A Guide to Advanced Skywatching

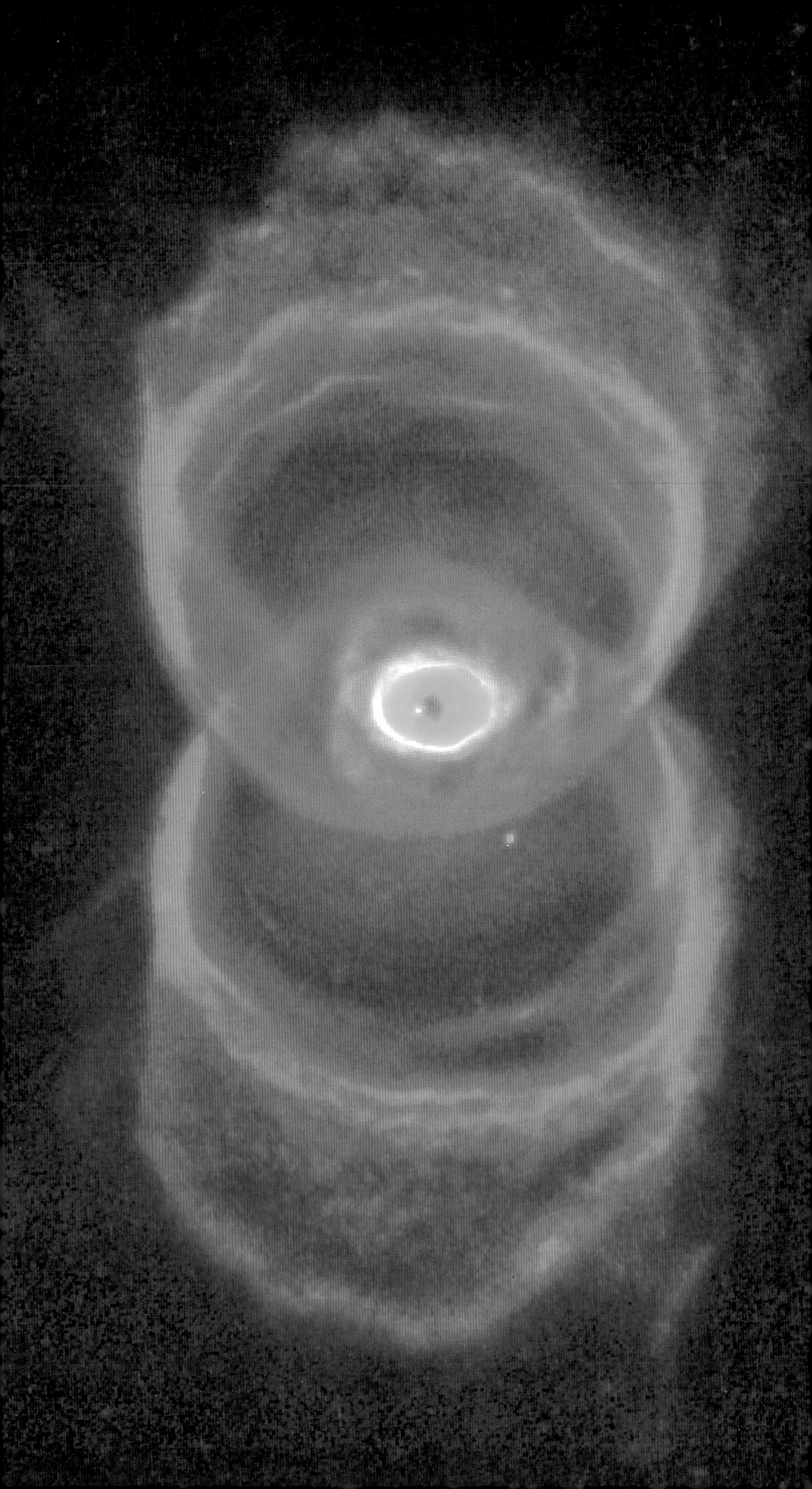

A GUIDE TO

ADVANCED SKYWATCHING

The backyard astronomer's guide to starhopping and exploring the universe

ROBERT BURNHAM, ALAN DYER,
ROBERT A. GARFINKLE, MARTIN GEORGE,
JEFF KANIPE, DAVID H. LEVY

CONSULTANT EDITOR
DR. JOHN O'BYRNE

FOG CITY PRESS

Published by Fog City Press
814 Montgomery Street
San Francisco, CA 94133 USA

This edition 2002

Chief Executive Officer John Owen
President Terry Newell
Publisher Lynn Humphries
Managing Editor Janine Flew
Art Director Kylie Mulquin
Editorial Coordinator Tracey Gibson
Editorial Assistant Kiren Thandi
Production Manager Caroline Webber
Production Coordinator James Blackman
Business Manager Emily Jahn
Vice President International Sales Stuart Laurence
European Sales Director Vanessa Mori

Project Editors Jenni Bruce, Elizabeth Connolly
Copy Editor Lynn Cole, Helen Cooney
Editorial Assistants Vesna Radojcic, Shona Ritchie
Project Designer Clare Forte
Picture Research Karen Burgess, Jenny Mills

A catalog record for this book is available from
the Library of Congress, Washington, DC.

ISBN 1 877019 32 1

Color separations by Leo Reprographics (Hong Kong)
Printed by Kyodo Printing Co. (S'pore) Pte Ltd
Printed in Singapore

A Weldon Owen Production

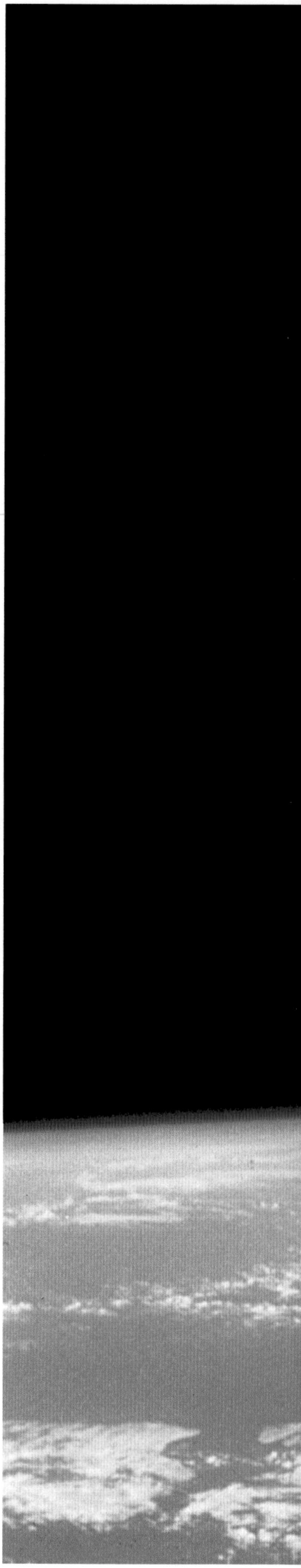

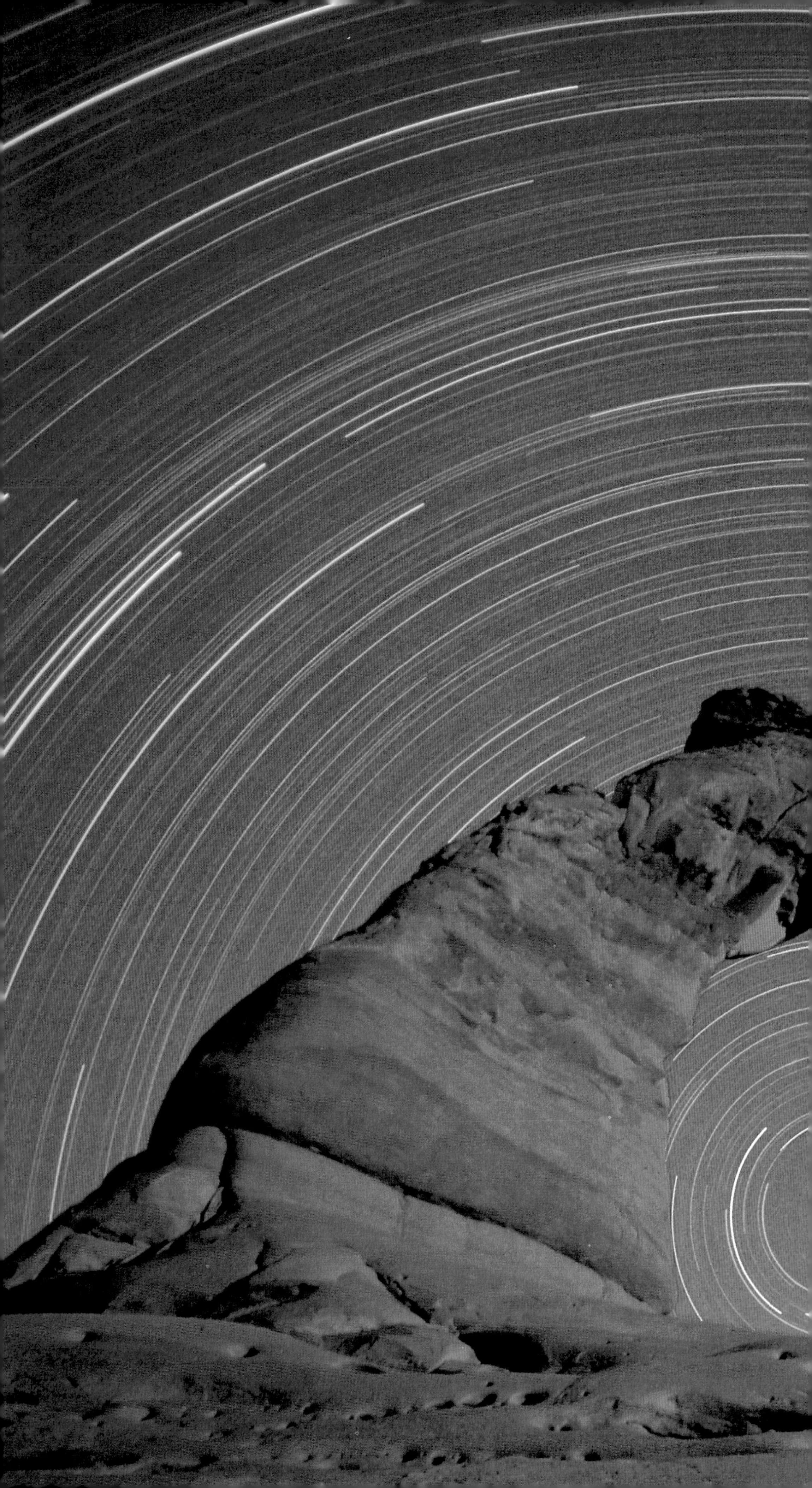

I open the scuttle at night and see the far-sprinkled systems,
And all I see multiplied as high as I can cipher
edge but the rim of the farther systems.

Wider and wider they spread, expanding, always expanding,
Outward and outward and forever outward.

Song of Myself,
WALT WHITMAN (1819–92), American poet

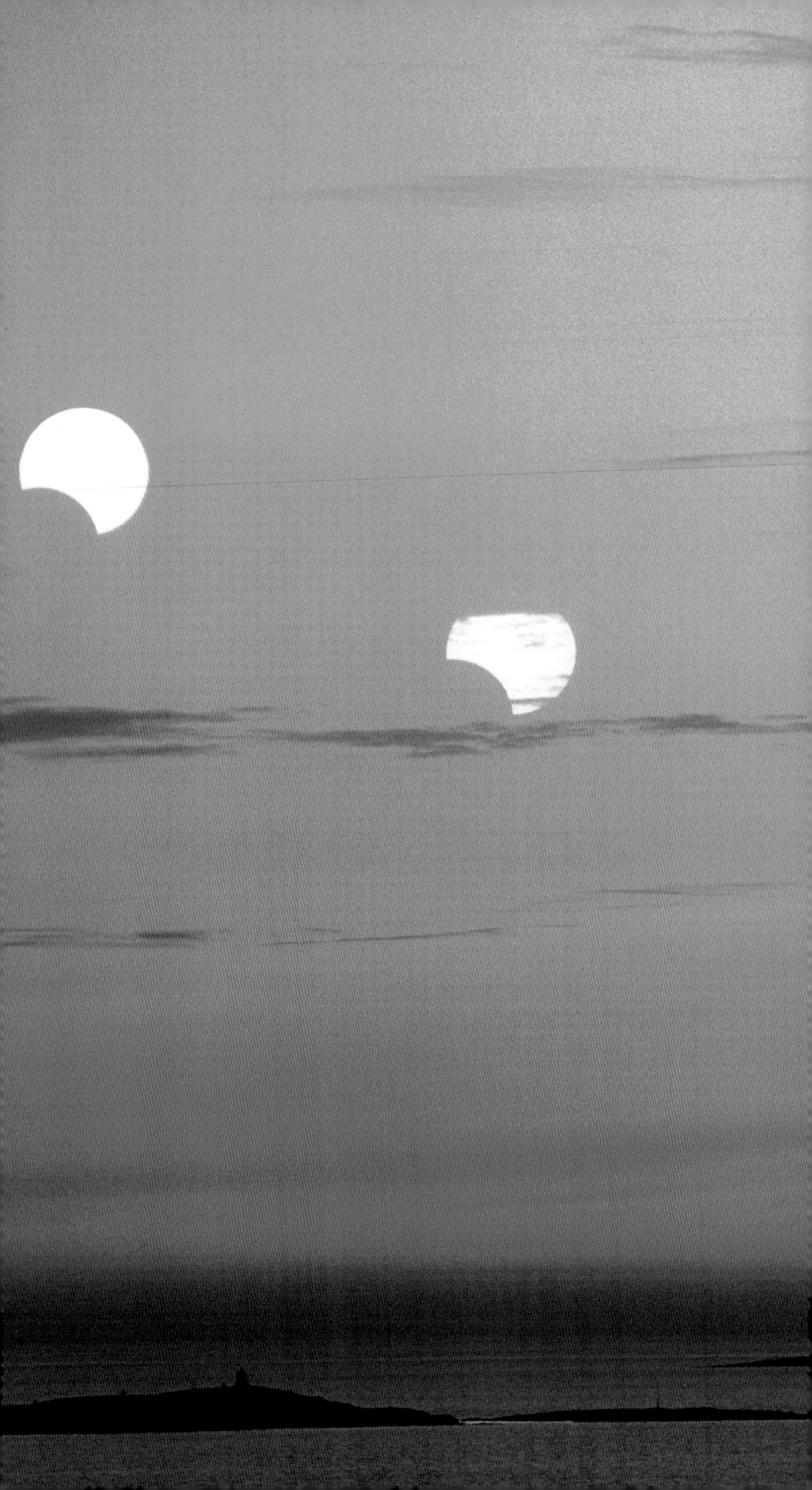

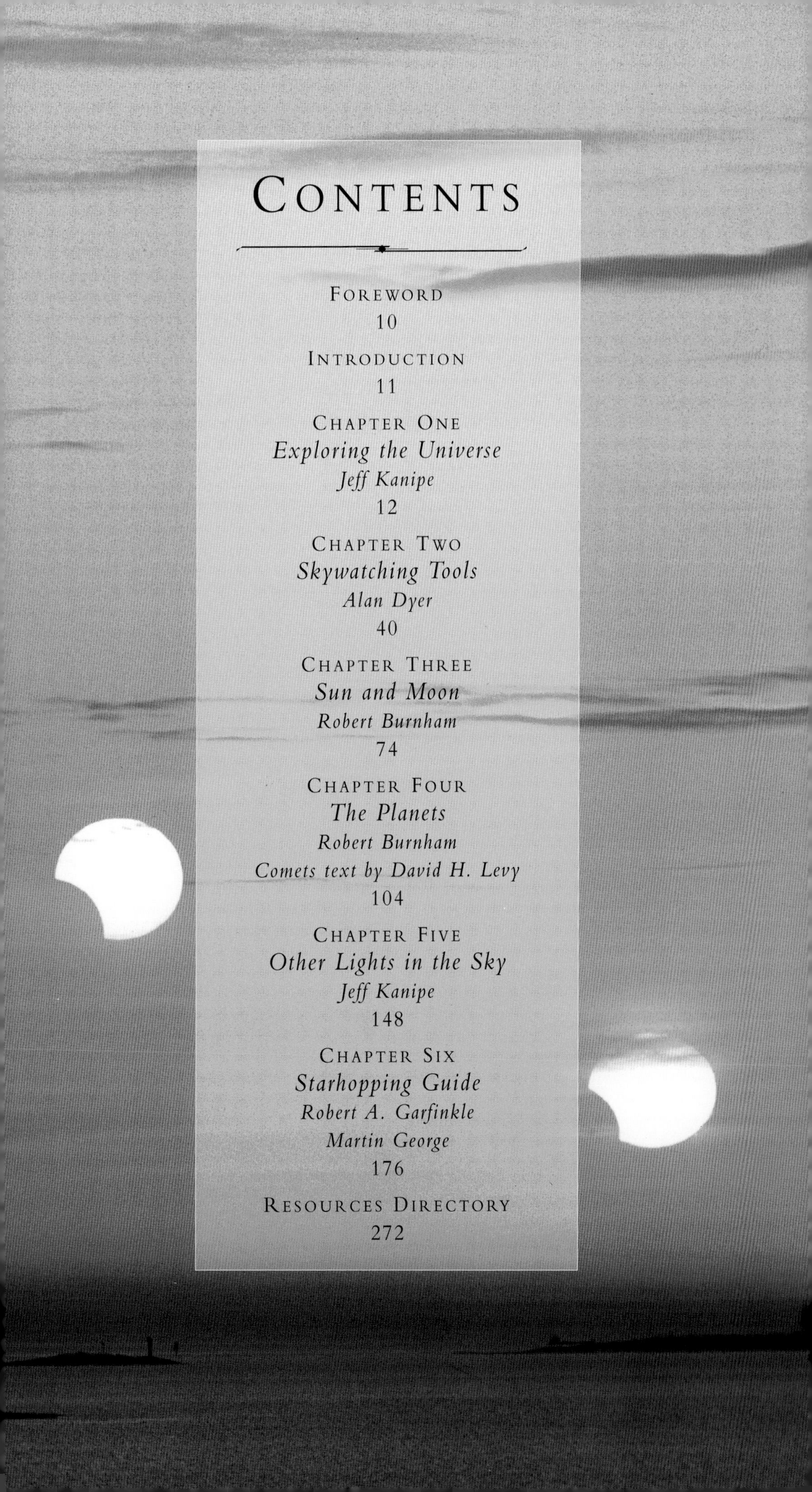

Contents

FOREWORD

When *Skywatching* first appeared, its simple goal was to promote an enjoyment of the night sky. *A Guide to Advanced Skywatching* begins where the earlier book left off. Designed as a perfect complement, *A Guide to Advanced Skywatching* extends the journey, peering more deeply into the universe that beckons above your backyard.

Einstein wrote that the most incomprehensible thing about the universe is that it is so comprehensible. The purpose of this book is to bring home that understanding with the help of superb color photographs and illustrations, as well as engaging text. The authors are experienced in their branches of astronomy and skilled at conveying the latest ideas and discoveries through easily understood explanations.

In a sense, the universe is merely an extension of our neighborhood. From the streets and parks of our youth, we can go outward still, to the planets, moons, comets, and asteroids that make up our Solar System. The Milky Way Galaxy, which is relatively nearby, lies within the Local Group of galaxies, itself a part of an immense confederation called the Local Supercluster. When we move on into the space between the superclusters, we truly leave our home and proceed to explore the great beyond—a universe where the stars grandly exceed all the grains of sand on Earth.

I beckon you to open *A Guide to Advanced Skywatching* and begin a personal voyage through space and time, to an understanding of the great universe of which you are a part.

David H. Levy

DAVID H. LEVY
Comet hunter and astronomy author

Introduction

Only a few centuries ago, humans believed that Earth lay at the center of the cosmos and that the movements of heavenly bodies influenced human destinies. Now we know that our planet is but a minute part of a vast, constantly expanding universe that functions according to physical laws. But despite the technological advances that have allowed us to probe deep into space and to peer at galaxies far way, the universe still contains many more tantalizing mysteries yet to unfold.

We can all delight in the wonders of the night sky just by gazing upward, but greater enjoyment lies in knowing what to look for and how. The possibilities are inspiring. With the naked eye you can chart the wanderings of five of the planets; binoculars reveal Jupiter's four Galilean moons; and a telescope allows you to view distant galaxies. *A Guide to Advanced Skywatching* provides all that you need to familiarize yourself with the sky's features. It profiles the Sun, Moon, planets, and other celestial bodies and sky phenomena, and gives expert advice on choosing and using binoculars, telescopes, and astrophotography equipment. A starhopping guide is also included, to help you traverse the night sky and identify the bodies in it.

The book also covers space exploration, from the early days of the space race to recent probes to the outer reaches of the solar system. Future explorations may well answer the eternal question: Are we alone in the universe?

THE EDITORS

US
United States
NASA
Atlantis

Chapter One
Exploring the Universe

Nothing puzzles me more than time and space; and yet nothing troubles me less.

Letter to Thomas Manning, 2 January 1810,
Charles Lamb (1775–1834), English essayist

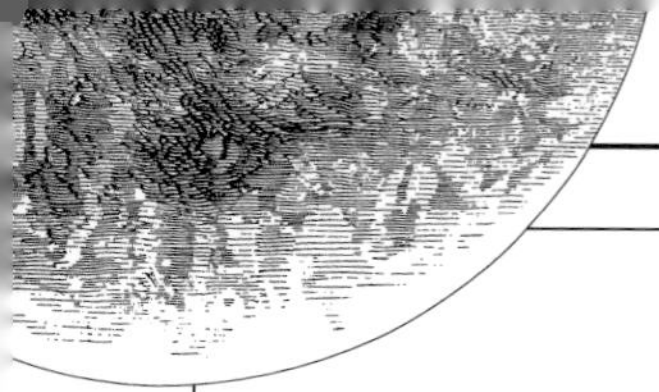

OUR VIEW *of the* UNIVERSE

The technological progress made in just one century has opened wide our windows on the universe.

For millennia, humans have gazed at the stars and planets and wondered how far away they lay, what they were made of, and how they got there. Skywatchers have charted and analyzed these celestial bodies—first with quadrants and armillaries, then using telescopes, spectroscopes, and spacecraft. The Moon and planets have been mapped, and thousands of star clusters, nebulas, and galaxies have been cataloged.

Today, huge telescopes and sensitive satellites examine the stars, and probes have been sent to some planets, but the goal remains the same: to discover what the universe is and why it is the way it is.

THE STARS FROM EARTH

Astronomy has developed exponentially since the mid-1800s, when the universe was a fairly small and uncomplicated place, and the view of the cosmos was Sun-centric. The planets were simply little-known cousins of Earth. The stars were thought to be immutable, and galaxies did not exist. Today, we know the planets as separate worlds, each with a unique geology and meteorology. Each star, too, displays a complex evolutionary history, and galaxies are recognized as giant stellar systems independent of our own.

In the 1860s, the development of spectroscopy—the breaking down of light into its constituent colors—started to expand astronomy from the study of the positions of celestial objects to the study of their physical properties. For the first time, astronomers could extract information from starlight. This allowed them to classify the stars and chart them from birth to death.

Giant reflecting telescopes at California's Mount Wilson and Palomar observatories were constructed in 1917 and 1947, and for decades, they were the largest optical telescopes in the world. Now, new mirror-making technologies are producing grander instruments. Each of the Keck telescopes on Mauna Kea, Hawaii, has a mirror 400 inches (10 m) in diameter.

Just as importantly, photographic cameras have now

OBSERVING *in 1925 at the Royal Greenwich Observatory (left) was still mostly concerned with determining the positions of objects.*

TRUTH & FICTION *In the nineteenth century, Jules Verne imagined this train of capsules (top) on their way to the Moon. A hundred years later, astronauts were actually floating in space (right).*

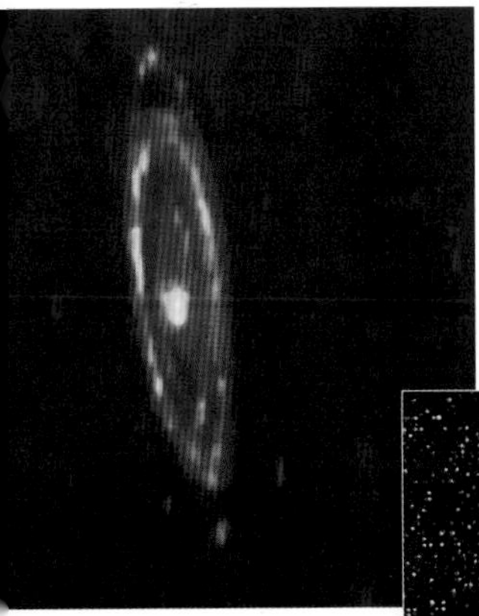

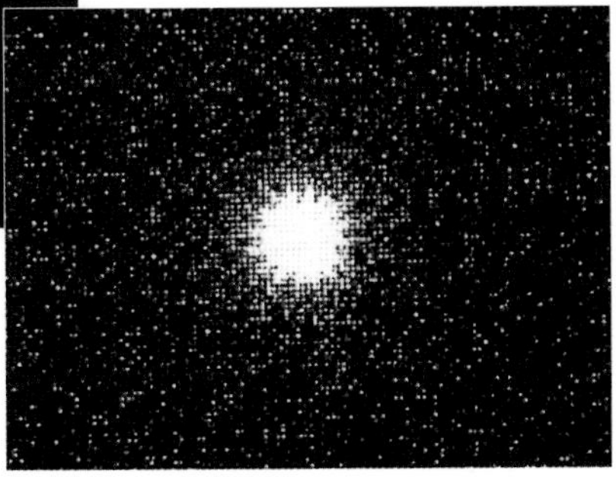

NON-VISUAL RAYS *Infrared image of Andromeda Galaxy (left). Cygnus X-1 (below), photographed by ROSAT, is a powerful X-ray source. It is a binary star system that probably hides a black hole.*

been largely replaced by electronic detectors called charge-coupled devices (see p. 68). CCDs capture images quickly, and are sensitive over a wide range of visible light.

Invisible Signals

Visible light waves, however, reveal only a fraction of the big picture. Just after World War II, developments in radio technology enabled scientists to measure energetic radio waves emitted by the Milky Way, other galaxies, and nebulas. Advances in electronics and solid-state physics in the 1960s and 1970s made radio astronomy one of the most powerful means of surveying the cosmos. It has been especially effective in probing distant galaxies, particularly their active centers, which are thought to harbor supermassive black holes.

Astronomers who were eager to explore the universe at other invisible wavelengths (see p. 21) had to get above the distorting effects of Earth's atmosphere. Since the late 1950s, Earth-orbiting satellites have allowed them to do this.

On the low-energy end, infrared wavelengths reveal the warm-to-cold universe. With infrared cameras, star clusters still shrouded in dense dust can be "seen" despite being visually obscured, as can the heart of our own galaxy. On the more energetic end of the spectrum, ultraviolet and X-ray wavelengths show the hot, active universe. Here, astronomers study novae and supernovae, quasars and the cores of active galaxies, and the disks of gas around black holes and neutron stars.

All-sky surveys from space in the 1970s and 1980s revealed thousands of new objects, including new classes of infrared-bright galaxies, high-energy binary stars, black-hole candidates, and enigmatic objects that emit bursts of gamma-ray energy.

Nothing better illustrates how rapidly the frontiers are advancing than the stream of discoveries from the Hubble Space Telescope (see p. 24). Its detailed imaging and spectroscopic observations are of incalculable value.

The twentieth century has also seen amazing leaps in the study of the universe from space itself. The dream of venturing into space is an old one. As far back as AD 160, a Greek named Lucian of Samosata wrote a story about traveling to the Moon. In the early seventeenth century, a journey to the Moon and speculations about weightlessness figured prominently in a story called *Somnium* ("Dream") by the German astronomer Johannes Kepler. Imagine what such visionaries might say about the Apollo missions to the Moon, the Viking landers on Mars, or the Voyager probes to the Solar System's outer planets.

THE HUBBLE SPACE TELESCOPE *looked deep into space to produce this image—each speck of light is a galaxy.*

The Birth *of* Astrophysics

The heavens have been a source of wonder since time immemorial. With the development of new tools, they began to give up their secrets at an unprecedented rate.

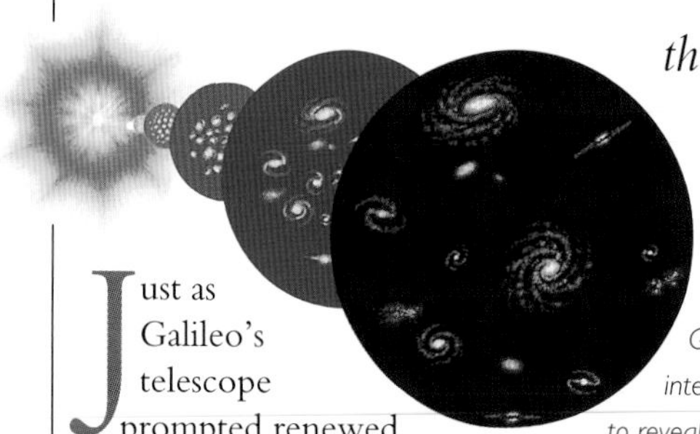

Just as Galileo's telescope prompted renewed enthusiasm for observational astronomy in the early seventeenth century, the development of the spectrometer two centuries later spawned a vigorous new science that focused on the dynamics, chemical properties, and evolution of celestial bodies.

The study of astrophysics, as it was called, began modestly in Munich in 1814, when physicist and telescope-maker Joseph von Fraunhofer passed the light of the Sun and stars through a spectrometer—essentially a prism that broke the light down into the colors of the spectrum. He noticed that the spectra had hundreds of dark lines, like fence slats, running across them. Although Fraunhofer realized that the lines were important, their meaning eluded him and other scientists for 45 years.

Reading the Signals

In 1859, German physicist Gustav Kirchhoff showed that each chemical element produced characteristic patterns of lines in the spectrum of the Sun. He correctly interpreted the bright spectral lines as the emission of light by particular chemical elements, and the dark lines as the absorption of light by the same elements. This meant spectral analysis could reveal, for instance, that the Sun contains sodium.

By 1863, spectroscopy had advanced enough to allow English astronomer William Huggins to publish lists of stellar spectral lines. He confirmed that the Great Nebula in Orion was gaseous, and detected hydrogen, which had already been noted in the Sun, in the spectrum of a nova—the first evidence that all stars contained hydrogen. Huggins also used spectroscopy to determine the speed and direction of the star Sirius's movement (see p. 19).

Progress in photography brought a further leap in astronomical research. Until the early 1800s, cameras were primitive pinhole types, but by the 1840s, lenses, mirrors, and better film emulsions were available. Successful photographs of the Moon galvanized astronomers into photographing the Sun, which became a routine matter by the 1870s.

SECRETS REVEALED *Gustav Kirchhoff (right) correctly interpreted spectroscopic data to reveal the composition of the Sun, while Edwin Hubble's work on galactic motions led to the Big Bang theory (left).*

Distant Neighbors

As powerful as spectroscopy and photography were, they could not help astronomers work out the distances to the stars. Industrial advances in the early 1800s, however, made possible more accurate instrumentation and optics. The German astronomer Friedrich Bessel was

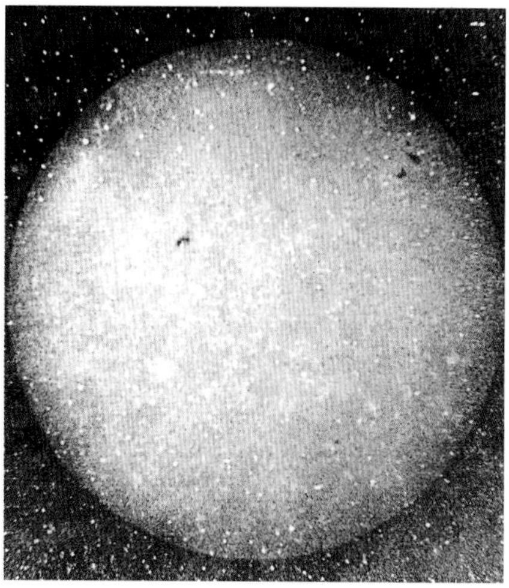

SUNSPOTS *can be seen on this daguerreotype (a kind of early photograph) of the Sun, taken in 1845 by Foucault and Fizeau.*

the first to capitalize on these advances. In 1838, by measuring the slight positional shift of the star 61 Cygni with respect to more distant background stars—an effect known as parallax—he estimated a distance for that star of 35 trillion miles (56 trillion km), or about 6 light-years.

Within two years, the distances to two other stars, Alpha Centauri and Vega, were determined using the parallax effect. In the decades that followed, many more distances were calculated and scientists realized that the universe was far more vast than once thought. No one, however, was prepared for how big it would "grow."

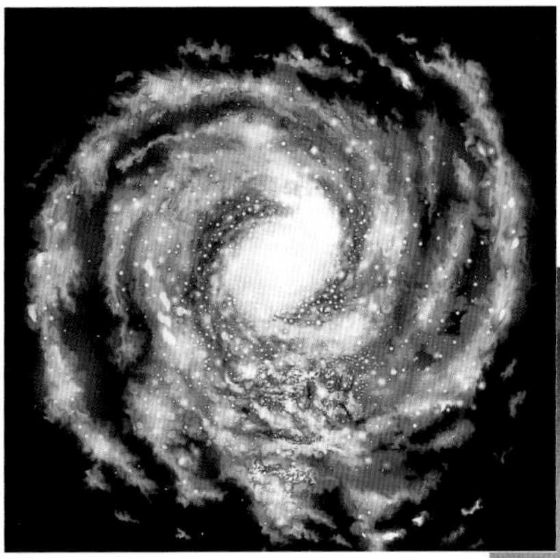

THE MILKY WAY *(illustrated above) is a spiral galaxy, possibly with a short bar (yellow-white) at the center. The spiral pattern is 100,000 light-years across.*

ASTRONOMER *Edwin Hubble (below) revised our ideas of the universe's scale.*

The Incredible Expanding Universe

The first decade of the twentieth century saw several discoveries that would profoundly affect astrophysics. In 1905, Danish astronomer Ejnar Hertzsprung and American astronomer Henry Norris Russell investigated the relationship between a star's color and temperature and its brightness. Russell's work on the evolution of stars led to the notion of red giants and white dwarfs. Charts that plot a star's color against its brightness are now called Hertzsprung-Russell, or H-R, diagrams in their honor (see p. 162).

In 1912, Henrietta Leavitt at the Harvard Observatory discerned a relationship between the period of brightness fluctuations and the true brightness of a class of stars called Cepheid variables (see p. 167). She noticed that Cepheids with slow cycles were brighter than those with fast cycles. Using Leavitt's period-luminosity law, Harlow Shapley calculated the distances to Cepheids in globular clusters. His results proved that the Milky Way was far larger than previously thought and that the Sun was not at its center.

Still under debate were the distances to, and the nature of, the "spiral nebulas," which most astronomers believed belonged to our galaxy. A series of observations, begun in 1912, showed that most spiral nebulas were receding from us. In 1924, American astronomer Edwin Hubble, scrutinizing the Great "Nebula" in Andromeda (M31), was able not only to make out individual stars in the nebula, but to determine that some of them were Cepheid variables. He then calculated that the nebula lay almost a million light-years away—a vast underestimate, as it turns out, but unquestionably beyond the bounds of our own galaxy.

But the biggest surprise came when Hubble used spectroscopy to analyze the velocities of galaxies, finding that the farther away a galaxy lay, the faster it was receding. This could best be explained, Hubble concluded, by presuming that the universe was expanding like a balloon. As time progresses, the universe's size increases, a possibility that had been predicted by Albert Einstein in his 1917 General Theory of Relativity.

Age-old Questions

In an expanding universe, astronomers realized, it should be possible to determine how and when the expansion began. Hubble's initial calculations indicated that the universe was only two billion years old, younger than the age of the Earth as determined by geologists. Once discrepancies in Hubble's calculations had been adjusted, a more geologically agreeable age of six billion years emerged.

Astronomers now believe the universe probably began 10 billion to 15 billion years ago in an immense explosion of matter and energy known as the Big Bang. Since then, the expansion of space itself has carried the galaxies away from each other, in accord with Hubble's Law of expansion.

FROM *the* GROUND

Our view of space from the ground is improving all the time as telescopes get bigger and more sophisticated, and computers provide new ways to gather and interpret astronomical data.

Visible light is radiation that the human eye can see. Compared to the total range of the electromagnetic spectrum, visible light constitutes a very narrow span (see p. 21). Yet much of what we know about the universe has come to us through this slim window.

The faintest star visible to the naked eye is about 6th magnitude, but binoculars and telescopes greatly extend this reach. Even a 6 inch (150 mm) scope gathers about 500 times more light than the human eye can—no wonder astronomers always want larger telescopes.

BIGGER AND BETTER

One of the first giant telescopes was built in 1845, by William Parsons on his Irish estate (see p. 71). Its mirror spanned 72 inches (1.8 m), and until 1908, it was the largest telescope in the world. In 1917, the 100 inch (2.5 m) Hooker Telescope was built by American astronomer George Hale on Mount Wilson, California. Hale started work on an even bigger telescope, but died before it was finished in 1948. The 200 inch (5 m) Hale Telescope on Mount Palomar, California, was the world's largest for nearly 30 years.

In 1977, the Russians completed the 240 inch (6 m) Bolshoi Telescope in the Caucasus Mountains. Although larger than the Hale, design flaws prevented it realizing its full potential. In 1992, the 400 inch (10 m) Keck I Telescope on Mauna Kea, Hawaii, became the world's biggest. It was followed by Keck II (also 400 inches) in 1995. Because such large telescope mirrors are very difficult to make in one piece, each was created from 36 hexagonal segments.

The resolution capabilities of ground-based telescopes have been limited by the effects of atmospheric seeing. This, however, is becoming less of a problem. Computer-controlled regulators can rapidly adjust the shape of a telescope's optics to reduce the turbulent effects of Earth's atmosphere and produce sharper images. Known as adaptive optics, the technique is being applied to many of the world's large telescopes.

CCD TECHNOLOGY

In the 1970s, a light-sensitive electronic charge-coupled device, known as the CCD, increased the telescope's light-gathering capability a hundredfold and revolutionized astronomical imaging

BIG SCOPES *Getting ready to photograph with the Hale Telescope (top). The Anglo-Australian (left) was one of the first electronically controlled telescopes. Observatory domes on Mauna Kea (below).*

(see p. 68). Today, CCDs are popular with amateurs and professionals alike, and are widely used in all aspects of astronomical research, from studying variations in starlight to imaging the centers of active galaxies. Although their resolution is lower than fine-grained film, they are far more efficient at gathering light and exposure times are shorter.

SPECTROSCOPY

Telescopes and CCDs may gather and record starlight, but only a spectrograph can extract its secrets. A spectrograph separates white light into its component wavelengths—a spectrum—crossed by numerous dark and bright spectral lines. It then records these lines as a kind of bar code of the object's physical properties. Spectroscopy is the basis of modern astrophysics—most of what we know about the chemical composition, temperature, and pressure in any astronomical object is encoded in its spectral lines.

One of spectroscopy's most useful applications is in determining an object's motion toward or away from Earth. This was discovered in 1868 by William Huggins, who found that dark lines in the star Sirius were shifted toward the red end of the spectrum, implying a decrease in frequency. Based on work by Austrian physicist Christian Doppler, who theorized that a source exhibits a change in frequency if it is moving toward or away from the observer, Huggins determined that Sirius was moving away from the Sun at about 25 miles per second (40 km/s). Had the spectral lines been blueshifted, increasing in frequency, it would have indicated that Sirius was approaching Earth.

Edwin Hubble later applied this method to galaxies, demonstrating that most galaxies are redshifted and their recessional velocities increase with distance (see p. 17).

Fiber-optic technology is improving the efficiency of spectroscopy. Many optical fibers can be arranged so that each receives light simultaneously from a different star or galaxy. The opposite ends of the fibers can then be aligned so that they feed into the spectograph at once, allowing a hundred or more spectra to be gathered at the same time.

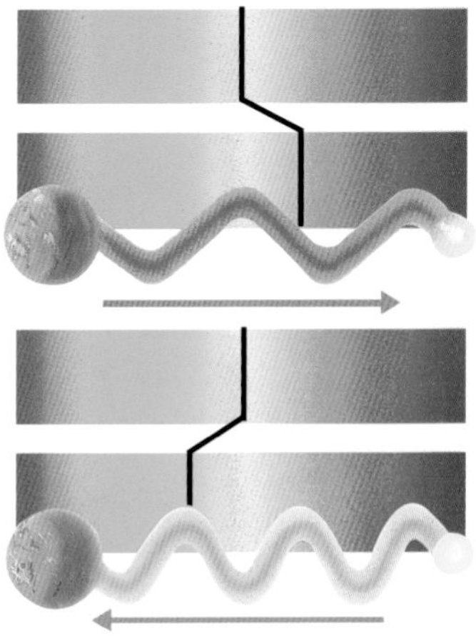

SPECTRAL SHIFT *William Huggins (above left) based much of his work on the idea that if a star's spectral lines shift toward the red end of the spectrum (right, top), the star is moving away from Earth. If it shifts toward the blue end (right, bottom), the star is approaching.*

TRUE AND FALSE COLOR IMAGES

The beautiful photographs of nebulas, galaxies, and star clusters that appear in astronomy books and magazines often exhibit a riot of color: hot pinks, lime greens, deep reds, and neon blues. As striking as they are, they usually do not represent the object's true colors. Astronomers generate computer-colored images, purposely assigning colors to different levels of intensity so as to bring out subtle details. Images taken at radio and optical wavelengths can also be represented by different colors and superimposed to form a composite. Such composites are helpful in correlating radio hot spots to optical sources.

False-color images are often used to map temperature differences and radiation intensity across an object. For example, both these close-up images of Halley's Comet are optical photographs taken by the Giotto spaceprobe in 1986. While the top image uses natural colors, the bottom image is computer processed to enhance relative brightnesses. The most brilliant part of the comet is coded white, with other bright regions in pale colors. The dark shades of pink, orange, yellow, blue, and green indicate successively fainter areas of the comet.

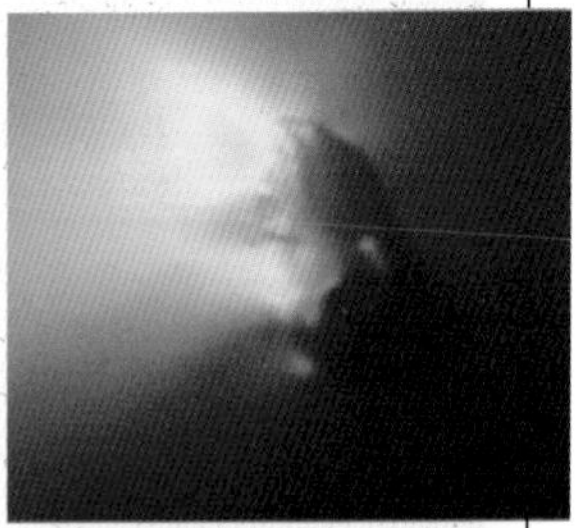

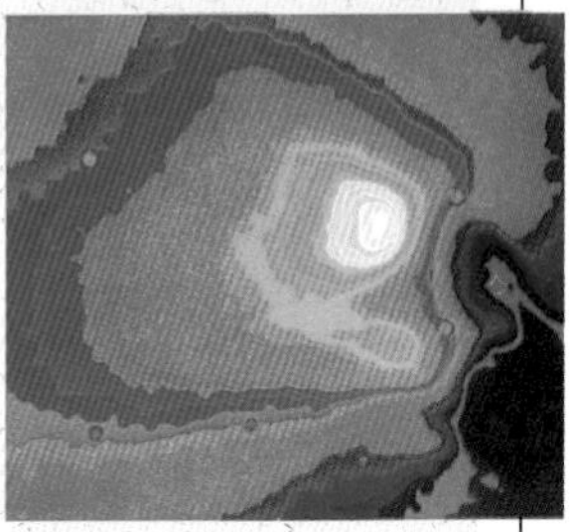

OTHER WINDOWS *on the* UNIVERSE

There is a wealth of information about the physical universe that lies beyond the realm of visible light.

THE VERY LARGE ARRAY *Telescope in New Mexico consists of 27 radio dishes linked to a central receiver.*

The physical universe is represented by more than what our eyes are able to see. Radiation spanning the full electromagnetic spectrum (see diagram, facing page) can tell us a great deal about celestial objects.

Signals from space—such as visible light, X-rays, and ultraviolet, gamma, and infrared rays—all reach Earth as waves, and the various "wavelengths" indicate the distances between the crests of the waves.

The features in these other regions of the spectrum are picked up by telescopes designed specifically to "see" them. The air above us absorbs and scatters particular wavelengths; to pick these up, some telescopes are placed beyond Earth's atmosphere.

THE LOW END

The low-frequency, low-energy part of the spectrum is the domain of infrared, millimeter, and radio radiation.

Infrared rays are the radiation we feel as heat. Most infrared sources radiate at temperatures between –430 and 1800 degrees Fahrenheit (–260° and 1000° C). Infrared observations can pierce dense dust clouds to reveal young stars that are invisible at optical wavelengths, accentuate dusty disks around stars, and even pinpoint remote dusty galaxies. Some near- and mid-infrared observations can be ground-based, but longer-wavelength infrared observations must be conducted from balloons, rockets, or satellites.

MILLIMETER WAVES

Millimeter and submillimeter telescopes address wavelengths corresponding to the lower temperatures. Optical lenses and mirrors are replaced with radio "mirrors"—radio dishes and receivers specifically tuned to receive short-wavelength radio signals. One of the largest millimeter-wave telescopes, with a dish 50 feet (15 m) in diameter, is the James Clerk Maxwell Telescope on Mauna Kea, Hawaii.

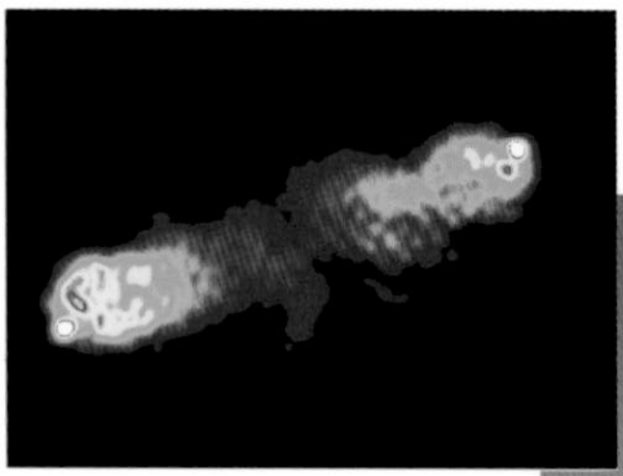

GROUND-BASED WORK *includes the MERLIN radio telescope array in England, which took this image of galaxy Cygnus A (above); the UK Infrared Telescope (above right); and the James Clerk Maxwell Telescope (right).*

The millimeter window is ideal for observing giant molecular clouds in which stars are likely to form.

RADIO SIGNALS

Longer-wavelength radio radiation reveals the processes occurring in gaseous nebulas, pulsars (spinning neutron stars), supernova remnants, and the active cores of distant galaxies. A radio telescope consists of an antenna, or antenna-array, connected to a receiver. The antenna

KARL JANSKY'S STATIC FROM SPACE

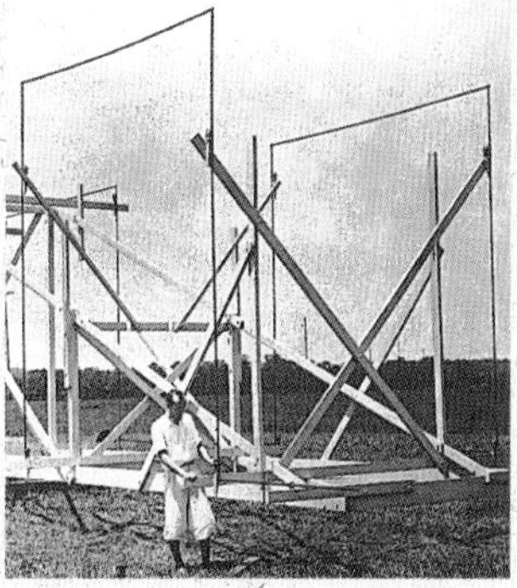

In 1928, just a year after graduating from the University of Wisconsin, radio engineer Karl Jansky (1905–50) accepted a job with Bell Telephone Laboratories. One of his first assignments was to track down the source of static that interfered with radio reception and ship-to-shore communications. Although much of the static could be attributed to storms, aircraft, or local electrical equipment, Jansky detected a fainter type of background static that appeared to rise and set with the Sun. As he continued to track the radio noise with his directional radio aerial system (above), he noted that the noise moved ahead of the Sun by about four minutes each day.

Realizing that this is the same amount that the stars gain on the Sun each day, Janksy reasoned that the source of the static lay beyond the Solar System. By 1932, he had determined that its source lay in the constellation Sagittarius, toward the center of the Milky Way.

Although Jansky did not pursue the science of celestial static any further, his discovery marked the birth of radio astronomy, which by the 1970s, had taken its place alongside optical astronomy as a highly effective means of studying the universe.

may be as simple as a dipole-type antenna, which looks something like a TV antenna. More often, though, it is a parabolic reflector dish, or an array of dishes. The radio waves are focused and collected by a secondary antenna, known as a feed. They are then transferred to a receiver for amplification and analysis. As with an optical telescope, the sensitivity of the radio telescope increases with the surface area of the antenna.

When two or more radio telescopes are linked so that their signals are fed to a common receiver, they are jointly termed an interferometer. The greater the distance between the radio telescopes, called the baseline, the finer the detail that can be resolved. The Very Large Array, near Socorro in New Mexico, has baselines up to 22 miles (36 km) long, while the Very Long Baseline Array consists of 10 dishes spread over about 5,000 miles (8,000 km)—from Hawaii across the USA to the Virgin Islands in the Caribbean Sea.

The High End

At the high-frequency, high-energy end of the spectrum, ultraviolet, X-ray, and gamma-ray radiation unveil the physical and chemical properties of objects that are incredibly hot and energetic. These wavelength regimes are particularly useful for studying the hot gas swirling around black holes and neutron stars; galactic haloes; interactions between cosmic rays and interstellar gas; and the rarefied outer atmospheres of stars.

The Universe is wider than our view of it.

Henry David Thoreau (1817–62), American naturalist and writer

ELECTROMAGNETIC SPECTRUM

This illustration shows the range of radiation wavelengths and some of the instruments designed to detect them. Only optical, radio, and some infrared rays can pass through Earth's atmosphere; satellite-based telescopes are used for detecting shorter wavelengths.

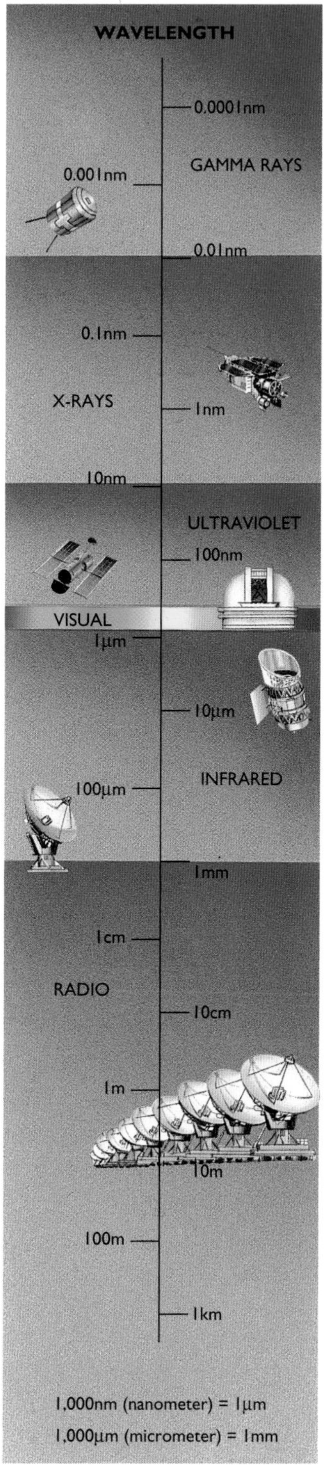
WAVELENGTH
0.0001nm
GAMMA RAYS
0.001nm
0.01nm
0.1nm
X-RAYS
1nm
10nm
ULTRAVIOLET
100nm
VISUAL
1µm
10µm
INFRARED
100µm
1mm
1cm
RADIO
10cm
1m
10m
100m
1km
1,000nm (nanometer) = 1µm
1,000µm (micrometer) = 1mm

AROUND *the* EARTH

Infrared, ultraviolet, X-ray, and gamma-ray signals have always been there, but until the late 1950s, most of them were hidden behind the curtain of Earth's atmosphere.

Observing the universe from space became a real possibility on 4 October 1957, when Russia launched Sputnik 1, the first artificial satellite. The United States lofted Explorer 1 into orbit on 31 January 1958, and between then and 1984, 65 Explorer-class satellites were successfully launched. These missions studied such scientific phenomena as Earth's magnetic fields, the solar wind (see p. 79), and ultraviolet radiation from stars and galaxies.

GETTING INTO SPACE

The most direct means of overcoming Earth's gravity and putting a payload, or cargo, into space is to launch it on a rocket. The velocity required is about 17,500 miles per hour (28,000 km/h). The rocket's thrust must be greater than its weight, so even a small cargo uses a lot of fuel. One way of lessening gravity's effect and conserving fuel is to launch the rocket in an easterly direction near the Equator. Earth's rotation acts like a sling, providing extra lift. Cape Canaveral, Florida, is an ideal location, boosting a rocket's velocity by almost 900 miles per hour (1,450 km/h).

With enough speed, a rocket can place satellites or other spacecraft into one of three types of orbit: equatorial, polar, or geostationary.

Equatorial satellites usually orbit at low altitude—about 100 miles (160 km) above Earth—and, because they are launched eastward, they travel from west to east.

Polar-orbiting satellites pass over the entire Earth in high-altitude orbits at a minimum of about 1,000 miles (1,600 km). Most of these scan geological or weather phenomena, but some are dedicated to military reconnaissance. They travel north to south, or south to north. From Earth, they appear to move more slowly because of their greater altitude.

Geostationary, or geosynchronous, satellites circle Earth once a day at an altitude of 22,000 miles (36,000 km). Because their speed matches that of Earth's rotation, they remain over a fixed point. This makes them ideal for gathering weather-forecasting data and for communications such as television transmissions.

WAVELENGTH WINDOWS

Earth's atmosphere often interferes with the reception of radiation, so once receivers could be positioned beyond the atmosphere, a new era in the study of the universe began.

A PRE-LAUNCH CHECK *of the ROSAT X-ray and extreme ultraviolet telescope (right). A model of IRAS, the Infrared Astronomical Satellite (top).*

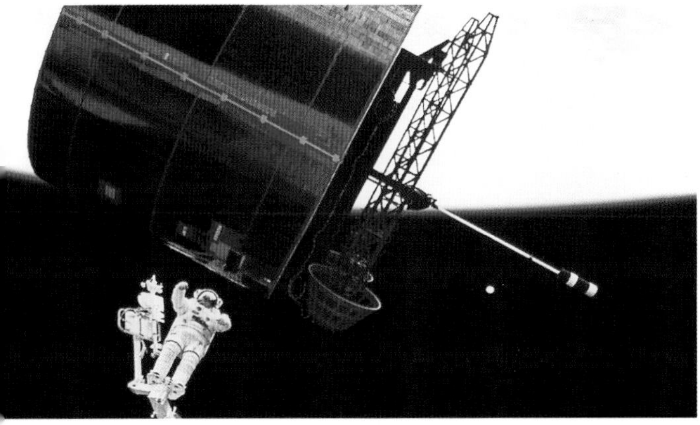

SPACE EYE *The Astro Observatory aboard the shuttle (below). A burst of gamma radiation recorded by the Compton Observatory (bottom). Repairing Syncom, a communications satellite (left).*

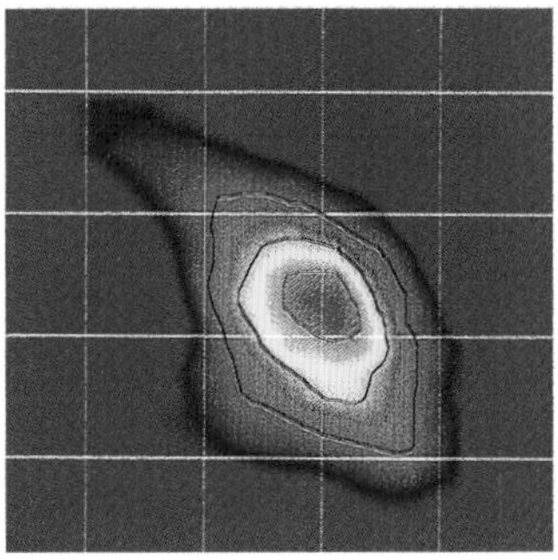

The first cosmic X-rays were picked up in 1962 by a detector aboard a small rocket. Their source was later found to be Scorpius X-1, a binary-star system that is the brightest X-ray source in the sky.

Satellites, including Japan's Ginga, the European Space Agency's EXOSAT, and Germany's ROSAT, have provided intriguing insights into X-ray sources, such as black holes, neutron stars, and supernova remnants. Scientists will image some of these sources in 1998 with the first Advanced X-ray Astrophysics Facility, AXAF-1.

Ultraviolet astronomy began in earnest in December 1968 with the Orbiting Astronomical Observatory-2, which returned data covering one-sixth of the sky. From this, a catalog of bright UV sources was compiled. Subsequent missions glimpsed sources of extreme ultraviolet radiation—hot stars, supernovae, the active cores of galaxies, and quasars (the intensely energetic centers of far-off galaxies)—and produced more than 90,000 ultraviolet spectra.

'Tis distance lends

enhancement to the view.

The Pleasures of Hope,
THOMAS CAMPBELL (1777–1844),
Scottish poet

In December 1990, an instrument package aboard the space shuttle *Columbia* provided spectroscopic observations in the far- and extreme-ultraviolet wavebands. The Astro mission also made images of UV sources and assessed how UV radiation is scattered by cosmic dust and strong magnetic fields. The Extreme Ultraviolet Explorer was launched in June 1992 and has surveyed the entire sky, mapping sources of extreme UV radiation. It is being used to study white and red dwarf stars, eruptive variable stars, and thin interstellar gas.

Exploring Further

Seeking and investigating sources of gamma rays, the most energetic form of radiation, was the objective of COS-B, launched in 1975 by the European Space Agency. The mission ran until 1982. In April 1991, the Compton Gamma Ray Observatory set out to explore further, particularly to identify sources of mysterious gamma-ray bursts.

The Infrared Astronomical Satellite (IRAS), a joint mission of the United States, Britain, and the Netherlands, discovered more than 200,000 infrared sources during an all-sky survey of IR objects in 1983. It also found a class of galaxies ultraluminous at far-infrared wavelengths that may be dust-embedded quasars.

The Big Picture

One of the most significant observations of the twentieth century was made in the early 1990s by the Cosmic Background Explorer. COBE measured the feeble background glow thought to be remnant radiation from the Big Bang—the explosion that created the universe.

In 1992, careful analysis of COBE data revealed slight temperature fluctuations in the background glow, which were interpreted as tiny variations in the smoothness of the universe 300,000 years after the Big Bang. These variations had been predicted by the Big Bang theory. Further study may reveal how these irregularities in the distribution of matter resulted in the formation of the galaxies.

The Hubble Space Telescope

No other orbiting telescope has generated as much excitement or so astounded astronomers.

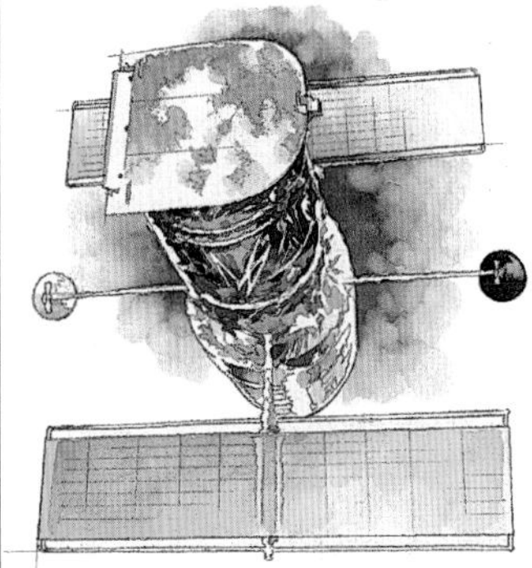

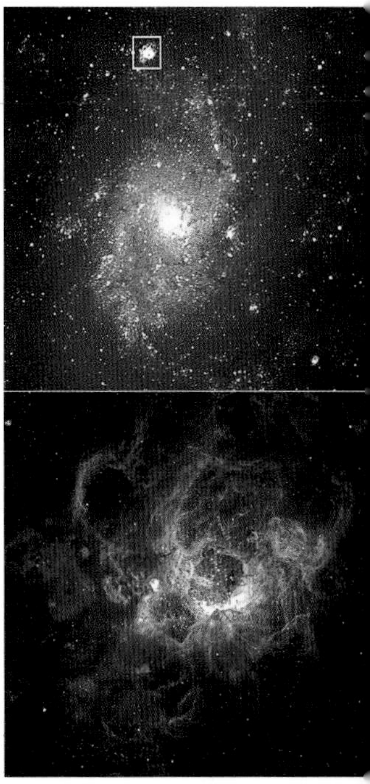

UP CLOSE *The HST (left) has vastly improved our view of space. The image of the galaxy M33 (right, top) was taken with the Hale Telescope, but a small part of it, NGC 604, inset, was captured in greater detail by the HST (right, bottom).*

Launched in April 1990, the Hubble Space Telescope (HST) has provided spectacular observations in visual, near-infrared, and ultraviolet wavelengths, as well as spectroscopic studies of stars, the thin interstellar gas, and galaxies. Consisting of a 95 inch (2.4 m) mirror and a suite of sensitive scientific instruments, it is controlled by some 400 astronomers, computer scientists, and technicians.

The Wide Field/Planetary Camera II, the most often used of HST's instruments, can detect objects as faint as 28th magnitude (about a billion times fainter than can be seen with the naked eye). The Faint Object Camera can also record 28th magnitude stars, but it offers higher resolution and a wider choice of viewing angles. It can distinguish between objects a mere 0.05 arcsecond apart (the naked eye can only split objects 60 arcseconds apart).

Two newer instruments, installed in February 1997, are the Near Infrared Camera and Multi-Object Spectrometer (NICMOS), and the Space Telescope Imaging Spectrograph (STIS). NICMOS handles both imaging and spectroscopic observations of objects at near-infrared wavelengths. Scientists hope to learn much from these observations about the birth of stars in dense, dusty globules, the infrared emission produced by the active centers of distant galaxies, and the nature of a class of galaxies as bright as quasars at infrared wavelengths.

The STIS is considered the most complex scientific instrument ever designed for space. It covers a broader wavelength range than its predecessor, and can block out, or occult, the light of distant stars to search for black holes and Jupiter-size planets in other galaxies.

Finally, HST's fine-guidance sensors, necessary for pointing the telescope and locking onto its target, can measure the positions of stars to 0.002 arcsecond. Such precision enables astronomers to note slight wobbles in a star's

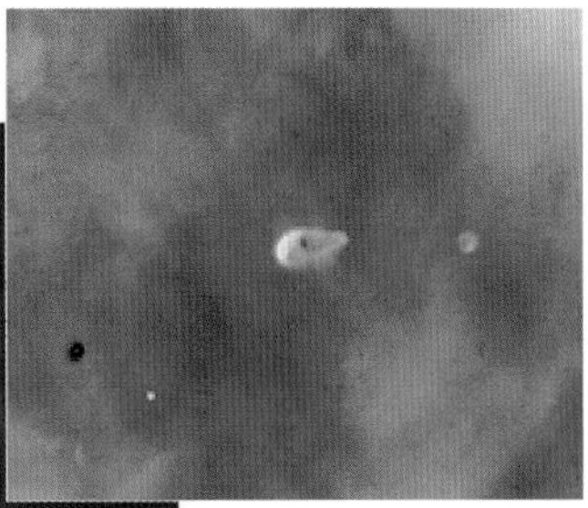

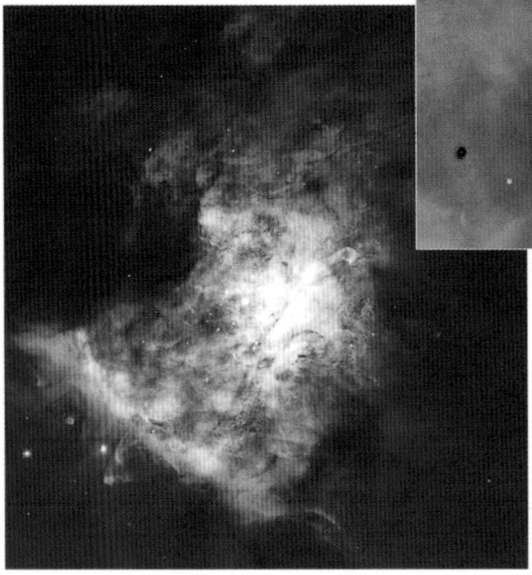

THE GREAT NEBULA *in Orion. The close-up (above) shows an area of dusty disks where young stars form. The larger composite image (left) is made up of 15 separate fields.*

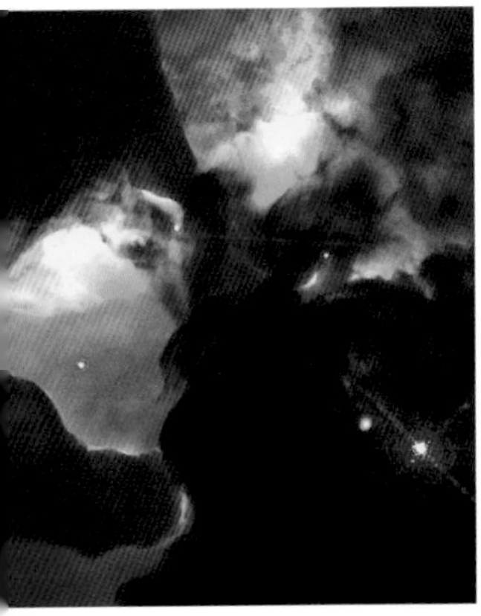

LOOKING INTO THE HEART *of the Lagoon Nebula (left) in Sagittarius, as seen by HST's Wide Field Camera.*

position, which might indicate the gravitational tug of an unseen planet—and the existence of other solar systems.

What HST Has Seen

As of 1997, HST has looked at more than 10,000 objects and made more than 100,000 exposures, yielded significant insights into the formation of stars and stellar disks, disclosed important evidence for the existence of black holes in galaxies and quasars, increased our knowledge of the size and age of the universe, and detected galaxies that formed a billion years after the Big Bang.

Its high-resolution images of Mars, Jupiter, Saturn, and Neptune are yielding details surpassed only by space-probe photographs. The world was astonished by the spectacular images it produced in July 1994 when 21 fragments of the comet Shoemaker-Levy 9 collided with Jupiter.

HST images of star-forming regions have shown nascent stars embedded in globules of dust and gas called EGGS (evaporating gaseous globules). In the Great Nebula in Orion, dusty disks visible around protostars have been interpreted as solar systems in the making. At the other end of stellar evolution, HST has produced a stunning image of Eta Carinae, a star/nebula system 8,000 light-years away (see p. 167). This is expected to explode one day as a supernova.

We can now see great disks of matter swirling around supermassive black holes at the centers of galaxies and quasars, as well as structural details in the spiral arms of nearby galaxies. Looking at the most remote corners of the universe, a 10-day series of HST exposures revealed an assemblage of galaxies of various sizes completely filling a single speck of sky in Ursa Major. Many of these may date back almost to the beginning of the universe.

The HST has established itself in astronomical history and will no doubt continue to make dramatic observations and discoveries. Nevertheless, astronomers are already planning a more powerful Next Generation Space Telescope. This will be able to look in even greater detail at a period in the universe when the primordial seeds of the galaxies began to evolve—just a few million years after the Big Bang. Astronomers long to study this epoch because it may help explain the origin and fate of the universe.

The new telescope would be much more sensitive than any existing telescope. The 240 to 320 inch (6 to 8 m) mirror would soak up light from remote proto-galaxies as well as study nearby objects in the universe. With this new technology, we may finally be able to address such burning questions as: How did galaxies form? and, What were the first generations of stars like?

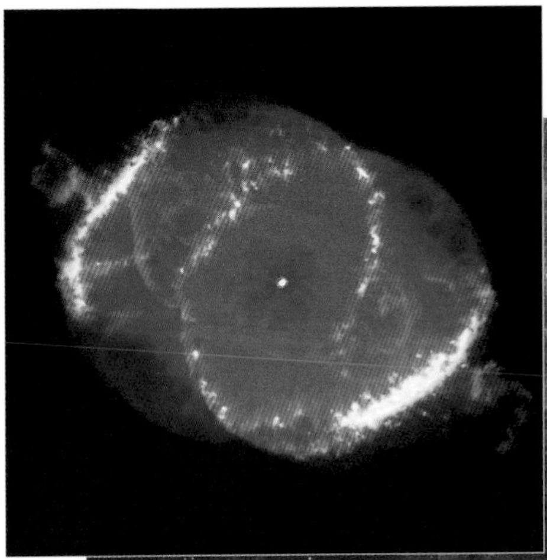

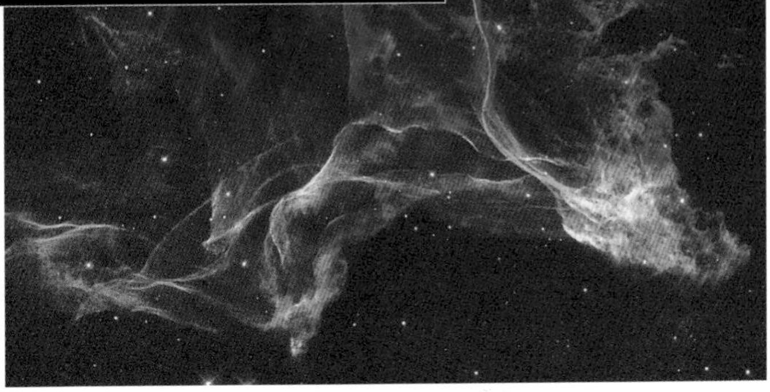

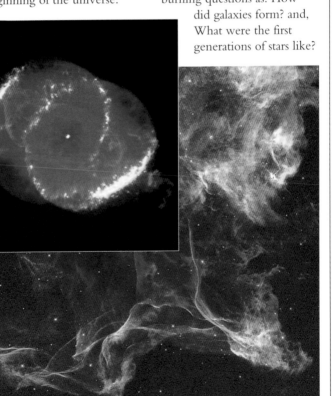

CELESTIAL BODIES *photographed by HST: NGC 6543 (above right), one of the brightest planetary nebulas, and a portion of the Cygnus Loop (right).*

Space Travelers

The fantasies of imaginative writers such as Jules Verne, H. G. Wells, and Cyrano de Bergerac suddenly became reality in the 1960s with the first human steps in space.

RECORDS *Mercury 2 sent a chimp (top) into space. The descent module (above) of Vostok 1, in which Yuri Gagarin became the first man to orbit the Earth. The first woman in space was Valentina Tereshkova (right).*

In 1957, the steady beeping of the Soviet Union's Sputnik 1 satellite, the first to orbit Earth, flung down the gauntlet before United States policy-makers. Who would control the high ground of space? On 1 October 1958, the United States government established the National Aeronautics and Space Administration (NASA). So began the space race, which, in retrospect, may be more accurately described as a race of political ideals embodied in the technological developments achieved by the United States and the Soviet Union. At any rate, the next 12 years saw phenomenal accomplishments, illustrious glory, and tragedy.

The Space Race

The Soviet Union leaped ahead first, with two Sputnik launches—the second of which carried Laika, a live dog. Laika survived for several days before her air supply ran out. The Soviets then placed the first human into Earth orbit, on 12 April 1961. The spacecraft, Vostok ("east"), was little more than a cabin in a cannonball, covered entirely by a heat shield. The cosmonaut, Yuri Gagarin, sat in an ejection seat, which he activated to safely parachute back to Earth.

Ostensibly, the goal of America's incipient space program, called Project Mercury, was to place astronauts into space, test their reactions and the spacecraft's maneuvering abilities, and return them safely to Earth. The long-term goal, however, as proclaimed in no uncertain terms by President John F. Kennedy in May 1961, was to land a man on the Moon by the end of the decade. It was a bold challenge, not only to the Soviets, but to the American people and their science and technology.

In contrast to Russia's daring steps into space, the Mercury program proceeded more cautiously, testing the footing each step of the way. There were 25 flights in all, only some of which were piloted. To ensure the safety of spacecraft for humans, NASA, like Russia, placed several animals into orbit. In November 1961, a chimp named Enos rocketed into orbit and safely returned. This inhumane use of chimps and monkeys would continue into the Apollo program.

Into Orbit

A month after Gagarin's historic mission, the first piloted Mercury flight was made by astronaut Alan Shepard, in space capsule Freedom 7. This was a suborbital flight, traveling beyond Earth's atmosphere without going into orbit. The second Mercury flight was also suborbital. By the time John Glenn earned the distinction of being the first American astronaut to orbit Earth, on 20 February 1962, the Soviet Union had already completed four piloted missions—one lasting four days. There were six Vostok flights in all, including one that carried Valentina Tereshkova, the first woman in space, in June 1963.

Later Mercury flights tested maneuvering systems, satellite deployment, and human performance in zero gravity.

Two's Company

The success of the Soviets' Vostok flights led them to embark on a brief but more ambitious program called Voskhod ("sunrise"). This entailed just two flights using a redesigned Vostok spacecraft, but each carried more than one cosmonaut at a time. In October 1964, five months before the first two-person Gemini mission, Vladimir Komarov, Boris Yegorov, and Konstantin Feoktistov rode Voskhod 1 into space and successfully completed 17 orbits. The first spacewalk was made by Aleksei Leonov from Voskhod 2 in March 1965.

In the same month, the first manned Gemini launch—Gemini 3—took place, piloted by Virgil Grissom and John Young. The astronauts performed orbital changes and other maneuvers that were important practice for future flights. Gemini 4, launched in June 1965, featured the first spacewalk by a United States astronaut. While pilot James McDivitt manned the spacecraft, Edward White drifted in space for 22 minutes, propelling himself on the end of a tether with a hand-held maneuvering device.

The next two years saw a succession of Gemini flights, which tested spacecraft maneuvers, refined spacewalking techniques, and studied the effects of spaceflight on humans. Perhaps the most important accomplishment of the program was the world's first docking maneuver in space, completed by Neil Armstrong and David Scott during the Gemini 8 flight in March 1966. Docking, a procedure in which one spacecraft joins another in space, was crucial to the success of future Moon missions. Other dockings followed with Geminis 10, 11, and 12.

Tragic Outcome

The Soviet's new Soyuz ("union") program dovetailed with the conclusion of Project Gemini. The Soyuz craft was designed for docking and for taking cosmonauts to the Moon. In April 1967, Soyuz 1 was to rendezvous with a second vehicle, Soyuz 2. The flight, piloted by Vladimir Komorav, ended in tragedy when Soyuz 1 developed serious control problems early in the flight. Its landing parachute failed to open, and Vladimir Komorav was killed.

Docking attempts continued with the launch of Soyuz 2 and 3 in October 1968. This attempt failed, but both spacecraft returned safely to Earth. Not until January 1969 did Soyuz 4 and 5 achieve a successful docking and crew transfer from one spacecraft to the other. By then, however, the Americans were orbiting astronauts around the Moon.

MILESTONES *Walter Schirra is pulled from the Mercury capsule after his 1962 mission (right). Edward White spacewalks during the Gemini 4 mission (below).*

UNI
ST

The Apollo Program

APOLLO 11

The race to put a man on the Moon was the impetus of NASA's ambitious Apollo program, which, in just six short years, changed the boundaries of human aspirations.

After the Gemini program came Apollo, NASA's most ambitious project yet. It began with seven unpiloted test flights, mostly involving the Saturn I and the Saturn IB rockets. These test flights led to the design and building of the much larger Saturn V rocket, which would become the heavy-lift vehicle for the Apollo program. The Saturn V was capable of hefting 285,000 pounds (130,000 kg) into Earth orbit, and 100,000 pounds (45,000 kg) to the Moon.

The cone-shaped Apollo command module, designed to carry three astronauts, sat atop the service module. While in Earth orbit, the lunar module was docked with the command module. Once in lunar orbit, two of the astronauts would enter the lunar module for the trip to the Moon's surface, leaving the command and service modules in orbit.

Reach for the Moon

Like the Soviet's Soyuz series, the Apollo program had a disastrous beginning. On 27 January 1967, an electrical fire erupted during an Apollo 1 training exercise just a month before its scheduled launch, killing astronauts Virgil Grissom, Edward White, and Roger Chaffee. The Apollo program was grounded for more than

a year until modifications made the spacecraft safer. Apollos 4, 5, and 6 were unpiloted flights designed to test the new systems and refine procedures for putting hardware into orbit. Their success led to the piloted Apollo 7 mission in October 1968, and the first lunar orbit by humans, in Apollo 8 in December 1968, which returned startling images of Earth from space. At least part of President Kennedy's goal had been realized.

On 3 March 1969, Apollo 9 tested rendezvous and docking

GIANT STEPS *Apollo 4, the first unmanned craft in the program, on the launchpad at the Kennedy Space Center (left). Walter Schirra (above) aboard the first piloted craft, Apollo 7. The Apollo 11 logo (top left) echoes Neil Armstrong's message that "the Eagle has landed."*

procedures with the lunar module in Earth orbit. Two months later, Apollo 10 returned to lunar orbit to check the module's guidance and navigation systems. Astronauts Thomas Stafford and Eugene Cernan brought the lunar module to within 9 miles (15 km) of the first landing site, the Sea of Tranquillity, before firing their ascent engine and returning to the command module, piloted by John Young. The lunar surface was now within reach.

Mission Accomplished

Less than a month later, on 20 July 1969, astronauts Neil Armstrong and Edwin "Buzz" Aldrin separated Apollo 11's lunar module, the *Eagle*, from

APOLLO 11 *"Buzz" Aldrin steps onto the lunar surface (above) as part of the 1969 mission. The command module in orbit over the Moon (right).*

the command and service modules and began the descent to the surface of the Moon. Soon after, the *Eagle* touched down on a flat area in the Sea of Tranquillity. The two astronauts checked the equipment, then Armstrong stepped out of the lunar module onto the powdery surface of the Moon, uttering the famous words: "That's one small step for a man, one giant leap for mankind." (A gap in transmission obscured the word "a" for most Earth-bound listeners.)

This historic mission was followed by five successful lunar landings: Apollos 12, 14, 15, 16, and 17. Apollo 13 was aborted on the way to the Moon because of an explosion that ruptured an oxygen tank and damaged other systems. Even this mission, however, was considered a victory of sorts—a catastrophe averted by the skill of the NASA team.

The Apollo Moon missions returned 844 pounds (382 kg) of rock and soil samples, amazing photographs of lunar features, and a vast quantity of data generated by instruments set up on the surface. But that was not all it accomplished. As one of the greatest technological feats of our species, it elevated our view of the place of humankind in the universe to unprecedented heights.

WEIGHTLESSNESS

Bungee jump from a bridge or stall an aircraft, and you will feel as if your weight is reduced to zero. Astronauts experience the same sense of apparent weightlessness during space missions, where it is a source of both enjoyment and discomfort. The term "zero gravity" is sometimes applied to this state. Like the term "weightless," however, it is somewhat misleading. Gravity is still present—gravity is what keeps the spacecraft in orbit. And although astronauts do not feel their own weight in space, they are still attracted by Earth's gravitational field. What causes the sense of weightlessness is the fact that the astronauts and the spacecraft are falling together toward the Earth.

Under such conditions, you do not walk, but float from place to place. Objects drift about (right) and heavy items are easily lifted. Astronauts may experience temporary nausea, called space sickness, but more serious is a loss of muscle tone and bone strength, and a decrease in white blood cells, which fight diseases. To deal with these effects, astronauts must perform special exercises during long space flights.

Beautiful, beautiful ...

Magnificent desolation.

Edwin "Buzz" Aldrin (b. 1930), Apollo 11 astronaut, as he followed Neil Armstrong to the surface of the Moon.

SHUTTLES *and* STATIONS

The space race is over, replaced by a spirit of cooperation where expertise is shared.

With the Moon race won, NASA began work on a new type of space vehicle that could be launched conventionally via rockets, but would also be reusable. On 12 April 1981, the space shuttle *Columbia* rocketed off the launch pad at Cape Canaveral, piloted by astronauts John Young and Robert Crippen. After a 54-hour test flight, it glided safely back to Earth.

The shuttle has three components: the orbiter, the external tank, and the solid rocket boosters. The orbiter is the distinctive, delta-wing vehicle that carries the payload into orbit. Its underside is covered with 23,000 tiles that protect it from burning up during reentry. The payload bay, at the center of the orbiter, is large enough to hold a passenger bus. The external tank fuels the shuttle engines during launch and burns up in the atmosphere, while the two solid rocket boosters that help propel the shuttle off Earth's surface are ejected prior to orbit. These parachute into the Atlantic Ocean, where they are retrieved and refurbished.

Once in orbit, the shuttle is maneuvered by means of two onboard engines, which are fired at mission's end to slow the orbiter's speed. As the shuttle reenters the atmosphere, aerodynamic controls take over. Moving into a shallow glide path, it makes an unpowered landing.

Today, NASA maintains four space shuttles: *Columbia,*

IN SERVICE *The space shuttle (right) has proved invaluable for making adjustments to satellites, including the very successful repair of the Hubble Space Telescope in 1993 (below).*

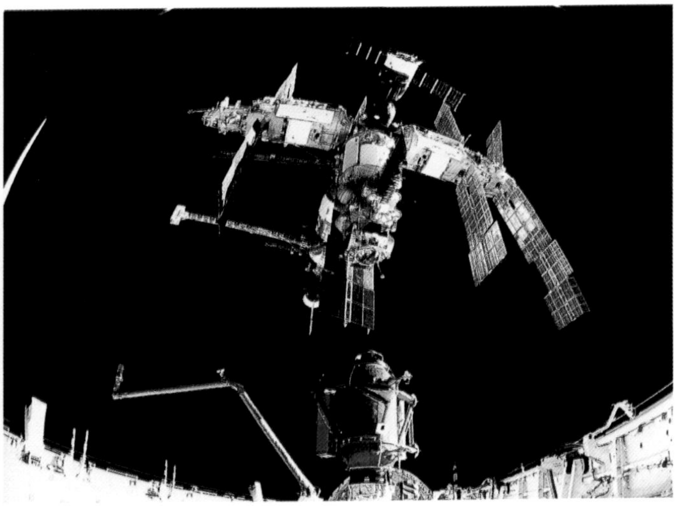

VISITORS *The Mir Space Station (left) about to dock with the space shuttle. A shuttle astronaut's view of a Mir cosmonaut looking out of his window (below).*

Discovery, Atlantis, and the newer *Endeavor,* which was built to replace *Challenger,* destroyed in a catastrophic explosion (see Box). *Endeavor* can remain in space for 16 days, while the others can make 14-day flights.

The shuttle has racked up many accomplishments. Reusable space laboratories have flown aboard it many times, conducting scientific experiments and astronomical observations. The shuttle's ability to retrieve and deploy satellites from the payload bay with the Remote Manipulator System, a huge mechanical space arm, has made it a valuable asset in the launching, servicing, and repair of satellites of all types. The Hubble Space Telescope has been successfully serviced twice from the shuttle's payload bay. Undoubtedly, the space shuttle will continue to serve science well into the twenty-first century.

Space Stations

In the early 1970s, the Soviets launched Salyut ("salute"), the first manned space station. Although rather cramped, Salyut established the viability of operating and maintaining laboratories in space. Six successful Salyut missions were made between 1971 and 1986.

The United States launched their space station Skylab on 14 May 1973. Hundreds of experiments and observations were made during its three missions. Skylab eventually fell back to Earth, burning up over Australia in 1979.

The most successful space outpost has been the Russian Mir ("peace") space station, launched in February 1986, and occupied almost continuously since. It has been visited several times by United States shuttle astronauts in a prelude to the construction of an international space station.

THE CHALLENGER DISASTER

Supposed to be a day of "firsts" for the space shuttle, 28 January 1986 is, instead, one of the American space program's darkest. *Challenger* lifted off with seven crew members, including its first civilian, Christa McAuliffe, a schoolteacher. Less than two minutes into the flight, the huge external tank exploded. The two solid rocket boosters went spiraling away from the blast, and *Challenger* broke into pieces. Minutes later, its cabin and crew fell into the sea.

A team of scientists and engineers determined that because of the extremely cold weather (the temperature had dipped below freezing the night before), a rubber O-ring seal in one of the solid rocket boosters was less pliable than it should have been. Instead of expanding to fill the joint in the booster, it remained stiff. Hot gases ruptured the joint and flames shot out, igniting the thin-skinned external tank, which was filled with liquid oxygen and liquid hydrogen.

Engineers labored for two years, not only to redesign the joints of the boosters, but to rethink all critical items that went into a shuttle launch. By the time *Discovery* lifted off on 29 September 1988, safety was the number-one priority, making the shuttle a reliable, but complex and expensive, way into orbit.

Flight directors in Houston Mission Control watch, horrified, as Challenger *explodes.*

To *the* Planets

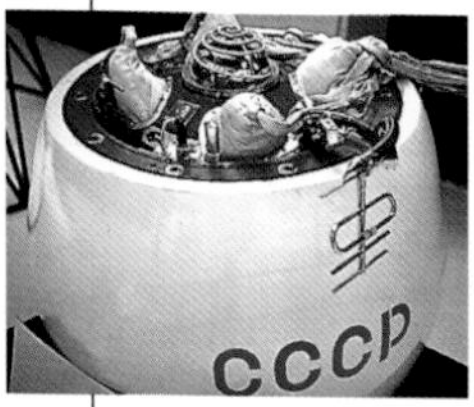
CCCP

Even before the first person was put in orbit, space visionaries were looking to more distant and tantalizing horizons—the planets.

The space race of the 1960s may have had a political agenda, but it also served to inspire the exploration of an entirely new frontier. Expeditions to the planets were made a few cautious and calculated steps at a time.

The First Journeys

Venus, the closest planet and a sister world to Earth, being similar in size and mass, beckoned as a reasonable and pragmatic first objective. Success, however, was literally hit and miss. Venera 1, launched by the Soviets on 12 February 1961, failed when radio contact was lost two weeks after launch. The probe missed the planet by 60,000 miles (97,000 km) and went into orbit around the Sun. A similar fate awaited Venera 2, while Venera 3 crashed into Venus. Not until Venera 4 would the Soviets successfully rendezvous with Venus, in October 1967, but the capsule that was supposed to make a soft landing was lost.

The Americans fared better with Mariner 2, a flyby mission to Venus, launched in August 1962. Mariner's instruments studied the planet's atmosphere and measured its lead-melting surface temperatures. The probe also discovered that Venus did not have a magnetic field or radiation belts.

ASSIGNMENT VENUS *The Mariner 2 spacecraft (right) made a successful flyby of Venus in 1962, five years before a Venera probe (left) reached the planet.*

The Soviets continued their efforts to land a spacecraft on Venus with Venera 7 and 8, launched in 1970 and 1972. Both capsules made it to the surface and transmitted data confirming Venus's extreme temperatures and pressures. A few years later, in 1975, Venera 9 and 10 were the first spacecraft to return images successfully from another planet's surface. Like the other kinds of data, the images were transmitted via radio beams.

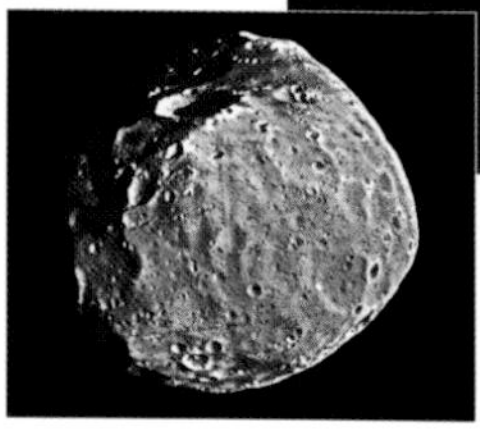

MARINER 9 *returned images of Mars and its moon Phobos (left).*

Mariner Missions

While the Soviets were busy with Venus, United States scientists were conducting a series of flybys of the planet Mars. Mariner 4 was launched in November 1964 and yielded a total of 22 photographs that revealed a heavily cratered terrain. Mariners 6 and 7 produced 201 photos, including the first close-up views of a Martian polar cap.

The most successful of the early Mars missions was Mariner 9, which entered orbit around the Red Planet on 3 November 1971, and spent the next 349 days photographing the surface. The probe's two video cameras returned more than 7,000 images and obtained fascinating close-up views of the planet's two moons, Phobos and Deimos.

Both Venus and Mercury were the targets of Mariner 10, launched 3 November 1973. Its cameras, equipped with ultraviolet filters, took photos of the layered cloud tops of Venus and revealed a complex global circulation pattern. Then the probe flew on to Mercury, where it captured more than 10,000 close-up photos of the Moon-like surface and discovered the planet's magnetic field.

Nature … Doth teach us

to have aspiring minds …

And measure every

wandering planet's course …

Tamburlaine the Great,
CHRISTOPHER MARLOWE
(1564–93), English dramatist

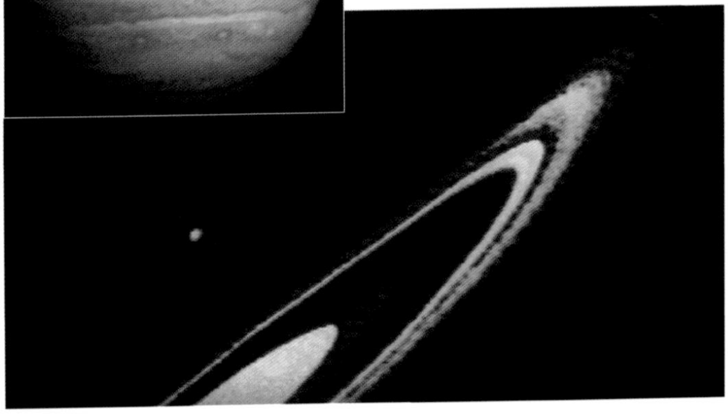

PIONEER 11 *was launched in 1973. It returned some of the first close-up images of Jupiter (left), before visiting the rings of Saturn (below) in 1979.*

THE PIONEER SPIRIT

The United States reached farther out into the Solar System with the Pioneer launches in 1972 and 1973. Pioneer 10 made the first flyby of Jupiter on 4 December 1973. Its closest pass was within about 80,000 miles (130,000 km) of the cloud tops. The photos it relayed showed Jupiter's churning atmosphere, including the Great Red Spot, and three of the planet's four large moons—Europa, Ganymede, and Callisto.

Almost exactly a year later, Pioneer 11 flew even closer to Jupiter—to within about 30,000 miles (48,000 km). Five years later, the probe passed Saturn, making a close approach of 18,600 miles (30,000 km) and returning 440 images and new data about the planet, its moons, and its ring system.

Today, both Pioneers 10 and 11 are heading into deep space in opposite directions. One day thousands of years from now, one or both of them may serve as an emissary from Earth to other intelligent life forms in the galaxy. Mounted on each spacecraft is a plaque that will convey information to anyone out there about our Solar System, the Earth, and the human race.

THE VIKINGS

In the middle to late 1970s, the United States embarked on a bold exploration of the outer Solar System. It began with the Viking probes, two identical spacecraft sent to Mars in the middle 1970s. Each consisted of an orbiter and a lander. The lander was designed to photograph the surface, collect and analyze soil samples, and make meteorological observations. The orbiter's duty was to image and map the planet's surface in detail. By the summer of 1980, the two Viking landers had sent back 4,200 photos of the surface and three million weather reports. The images showed a desolate rock and sand-dune terrain stretching to the horizon and a dusty pink sky. Viking 1 measured light winds, with gusts of up to 30 miles per hour (50 km/h). From afternoon to night, temperatures fell from –20 to –120 degrees Fahrenheit (–30° to –85° C).

Each lander outlived its expected 90-day life span, Viking 2's functioning until April 1980, and Viking 1's operating until late November 1982. The orbiters, meanwhile, sent back more than 52,000 photographs.

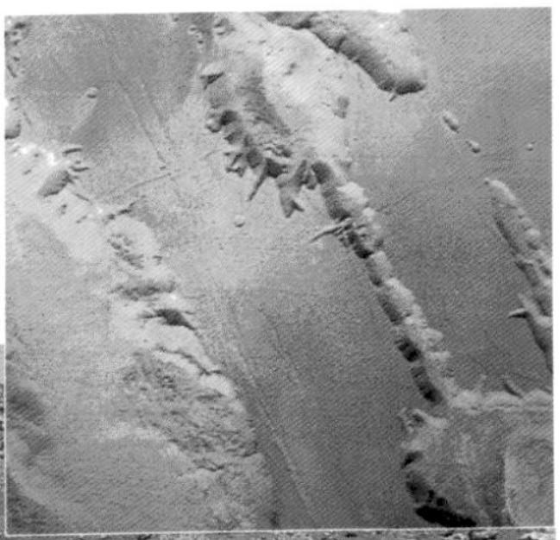

ROUGH SURFACE *The Viking landers returned images of Mars's rocky terrain (left), while the orbiters showed features such as huge canyons (above left).*

Voyager *and* Beyond

Recent years have seen even more ambitious explorations of the Solar System, from the outer planets to the Sun at its center.

In March 1979, Voyager 1 flew past Jupiter, sending back thousands of high-resolution photos of the cloud tops as well as intriguingly detailed images of the moons Callisto, Ganymede, Europa, and Io. Jupiter's gravity gave Voyager 1 the velocity needed to reach its ultimate destination, Saturn, in November 1980. There, it discovered six new moons and returned spectacular pictures of the planet's rings.

Perhaps the most successful probe of all was Voyager 2, which flew past all four gas-giant planets. It reached Jupiter in July 1979, where it skimmed past the moons Callisto, Ganymede, Europa, and Amalthea, then went on to Saturn, traversing the planet's ring plane and passing closer to its moons than had Voyager 1. In January 1986, Voyager 2 reached Uranus, where it discovered several new moons. Finally, in August 1989, Voyager 2 made a close approach to Neptune, revealing an aqua-blue atmosphere punctuated by a huge dark spot and fleecy white cirrus clouds. It also discovered six new moons and four dark rings encircling the planet.

Galileo

Intrigued by the Voyager findings at Jupiter, planetary scientists once again turned their attention toward that planet in the early 1980s. They designed a mission to orbit Jupiter and monitor the planet and its satellites for at least two years. In addition, a probe was to be deployed from the spacecraft into Jupiter's atmosphere.

The launch was scheduled for May 1986 from a space shuttle, but the *Challenger* disaster (see p. 31) in January of that year put all shuttle missions on indefinite hold. By the time shuttles were flying again, nearly two years later, the launch window that had afforded a direct path to Jupiter had come and gone.

Galileo did not have enough power for a direct flight, so a three-year flight path was designed to "sling" Galileo to Jupiter using gravity assists from Venus and Earth. The flight path also brought it near asteroids Gaspra in 1991 and Ida in 1993. High-resolution images of Gaspra indicated that it was possibly a fragment of a larger body, while those of Ida revealed that it was orbited by a small moon.

The spacecraft finally made it to Jupiter in December 1995.

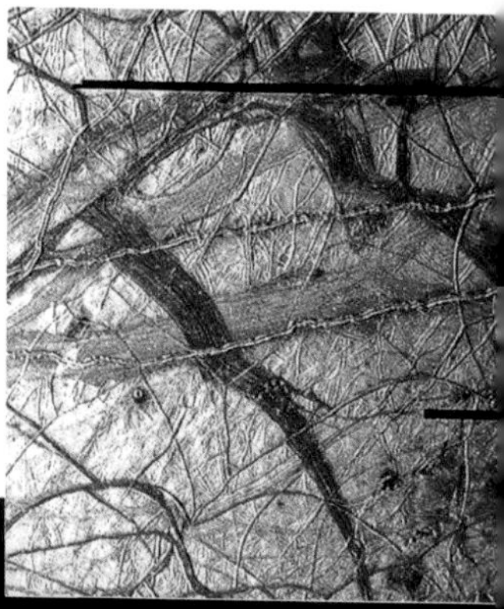

WORTH THE WAIT *Voyager 2 (top left) returned the first close-up images of Neptune. Since its launch from the space shuttle* Atlantis *in 1989 (below), the Galileo probe has returned amazing images, such as this (right) of the surface of Jupiter's moon Europa.*

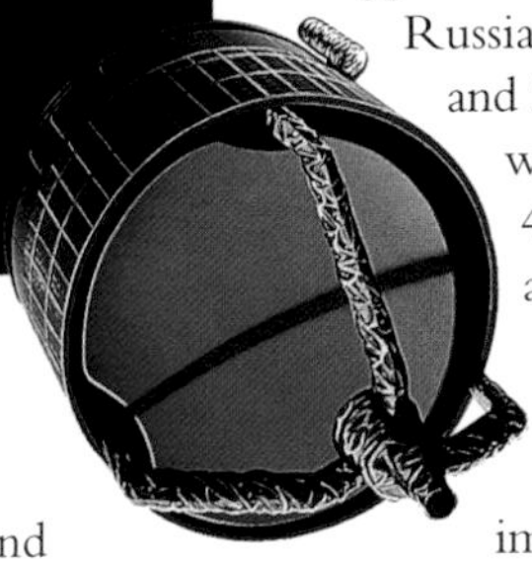

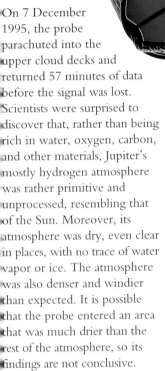

LONG-TERM PROJECTS *The Ulysses spacecraft (left) on its way to the Sun; and the ESA's Giotto space probe (below) heading toward Halley's Comet.*

On 7 December 1995, the probe parachuted into the upper cloud decks and returned 57 minutes of data before the signal was lost. Scientists were surprised to discover that, rather than being rich in water, oxygen, carbon, and other materials, Jupiter's mostly hydrogen atmosphere was rather primitive and unprocessed, resembling that of the Sun. Moreover, its atmosphere was dry, even clear in places, with no trace of water vapor or ice. The atmosphere was also denser and windier than expected. It is possible that the probe entered an area that was much drier than the rest of the atmosphere, so its findings are not conclusive.

Meanwhile, the orbiter has returned tantalizing data on the Galilean satellites (so called because they were first seen from Earth by Galileo), discovering, among other things, that Europa may have a relatively thin icy crust covering a mantle of liquid water or slush. This concoction may contain chemicals that could nurture life.

Comet Missions

Although the planets have usually been the destinations of space probes, minor members of the Solar System have not been ignored. In March 1986, several space probes passed near Halley's Comet as it approached Earth. Two Russian probes, Vega 1 and 2, approached within 5,520 and 4,990 miles (8,890 and 8,030 km) respectively on 6 and 9 March. Although the images from their cameras were blurry, onboard spectrometers successfully measured the composition of the comet's dust.

Four days later, the European Space Agency's probe, Giotto, intercepted Halley's Comet. It penetrated the coma, or head, of the comet, passing within 370 miles (600 km) of the nucleus and sending back astonishing images of bright geyser-like jets erupting from a black potato-shaped body. The composition and mass of the coma and the dust tail were measured, as was the interaction between the comet's gas and the solar wind. Halley's solar-wind interactions were further studied by the International Sun-Earth Explorer and the Japanese probes Sakigake and Suisei.

Solar Probes

Ulysses, a joint mission by NASA and the European Space Agency (ESA) to study the Sun's environment and the solar wind, was launched in October 1990. This was the first time a spacecraft had been deliberately sent out of the ecliptic (the plane of the Solar System). After receiving a gravity assist from Jupiter in February 1992, the Ulysses probe went on to pass over the Sun's south pole in May 1994, then over its north pole a year later. As well as gathering data about the solar wind, Ulysses' instruments have returned important information about the interplanetary magnetic field, interplanetary dust, and cosmic rays.

The Japanese have sent two solar observatories designed to study extremes of solar energy. Hintori ("firebird"), launched in 1981, collected X-ray data on the 11-year cycle of solar activity. Yohkoh ("sunbeam"), launched in 1991, carries X-ray and ultraviolet telescopes to image the Sun's rarefied corona and to study solar flares.

The Solar Heliospheric Observatory (SOHO) was launched in December 1996. A joint project of the ESA and NASA, SOHO was placed into orbit around a Lagrangian point—a region about 930,000 miles (1.5 million km) sunward from Earth where the gravitational pull of the Sun and Earth are in balance. SOHO's instruments are studying the temperature and structure of the Sun's interior, the corona, and the solar wind, and how these change as the Sun becomes more active.

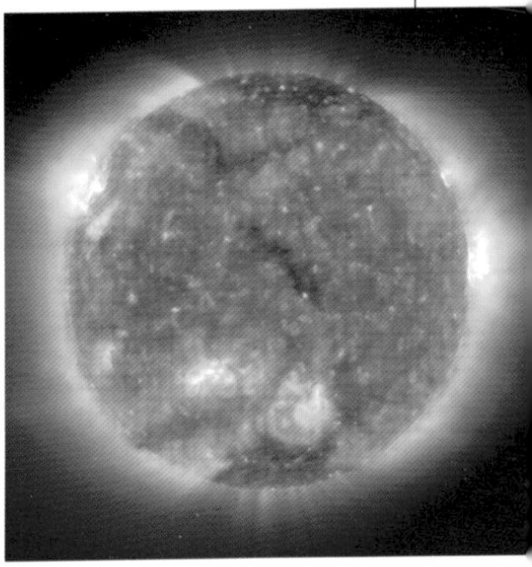

SOHO *returned this ultraviolet image of the Sun's corona in 1997. Scientists hope SOHO will make observations until 2002.*

WHERE *to from* HERE?

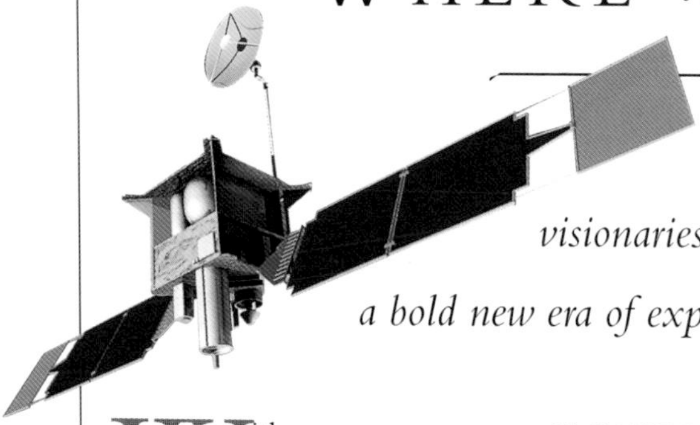

If just some of the dreams of the science visionaries come true, the next century promises to be a bold new era of exploration and discovery in astronomy.

With ever more sophisticated technology and mountains of data pouring in, it is hard to imagine what revelations even the next 25 years will bring. We will never understand everything about the universe because each answer poses new questions, but astronomers are closing in on some notable controversies: the existence of fossilized forms of microbiotic life on Mars; the intriguing conditions on the moons of Jupiter; the ages of the oldest stars; the presence of Earth-like planets around other stars; and the distances to remote galaxies. Still ahead lie the daunting tasks of ascertaining the true size and age of the universe, the physical characteristics of the earliest galaxies, and the composition of the mysterious "dark matter," said to constitute more than 90 percent of the universe's matter.

In some respects, the future is already here. Ground-based astronomy is using new mirror-making technology to create larger, less-expensive telescopes. With computer-controlled optics to counteract the effects of Earth's atmosphere, these instruments will be almost like space telescopes on the ground. Radio telescopes linked to form the equivalent of one huge telescope will look into the hearts of galaxies and quasars, and map the universe for unknown sources of radio energy.

Space missions currently underway include NASA's Cassini-Huygens mission, a seven-year, 2 billion mile (3.2 billion km) journey to Saturn. Soon after arrival, it

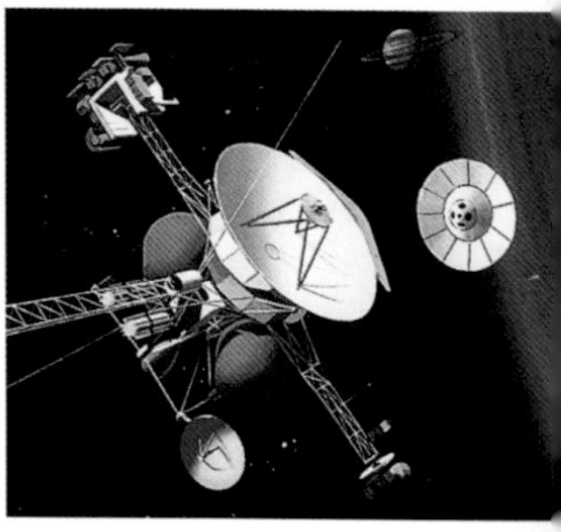

IN FLIGHT *The Mars Global Surveyor (left) on its way to the Red Planet. The Cassini orbiter (right) will study Saturn.*

MARS MISSION *The Mars Global Surveyor was assembled at Cape Canaveral (left). The Pathfinder lander carried the solar-powered Sojourner rover (below), which began its work by inspecting two rocks dubbed "Barnacle Bill" and "Yogi."*

will deploy the European-built Huygens probe to explore the atmosphere and surface of the giant moon Titan. Then the Cassini orbiter will embark on a four-year tour of Saturn, sending back data about the planet, its 18 known moons, and its spectacular ring system.

Scientists have ambitious plans for Mars, too. Leading off is the Mars Pathfinder, which arrived at the planet on 4 July 1997. Its surface-rover, Sojourner, studied Martian geology, analyzed soil, and transmitted exciting images of surface features. Hot on its heels is the Mars Global Surveyor, which will orbit the planet to map surface topography, mineral distribution, and global weather.

In February 1996, NASA launched its Near Earth Asteroid Rendezvous mission. The NEAR spacecraft's trajectory first takes a flyby look at asteroid 253 Mathilde, but its ultimate destination is 433 Eros, in 1999. It will orbit and study Eros's surface for a year, giving scientists their first look at an asteroid of a type that has collided with Earth.

COOPERATIVE EFFORT *The space shuttle approaches the proposed International Space Station (right).*

Missions in View

A number of other expeditions are planned, subject to available funding. With the Lunar Prospector, NASA hopes to map the Moon's chemical composition and magnetic and gravity fields. The Moon is also being considered as a location for both radio and optical telescopes. It has a stable surface, low gravity, and a negligible atmosphere, and its slow rotation offers continuous viewing for nearly two weeks at a time.

SPACE JUNK

There are literally thousands of artificial "satellites." The US Space Command's Space Surveillance Center near Colorado Springs, Colorado, monitors the whereabouts of more than 7,000 artificial objects. A few hundred are working satellites and spacecraft, but thousands more are spent rocket boosters, shards from exploded rockets, chunks of solar panels, screws and tools from previous space missions, and several thousand defunct satellites. This space debris (depicted below as yellow dots) is becoming a big safety issue. A fleck of paint moving at speeds of 17,000 miles per hour (27,000 km/h) has the energy to chip a space shuttle window or pierce the skin of a satellite. Imagine how damaging, and dangerous, an errant screw or a large piece of metal could be moving at that velocity.

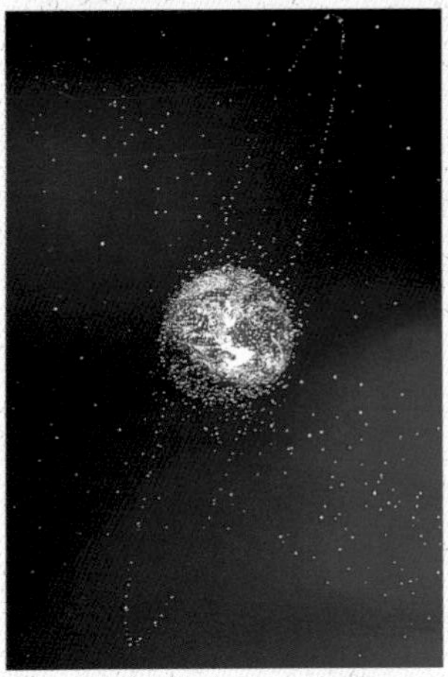

Objects in low Earth orbit reenter the atmosphere and burn up after a few months. Objects at greater altitudes could remain in orbit for centuries. Each piece of debris must be diligently tracked to avoid harm to spacecraft.

The Venus Multi-Probe Mission spacecraft would deploy 14 probes to measure temperature and wind simultaneously at different altitudes, revealing more about Venus's global circulation pattern.

The three-year Suess-Urey mission would travel beyond Earth's magnetosphere and use huge collectors to gather particles from the solar wind and return them to Earth for study.

The European Space Agency (ESA) hopes to examine a comet with Rosetta, which would spend eight years maneuvering among the planets. By 2011, it would rendezvous with periodic Comet Wirtanen, and spend 17 months monitoring the comet's changes as it nears the Sun.

NASA and Japan's Institute of Space and Astronautical Science are cooperating on MUSES-C, which would be the first mission to return samples from an asteroid.

People in Space

Will humans ever live in space? An International Space Station is being developed by the United States, Canada, Japan, Russia, and the ESA. This would be an orbiting laboratory for scientific experiments, astronomical observations, and biomedical research. Some forward-looking scientists and engineers foresee human bases on the Moon and even on Mars.

WHERE *in the* WORLD?

You may never travel in a spacecraft, but you can visit one of the many professional observatories and space centers on Earth.

This map marks just some of the world's largest observatories and space centers. Many of these provide tours, brochures, and programs for the public. Space centers tend to be located in major cities, but functioning optical and radio observatories will be found in more remote regions, well away from city light pollution and radio noise. Old observatories in urban areas, such as the Royal Greenwich Observatory in London, are often kept as museums and are well worth a visit.

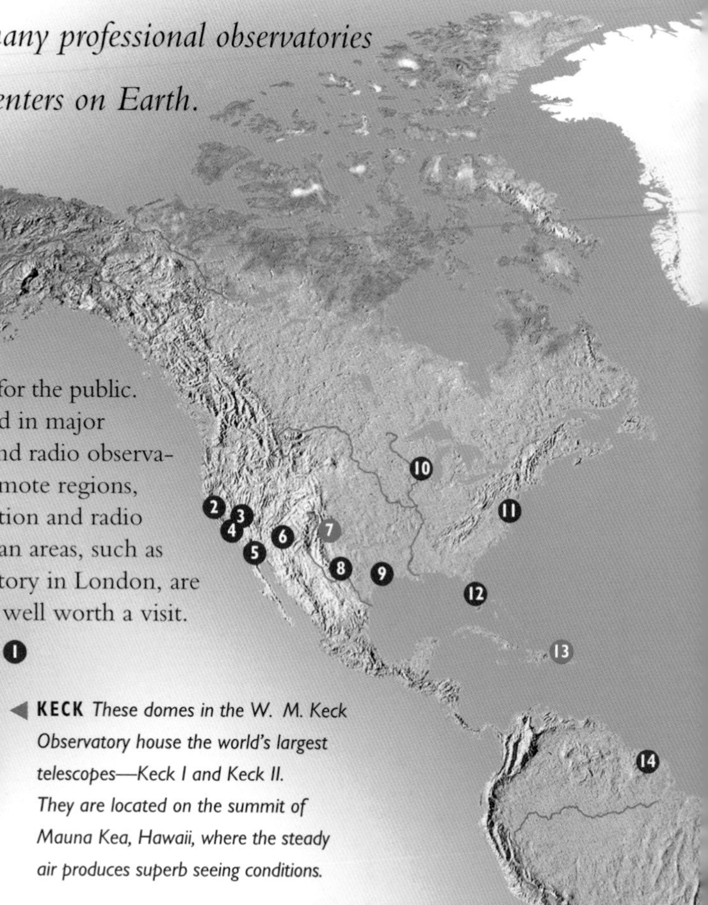
1
2
3
4
5
6
7
8
9
10
11
12
13
14
15
16

◀ **KECK** *These domes in the W. M. Keck Observatory house the world's largest telescopes—Keck I and Keck II. They are located on the summit of Mauna Kea, Hawaii, where the steady air produces superb seeing conditions.*

▲ **THE VERY LARGE ARRAY** *(VLA) near Socorro, New Mexico, consists of 27 dish antennas, each 80 feet (25 m) across, arranged in a huge Y-shape. The antennas can be linked to form a powerful radio telescope—a technique known as interferometry. The VLA is operated by the National Radio Astronomy Observatory (NRAO).*

WORLD OBSERVATORIES AND SPACE CENTERS

1. W. M. Keck Observatory, Hawaii
2. Lick Observatory, California
3. Jet Propulsion Lab, California
4. Mt Wilson Observatory, California
5. Palomar Observatory, California
6. Kitt Peak Observatory and Mt Hopkins Observatory, Arizona
7. Very Large Array, New Mexico
8. McDonald Observatory, Texas
9. Johnson Space Center, Texas
10. Yerkes Observatory, Wisconsin
11. Goddard Space Center, Maryland
12. Kennedy Space Center, Florida
13. Arecibo Telescope, Puerto Rico

esa
esoc

ESOC *Banks of monitors stretch across the mission-control room of the European Space Operations Centre (ESOC). Located in Darmstadt, Germany, ESOC oversees satellite operations, data reception, and processing for space flights conducted by the European Space Agency (ESA).*

18
19
20
21
22
23
24
25
17
26
27
KEY
optical telescope
radio telescope
space center

AAO *The Anglo-Australian Observatory at Siding Spring in New South Wales, Australia, includes the UK Schmidt Telescope. This 70 inch (1.8 m) instrument is used mainly for astro-photography and sky surveys.*

ARECIBO *Cornell University operates a spherical 1,000 foot (305 m) radio telescope near Arecibo, Puerto Rico. The enormous dish is formed by means of wire mesh suspended above the ground in a natural hollow.*

14 European Space Agency launch site, French Guiana
15 European Southern Obs., Chile
16 Cerro Tololo Observatory, Chile
17 La Palma Observatory, Canary Is.
18 Jodrell Bank, England
19 Royal Greenwich Obs., England
20 European Space Agency, France
21 Effelsberg Telescope, Germany
22 European Space Operations Centre (ESOC), Germany
23 Bolshoi Telescope, Russia
24 Baikonur Cosmodrome, Kazakhstan
25 Kagoshima Space Center, Japan
26 Anglo-Australian Observatory
27 Australia Telescope

At length, by sparing neither labor nor expense, I succeeded in constructing for myself an instrument so superior that objects seen through it appear magnified nearly a thousand times …

GALILEO GALILEI (1564–1642),
Italian astronomer, mathematician, and physicist

CHAPTER TWO

SKYWATCHING TOOLS

BECOMING *a* BETTER OBSERVER

Skywatchers can spend a fortune on elaborate equipment, but the sky comes free of charge.

Armed with no more than a good astronomy guidebook, you can enjoy skywatching from any location. A beautiful grouping of the Moon and planets, the swirling display of an aurora, or the progress of stars and constellations through the seasons—these are the simple pleasures of low-cost, naked-eye astronomy.

A pair of binoculars, which you may already own, opens up new vistas—from the Milky Way resolved into thousands of stars, to a glowing comet or the shuttling moons of Jupiter.

While telescopes are seen as the entry tickets to the hobby, many people buy one far too soon. If you do not yet own a telescope, wait until you know your way around the sky. Take the opportunity at star parties and observatories to look through a variety of telescopes. When you are ready to buy one of your own, consult the guidelines in this chapter. If you already have some skywatching equipment, the advice in this chapter will help you make the most of it.

RECORDING THE SKY

Some amateur astronomers specialize in particular kinds of observations: perhaps mapping storm features on Jupiter, tracking down an asteroid, or estimating the brightness of a variable star. Submitting formal reports of these observations to organizations can provide a great sense of accomplishment. Indeed, astronomy is one of the few sciences in which amateurs still regularly make genuine contributions.

Even if your observations go no farther than home, you will find keeping some type of sky diary or logbook very satisfying. It can be simply a list of what you see each night out, embellished with some descriptions. Many people like to add sketches of eyepiece views. Others use cameras to record the sky (see pp. 60–69).

CHOOSING A SITE

Your observing site will have a greater effect on your skywatching than any piece of equipment. Of course, a dark site far from city lights is ideal, but few of us have that luxury. City-bound observers may not be able to find faint galaxies, but the Moon and planets can still look splendid.

If a bright light is glaring at you, your eyes will not fully adapt to the dark and it will be difficult to see fainter targets. Try to find a site shadowed from street and yard lights. A black cloth thrown over your head, in the style of old-time photographers, can also help.

Some enthusiastic amateurs build backyard observatories (check back issues of astronomy magazines for instructions). Some are simple garden sheds with flip-top or roll-off roofs. Others have rotating domes and heated rooms. No matter how modest, the convenience of a home observatory is hard to beat, even if the sky conditions are less than ideal.

NOVICE SKYWATCHERS *(left) should learn key stars and constellations before buying a telescope. Keep a sky diary (top) with notes and sketches as your personal record.*

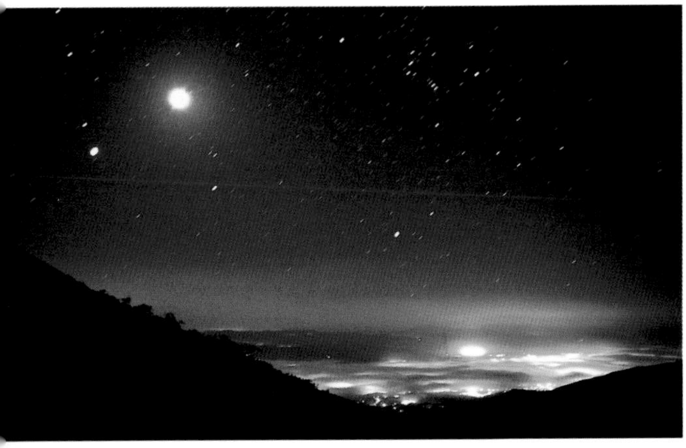

URBAN SKIES *have much to offer stargazers, with the Moon and planets unaffected by skyglow (left). However, the spread of light pollution from cities (indicated by white on the satellite image below) means many astronomers must travel farther to find dark skies.*

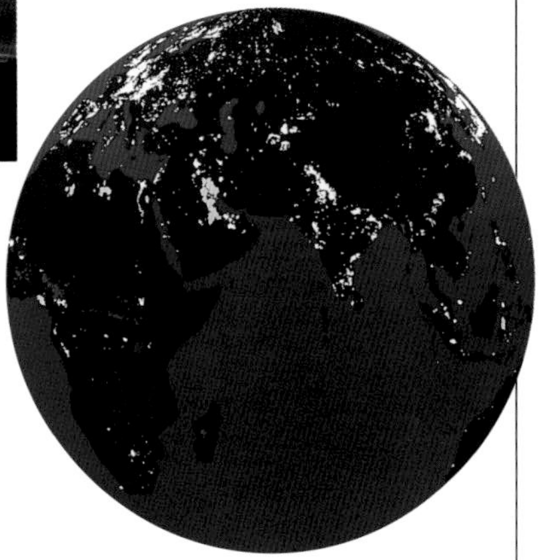

As with most naturalist pursuits, however, the beauty of the night sky is enhanced if you leave the urban sprawl behind. A drive of 30 to 45 minutes beyond city limits is often enough to get out from under the dome of light that covers every city and town.

Particularly for observers in the Northern Hemisphere, the best direction to drive from a city, if possible, is south. This puts the city glow to the north and places the southern sky—the most interesting part, which includes the Milky Way—above the darkest horizon. Parks are good destinations, but introduce yourself to the officials in charge or they might think you are a late-night vandal. Local astronomy clubs often maintain rural observatories on private land for their members. Observing with a group provides security and companionship.

Clear Skies

Amateur astronomers are also weather-watchers, paying close attention to daytime cloud patterns and TV weather maps. They soon learn that a summer sky filled with puffy cumulus clouds during the day often heralds a clear night sky.

If the sky is clear, the Milky Way may be seen even from suburban locations, and you are more likely to find faint objects such as nebulas.

Moisture of any kind in the air degrades the transparency of the sky. Clear nights usually come after a cold front sweeps out haze and humidity, then brings a dry high-pressure center (marked with an H on weather maps). Nights with a low dewpoint also tend to be transparent. The dewpoint is the temperature at which moisture in the air condenses, and a low dewpoint means moisture is unlikely to condense out of the atmosphere as the temperature falls during the night.

Good Seeing

Memorable nights of viewing can, however, come with many kinds of weather. For the Moon and planets, humid nights with stagnant haze or fog can actually bring the sharpest views. Even though little else is visible in the sky, planet disks appear rock steady, revealing astonishing detail. This is called good seeing—it happens when the layers of Earth's atmosphere are calm and stable, and not mixed up by winds at different altitudes. Under turbulent conditions, the poor seeing turns the disks of planets into boiling blobs with, at best, fleeting moments of sharp views.

A SKYWATCHING KIT *might include anything from a hot drink in a vacuum flask to a computerized tracking system.*

At most sites, nights of good seeing are often the least transparent, and vice versa. Only the finest mountaintop observatories boast the best of both worlds. Backyard astronomers soon learn to adapt their observing priorities to the conditions of the night, perhaps pursuing a galaxy hunt on a clear night, and a planet study on a night when the seeing is superb.

The Best Binoculars

Contrary to popular belief, the first piece of skywatching equipment to buy is not a telescope but a pair of binoculars.

Anyone with an interest in the sky should own a pair of binoculars. Many people mistakenly think that binoculars cannot be as good as a telescope because they do not magnify enough. True, binoculars magnify only 10 to 20 times at best—not enough to show details on the planets or resolve small clusters of stars. But their modest magnifying and light-gathering power is more than enough to reveal many of the sky's most interesting objects.

A three-day-old Moon set in the deep blue twilight and bathed in earthshine is an ideal binocular target. Using binoculars, you can see three or four of Jupiter's moons and even spot Uranus and Neptune as pale blue-green "stars." Dozens of asteroids come within reach of binoculars each year, while bright comets look spectacular.

No astronomical life is complete without a binocular tour of the fabulous dark lanes and star clouds near the center of our galaxy in Sagittarius and Scorpius. With binoculars, you can even range out past the Andromeda Galaxy to the bright galaxies of the Virgo Cluster, 50 million light-years away. Not bad for an instrument costing only about a quarter of the price of a standard beginner's telescope.

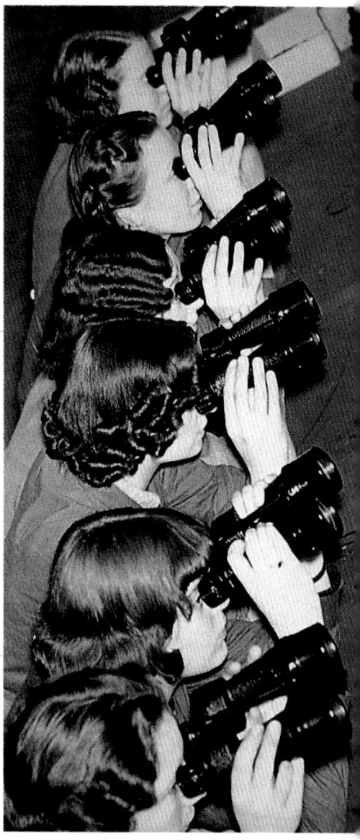

BINOCULAR TESTING *in 1939 (right). When trying out binoculars, check that the image looks sharp and bright.*

Why Binoculars?
The advantages of a pair of binoculars over a telescope are significant. Binoculars are inexpensive and readily available. Most are lightweight and require no effort to set up.

The twin barrels of binoculars are comfortable to look through. Because they are actually two small telescopes, they create a three-dimensional effect. Binoculars provide a

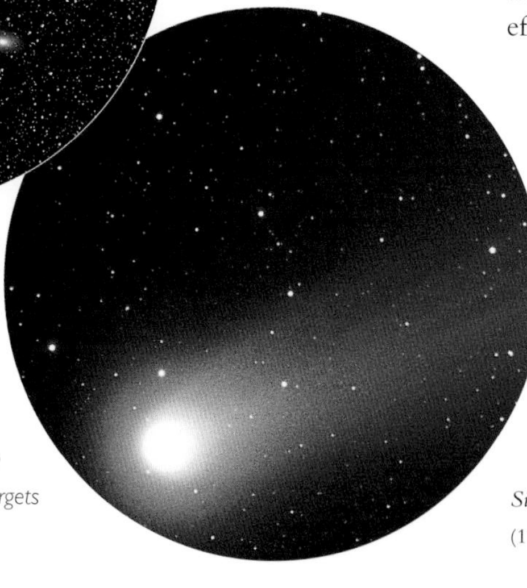

THE WIDE FIELD OF VIEW *provided by binoculars can sometimes be more suitable than a telescope view. Some galaxies, such as Andromeda Galaxy (above), and comets, such as Hyakutake (right), are large targets best seen through binoculars.*

... the sky
Spreads like an ocean hung on high,
Bespangled with those isles of light ...

Siege of Corinth, Lord Byron (1788–1824), English poet

wide field of view that makes it easy to find your target. It also helps that the images you see are right side up, unlike the upside-down images in many telescopes.

APERTURE *is the key feature in binoculars. Models with front lenses 50 mm across (left) are the most popular for astronomy, while an aperture of 35 mm (below) is a common size for daytime use.*

Which Binoculars?

The best binoculars for astronomy are 7 x 50 and 10 x 50 models. The 7 or 10 figure indicates the magnification. The 50 is the aperture—the diameter of each front lens in millimeters. Compared to 7 x 35 binoculars, a common size for daytime use, models with 50 mm lenses collect twice as much light, yielding brighter views of the night sky.

It is a tossup between a 7x and a 10x pair. The higher power of the 10x model is better for revealing lunar craters, star clusters, and faint stars. High-power binoculars, however, are harder to hold steady and can be less comfortable to look through because your eyes must be closer to the eyepieces. In low-cost models, the extra power may reveal flaws in the optics that would be less apparent in 7x binoculars.

The Exit Pupil

The exit pupil of a pair of binoculars is the diameter of the beam of light leaving each eyepiece and entering each eye. It is easy to calculate—simply divide the aperture by the power. For a 7 x 50 model, the 50 mm aperture divided by the 7 power equals a 7.1 mm exit pupil.

At night, under ideal conditions, the pupil of a dark-adapted human eye opens to 7 mm. Binoculars with a 7 mm exit pupil—a 7 x 50, an 8 x 56, or a 9 x 63 model—produce as wide a beam of light as the average human eye can accept, yielding maximum image brightness. Such models are commonly recommended for astronomy—indeed, the 7 x 50 model is sometimes called a night glass.

So why consider a 10 x 50 model? The nighttime viewing conditions most of us contend with are far from pitch black, so our eyes never fully dilate to 7 mm. Age also takes its toll. The pupils of most people over 30 open to only 6 mm in the dark. By the age of 50, our pupils' maximum aperture may be no more than 4 to 5 mm, so the light from binoculars with a 7 mm exit pupil is simply wasted, reducing the effective aperture of the binoculars. For many skywatchers, especially city dwellers, binoculars with a 5 or 6 mm exit pupil (a 10 x 50, or a lighter-weight 7 x 42) are a better choice.

YUJI HYAKUTAKE *used binoculars, albeit very large ones, to discover the brightest comet of 1996.*

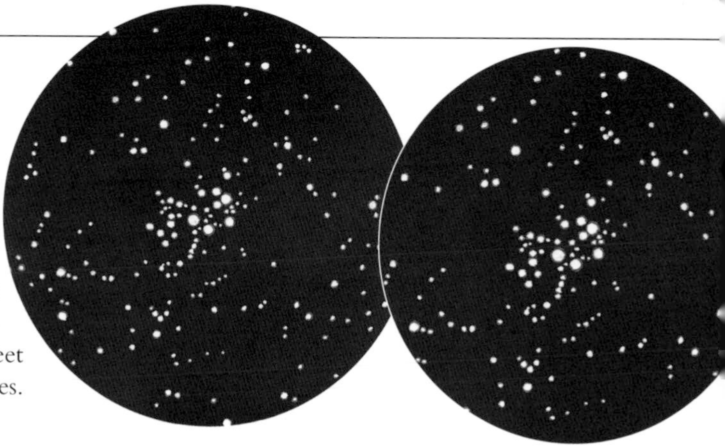

COMPARE THESE VIEWS *of the Pleiades (M45) in Taurus. Most 7x binoculars show about 7 degrees of sky (above left). Higher-power 10x models magnify the image more, but usually show only 5 degrees of sky (above right).*

Fields of View

The field of view is usually stamped on binoculars, but may be given as "feet at yards." For example, 367 feet at 1,000 yards equals a 7 degree field of view, and 262 feet at 1,000 yards equals 5 degrees.

Most 7x binoculars take in about 7 degrees of sky—roughly 14 times as big as the Moon—while 10x models see a smaller area, about 5 degrees.

Special wide-angle models provide impressive views of 8 to 10 degrees. But there are trade-offs: stars at the edge of the field tend to distort, and your eyes need to be close to the eyepieces.

Zoom binoculars sound attractive at first, but in almost all cases they are crippled by inferior optics and restricted fields of view.

Eye Relief

Eye relief is the distance that your eyes must be from the binocular eyepieces to see the whole field. An eye relief of less than 9 mm makes for uncomfortable viewing.

If you wear eyeglasses, especially ones that correct for astigmatism, you will get sharper views if you keep them on when using binoculars. In this case, an extra-long eye relief of at least 15 mm and rubber eyecups that roll back are essential.

Types of Prisms

Binoculars are made with one of two types of internal prism: porro prisms or roof prisms. Porro-prism binoculars, the most common type, have a zigzag shape. They are the best choice for skywatchers.

Roof-prism binoculars have a compact "straight-through" shape and cost more than porro-prism models. In all but the most expensive models, however, binoculars with roof prisms exhibit spikes of light on bright stars, making them unsuitable for astronomy.

Lens Coatings

A key difference in otherwise similar binoculars is the lens coating. Optical coatings give the lenses a blue, green, amber or red tint. They improve image brightness and contrast, without affecting the color.

Low-cost models have only single-layer coatings, often on just the outside lenses. In better models, all optical surfaces are coated, and the best ones feature multi-layer coatings.

Some binoculars have rubber armor to cushion the optics against minor bumps. Premium models are often waterproof, which prevents dew and contaminants from seeping into the innards.

In Focus

Waterproof binoculars often have two focus adjustments, one on each eyepiece. These individual-focus models are awkward for daytime use, but are fine for astronomy because every subject is located at infinity, requiring only one focus setting.

PORRO-PRISM BINOCULARS *use prisms made with either BK7 or BAK4 type glass. BAK4 prisms are more expensive, but they yield brighter images.*

IN-STORE TESTS

Try before you buy is a good rule with binoculars. When comparing the models available, carry out these simple tests.

- Do the images look sharp in the center? At what point do the images start to fuzz toward the edge of the field? Better models exhibit less edge distortion.
- Hold the binoculars at a distance. Look down the eyepieces. The circles of light should be evenly illuminated. Dark squared-off edges are a sign of the lower-quality BK7 prisms.
- Look into the main lenses. Lots of white reflections indicate poorly coated optics, while dark lenses with a few deep purple or green reflections indicate high-quality multi-coated optics.
- Are the binoculars comfortable to hold? Can you see the entire field without pressing your eyes right up to the eyepieces?
- Pass your hand across the front lenses so you look through just one side of the binoculars, and then the other. Does the image jump back and forth? If so, the binoculars are "cross-eyed," or out of collimation—a serious defect.

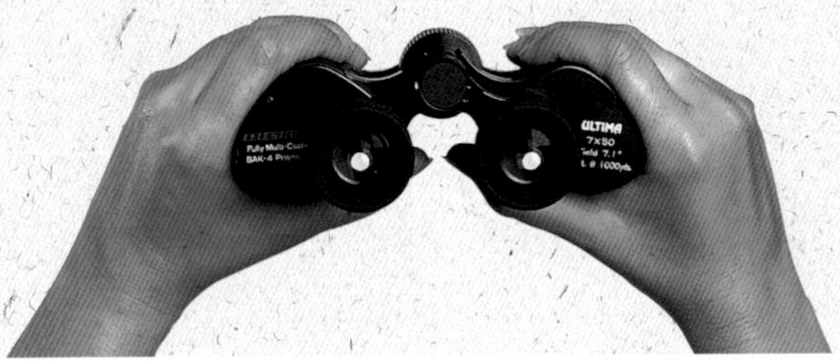
ULTIMA
7x50

Even so, most people prefer o use center-focus models. Ihese have a single focus adjustment that controls both eyepieces, as well as a "set-once-and-forget" adjustment on one eyepiece that compensates for focus differences between your right and left eyes.

In recent years, some manufacturers have sold binoculars hat have no focus adjustments. Ihese fixed-focus models are unusable for astronomy. They do not focus at infinity and provide no means to compensate for the focus variations among different people's eyes.

BINOCULAR MOUNTS

f you mount binoculars on a steady tripod, you will see much more detail in everything you look at. Many binoculars feature a centrally located threaded socket. This accepts an L-shaped bracket, which in turn bolts to any camera tripod.

To make it easier to get under the binoculars and look through them when they are aimed high in the sky, some manufacturers offer cantilevered swing arms that suspend the binoculars away from the tripod.

Another way to steady your view is to lie back in a deck chair or recliner, and use the chair's arms for support. What must be the ultimate luxury, however, is a rotating binocular-observer's chair.

BIG BINOCULARS

Every skywatcher, novice or experienced, should own a pair of binoculars in the 50 mm aperture league. Serious binocular users may want to graduate to a pair in the 70 to 80 mm class. With 11 to 20 power, these big binoculars cost as much as a small telescope and their hefty weight demands tripod mounting. Binoculars in this class can provide bright, detailed views of Milky Way star fields, large nebulas, comets, and lunar eclipses unmatched by any other type of instrument.

CARE OF BINOCULARS

With care, binoculars can last a lifetime. Avoid touching the lenses with your fingers and clean them as you would telescope eyepieces (see p. 55). Most importantly, try not to drop them. A sharp jolt can loosen the internal prisms or knock the two halves out of collimation (alignment). It can cost as much to repair binoculars as to buy a new pair of similar quality.

ANTILEVER STANDS, *shown here upporting big binoculars, make it easy to ook anywhere in the sky, even straight up.*

TELESCOPE TYPES

From a no-frills beginner's telescope to an advanced computerized model, a superb selection of tempting equipment awaits the telescope shopper.

For the first-time buyer, the extraordinary variety of telescopes available can be overwhelming. There are three main optical designs. Reflecting telescopes use a mirror to gather light; refracting telescopes use a lens. A third type of telescope, the catadioptric, uses a combination of a mirror and a lens. Schmidt-Cassegrains and Maksutovs are examples of catadioptric telescopes.

Each of these three types has its selling points (see p. 50). All of them can show you details on the Moon as tiny as a mile across, the stunning rings of Saturn, the changing clouds of Jupiter, and the brightest galaxies, nebulas, and star clusters of deep space.

When shopping for a new telescope, the choice of optical design is less critical than the features listed below in our Telescope Shopper's Checklist. Most importantly, avoid any telescope sold by how much it can magnify. Low-quality telescopes are often promoted by claims such as "450 power" or "high-power professional model." Many of the 2.4 inch (60 mm) refractors and 4.5 inch (110 mm) reflectors commonly seen in camera shops and department stores fall into this category. They feature wobbly mounts, eyepieces and finderscopes with poor optics, and plastic fittings. The views through these "toy" telescopes are disappointing, to say the least. However, there is nothing inherently wrong with smaller telescopes. A good 2.4 inch (60 mm) refractor with the features in our checklist can make a fine starter scope.

Most telescopes are sold with at least one or two eyepieces. Try to find a telescope with quality eyepieces—such as Kellner, Orthoscopic, Modified Achromat, and Plössl types—giving no more than 75 to 100 power. Additional eyepieces can be purchased separately (see p. 54), but if the telescope provides only high-power (150x to 300x) eyepieces, it is probably of poor quality

TELESCOPE SHOPPER'S CHECKLIST

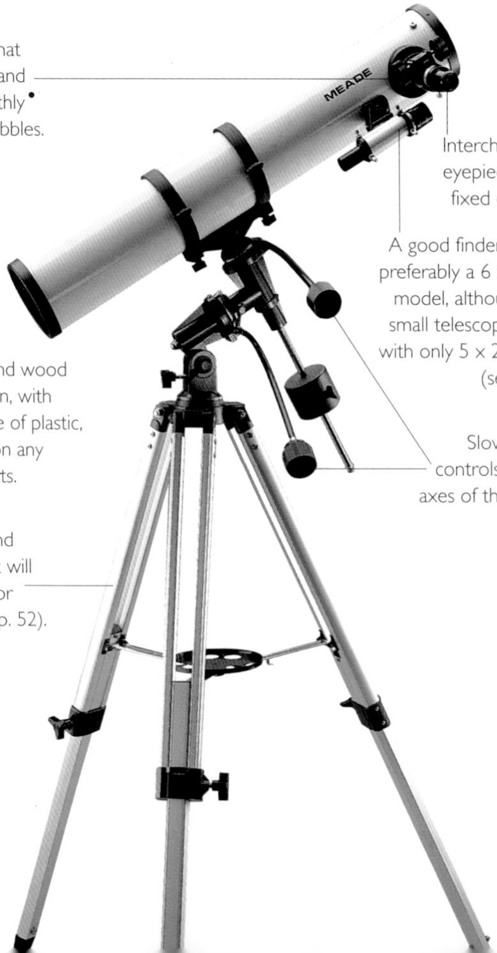

MEADE
A focuser that slides back and forth smoothly with no wobbles.
Interchangeable eyepieces—not fixed or zoom.
A good finderscope—preferably a 6 x 30 mm model, although most small telescopes come with only 5 x 24 finders (see p. 57).
All metal and wood construction, with minimal use of plastic, especially on any moving parts.
Slow-motion controls on both axes of the mount.
A tripod and mount that will not shake or bend (see p. 52).

OUR CHECKLIST *of features (left) is marked on a Newtonian reflector, but applies to all types of telescope. Almost any telescope you buy today will be more effective than Galileo's early telescopes, which were refractors (top).*

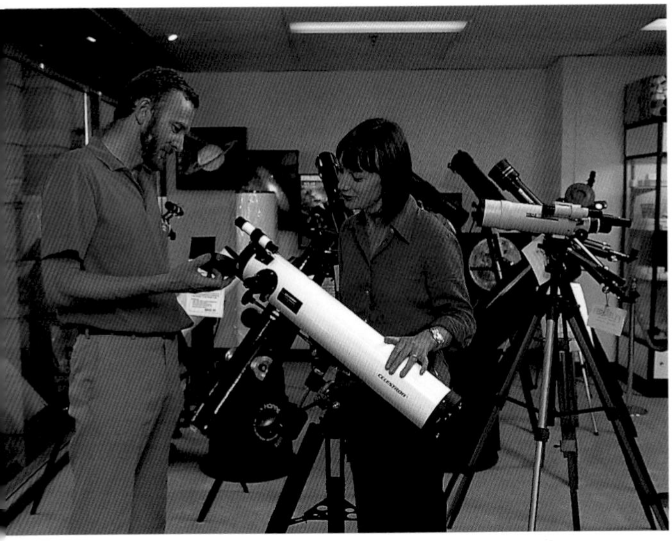

BEFORE YOU GO SHOPPING, *read this chapter carefully and compile a list of questions based on the skywatching activities you plan to pursue.*

and should be avoided. Most basic telescopes accept only 0.965 inch (24.5 mm) diameter eyepieces, but if you are prepared to spend a bit more, look for a telescope that takes the better grade of 1.25 inch (31.8 mm) diameter eyepieces.

Aperture not Power

The key specification of any telescope is its aperture—the diameter of the lens or mirror that collects the light. The magnification is unimportant—by changing eyepieces it is possible to make any telescope magnify any amount. But the maximum power any telescope can deliver is equal to about 50 times its aperture in inches. A 2.4 inch (60 mm) telescope cannot operate much higher than 120 power. The limit for a 4.5 inch (110 mm) scope is 225 power. Exceed these limits and the image will indeed get bigger, but it will also look faint and fuzzy.

The critical factor is the light-gathering power of the instrument. The greater the aperture of the telescope, the more light it can collect. In fact, for every doubling of aperture, light-gathering power goes up by a factor of four. For example, a 6 inch (150 mm) mirror has four times the surface area of a 3 inch (75 mm) mirror. It collects four times the light, making the images appear four times brighter.

Doubling the aperture of a telescope also doubles its resolving power—the ability to see fine details on planets or split closely spaced stars. Under excellent seeing conditions, a 4 inch (100 mm) telescope can resolve stars about 1 arcsecond apart, while an 8 inch (200 mm) telescope's resolving power is 0.5 arcsecond. (By comparison, the human eye can resolve only about 60 arcseconds.)

So when you are shopping, consider the telescope in your price range with the largest aperture. Then again, the biggest telescope you can afford is not necessarily the best for you. Think about where you will store it, and where and how you will use it. A smaller, portable instrument may get used more frequently than a large, unwieldy one. The best telescope for you is the one you will use most often.

BUYING FOR A CHILD

With their sharp eyesight and insatiable curiosity, children make excellent skywatchers, and the gift of a telescope can lead to a lifelong hobby—or even a career.

When buying a scope for a child, avoid cheap "high-powered" instruments. Even adults have a difficult time using these shaky, fuzzy telescopes. Purchase a good-quality 2.4 inch (60 mm) refractor or, better yet, move up to a larger refractor or a 4 to 6 inch (100 to 150 mm) reflector. Choose one with an altazimuth or Dobsonian mount, rather than a complicated equatorial mount (see p. 52). It is important that the child can use the telescope unsupervised, although you will both get much more out of the purchase if you spend time skywatching together.

If you are concerned that a good-quality telescope is too expensive to start with—or if the child is still too young for a delicate instrument—purchase binoculars instead. Buy a set of star charts (see p. 70) and spend a year with the child learning the constellations and picking out binocular targets. You can sample telescopic views at star parties and public observatories.

STAR HOPPER

A 6 inch (150 mm) reflector on a Dobsonian mount.

REFRACTORS *use a lens to collect light, which is focused to an eyepiece found at the bottom of the tube.*

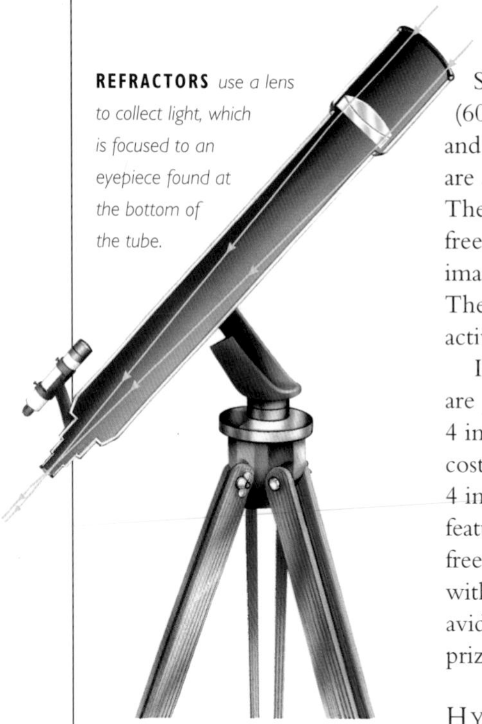

Lens or Mirror?

The choice of serious but budget-minded skywatchers is usually a Newtonian reflector, a design relatively unchanged since Isaac Newton invented it in 1668. As a rule, reflectors provide much greater aperture for the money than refractors do, and an economical 4 or 4.5 inch (100 or 110 mm) Newtonian can serve as a good starter scope. Several manufacturers now offer very affordable 6 inch (150 mm) reflectors on simple Dobsonian mounts (see p. 52). These provide views of planets and deep-sky targets so superior to those in smaller telescopes that it is worth making the jump up. The main drawback of a Newtonian reflector is that it requires some maintenance—the exposed and delicate mirrors need occasional cleaning and collimation.

Small refractors—in 2.4 inch (60 mm), 3.1 inch (80 mm), and 3.5 inch (90 mm) sizes—are also popular starter scopes. These rugged, maintenance-free instruments offer crisp images and are easy to aim. They can double for daytime activities, such as birding.

In larger sizes, refractors are premium instruments—a 4 inch (100 mm) refractor can cost five times as much as a 4 inch reflector. They often feature apochromatic (color-free) lenses, and are popular with optical connoisseurs and avid astrophotographers who prize their ultra-sharp images.

Hybrid Scopes

Catadioptric telescopes—the Schmidt-Cassegrains and Maksutovs—are hybrids, combining a reflecting mirror with a large corrector lens that eliminates optical distortions.

The Schmidt-Cassegrain design employs optical technology invented by Guillaume Cassegrain in the seventeenth century and Bernhard Schmidt in the 1930s. First introduced to amateur astronomers in the early 1970s, this telescope's key selling point is its portability: an 8 inch (200 mm) model has a tube only one-third as long as an 8 inch Newtonian reflector. Extensive systems of accessories make Schmidt-Cassegrains good choices for astrophotography. They cost at least 50 percent more than reflectors—but far less than refractors—of the same aperture.

The Maksutov, a similarly compact instrument, was invented independently by Bouwers and Maksutov in the 1940s. Also known as "Maks," these scopes come in apertures of up to 8 inches (200 mm), the most popular being the very portable 3.5 inch (90 mm) models. Maksutovs provide sharp, high-contrast images, making them favorites for planetary viewing.

Which Way Up?

All telescope designs produce upside-down images, but the refractors and the Schmidt-Cassegrains are usually used with an accessory called a star diagonal, which fits just in front of the eyepiece. This turns the image right way up but also swaps it left to right, so that you see a mirror image. The images in reflectors are generally upside down.

NEWTONIAN REFLECTORS *(right) use a mirror to collect light, which is reflected and focused back up the tube to an eyepiece near the front of the telescope.*

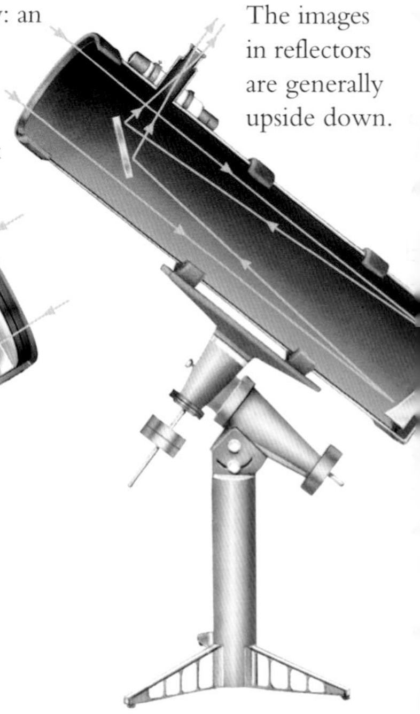

CATADIOPTRIC TELESCOPES, *such as the Schmidt-Cassegrain (near right), use a mirror and a corrector lens. The light travels back and forth then exits through a hole in the mirror to an eyepiece at the back of the telescope.*

League
of
Champions

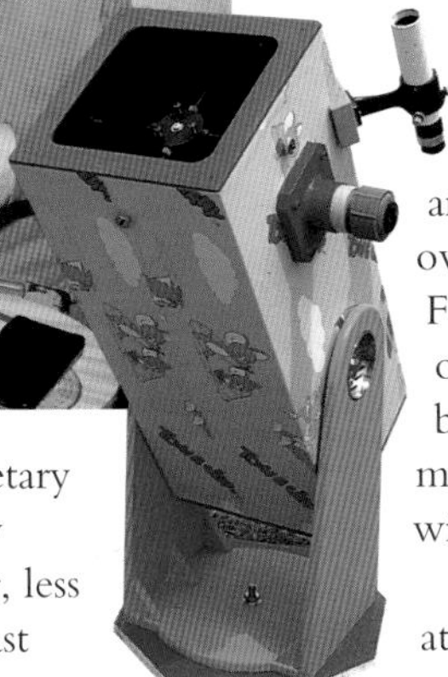

TELESCOPE BUILDING *can become a serious and engrossing hobby (left), but a simple Dobsonian mount (below) can be built in a couple of weekends.*

Scope Speed

A specification usually marked on the telescope's tube or given in its instruction manual is the focal length, almost always given in millimeters. This is the length of the light path from the lens or mirror to the focus point. With refractors and reflectors, the focal length is roughly equal to the physical length of the tube. But in Maksutovs and Schmidt-Cassegrains, the light path is folded back and forth several times, allowing long focal-length optics to reside in a shorter, more compact tube.

Dividing the focal length of a telescope by its aperture in millimeters gives its f-ratio. A telescope with a focal length of 2,000 mm and an aperture of 200 mm (or 8 inches) is an f/10 telescope. If the focal length were 1,000 mm, it would be an f/5 scope.

The smaller the f-ratio, the faster the telescope. Fast telescopes are an advantage for deep-sky photography—an f/5 telescope will record a nebula in a quarter of the exposure time required by an f/10 telescope. Hence, the term "fast" telescope. For visual use, however, there is little advantage of one f-ratio over another. Slower telescopes are sometimes sharper, making them better suited to high-power planetary viewing, but they often have longer, less portable tubes. Fast telescopes give lower powers and wider fields with any given eyepiece, making them better suited to deep-sky viewing, but they also tend to exaggerate flaws in the optics.

There is no single ideal telescope, but by taking into account your skywatching interests, you can certainly make an informed choice.

Building Your Own

In the past, many amateurs ground their own telescope mirrors. Few pursue this aspect of the hobby today because quality commercial optics are now widely available.

Building a telescope at home, however, is as popular as ever. You simply buy a mirror and other parts, then place them into a thick cardboard tube. This sits on top of a Dobsonian mount, which you can construct from plywood. There are some excellent references on telescope building. Check the recommendations in the Resources Directory (see p. 274).

ALVAN CLARK AND SONS

At the end of the nineteenth century, it was the well-heeled and fortunate amateur astronomer indeed who owned a telescope made by Alvan Clark and Sons. Based in Massachusetts, this famous company built refractors in sizes from 3 inches (75 mm) on up. Way up! In 1897, Clark's son Alvan Graham (shown here, on the left, with assistant Carl Lundin) completed the optics for the 40 inch (1 m) Yerkes refractor in Williams Bay, Wisconsin. It remains the largest working refractor ever made and is still in use. Clark and Sons' smaller refractors for amateurs provide views surpassed by few telescopes today, and are much sought after by collectors of fine optics.

Choosing *a* Telescope Mount

The best optics in the world are worthless without a sturdy mount—it is half the telescope.

When a telescope has a poorly made mount, the image in the eyepiece jumps around—finding objects becomes a frustration, while seeing details within objects is nearly impossible. Fortunately, many telescopes sold today come with excellent mounts that move smoothly across the sky, and also remain firmly on target when released.

Small "department-store" telescopes, however, often have shaky mounts and flimsy tripods that flex and bounce with every touch and breath of wind. Do not be fooled by the high-tech dials and cables on some cheap mounts. Here is an easy showroom test: give the telescope tube a sharp rap. All telescopes will shake for a moment, but in a well-mounted telescope vibrations should die out within three to four seconds.

One or Two Motions

There are two main designs to consider—altazimuth and equatorial. The simplest and most common is the altazimuth mount, which swings up-and-down and left-to-right. Moving an altazimuth scope around the sky to find a target is easy. But as the Earth rotates, the target will also move. To keep it in view, you must move the telescope a little in both directions every few seconds, which can be awkward. It is now possible to motorize altazimuths, but only with the aid of expensive computer controls (see p. 59).

The equatorial mount is the usual solution for tracking objects. One of its axes—the polar axis—must be lined up so that it points to the celestial pole, the point in the sky around which the stars appear to turn during the night.

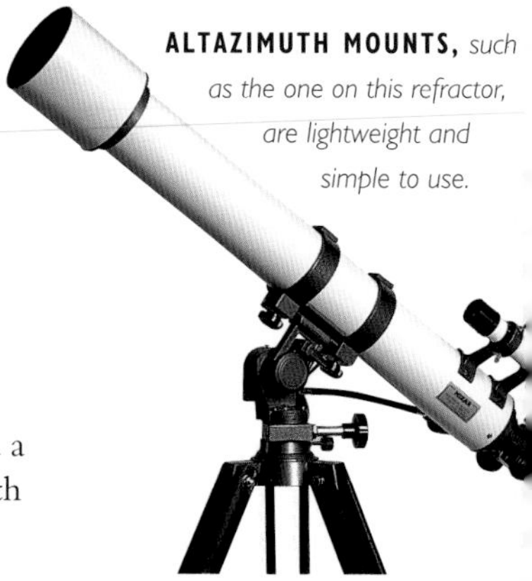

ALTAZIMUTH MOUNTS, *such as the one on this refractor, are lightweight and simple to use.*

When you rotate the polar axis, the telescope moves in the same east-to-west direction as the stars, allowing it to track a particular object with a single motion. Add a motor to the polar axis and the telescope can follow objects automatically, leaving your hands free.

Which Mount?

Altazimuth mounts are inexpensive and quick to set up. A special type of altazimuth is now one of the most common for reflector telescopes. Popularized by the Californian telescope-maker John Dobson, this wooden mount uses simple Teflon pads to provide smooth motions in both axes. It moves effortlessly, yet stays firmly on target. A gentle nudge every minute or two is all you need to keep objects in view.

THIS DOBSONIAN MOUNT *(left), built by John Dobson in the late 1970s, certainly has a much simpler design than the complex mount photographed in 1912 (top left).*

Their simplicity, economy, and stability make Dobsonian mounts an ideal choice for anyone looking for the best performance for the least money. Even with a computer-tracking system, however, altazimuths are only for visual observing and will not work for most astrophotography.

A motorized equatorial mount certainly makes it easier to study objects such as planets, and can be used for tracked astrophotography (see p. 64). But equatorial mounts are much heavier, less portable, and more complex than altazimuth mounts. They require polar alignment (see Box) and can be confusing for beginners to move around the sky. And they are more expensive—an equatorial version of a 3.5 inch (90 mm) refractor or a 6 to 8 inch (150 to 200 mm) reflector is likely to cost twice as much as the altazimuth version.

There are several varieties of equatorial mount, but the most popular ones are the fork and the German equatorial mount. Fork mounts work well for short-tube telescopes such as Schmidt-Cassegrains, while German equatorials are commonly supplied with refractors and Newtonian reflectors—telescopes that have longer tubes.

Motor Drives

Motor drives come in either AC or DC models. AC models run from household power, or from a power inverter connected to a car battery. DC models have a built-in battery or a separate battery pack, so you can use them anywhere. Many DC drives also have speed controls, which are essential for advanced astrophotography.

GETTING ALIGNED

For an equatorial mount to work properly, it must be polar aligned to the celestial pole. For Northern Hemisphere observers, this means aiming the polar axis toward Polaris, the northern pole star. For Southern Hemisphere observers, the pole star is the faint Sigma Octantis. You need not be too precise—unless you are taking photographs (see p. 65), getting within 5 degrees of the pole will be fine.

To polar align a fork mount, place it so the fork arms, which parallel the polar axis, aim due north (true north, not magnetic north)—or due south in the Southern Hemisphere. Level the top of the tripod, then adjust the tilt of the wedge so that the angle between the ground and the fork arms equals your latitude. For example, an observer in New York, which is at 40 degrees latitude, would tilt the wedge to a 40-degree angle. There may be a scale on the mount to help set the angle, or you can use a protractor. This angle needs to be set only once—on subsequent nights, simply aim the fork arms toward the pole star.

For a German equatorial mount, the procedure is the same, but you aim and tilt the mount's polar-axis housing (indicated below).

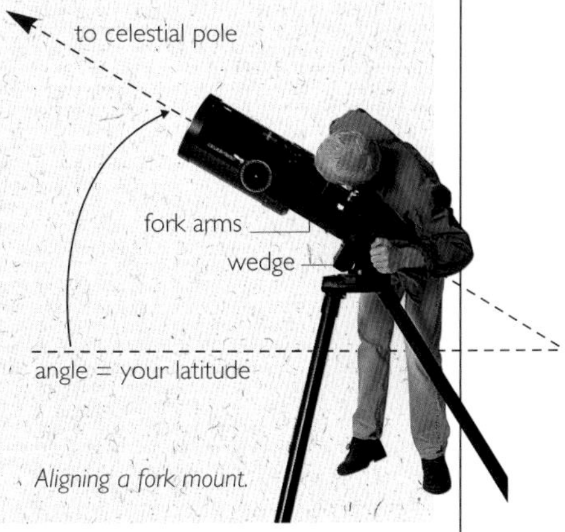
to celestial pole
fork arms
wedge
angle = your latitude

Aligning a fork mount.

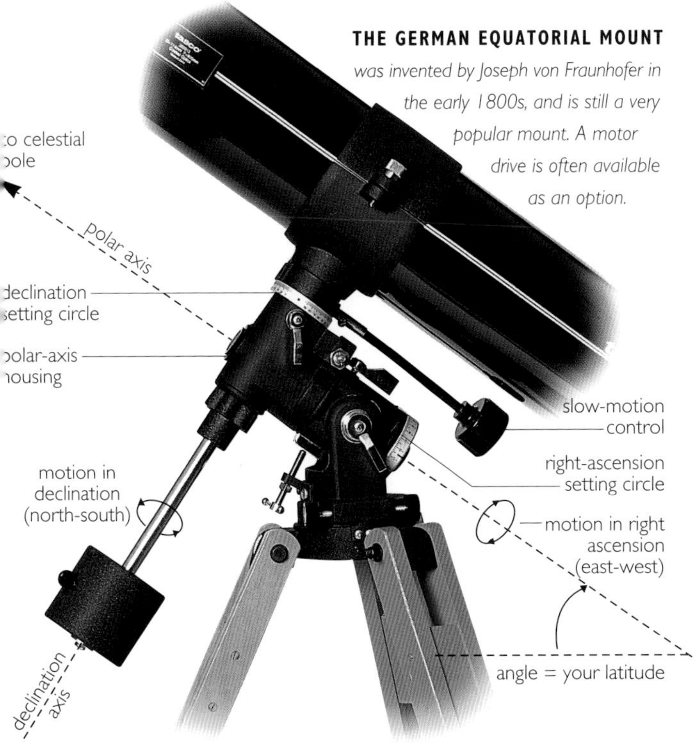
to celestial pole
polar axis
declination setting circle
polar-axis housing
motion in declination (north-south)
declination axis
slow-motion control
right-ascension setting circle
motion in right ascension (east-west)
angle = your latitude

THE GERMAN EQUATORIAL MOUNT *was invented by Joseph von Fraunhofer in the early 1800s, and is still a very popular mount. A motor drive is often available as an option.*

The heavens call to you, and circle around you, displaying to you their eternal splendors ...

The Divine Comedy, DANTE (1265–1321), Italian poet

SELECTING EYEPIECES

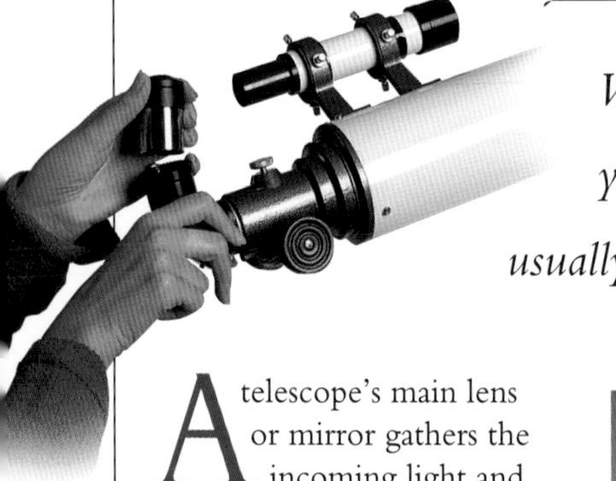

Whenever you look through a telescope, you look through an eyepiece. Eyepieces are usually the first accessories telescope owners buy.

A telescope's main lens or mirror gathers the incoming light and focuses it into an image, but it is the eyepiece that magnifies the image. To change magnifications, you need to change eyepieces. Even the best telescopes may come with only one general-purpose eyepiece as standard equipment. Two or three additional eyepieces are essential.

EYEPIECES *slide into the telescope's focuser (top). The most common barrel sizes are 0.965 inch (left, above) and 1.25 inch (right, above). Premium models feature 2 inch barrels (center, above).*

EYEPIECE FOCAL LENGTHS

Eyepieces are sold not by their magnifications but by their focal lengths (always given in millimeters). Look for numbers such as 25 mm, 12 mm, or 9 mm on the barrel. This indicates the focal length of the miniature optics inside the eyepiece. To calculate how much power a given eyepiece will produce on your telescope, divide the focal length of the telescope (see p. 51) by the focal length of the eyepiece.

A 20 mm eyepiece inserted into a telescope with a focal length of 2,000 mm, for instance, would give 100 power. The same eyepiece inserted into a 1,000 mm focal length scope would give 50 power.

The shorter the eyepiece's focal length, the higher the power it gives. Any observer can use at least three eyepieces: a low power (35x to 50x), a medium power (80x to 120x), and a high power (150x to 180x) eyepiece. This means that if your scope has a focal length of 1,000 mm, for example, you might select 25 mm, 12 mm, and 6 mm eyepieces. Low power provides wide fields for locating targets and for panoramic views of star fields; medium power resolves clusters and double stars; and high power reveals details on the planets. Magnifications much higher than 180x are rarely of any use—the image gets bigger but becomes blurry in the process. Also avoid zoom eyepieces—they may promise to combine a range of powers, but in practice they deliver poor images.

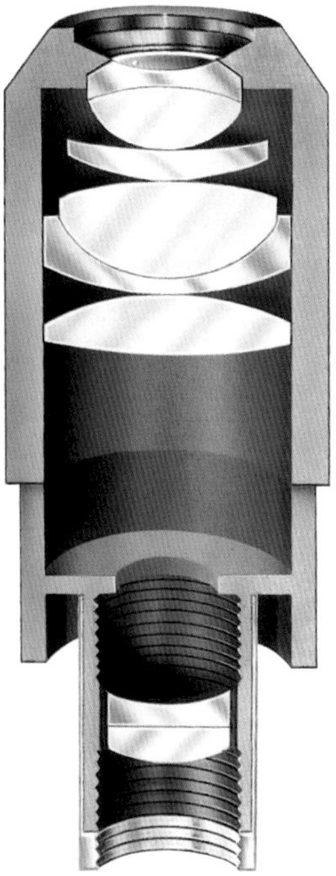

NAGLER
(8 lens elements)

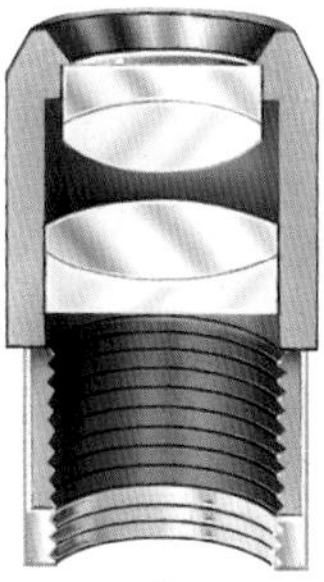

PLÖSSL
(4 lens elements)

KELLNER
(3 lens elements)

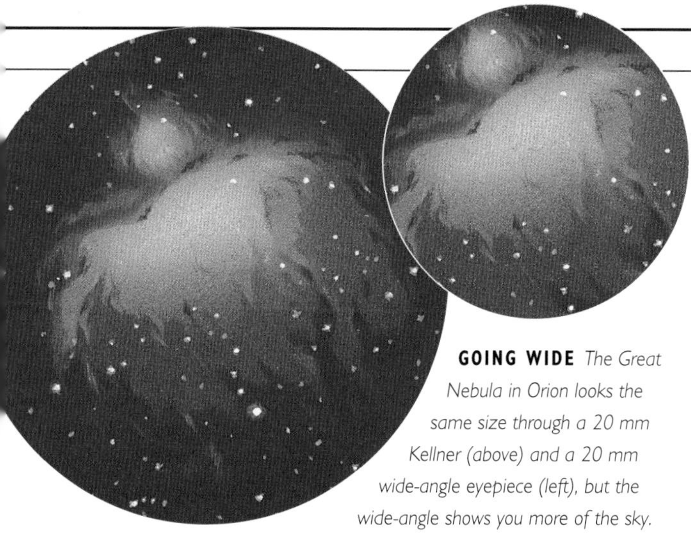

GOING WIDE *The Great Nebula in Orion looks the same size through a 20 mm Kellner (above) and a 20 mm wide-angle eyepiece (left), but the wide-angle shows you more of the sky.*

Barrel Sizes

Eyepieces come in three barrel sizes: 0.965 inch (24.5 mm), 1.25 inch (31.8 mm), and 2 inch (50.8 mm). Many small starter telescopes accept only 0.965 inch eyepieces. Unfortunately, the Huygenian and Ramsden models of eyepieces usually packed with these telescopes are of mediocre quality. Better models—such as Kellners, Orthoscopics, and Modified Achromats—are available for this smaller barrel size. Upgrading to higher-quality eyepieces is one of the best things you can do to improve the performance of a low-cost telescope.

A telescope that accepts the larger 1.25 inch eyepieces is a definite plus. These eyepieces are consistently better quality and are available in a vast selection of models. A small number of eyepieces, however, have even larger, 2 inch barrels. These are specialized wide-angle eyepieces that fit only some of the premium telescopes on the market.

Optical Designs

Eyepieces are available in various optical designs. Some manufacturers have exclusive rights to certain designs—such as Edmund Scientific's RKE models, Meade's Modified Achromats, and TeleVue's Nagler and Panoptic series. But the most common designs are sold by nearly every dealer.

Kellner, RKE, and Modified Achromat eyepieces use three lens elements for good image quality at low cost.

A step up in quality and price are the four-element Orthoscopic and Plössl models. Most manufacturers have matched sets of Plössls in their catalogs—they represent the best buys in today's marketplace. Every telescope owner can use a set of three or four Plössls.

In the premium price bracket are eyepieces known by generic names such as Erfles, and brand names such as Wide Fields, Ultra Wide Angles, Panoptics, and Naglers. These five- to eight-element models come in focal lengths from 35 to 4.7 mm. Compared to a standard Plössl, these models all show a much wider field, providing a wonderful "picture-window" view of the universe.

Light from distant places
has made the journey
to earth, and it falls on
these new eyes of ours,
the telescopes ...

Cosmic Landscape,
Michael Rowan-Robinson
(b. 1942), British mathematician

KEEPING OPTICS SPARKLING

Nothing can smear your view of stars and planets better than an eyepiece coated with dust, eyelash oil, and fingerprints. To clean an eyepiece, first blow off loose dust with a lens blower. Then lightly moisten a Q-tip cotton swab with camera-lens cleaning fluid. Wipe the lens gently with the wet swab, then again with a dry one to remove the streaks. Never pour the fluid directly onto the lens—it can seep into the eyepiece, staining interior lenses. And do not try to disassemble an eyepiece—the lenses could all tumble out.

The main lenses and mirrors of telescopes should be cleaned with caution and as seldom as possible. To clean refractor lenses, blow off as much dust as you can, then moisten lens cleaning tissue with lens cleaning fluid. Wipe the lens with gentle strokes. Do not rub hard.

Reflecting mirrors are the most delicate of all. They must be cleaned only with gentle swipes of cotton balls moistened with a solution of distilled water and mild detergent, then rinsed with pure distilled water. Other methods can leave scratches, which do far more harm than dust. To keep the dust off in the first place, never store telescopes with their lenses or mirrors exposed.

The Best Accessories

A few well-chosen accessories can improve the performance of your telescope and increase your enjoyment of skygazing.

MEADE
Series 4000

Like photography, backyard astronomy is a hobby filled with gadgets and accessories. Some are frills, but the best ones can enhance your view of the universe.

Filtering Out Light

You can select from a wide range of filters according to your skywatching interests. Filters usually screw into the base of the telescope's eyepiece.

Some filters minimize the effects of light pollution. Known as LPR (light pollution reduction) filters, they block the green and yellow wavelengths emitted by street lights, but transmit the red and blue-green colors of nebulas. While these filters can marginally improve views of star clusters and galaxies, they work best enhancing emission nebulas (ones that produce their own light).

LPR filters come in several varieties. Broadband, or deep-sky, filters transmit the widest range of colors, providing a modest enhancement of a range of deep-sky objects. Narrowband, or nebula, filters are much more effective on emission nebulas and are the best choice if you are planning to buy just one filter. Line filters, such as Oxygen III and Hydrogen-beta filters, transmit only a single color, providing excellent results on a limited number of nebulas.

Solar System Sights

In the past, many entry-level telescopes came with "sun filters" that screwed into an eyepiece. These filters are very dangerous and should never be used. Thankfully, they are rarely supplied today.

The *only* safe way to view the Sun directly through a telescope is with a filter that covers the entire front aperture of the scope. Mylar filters are the least expensive and are available in sizes to fit most telescopes. They produce a blue-tinted view of the Sun. Glass filters coated with metal show a

MEADE #911B
BROADBAND
MEADE #911N
NARROWBAND
Series 4000 NEBULAR
Series 4000 NEBU

FILTERED LIGHT *Colored filters (top) and LPR filters (right) provide clearer images of some objects. A light-polluted view of the Veil Nebula (below left) shows a dramatic improvement (below right) when a narrowband LPR filter is used.*

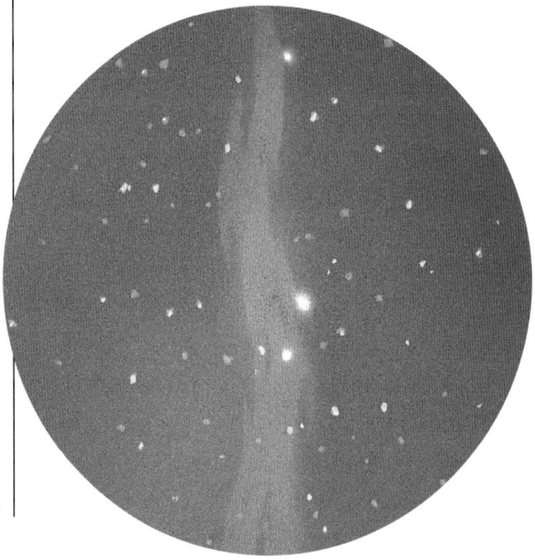

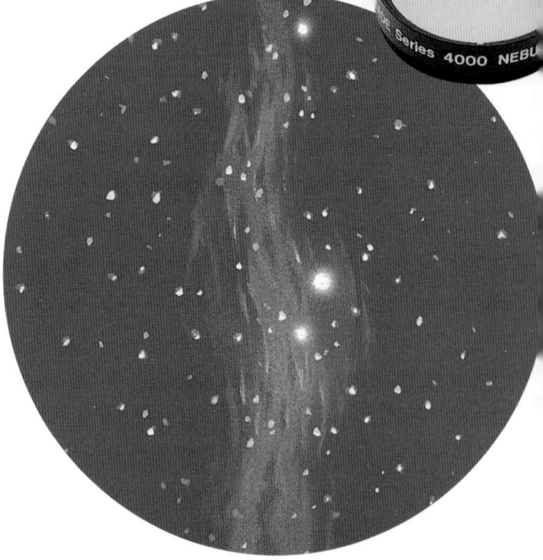

Tele Vue
BIG
2x
Barlow
Made in U.S.A.
TELE VUE 2x BARLOW

BARLOW LENSES *expand the use of your eyepieces and the range of your magnification choices. They come in 2x and 3x models.*

natural-looking yellow Sun, but cost about 50 percent more. Both Mylar and glass filters give "white-light" views of features such as sunspots and faculae (see p. 80). To see towering prominences on the edge of the Sun, however, you need a Hydrogen-alpha solar filter. Unfortunately, these can cost as much as a good telescope.

Colored filters can accentuate particular features on the planets: a green filter brings out Jupiter's Red Spot; a red filter improves views of the elusive dark markings on Mars. A set of four to six colored glass filters often costs no more than an eyepiece. An even less expensive option is to purchase colored gelatin filters from a camera store (see pp. 111, 119, and 125).

Easy Power

Eyepieces with a short focal length (4 to 8 mm) provide high power, but, as a rule, the higher the power of the eyepiece, the shorter its eye relief. The shorter the eye relief, the closer your eye must be to the eyepiece, making it less comfortable to use.

An alternative is a Barlow lens inserted between an eyepiece and the telescope. A 2x Barlow, which doubles the power of the eyepiece, is the most common size. A 12 mm eyepiece with a 2x Barlow produces the same high magnification as a 6 mm eyepiece, but without giving up the comfortable eye relief of the longer 12 mm eyepiece.

Dewcaps

When exposed to the cool night air, the front lens of a refractor or Schmidt-Cassegrain can readily attract dew or frost. One remedy is a dewcap—a tube extending in front of the telescope. Dewcaps come with most refractors, but they must be bought separately for Schmidt-Cassegrain telescopes. Dewcaps are fine for light dew. On very humid or frosty nights, however, a heater coil that wraps around the telescope tube is much more effective. Heater coils are also available for eyepieces and finderscopes.

Creature Comforts

The best accessories of all are items that increase your comfort during long skywatching sessions. In the cold night air, warm boots and clothing are essential. In warm climates, you may need insect repellent. An adjustable-height stool or chair lets you sit comfortably at the eyepiece of your telescope. A folding table provides a place to lay out charts and accessories, while a foam-lined camera case is good for storing them. A flashlight with a red filter helps you find things in the dark without affecting your night vision.

FINDING THINGS

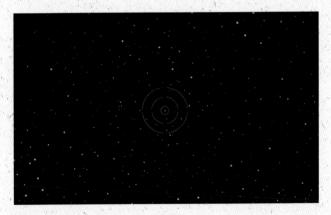

Finderscopes are small, wide-field secondary telescopes mounted on the main telescope. They make it easy to aim the main telescope and conduct one of the starhops detailed in Chapter 6.

Many entry-level telescopes come with 5 x 24 finders—these give 5x magnification and have apertures of 24 mm. Their small lenses are fine for centering a telescope on a bright planet. If you want to locate fainter targets, however, consider upgrading to at least a 6 x 30 finder. For owners of 6 to 8 inch (150 to 200 mm) telescopes, an 8 x 50 finder is a better choice.

An accessory some people find wonderfully useful, either instead of or as well as a finderscope, is one of the new reflex finders, such as the Telrad. Looking through a reflex finder, you see a naked-eye view of the sky with a red dot or bull's-eye floating among the stars (above).

Computer Control

Computer technology now makes it possible to find even the most obscure galaxy at the mere push of a button.

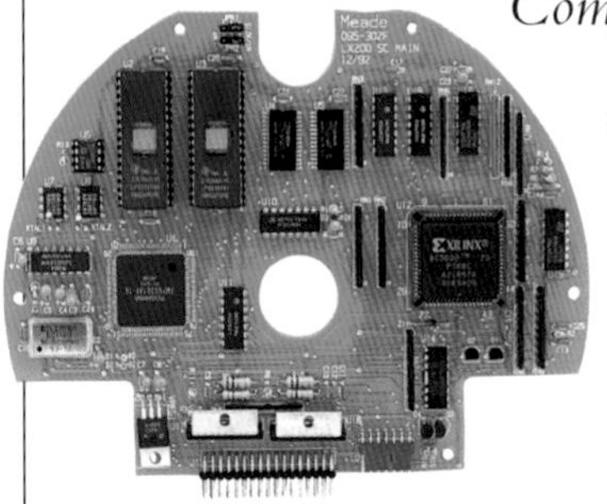

COMPUTER CHIPS *are everywhere, even inside astronomical telescopes.*

Finding objects in the night sky is a challenging task for most newcomers to the hobby. They stand in awe as veteran skygazers aim their telescopes at seemingly blank areas of the sky to reveal extraordinary views of galaxies and nebulas. How do the experts do it?

Starhopping is the answer—using a good finderscope to hop from a bright, easy-to-find star over to fainter stars and then to the target. Chapter 6 contains starhopping charts for many of the sky's best and brightest deep-sky wonders. Conduct a starhop a few times and you will probably remember the star pattern, making it a snap to locate a particular object again. You will have graduated from a novice lost in the stars to an expert trekking across the sky with confidence. But is there another way to navigate the sky?

Setting Circles

Most equatorial mounts have numbered dials on each axis. These are manual, or analog, setting circles, and they can be used to locate an object using its right ascension and declination coordinates (see Box). However, newcomers to astronomy rarely find setting circles very helpful. The dials require careful adjustments with every use and can quickly lose accuracy as the sky moves during the night.

Computerized, or digital, setting circles, are generally more useful, though more expensive. These devices are programmed with the locations of hundreds, if not thousands, of stars, nebulas, clusters, and galaxies—and even planets on some models. Add-on digital setting circles come with a pair of encoders, one for each axis of the mount. The encoders sense the motion of the mount and keep track of how far it has moved.

Digital setting circles are easy to use and provide a highly accurate method of locating objects. You begin by aiming the telescope at two bright stars on either side of the sky. These two positions are all the computer needs to know to guide you to any of the objects in its memory. Call up one of those objects and

Schmidt-Cassegrain telescope
analog setting circle
fork mount set up in altazimuth position
stepper motors (inside base)
computerized hand controller

COMPUTERIZED TELESCOPES *provide easy access to the thousands of deep-sky objects on the Messier and NGC lists at the push of a button.*

THE POSITION *of Alcyone in the Pleiades (M45) can be read from this sky chart as 3 hours 47 minutes 30 seconds RA and +24 degrees 7 minutes dec.*

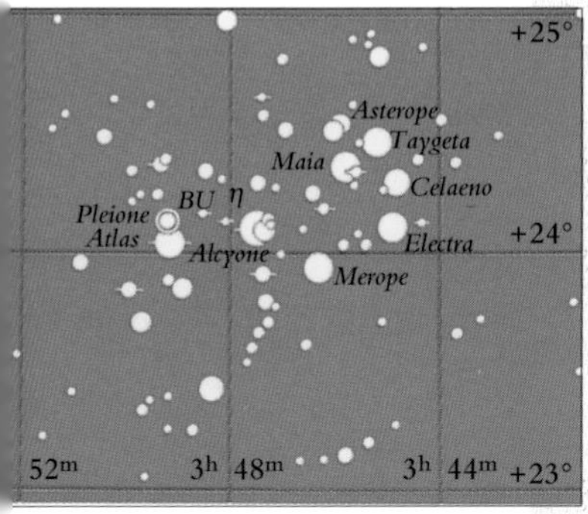

+25°
+24°
+23°
Asterope
Taygeta
Maia
Celaeno
BU
η
Pleione
Atlas
Alcyone
Electra
Merope
52m
3h 48m
3h 44m

ı display shows which way :o swing the telescope and ndicates when you are on :arget. Look in the eyepiece ınd there the object is!

Digital setting circles can be ıdded to most of today's ɔopular telescopes. Re-narkably, they do not ıeed a polar-aligned ·quatorial mount, ınd will work ɔn altazimuth or Dobsonian nounts. Most nodels sell for the ɔrice of a beginner's :elescope.

HANDS-OFF ASTRONOMY

A step up in sophistication is ı telescope with high-speed tepper motors on each axis ınd a computerized hand :ontroller to drive them. 5elect an object on the hand :ontroller's display, press "Go," and the telescope ıutomatically slews itself across he sky to the target. You can ılso operate these telescopes rom a laptop computer with a ky-charting program. Point at ı galaxy on the screen, click the nouse, and away the telescope goes to find the real galaxy.

This technology is most ɔopular on fork-mounted Schmidt-Cassegrain telescopes. A side-benefit is that you can track the stars without polar-aligning the mount. When the mount is set up in altazimuth fashion, the computer will pulse each of the motors by the correct amount to track an object moving across the sky.

The ability to find and follow objects easily without polar alignment has made computerized Schmidt-Cassegrains very popular. Of course, there is a price to pay—a computerized model will cost about twice as much as a similar telescope on a manually operated mount.

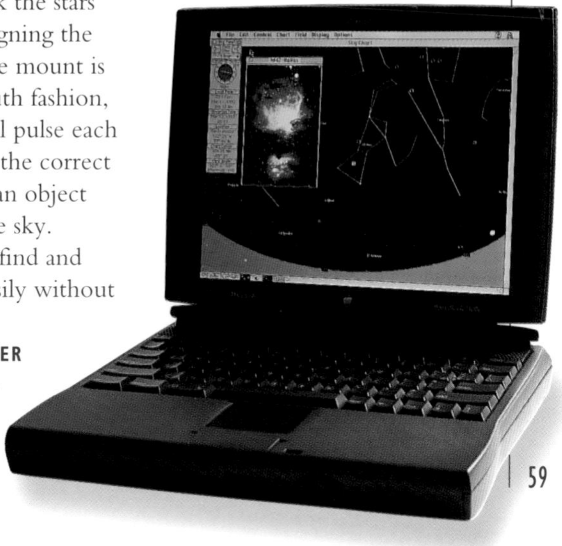

A LAPTOP COMPUTER *can control a high-tech telescope from up to hundreds of feet away.*

SKY COORDINATES

Every location on Earth has a latitude and longitude coordinate. Latitude measures how far north or south of the Earth's equator a place is. Longitude measures how far west or east a place is from a north-south line called the prime meridian, which runs through Greenwich, England.

Similarly, every object in the sky can be mapped with two coordinates: declination and right ascension. Declination (dec.) measures the object's position north (+) or south (–) of the celestial equator, a circle that divides the sky into northern and southern hemispheres. Declination is measured in degrees, minutes, and seconds of *arc*. There are 60 arcseconds in an arcminute, and 60 arcminutes in a degree. An object on the celestial equator has a declination of 0 degrees. The distance from the celestial equator to either of the two celestial poles is 90 degrees.

Right ascension (RA) measures how far east an object lies from the sky's prime meridian. The prime meridian is an imaginary line running due north and south through the vernal equinox—the point where the Sun crosses the celestial equator on 20 March each year. Right ascension is given in hours, minutes, and seconds of *time*. It is measured eastward from the prime meridian, which has an RA of 0 hours, until we reach 23 hours, 59 minutes.

While all celestial objects have right ascension and declination coordinates, these numbers slowly change because of a motion of the Earth called precession. As a result, catalogs and star charts specify which "epoch" they are designed for—most currently in use give coordinates accurate for the year 2000.

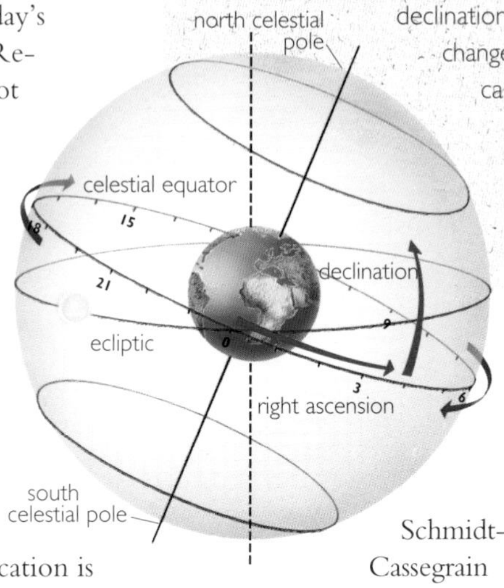

north celestial pole
celestial equator
18
15
21
declination
ecliptic
0
9
3
6
right ascension
south celestial pole

Becoming *an* Astrophotographer

Kodak
Flexi
AGFACOLOR
ULTRA 50
Sensia
100
400

Some astrophotography techniques require no more than the camera gear you may already own.

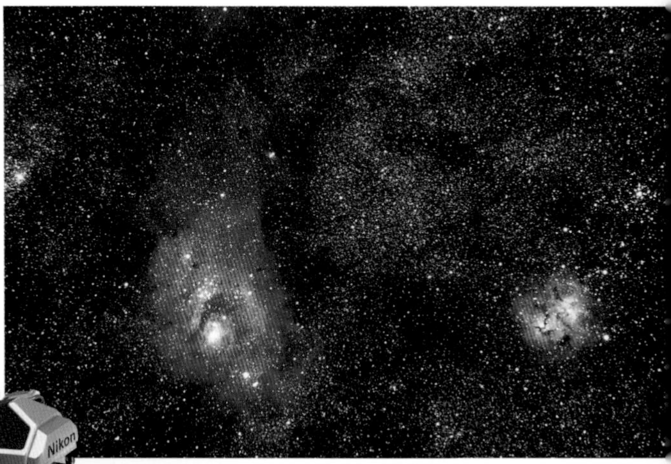

Nikon
Meade
CAMERA ADAPTER

LONG EXPOSURES *show up colors in the Lagoon and Trifid nebulas (above) that remain invisible to the eye. With a camera adapter (left), you can easily attach a camera to a telescope.*

Hardly anyone who has a camera and an interest in astronomy can resist taking photos of the night sky. Snapshot astrophotos will record a memorable night-sky scene just as your eye saw it. Long-exposure photos can actually capture far more than meets the eye, recording striking colors and faint objects that are invisible by eye even through a telescope. These images may serve as valuable scientific records or simply as beautiful "paintings" of celestial light.

Cameras

For casual sky shooting, almost any design of camera will do, but it must have a B (for Bulb) setting for the shutter. This setting holds the shutter open for as long as the shutter button is pressed—an essential feature because most night-sky photos require exposures of several seconds, if not minutes. To hold the button down during these long exposures, you use a locking cable release, which screws into the shutter button.

The drawback of most new cameras is that their shutters are battery-operated. Relying on batteries is risky—they may fail in the cold night air or during long exposures. This is why many astrophotographers seek out mechanical cameras on the secondhand market. (Good models include 35 mm single-lens-reflex, or SLR, cameras such as the Canon FTb and F1, Nikon FM2 and F3, Olympus OM-1 and OM-4T, and Pentax K1000 and LX.) It might seem strange to forgo the electronic features that adorn today's cameras, but none of them is needed for astrophotography.

While a close-up of the Moon or a clear image of a faint nebula requires shooting through a telescope (see pp. 66–67), many sky subjects can be captured with standard camera lenses (see pp. 62–65). Lenses that have fixed focal lengths are usually of better quality and faster than zoom lenses. For 35 mm cameras, a set of three fixed lenses is useful: a wide-angle lens (one with a focal length of 24 to 35 mm), a normal lens (50 to 55 mm), and a short telephoto lens (85 to 200 mm). Try to choose fast lenses—these have a maximum aperture of f/2.8 or f/2.

For photographing the night sky, especially with color film, most of the filters employed by photographers and amateur astronomers are of limited value. Broadband LPR, or deep-sky, filters (see p. 56) can help darken sky backgrounds but usually shift the color balance to green. A dark red filter will enhance nebulosity when shooting with the favored black-and-white film, Kodak Technical Pan 2415, or Tech Pan.

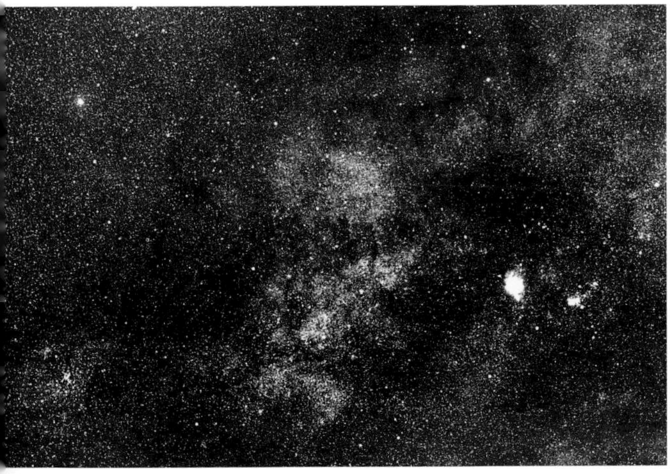

THIS EARLY ASTROPHOTO *of the region surrounding the Lagoon and Trifid nebulas was taken by E. E. Barnard.*

Fast Films

The speed of a film—its ISO rating—indicates how quickly it reacts to light. Faster films allow shorter exposure times, but they also tend to produce grainier images.

Twilight scenes or bright objects such as the Moon are best captured on slow, fine-grained films, which have an ISO rating of 50 to 100.

Most celestial shots require the extra light-gathering ability of at least an ISO 400 film. In recent years, ISO 400, 800, and 1000 print films from Kodak and Fuji have been popular. For ultra-high speed, Konica offers an ISO 3200 print film. For slides, high-speed films such as Kodak's Ektachrome 400 and Ektachrome P1600 have proved suitable for astrophotography.

For extra speed, most color slide films and some print films can be "pushed," a procedure that extends the film's development time. Most photolabs offer push-processing for a small extra charge. An ISO 800 film pushed "one stop" has the recording power of an ISO 1600 film, though with an increase in graininess.

For very detailed images, some astrophotographers shoot Tech Pan, a fine-grained black-and-white film much sharper than any color film. But for Tech Pan to pick up faint stars and nebulas, it must be hypersensitized—a baking process that increases its speed. Color film can also be hypered, but this is not essential. You can take superb photos with today's ISO 400 and 800 films right out of the box.

While a one-hour photolab can provide fine results, your photographs will be safer with a specialized professional lab. Always shoot a daylight scene at the start of the roll so the lab can work out where the frames are on your film.

Keeping Records

The best tip for improving your astrophotos is to keep records of the exposure data, sky conditions, and equipment used for every shot. This will help you learn which exposures and methods work best. Always shoot a range of exposures—a technique called bracketing. A shorter or longer exposure will often produce a better result than an estimated "best" exposure.

E. E. BARNARD

Imagine painstakingly guiding a telescope for three hours on a cold Wisconsin winter night. "How do you keep warm?" visitors to the Yerkes Observatory would ask. "We don't," replied Edward Emerson Barnard (1857–1923). For this pioneer of astrophotography, the arduous conditions were part of a normal night's work.

From the 1840s on, astronomers had been among the first to put the new technique of photography to work. Until Barnard, however, nobody had used it to map the Milky Way. From 1889 to 1895, he used an astrograph—a device with a 6 inch (150 mm), f/5 lens—and multi-hour exposures to photograph the Milky Way. By 1905, he had switched to the 10 inch (250 mm) Bruce Photographic Telescope, which used glass plates. Images taken with these two instruments formed the basis of Barnard's 1927 *Atlas of Selected Regions of the Milky Way,* a work that proved our galaxy is riddled with dark dust and nebulosity. Many of the famous star fields portrayed in this stunning atlas were first photographed, and in some cases discovered, by Barnard. And while our optics and films have improved, few astrophotographers have surpassed the quality and quantity of E. E. Barnard's century-old images.

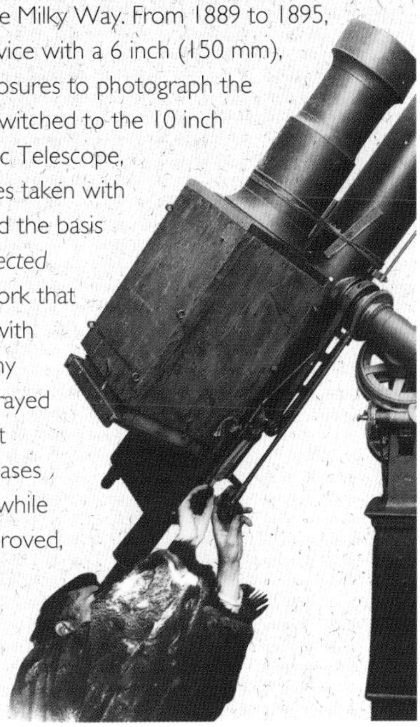

CAMERA *on* TRIPOD

With today's fast films, the only equipment you need to start taking spectacular photographs of the night sky is a camera and a tripod.

Taking pictures of the night sky is back-to-basics photography. With today's automated cameras you need only point and shoot to get technically perfect photos—at least during the day. At night, especially with faint subjects such as stars, old-fashioned skill has to replace the microchip circuits of modern cameras. Light meters and autofocus sensors will not function in the dim light.

An astrophotographer must manually set the f-stop and the shutter speed. The f-stop is the adjustable aperture of a lens, which controls how much light enters the camera; the shutter speed is the length of time the shutter remains open to expose the film. Simple astrophotography—shots done with a camera on a tripod—usually employs fast apertures of f/2 to f/2.8. A lens set to an aperture of f/2.8, for example, lets in twice as much light as a lens set to f/4.

With a good-quality fast lens and one of today's fast films, you can capture striking portraits of constellations and Milky Way star clouds without a tracking system. No guiding, no polar alignment, no elaborate equipment to buy or set up. Just place the camera on a sturdy tripod, focus its lens at infinity, and, with the camera in manual-exposure mode, set the lens aperture to f/2.8. Frame the scene, then lock the shutter open on its B setting for 10 to 80 seconds. Try a range of exposures at first.

SHORT EXPOSURES *and fast films keep stars looking like points. This portrait of Moon, planets, and stars was taken with an exposure of 4 seconds.*

LONG EXPOSURES *allow the stars to trail across the film. This photo shows the same scene as the portrait does, but was taken with an exposure of 90 minutes.*

The secret of success is the film. Use at least an ISO 400 film. Better still are the latest ISO 800 to 3200 films. With exposures of less than 60 seconds, these films can record stars far fainter than you can see with the naked eye.

CONSTELLATIONS

With this simple technique, it is possible to capture constellation patterns, the Milky Way, and even bright nebulas such as the Eta Carinae Nebula and the Great Nebula in Orion. Only the brightest stars will record in light-polluted skies, so the darker the sky, the better.

Of course, the sky is always moving. How long an exposure you can use before the stars start to streak depends on the lens's focal length and

CONSTELLATION EXPOSURES

To avoid star trailing, use exposures no longer than these.

Lens focal length	Near 90° dec. (cel. poles)	Near 45° dec.	Near 0° dec. (cel. equator)
28 mm	80 seconds	50 seconds	30 seconds
35 mm	60 seconds	35 seconds	20 seconds
50 mm	40 seconds	25 seconds	15 seconds

THIS LUCKY SHOT *captures a meteor streaking across a sky that is lit up by a red and green aurora.*

TOP CAMERA-ON-TRIPOD TARGETS

1. **Total solar eclipse**
2. **Rare all-sky aurora**
3. **Crescent Moon near Venus in twilight**
4. **Milky Way in Sagittarius**
5. **Circumpolar star trails**

n where in the sky it is ointed (see Table). Stars ear the celestial equator, uch as those in Orion, will rail more quickly than stars ear the celestial pole, such as hose in the Big Dipper.

TAR TRAILS

3ut why not let the stars trail cross the frame? Star-trail ictures are always impressive. 'or a dramatic composition, nclude a landscape or a fore-round object such as a tree.

Star trails are best taken vith an ISO 100 to 400 film. 'or first attempts, try expo-ures of 5 to 30 minutes at f/4. Check the results—if the sky s brightly lit, the exposures vere too long, but if the sky is lark, your site is good enough o allow even longer expo-ures. At very dark sites, xposures can last all night, specially with the lens topped down to f/8 or so.

MOONLIGHT PORTRAITS

Astrophotography need not top when the Moon comes ıp. On moonlit nights, venture out to scenic loca-ions. During a Full Moon, xposures of 40 seconds t f/2.8 with ISO 400 film vill record a scene that ooks like daylight, complete vith blue sky. And yet there will be stars in the sky—a nightscape shown in literally a different light.

TWILIGHT SCENES

Few images portray the beauty and tranquillity of the night sky better than a photograph of the crescent Moon at twilight, especially when a bright planet such as Venus is nearby. Use exposures of 1/2 to 4 seconds at f/2.8 with ISO 50 to 100 film. With a sensitive light meter, you might be able to take an exposure reading off the twilit sky—this is the only astrophotograph where a light meter is of any value.

AURORAS

The Northern and Southern lights are photogenic, but elusive, subjects. During a bright display, use a fish-eye (8 mm to 16 mm), wide-angle (24 mm to 35 mm), or normal (50 mm) lens set at f/2.8 or faster. Use at least an ISO 400 film and expose for 10 to 40 seconds. There is no single correct exposure—a shorter exposure freezes the rippling curtains of light, but may not pick up the subtle shades of red, green, blue, and purple that film can record but the eye usually cannot see.

METEORS

Take enough constellation and star-trail shots, and chances are a meteor will eventually streak across a frame. It is, however, much more difficult to catch a meteor deliberately. The best times are during an annual meteor shower (see p. 156). Use ISO 400 film, a 24 mm to 50 mm lens set to f/2.8, and exposures as long as your site allows. For longer meteor streaks, aim away from the shower's radiant. Keep shooting all night—with luck, one frame will grab a meteor.

THIS NIGHTSCAPE *is illuminated by a Full Moon just outside the frame.*

PIGGYBACKED CAMERA

By using an equatorial mount to track the stars during an exposure, you can reach deep into space and record fainter targets.

A TRIPOD HEAD *attached to a telescope allows the piggybacked camer to aim independently of the telescope.*

The next step up in complexity still uses the camera's own lens to shoot the sky, but now the camera rides piggyback on a platform that follows the stars. The advantage is that stars will record as points, even in 10- to 30-minute exposures. Because the exposures can be so long, a piggybacked camera is able to capture stars and nebulosity far fainter than any stationary camera will record.

However, the platform must be aligned as closely as possible to the celestial pole. As the focal length of the lens goes up, so does the intolerance for poor alignment. For example, telephoto lenses with focal lengths of 100 to 300 mm demand alignment within 1/5 degree. Just aiming at Polaris or Sigma Octantis will not be sufficient (see Box).

TOP PIGGYBACK TARGETS

1 **Bright comets such as Hale-Bopp**
2 **Milky Way in Sagittarius**
3 **Constellation of Orion**
4 **Nebulas and clusters of Carina/Crux area**
5 **Large Magellanic Cloud**

THE PIGGYBACK PLATFORM

The simplest platform is the barn-door tracker, a home-built contraption made of two wooden boards bolted together. As you slowly turn the bolt, the "barn door" gradually opens up, swinging the camera from east to west across the sky. Exposures of up to 10 minutes are possible. Detailed plans for barn-door trackers can be found in back issues of most astronomy hobby magazines (see p. 276).

Any motorized equatorial mount can serve as a platform for a piggybacked camera. A few manufacturers market small "scopeless" mounts that serve as dedicated camera platforms, but most people attach the camera to the side of a telescope tube. Your telescope mount may already hav an attachment bolt, or you ca buy an adapter bracket.

FILM AND EXPOSURES

The ideal subjects for piggyback photos are constellations Milky Way panoramas, large nebulas and galaxies, and comets. These are best shot with an ISO 400 or 800 film and lenses set to an aperture o f/2.8 to f/4. Lenses with faste settings are available, but mos perform poorly when used "wide open" at f/1.4 or f/2.

A CONVENTIONAL CAMERA *riding piggyback can capture an expanse of the Milky Way. This image was taken with a wide-angle lens set at f/4.*

.. look, how the floor

f heaven

s thick inlaid with patines

f bright gold.

he Merchant of Venice,
VILLIAM SHAKESPEARE (1564–1616)

A TELEPHOTO LENS *works well for nebulous areas. This is a 20-minute exposure of the Eta Carinae Nebula, taken with a 300 mm lens set at f/4.*

tars appear bloated, especially t the corners of the frame.

Exposure times depend on ky conditions. In a dark sky, 0 to 20 minutes at f/2.8 with SO 800 film records an astonshing amount of nebulosity.

GUIDING OUT ERRORS

When shooting with short focal ength lenses (under 135 mm), ou can usually turn on the rive, open the shutter, and valk away. Longer lenses, however, are more demanding, and slight irregularities in the speed of the drive must be guided out. To do this, you monitor a star through the main telescope during the exposure. An illuminated reticle—an eyepiece with cross hairs illuminated by a dim red light—can help you detect any wandering of the guide star.

The mount should have drive motors on both axes, each with a push-button speed control. If the star wanders one way along the cross hairs, press the appropriate button to jog it back. Taking a long exposure demands constant vigilance at the eyepiece, or the use of an automatic CCD guider (see p. 67).

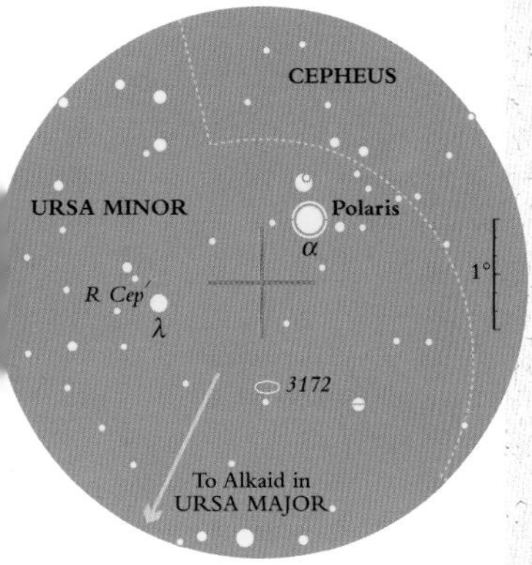

CEPHEUS
URSA MINOR
Polaris
α
1°
R Cep
λ
3172
To Alkaid in
URSA MAJOR

NORTH CELESTIAL POLE

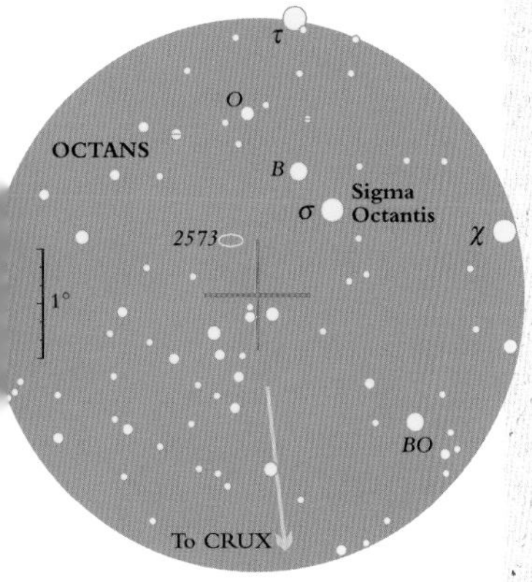

τ
OCTANS
B
Sigma
Octantis
σ
2573
χ
1°
BO
To CRUX

SOUTH CELESTIAL POLE

PRECISE ALIGNMENT

To record pinpoint stars in long exposures, you need to zero in on the *exact* location of the north or south celestial pole. One way to do this is to use these charts (left). Each chart has a field of view of 5 degrees, about the same as most finderscopes.

Step 1 Ensure that the finderscope's cross hairs aim at the same spot in the sky as your main telescope.

Step 2 Place the mount so that its polar axis (see p. 53) aims roughly at the celestial pole.

Step 3 Swing the tube so that it is aimed at 90 degrees declination according to the declination setting circle.

Step 4 By moving the *whole* mount, now aim the finderscope at the pole location shown on the chart. Take care not to move the mount in declination or right ascension. Use the star patterns as a guide—while peering through the finder with one eye, watch Alkaid or Crux with the other eye to work out which way to offset the telescope.

Trouble-shooting This method can go awry if your declination setting circle does not read a true 90 degrees. To calibrate it, set the telescope at 90 degrees, then rotate the telescope back and forth in right ascension (east-west). If the telescope is set accurately, the stars will appear to revolve around the center of the eyepiece field. If they do not, move the telescope slightly in declination and try again. Once you have the stars revolving around the center, loosen the declination circle and turn it until it reads 90 degrees. Then lock it down. You should not need to adjust it again. Now return to Step 4.

THROUGH *the* TELESCOPE

By shooting through a telescope, you can add frame-filling snapshots of the Moon and close-ups of glowing nebulas to your astro-album.

When newcomers to the hobby think about taking astrophotographs, this is what they have in mind—attaching a camera to the focus of a telescope, a technique often called prime-focus photography. The lens is now the telescope itself, so a single-lens-reflex (SLR) camera is essential—it allows you to view and focus through the same lens that will take the photograph.

To adapt a camera to a telescope, remove the lens from the camera and the eyepiece from the telescope. Attach a T-ring designed for your camera brand onto the camera's lens mount, then screw the T-ring onto a camera adapter. The tube of the camera adapter often just slides into the focuser of the telescope in place of an eyepiece.

Any telescope mount will do for snapshots of the Moon (and of the Sun—with a safe filter, see p. 56). However, to take long-exposure photographs of planets and deep-sky objects such as nebulas and galaxies, you will need the tracking ability of a sturdy equatorial mount equipped with an accurate drive and variable speed controls.

MOON SHOTS *at the prime focus of most telescopes record the entire lunar disk (above). Lunar close-ups (right) require eyepiece projection and exposures of 1 to 4 seconds. For good deep-sky images, you may need an off-axis guider between scope and camera (top).*

SHOOTING THE MOON

The best subject for trying out prime-focus photography is the Moon. Even with slow film, exposures are only a fraction of a second (see Table).

The size of the Moon in the frame depends on the telescope's focal length (see p. 51). A 2,000 mm focal length system produces a lunar disk just big enough to fill a 35 mm frame. With a 1,000 mm system, the Moon will be about 9 mm across—still impressive.

Because the viewing screens of most cameras go dark at

LUNAR EXPOSURES

Assuming the use of ISO 50 film. Exposure times given in fractions of a second.

f-ratio	Thin Crescent	Thick Crescent	Quarter Moon	Gibbous Moon	Full Moon
f/4	1/30	1/60	1/125	1/250	1/500
f/5.6	1/15	1/30	1/60	1/125	1/250
f/8	1/8	1/15	1/30	1/60	1/125
f/11	1/4	1/8	1/15	1/30	1/60
f/16	1/2	1/4	1/8	1/15	1/30

For ISO 100 film, use shutter speeds twice as fast (1/60 instead of 1/30).
For ISO 200 film, use shutter speeds four times as fast (1/125 instead of 1/30).

TOP PRIME-FOCUS TARGETS

1 Great Nebula in Orion
2 Ring Nebula
3 Omega Centauri
4 Moon with earthshine
5 Close-ups of First Quarter Moon

uch long focal lengths, the rincipal challenge is to get he Moon in focus. A camera vith an interchangeable ocusing screen helps. Matte creens made for telephoto-ens work provide a brighter mage, making it easier to get hat sharp lunar snapshot.

ZOOMING IN

illing a frame with a small ection of the Moon requires ots of focal length, more than nost telescopes offer. The nswer is eyepiece projection. Many camera adapters come vith an add-on extension ube that accepts an eyepiece, lthough Schmidt-Cassegrain elescopes require a separate dapter called a tele-extender. The eyepiece projects a highly nagnified view of the Moon nto the film.

Eyepiece projection is also ssential for photos of planets. Without the extra magnifi-ation, their disks will be far oo small to show any detail.

Lunar close-ups and planet ortraits usually require faster SO 200 to 400 film and expo-ures of 1 to 4 seconds. To nake the exposure, hold a lack card over the front of he telescope, open the camera hutter on Bulb, then flip the ard away and back again. Without this "hat trick," your hotographs may be blurred y the slight vibration of the amera's own shutter.

DEEP-SKY SPLENDORS

For faint deep-sky targets such as nebulas and galaxies, faster telescopes—ones with f-ratios of f/4 to f/7—are a definite advantage because they allow shorter exposure times. Alternatively, you can speed up a slower telescope with an accessory called a tele-compressor.

Even at fast f-ratios and with ISO 800 film, however, you will need exposures of 15 to 60 minutes. It would be wonderful if we could simply attach a camera at prime focus, open the shutter, and walk away. Try it at these focal lengths and the result will be horribly trailed stars, despite the most precisely aligned mount.

Errors in the drive gears are the chief culprit, but flexing mounts and tubes, atmospheric refraction, and wind all contribute to wiggly stars. Some method of guiding the telescope is called for.

Schmidt-Cassegrain owners usually turn to off-axis guiders. These devices contain a small prism that picks off a star image just outside the film frame, allowing the photographer to monitor the guide star's position while photographing through the same telescope. Owners of refractors and Newtonian reflectors often opt for a separate piggybacked guidescope—perhaps a small 2.4 or 3 inch (60 or 75 mm) refractor.

The least expensive but most difficult method of guiding is to do it by hand, tweaking speed controls to keep a star centered on the illuminated cross hairs of the guiding eyepiece. Today there is an alternative. Automatic guiders using CCD chips detect any wandering of the star's image and send pulses to the telescope's motors, which move the scope so the star returns to its centered position. The constant, unfailing corrections result in a photograph more perfectly guided than any human is capable of, especially while falling asleep at the eyepiece at 3 am.

PRIME FOCUS *Shot through an 8 inch (200 mm) Schmidt-Cassegrain, the Great Nebula in Orion fills the frame.*

A NORMAL EYEPIECE *inserted into a camera adapter will magnify the image—an essential technique for planet shots.*

Canon
EOS
630

VIDEO *and* CCD CAMERAS

With CCD technology, amateurs can capture images of planets and faint galaxies that were once available only through observatory telescopes.

SONY
CELESTRON

SIMPLE ASTRO-MOVIES *can be made by holding a home camcorder above the eyepiece of your telescope.*

Very few professional astronomers still use photographic emulsions to record the sky. Film has been replaced by electronic chips. This digital revolution is now happening in amateur astronomy.

PUTTING THE MOON ON TV

Going digital with astrophotography can be as simple as holding a camcorder up to a telescope eyepiece. Aim the telescope at the Moon, insert a low-power eyepiece, then hand-hold the camcorder above the eyepiece or, better yet, place the camcorder on a separate tripod so that it looks directly into the eyepiece. Focus the camcorder's built-in lens to infinity and zoom it in to its telephoto position to see highly magnified views of small areas of the Moon. You can also use a camcorder in this way to capture images of the planets.

DIGITAL IMAGING

Most home camcorders are not sensitive enough to pick up images of nebulas and galaxies. For such distant objects, astro-imagers have turned to specialized CCD cameras. CCD stands for charge-coupled device, a type of light-sensitive chip similar to the chips used in home camcorders.

Instead of movies, CCD cameras record single long exposures. During an exposure, light falls onto an array of pixels arranged in a grid pattern on the chip. At the end of the exposure, each pixel reads out a voltage that corresponds to how much light fell onto it. The readings are digitized (turned into binary numbers consisting of 0s and 1s), then sent to a computer that can re-create the image on its monitor.

CCD cameras have major advantages over conventional film. A CCD image is digital data, which can be manipulated, stored on computer disks, and even transmitted via modem. You can also see the results immediately at the telescope. Most importantly, CCDs are much more sensitive than film, requiring exposures of only 5 minutes to record what might take a fast film 30 to 60 minutes to pick up. With these shorter exposures, there is often no need for elaborate guiding. Because the images are digital, they can be enhanced on a computer, making it possible to discover faint objects even in light-polluted or moonlit skies.

The main drawback is the expense of a CCD camera—which can cost as much as a good telescope—and also the hardware and software that

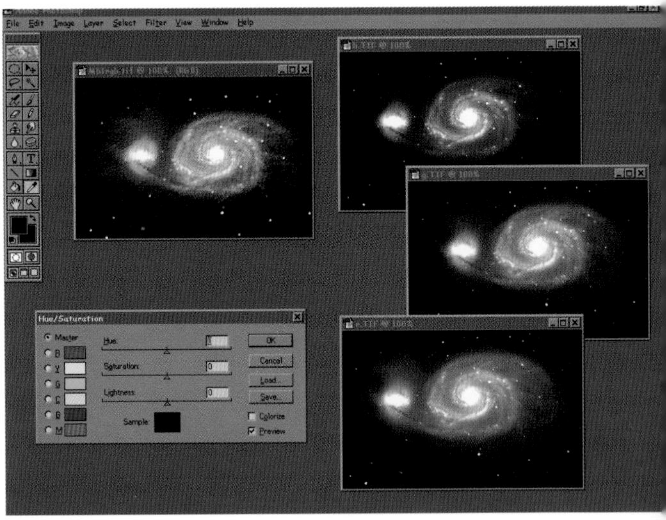
File Edit Image Layer Select Filter View Window Help
Hue/Saturation
Master
Hue:
Saturation:
Lightness:
Sample:
OK
Cancel
Load...
Save...
Colorize
Preview

IMAGE-PROCESSING SOFTWARE *such as* Photoshop *can merge three black-and-white CCD images taken through filters and create a color image.*

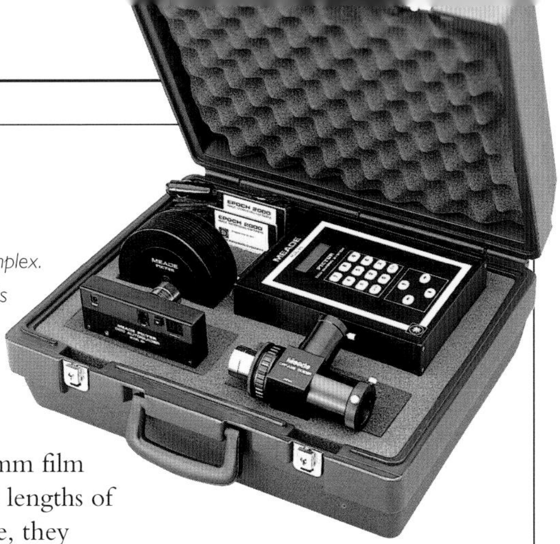

A CCD SYSTEM *is complex. The heart of the system is the round CCD head, which attaches to the telescope's focus.*

;oes along with it: you will ɪeed a personal computer oaded with plenty of RAM, pace on the hard drive, and mage-processing programs. Ɣou may also want a laptop ·omputer so you can easily ›perate the camera in the field.

ƆHOOSING A CCD

Ƭhe main difference between ƆCD cameras is the size of heir chips. The lowest-priced ·ameras use chips with a .96 x 165 pixel array, while ›op-end cameras have arrays of ,000 x 1,000 pixels or more.

Chips are getting bigger .nd cheaper each year. The .dvantage of a larger chip is hat it can record a larger area ›f sky. This is critical since ·ven large chips are much smaller than a 35 mm film frame, and at focal lengths of 2,000 mm or more, they "see" an area of sky only a few arcminutes wide, smaller than the disk of the Moon. Nevertheless, CCD cameras are superb devices for recording planets, galaxies, and smaller nebulas and clusters.

'HIS CCD IMAGE *of the Dumbbell ɴebula (M27), in Vulpecula, was ·aptured by Jack Newton.*

JACK NEWTON

If you see an astounding color CCD image, chances are it was taken by Jack Newton. From his home observatory in British Columbia, Canada, Newton pioneered color CCD techniques.

"The results I got with the first CCD cameras in 1990 blew me away," says Newton. "I couldn't believe what I was able to record. Now, with the latest cameras, it's possible to routinely take images that duplicate the best photos taken with the 200 inch Hale Telescope." Indeed, backyard telescopes can image objects never seen before from Earth. "With a 16 inch telescope, anyone can discover five or six new asteroids a night, every night." Newton predicts film will be all but gone in 10 to 20 years. "CCDs will replace film just as CDs replaced vinyl records. The next few years are going to be exciting times for amateurs."

It is important to match the CCD camera to a telescope of the appropriate focal length. Many of today's CCD cameras have tiny pixels just 9 microns, or 0.009 mm, across. Attached to a telescope with a focal length of at least 1,000 mm, these high-resolution cameras will produce images as sharp as the atmosphere allows. When used on shorter focal length telescopes, however, the pixels will be larger than the star images, yielding odd-looking square stars.

Most CCD cameras are black and white. To produce a color image, you need to take three exposures, one each through red, green, and blue filters, and then merge them in the computer using image-processing software.

Producing a clean, noise-free image requires taking two calibration frames. A "dark frame"—an image with the same exposure time as the main exposure but taken with the lens cap on—records the background noise inherent in all cameras. A "flat field" frame—a short exposure of a blank screen or bright light—records flaws such as dead pixels and dust specks. You can then use image-processing software to electronically subtract these unwanted defects from the main image.

CCDs are not devices for the technologically challenged, but if you love computers, you will love CCD imaging.

THE SKYWATCHER'S LIBRARY

VRANO
METRIA

Exploring the universe of information about astronomy can be almost as engrossing as exploring the universe itself.

Starting to learn about the universe requires no more than one or two good books. You can also keep in touch with celestial events and discoveries by subscribing to one of the popular astronomy magazines (see Resources Directory, p. 276). But every backyard stargazer soon discovers the need for a few standard reference works.

STAR ATLASES AND FIELD GUIDES

A star atlas is a road map to the sky. It can help you find hundreds of telescopic targets, but you must first know how to find key constellations and bright stars. This is where a planisphere can help. These "star wheels" show the entire sky on a disk that rotates to any date and time of night.

A step up in detail is one of the "field guides" to the night sky. These books feature month-by-month charts of the entire sky, supplemented by more detailed maps that are often arranged by constellation. A companion volume in this Nature Company series, *Skywatching* by David H. Levy, is an excellent example of this type of field guide.

More detail is provided by popular star atlases, such as *Norton's 2000.0 Star Atlas and Reference Handbook* and *The Cambridge Star Atlas 2000.0*. They contain excellent star charts that plot all stars down to the naked-eye limit of magnitude 6, as well as hundreds of other clusters, nebulas, and galaxies. Both atlases and field guides contain lists of the finest telescopic targets for each area of the sky, and any one of them is a valuable addition to an astronomy library.

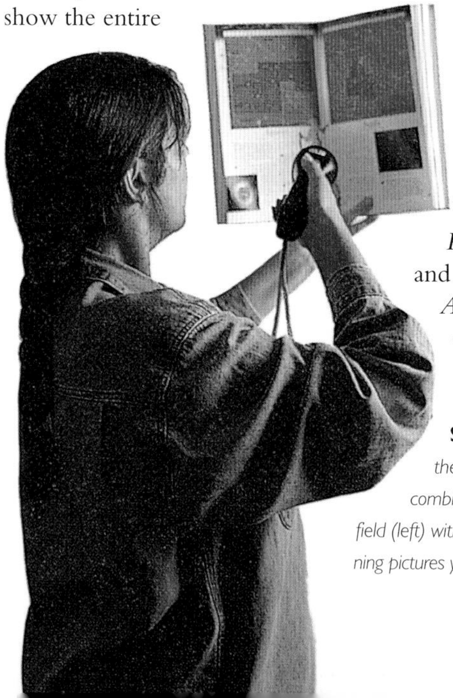

SKYWATCHING, *one of the Nature Company Guides, combines star maps to use in the field (left) with informative text and stunning pictures you can enjoy in an armchair.*

FANCIFUL FIGURES *adorn early star charts such as these celestial globes of 1630 (above). The first serious star atla was Bayer's* Uranometria *in 1603 (top*

More detailed still is the *Sky Atlas 2000.0*. This large-format atlas plots all stars down to magnitude 8 and includes 2,500 deep-sky objects—many more than the magnitude 6 atlases do. One of the most comprehensive atlases in print is the *Uranometri 2000.0*, a three-volume set (including a massive catalog, the Field Guide) that divides the sky into 473 charts, plottin all stars down to magnitude 9, as well as 14,000 deep-sky objects to magnitude 15.

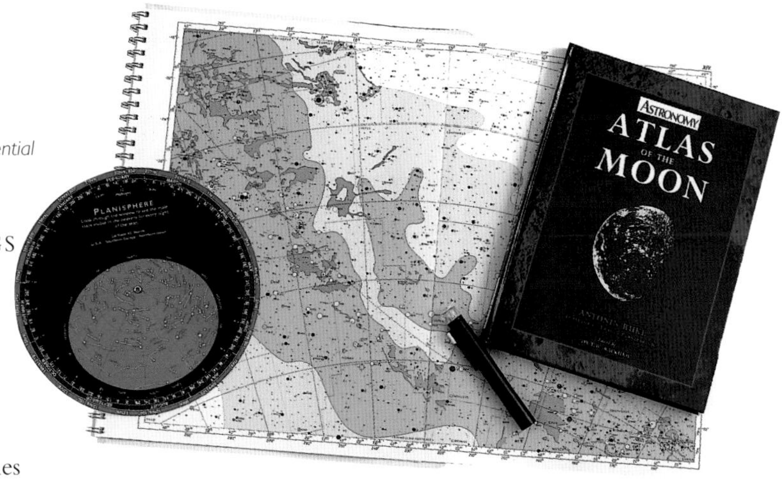
ASTRONOMY
ATLAS
OF THE
MOON
PLANISPHERE

ARTS AND ATLASES *of the oon, planets, and stars are essential ols" for skywatchers.*

EEP-SKY CATALOGS

glance at any star map veals cryptic references objects such as M31, GC 4565, and IC 434. 'hat do they mean?

In the late 1700s, the ench astronomer Charles essier compiled the *Catalogue Nebulous Objects and Star lusters*, which remains the ost popular list of bright eep-sky wonders. Number 31 Messier's list, for example, the Andromeda Galaxy 131). Modern versions of e list contain 110 entries, d seeing all of them is a allenging goal for amateur tronomers (see p. 231).

By the mid-1800s, astronomers had discovered many more fuzzy telescope targets in the night sky. In 1864, the English astronomer John Herschel published the *General Catalogue of Nebulae*, a list of more than 5,000 objects discovered mostly by him and his father, William Herschel (famous for discovering the planet Uranus).

In 1888, J. L. E. Dreyer published a master list, the *New General Catalogue of Nebulae and Clusters of Stars* (see Box). Supplementary *Index Catalogues* (IC) were added in 1895 and 1908.

Astronomers such as Wilhelm and Otto Struve and S. W. Burnham compiled catalogs of double stars in the nineteenth century. For example, stars designated with the Greek letter sigma (Σ) are from Wilhelm Struve's catalogs. Collinder, Melotte, Ruprecht, Stock, Trumpler, and many others in the twentieth century have cataloged star clusters and lent their names to clusters that Messier and the NGC authors missed. In the 1950s, George Abell cataloged clusters of galaxies—distant Abell clusters are challenging targets for owners of large telescopes.

THE NEW GENERAL CATALOGUE

In 1874, the largest telescope in the world was owned by an amateur astronomer. At his ancestral home of Birr Castle, Ireland, the third Earl of Rosse, William Parsons, built what became known as the "Leviathan of Parsonstown" (pictured below). For its day, it was a monster—a 72 inch (1.8 m) diameter telescope slung on cables between two massive brick walls. With this unwieldy instrument, Rosse discovered that many "nebulas" had a spiral shape. Today we know them to be spiral galaxies. Rosse himself rarely used the telescope, but many other keen observers did. Among them was Johann Louis Emil Dreyer. Dreyer served as Rosse's assistant from 1874 to 1878, recording the mysterious spiral nebulas, among others, with sketch pad and notebook.

By 1886, Dreyer and other observers had discovered so many nebulas and star clusters that a new catalog was needed. The Royal Astronomical Society assigned Dreyer the task. Published in 1888, the *New General Catalogue* (NGC) contained entries for 7,840 objects. The two *Index Catalogues* (IC) later added another 5,386 objects, some discovered with the newly invented process of photography. A century later, the 13,000-plus objects of the NGC and IC lists form the core of all of today's comprehensive databases of deep-sky wonders.

CELESTIAL HANDBOOKS

Many of these catalogs are now out of print or inaccessible to amateurs, but some excellent encyclopedic works are still available. The three-volume *Celestial Handbook* by Robert Burnham, Jr., originally compiled in the 1960s, remains the most comprehensive. Another good reference is Rev. T. W. Webb's *Celestial Objects for Common Telescopes*, first published in 1859 but regularly updated (see p. 171).

COMPUTERS *and* CLUBS

Computers have revolutionized the way amateurs find information, but nothing can replace the advice of fellow skywatchers.

Voyager II
the Dynamic Sky Simulator
REDSHIFT 2
THESKY

PLANETARIUM PROGRAMS *present even more detailed views of the sky than printed atlases do.*

Personal computers and astronomy seem made for each other. One of the most useful pieces of software is a planetarium program, which is like an electronic star atlas. Some versions can be used to print out customized star charts—perhaps showing a select region of the sky or a new comet's path, or printed in mirror-image fashion to match the view through some telescopes and finderscopes.

You can set planetarium programs for any date thousands of years into the past or future. They are indispensable for figuring out details such as what the sky will look like at a certain time, when Venus will be closest to the Moon, or where a comet will appear.

At the very least, these programs plot all Messier and many NGC objects and stars down to magnitude 9. More advanced programs contain databases of thousands more deep-sky objects from newer catalogs, as well as providing the option of loading the *Hubble Guide Star Catalog* from CD-ROM. This database contains more than 19 million stars and non-stellar objects.

BEYOND STAR CHARTS

Many planetarium programs go far beyond star-charting functions. For example, they may simulate flights around the Solar System, showing the sky from other planets. Some contain still images and movies of planets and deep-sky objects, or even allow you to create animated movies of celestial events.

Multimedia programs available on CD-ROM are like electronic, interactive books. You can tour the planets or fly over a landscape on Venus or Mars. Such programs are an excellent way to get your family and friends interested in astronomy.

ASTRONOMY ON-LINE

Professional astronomers were among the first to put the Internet to use, back in the days before the World Wide Web and graphical browsers made it easy for anyone to navigate "the Net." Today, astronomy-related sites on the Internet are among the most

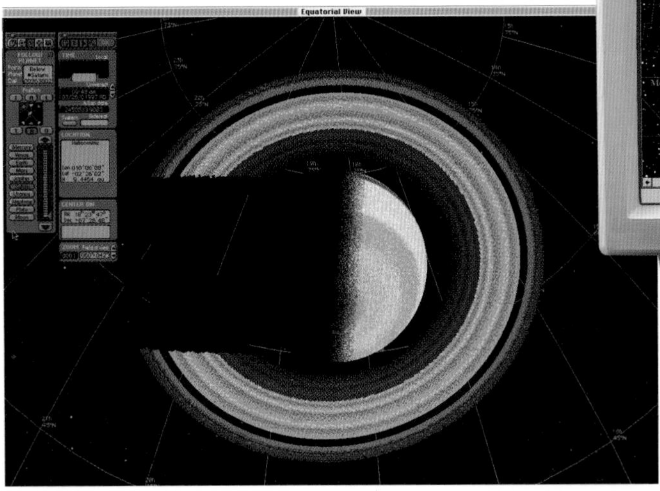
Equatorial View

Meade Epoch 2000sk
File Time Ephemeris Starmap Telescope Options Help
Starmap - Spherical Projection - FoV=70.000000°
Taurus
Orion
Monoceros

EXPLORE A VIRTUAL UNIVERSE *using programs that can show you various views of Saturn or help you to locate a particular star or galaxy.*

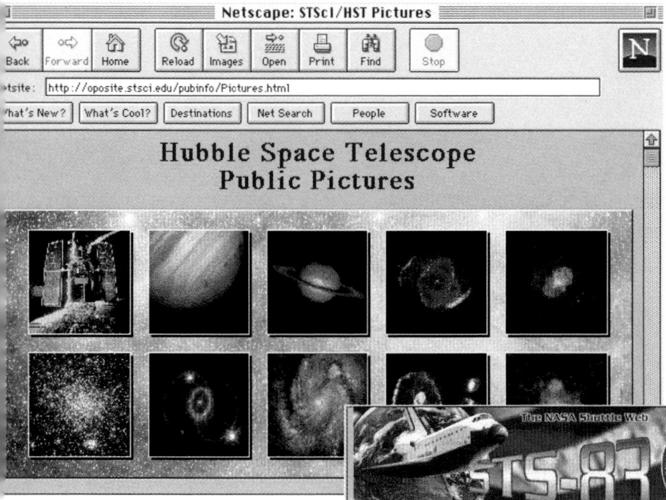
Netscape: STScI/HST Pictures
Back
Forward
Home
Reload
Images
Open
Print
Find
Stop
http://oposite.stsci.edu/pubinfo/Pictures.html
hat's New?
What's Cool?
Destinations
Net Search
People
Software
Hubble Space Telescope
Public Pictures

The NASA Shuttle Web
STS-83
Sneak Preview
Previous Missions ** Future Missions ** Shuttle-M

opular. Selected images rom the Galileo spacecraft at upiter, the Mars probes, or he Hubble Space Telescope ppear first on the Internet. NASA maintains a host of sites ich in images and information.

You can do your telescope hopping on-line. The web ites of equipment manufac-turers provide information and dvice on the latest models, nd allow you to order them lirect. You can even down-oad samples of new software.

Using the Internet, amateur stronomers can control tele-scopes halfway around the vorld, creating digital images hat they can then download nto their own computers. Current aurora predictions, he latest pictures of a new omet, this hour's satellite veather map—all this and nore are on the Web. There s so much to explore that ome amateurs conduct their obby via the computer creen, neglecting the real sky or the digital universe.

GETTING SOCIAL

Jsing the Web's newsgroups nd the special-interest forums ffered by commercial on-line ervices, amateur astronomers rom every nation now talk to ach other every day. New-omers can post questions bout what to buy or how to get started in the hobby. FAQ (Frequently Asked Questions) files provide useful advice on equipment and astrophotog-raphy. The world has become one big virtual astronomy club.

But nothing can replace meeting fellow skywatchers face to face. Most large towns have astronomy clubs that host regular talks, courses, and star parties. A club can be an excellent source of ad-vice and learning, and many clubs offer observing nights or courses for beginners.

A favorite event is the local star party. Most clubs will host monthly observing sessions at their club site or observatory in the country. At these gatherings, a newcomer can look through a variety of instruments and talk to the owners about what they like and dislike about their tele-scopes. To contact your local club, call a nearby planetarium, nature or science center, college, or univer-sity. Chances are the club is affiliated with one of those. You could also check the Internet for lists of clubs in your area.

Once a year, most regions also host an enormous star party, usually at a park or campground with dark skies. These events attract hundreds of amateur astronomers, and for many, the annual star party becomes their main chance to escape light-polluted skies. Swap-tables become popular places to pick up bargains, and vendors often set up sales booths. Check astronomy magazines or their web sites for dates of major star parties coming up in your area.

THE WORLD WIDE WEB *brings you the latest images from the Hubble Space Telescope and keeps you up to date on the space shuttle missions.*

STAR PARTIES *are popular events in every country. They are excellent opportunities to visit dark skies, find a bargain, and meet other skywatchers.*

Over the rim of the waiting earth the moon lifted with low majesty till it swung clear of the horizon and rode off, free of moorings …

The Wind in the Willows,
KENNETH GRAHAME (1859–1932), British author

CHAPTER THREE

SUN *and* MOON

OBSERVING *the* SUN

Life-giver, and even law-giver in the past, the Sun has a strong influence on the world that anyone can feel on a sunny day.

A Sun god has been a major feature in almost every human culture. Among the Hopi Indians of the arid American Southwest, for instance, the Sun is the "keeper of the ways." Each day it climbs out of its eastern kiva (an underground ceremonial chamber) and up into the sky to keep watch on the world.

Sun-watching in ancient times was inextricably mixed with astrology and divination. In the European Renaissance, however, the first telescopes were invented and studies of the Sun finally started to address its physical reality.

Amateur astronomers were responsible for many of these early studies. For example, an amateur determined that sunspots occur in cycles, and two amateurs discovered the existence of solar flares and their links to geomagnetic disturbances such as auroras.

During this century, most solar research has become "Big Science," with budgets and sophisticated instruments to match. Ambitious large-scale projects include the SOHO and Yohkoh satellite observatories (see p. 35).

But solar observation remains a rich and enjoyable field for the amateur, thanks in part to the Sun's ever-changing appearance. No other area of astronomy offers so many opportunities to see a celestial object change.

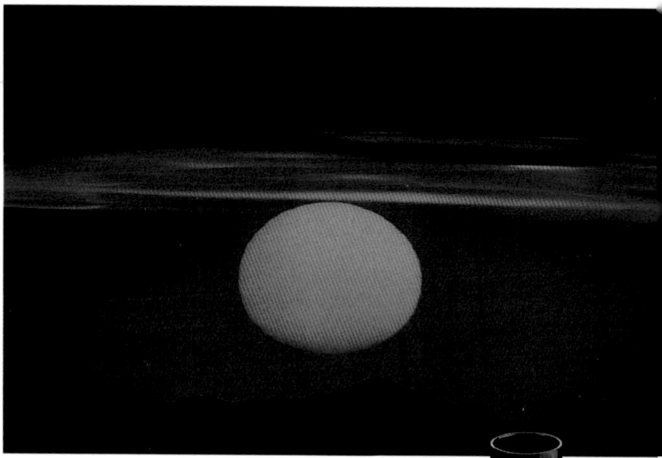

SUN GOD *An Egyptian pendant shows a Sun god in scarab-beetle form (top). The setting Sun (above). To match your telescope to an H-alpha filter, you may need a kit of extra accessories (right).*

Some amateurs also find that solar viewing fits more easily into their daily lives than does nighttime astronomy, with its late hours and long trips to find dark skies.

THE NEAREST STAR

The Sun is our closest star—in fact, it is about 270,000 times closer than the next nearest star, Alpha Centauri, and is the only star whose surface we can see in any detail.

The Sun's proximity makes proper safety precautions essential. It is possible to view the Sun with complete safety, but you must follow the procedures described in Chapter 2 (see p. 56) and below. Cutting corners can permanently damage your eyesight—Galileo's blindness in later life was probably caused by slapdash solar-observing methods

THE SOLAR SCOPE

A solar telescope does not need much aperture—6 inch (150 mm) is ample, and good observations can be made wit apertures of 2 to 3 inches (50 t 75 mm). Because sunlight usually creates poor seeing, yo cannot use the higher resolution afforded by a larger scope

To reduce the effects of ba daytime seeing, try observing in the early morning, before the Sun has had a chance to eat up the surroundings. Erect our telescope on a grassy area, voiding pavement, rooftops, or any other heat-absorbing urface. Another solution is to iew the Sun across a lake, vhere the water helps to keep he air relatively settled.

The only safe way to view he Sun directly is to use a olar filter that fits over the ntire aperture of the telescope see p. 56). An aperture filter educes sunlight to safe levels efore it enters the scope. This eeps the optics cooler and mproves the viewing.

Most solar filters are made rom Mylar plastic or glass; oth kinds will show sunspots nd faculae (see pp. 80–81). Expensive filters that isolate arrow wavelengths, such as Hydrogen-alpha (H-alpha), an show other surface details nd spectacular prominences.

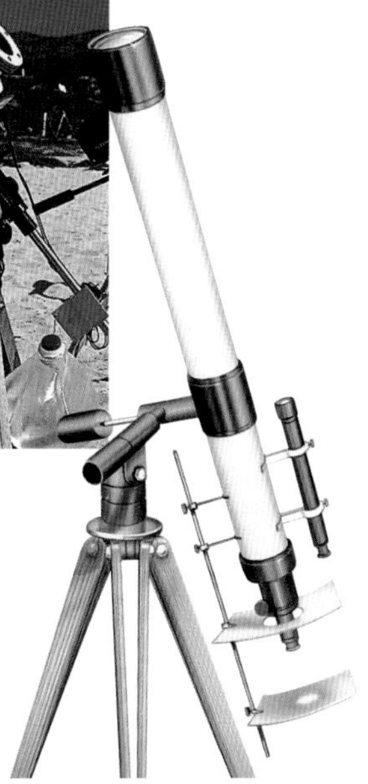

TWO SAFE METHODS *You can view the Sun directly through a telescope fitted with an aperture filter (above), or you can use a solar projection screen (right).*

Projecting the Sun

The other technique used for solar viewing is projection. It requires no costly equipment and allows a group of people to view the Sun at one time. Aim your telescope at the Sun and affix a sheet of white cardboard behind the eyepiece, focusing until the image looks sharp. Shield the projected solar image from any direct sunlight. Several observers can then gather around the telescope and study the Sun and its features. A projected solar diameter of 6 inches (150 mm) is easy to view and useful for sketching.

When aiming the telescope toward the Sun, never sight upward along the tube—instead, look at the scope's shadow on the ground. Cover the finderscope's aperture so that it cannot accidentally burn you and no one is tempted to look through it. Never leave the telescope unattended—and if your group includes children, watch them like a hawk.

The heat of the Sun can ruin the optical cement in a compound eyepiece, so do not use your best eyepiece. Opt for a cheaper design such as a Huygenian or a Ramsden. If your scope's aperture exceeds 4 inches (100 mm), place an opaque mask with a circular cutout over the aperture to reduce it to 4 inches, or even less. This reduces the heat buildup inside the instrument.

PHOTOGRAPHING THE SUN

Solar astrophotography can become a simple and enjoyable matter with the right equipment—an aperture solar filter, an adapter to couple your camera to the telescope, and, preferably, a telescope mounting that can track the Sun (see pp. 60–67).

Some amateurs take a photograph of the Sun every clear day, building up a valuable archive of images. Others photograph specific solar activity, such as sunspots, or use narrow-wavelength filters, such as H-alpha filters, to capture the active surface of the Sun.

The Sun's brightness means you can use slower films—with speeds of ISO 25 to 100—which provide better resolution. If you prefer black-and-white film, Kodak's Tech Pan is ideal for solar photography (see p. 61). Solar filters reduce the Sun's brightness to about that of the Full Moon—refer to the lunar exposure table on page 66 as a guide to exposure times for solar photographs, but take a few test rolls to see what works with your equipment.

Experienced solar photographers know that the toughest problem to contend with is poor seeing. Shooting in the early morning or across a body of water can help, but part of the challenge lies in discovering what works best for an individual site.

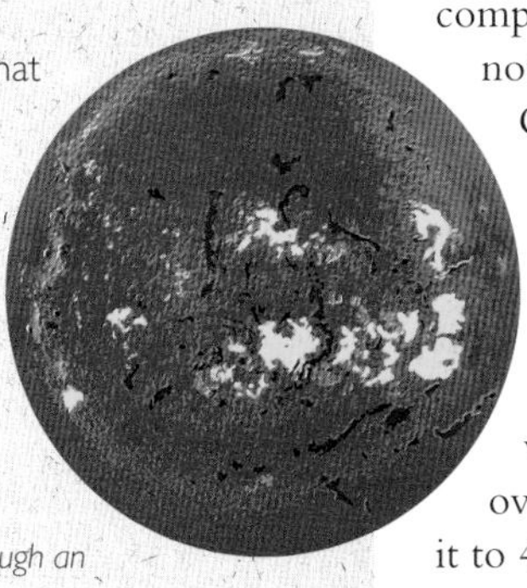

This false-color image of the Sun's surface was taken through an H-alpha filter. Regions of activity are shown as yellow.

THE STRUCTURE *of the* SUN

The Sun is a vast, layered structure, comprising all but a small fraction of the mass of the entire Solar System.

THE FACE OF THE SUN *This nineteenth-century painted woodcarving from Germany depicts "Mrs Sun."*

The Sun is a ball of gas about 865,000 miles (1,392,000 km) in diameter. If you could put the Earth at its center, the Moon would orbit about halfway to the Sun's surface. It would take 109 Earths, lined up side by side, to span the Sun's diameter. The Sun contains about 333,000 times as much matter as Earth does, and comprises more than 99.99 percent of the mass in the entire Solar System. The remaining 0.01 percent makes up the rest of the Solar System—that is, the nine planets and all the moons, comets, asteroids, and dust.

The Sun is a medium-size, medium-hot star in the middle of its life, which began some 4.6 billion years ago. We can be thankful that it is such an ordinary star. If the Sun were 10 times more massive, it would already have lived its life and exploded as a supernova (see p. 168). And had the Sun been 10 times smaller, it would not produce enough light and heat for life to exist in the Solar System.

A single element, hydrogen, accounts for 92.1 percent of all the atoms in the Sun. The next most common element is helium, which amounts to 7.8 percent. All the other elements together make up the remaining 0.1 percent. The Sun is fueled by thermonuclear reactions in its core, in a process where hydrogen atoms are fused to make heavier helium atoms, which results in an enormous release of energy.

A HOT HEART

Solar astronomers use a technique called helioseismology to look below the surface of the Sun. Using telescopes positioned around the Earth (to keep the Sun under constant watch), they study subtle pulsations in the Sun's surface. These pulsations—essentially sound waves—probe the Sun's interior in much the same way that seismic waves from earthquakes enable geophysicists to study our planet.

Such studies have shown that the Sun has a layered structure. The core extends outward to about a quarter of the Sun's radius, containing merely 7 percent of the volume, but half the mass, of the Sun. It

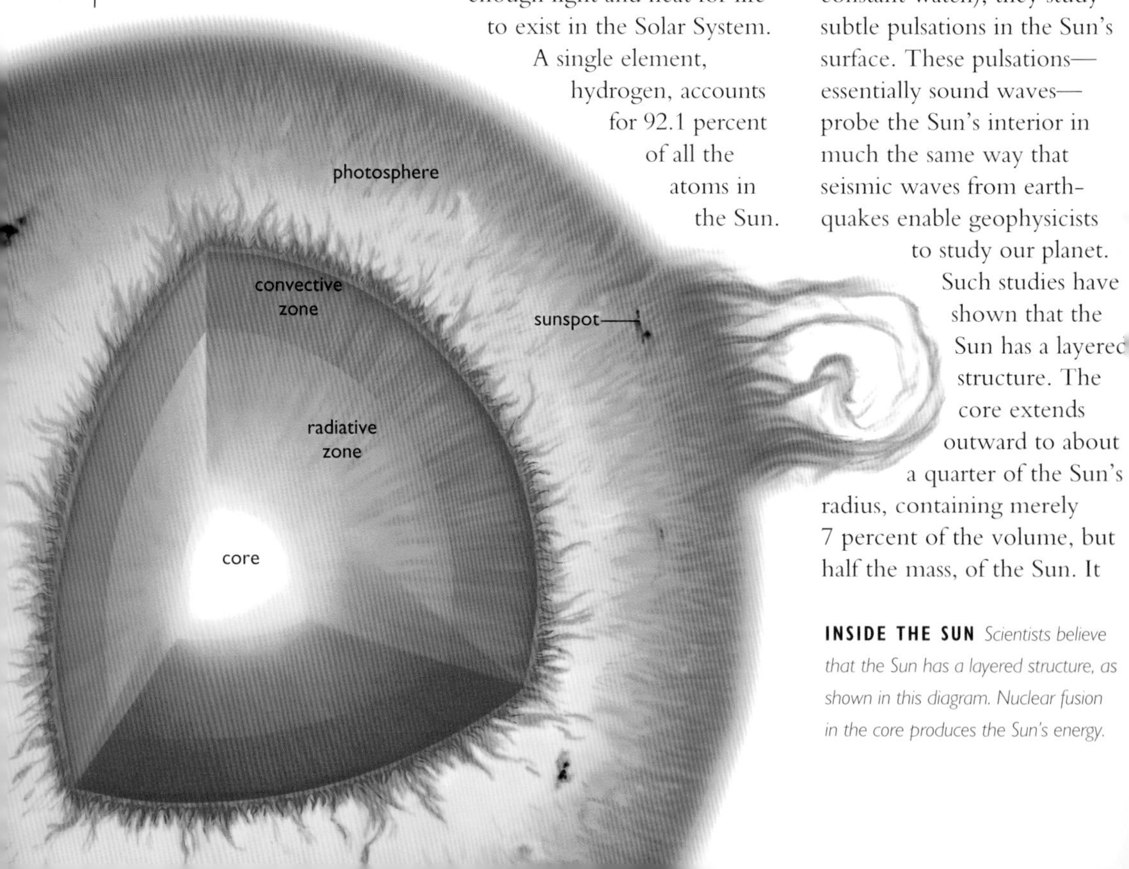
photosphere
convective zone
sunspot
radiative zone
core

INSIDE THE SUN *Scientists believe that the Sun has a layered structure, as shown in this diagram. Nuclear fusion in the core produces the Sun's energy.*

has a temperature of some 27 million degrees Fahrenheit (15,000,000° C). The core is the "engine room" of the Sun, where the energy released from nuclear fusion streams outward. This energy is carried by radiation in the form of photons, which are parcels of electromagnetic energy.

Three-quarters of the way to the surface, the solar gas becomes considerably cooler and undergoes a boiling, convective motion. Convection carries the core's energy along the last step outward. Just below the photosphere—the visible surface—the big upwelling cells of gas break down into smaller ones, resulting in the granules seen on the surface. By the time it reaches the photosphere, the gas has cooled to about 11,000 degrees Fahrenheit (6,000° C).

The Pearly Corona

Above the Sun's photosphere lies a thin, cool, pinkish layer called the chromosphere. It is visible only during solar eclipses or through an H-alpha filter. The top of the chromosphere merges into the corona, the Sun's vast outer atmosphere, which reaches outward at least 10 times the solar radius. It, too, can usually be seen from Earth only during solar eclipses, when its pearly light

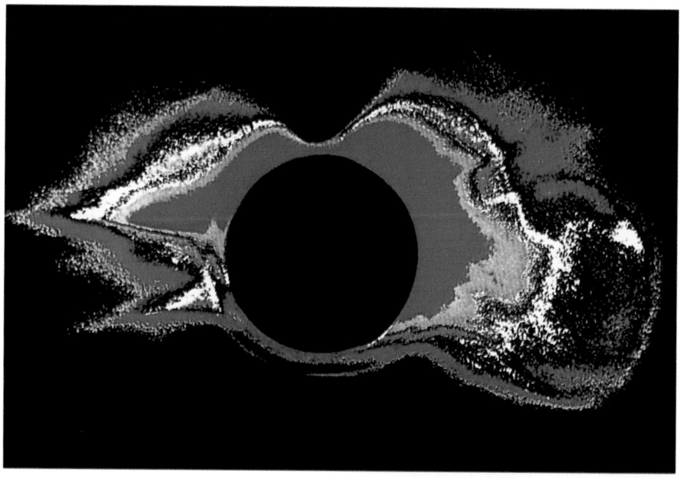

THE SUN'S CORONA—*its outer atmosphere—extends far into space in this infrared image (above), taken with a telescope that created an artificial eclipse.*

is a beautiful sight to behold. The coronal gas is very thinly spread but extremely hot—about 2 million degrees Fahrenheit (1,000,000° C). It is heated by processes that remain unclear.

Beyond the corona, the Sun sends out a stream of charged particles—mostly negatively charged electrons and positively charged protons—into space in what is called the solar wind. Near the Sun, this wind blows at 2,500 miles per second (4,000 km/s), but at Earth's distance, the velocity is a tenth of that figure.

The solar wind sometimes interacts with Earth's atmosphere and creates auroras. It also controls the shape and direction of a comet's ionized gas tail (see p. 143).

The Future of Our Star

Astronomers predict the future of the Sun by looking at what has happened to similar stars. Its life is now about half over. As it consumes hydrogen in its core, leaving an "ash" of helium behind, the Sun will grow hotter and brighter. When its hydrogen begins to run

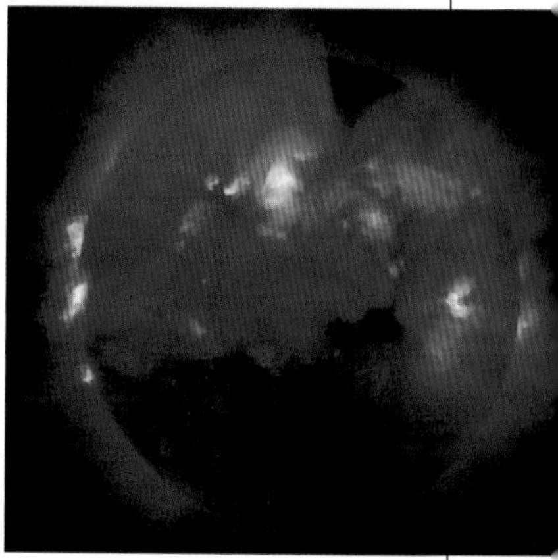

THIS X-RAY VIEW *shows the hot gases of the inner corona above the photosphere, which looks dark in X-rays.*

low, about five billion years from now, the Sun will grow into a red giant, expanding its diameter a hundredfold and swallowing the planet Mercury in the process.

The Sun will spend a few hundred million years as a red giant. Then, as nuclear reactions in its core start to sporadically fuse the helium "ash" into carbon, the Sun may begin to pulsate like the variable star Mira. Finally, as it exhausts its reserves, the Sun will throw off its outer layers to form a planetary nebula. This shell of gas will endure for a few tens of thousands of years before it dissipates into space. Left behind will be the white-hot core of the Sun, a white dwarf that will slowly cool and fade over billions of years.

SUN FACTS

Distance from Earth 92,960,000 miles (149,600,000 km)
Mass 333,000 × Earth's mass
Radius 109 × Earth's radius
Apparent size 32 arcminutes
Apparent magnitude –26.7
Rotation (relative to stars) 25.4 days at equator to 35 days near poles

Solar Activity

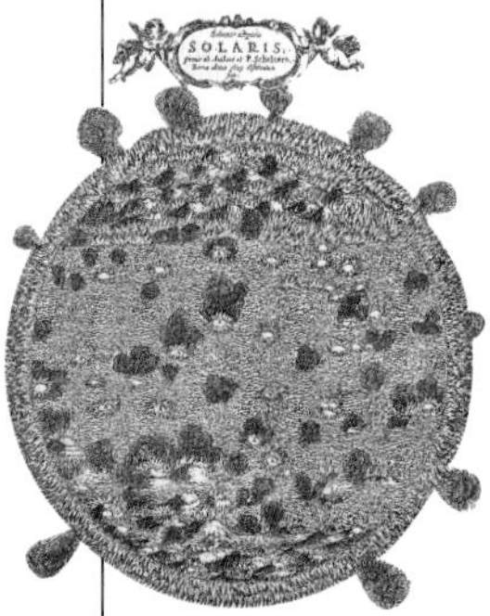

The surface of the Sun is a spectacular, turbulent place, and signs of intense solar activity, such as solar flares and sunspots, can be seen from Earth.

The visible surface of the Sun is called the photosphere ("sphere of light"). It looks solid, but in reality it is a shell of gas about 250 miles (400 km) thick. At about 11,000 degrees Fahrenheit (6,000° C), this outer layer is much cooler than the gas deep beneath it.

The edge of the photosphere looks dark compared to the center of the disk. This limb-darkening effect occurs because at the center of the disk we see deeper into the Sun where temperatures are higher and the gas shines more brightly. At the limb we are seeing cooler gas at higher elevations and it appears darker.

Viewed at high magnification, the photosphere shows many features. The smallest of these are individual cells of gas called granules, which average about 700 miles (1,100 km) across. Granules can last from several minutes to half an hour, but are usually difficult to observe because of the poor seeing during daytime.

The photosphere also has large, irregular patches of slightly hotter material called faculae (Latin for "torches"). Because they are not much brighter than the rest of the surface, they are often hard to see, except near the limb.

With a standard H-alpha filter, you may see bright loops called prominences extending from the Sun's disk. They can also be seen with a narrowband H-alpha filter as dark filaments on the disk itself.

Another kind of activity—a flare—is always worth looking for, although it is rare to see one. The first was discovered in 1859 by two English amateurs, Richard Carrington and Richard Hodgson, who were working independently. Flares are brilliant explosions of pent-up magnetic energy and radiation. They come and go, often within minutes, but most are invisible without an H-alpha filter. White-light flares, which can be seen with any solar filter, are rare but powerful events. They may release as much as 2 percent of the Sun's energy from a tiny area in a few minutes, and can disturb the electromagnetic environment of the whole inner Solar System. When the flare's particles arrive at Earth as a high-speed gust in the solar

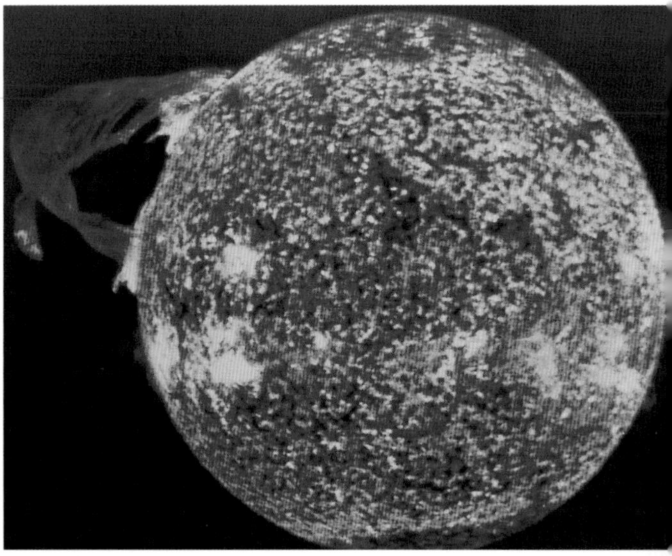

THIS SPECTACULAR PROMINENCE *is erupting more than 370,000 miles (600,000 km) into space.*

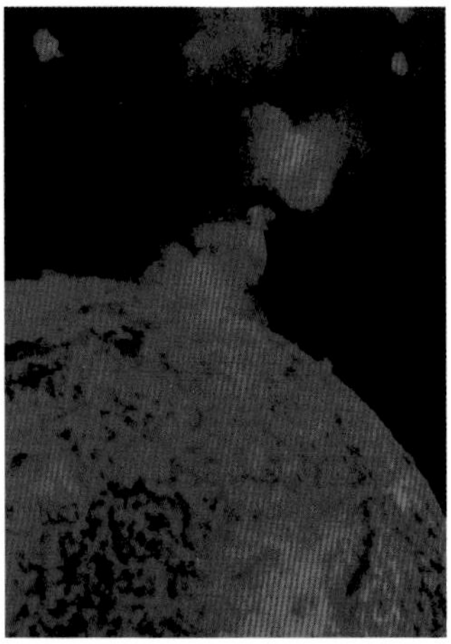

ERUPTIONS *on the Sun's surface (left) release an enormous amount of energy. A 1635 illustration (top) wrongly depicts the Sun as a burning mass of carbon.*

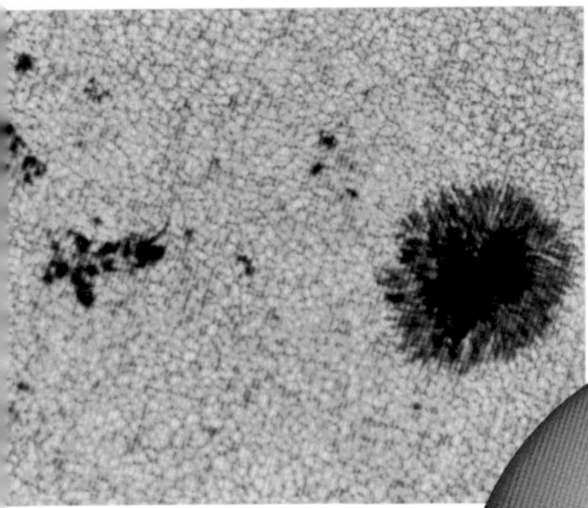

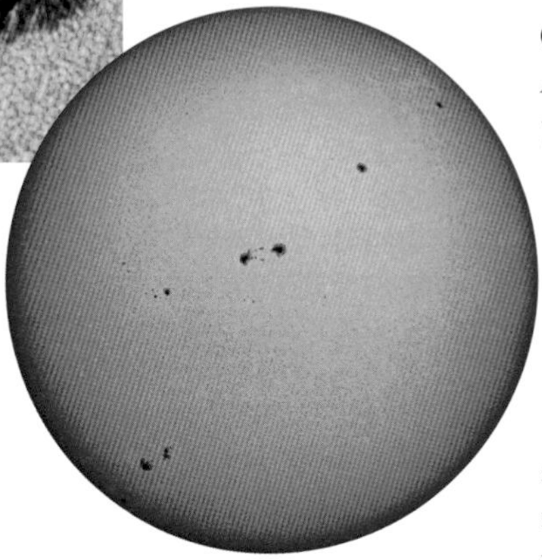

SUNSPOTS *can be observed on the Sun's disk (below). Up close (left), they show a dark umbra and a lighter penumbra.*

wind (see p. 79), they create geomagnetic disturbances such as auroras. In extreme cases, they can produce surges of current that cause power grids to fail, leading to widespread blackouts.

Sunny Side Up

The most famous solar features are the sunspots. Records from China show that since ancient times, observers have noticed dark spots on the Sun. (These naked-eye observations were made when the Sun was low in the sky or partly obscured by fog, mist, or clouds. Do not try to copy their technique, however—many astronomers in antiquity probably went blind.)

In the early 1600s, Galileo used his telescopes to show that sunspots were not clouds above the photosphere as some claimed, but features on the photosphere itself. Eventually, he and others established that sunspots occur within about 35 degrees of the solar equator. In the nineteenth century, Richard Carrington noticed that sunspots near the Sun's equator took about two days less to circle the Sun than did spots in higher latitudes. He concluded that different parts of the Sun rotate at different rates, and that this differential rotation occurs because the Sun is a gaseous body. It was an important new finding.

Sunspots are places where the photosphere is about 3,600 degrees Fahrenheit (2,000° C) cooler than the surrounding sunscape. This temperature difference makes the spots look darker. Sunspots occur where the Sun's magnetic field has emerged through the photosphere and stopped some of the rising energy from reaching the surface.

Sunspots have a dark central region, called the umbra, surrounded by a striated, lighter-colored halo, known as the penumbra. They vary greatly in size, shape, and complexity, and can appear as single, small, black dots or as complicated groups with so many discrete features that it is difficult to tally them. The largest groups can reach a size of 60,000 miles (100,000 km) across, many times the diameter of the Earth.

In 1769, the Scottish solar astronomer Alexander Wilson noted that the largest sunspots appeared saucer-like when seen near the limb. He proposed that these spots were depressions in the solar surface. Today's astronomers still call this the Wilson Effect, but ascribe it to changes in the transparency of the solar gas over a spot when it is seen at an extreme angle.

Observing Sunspots

As the Sun rotates on its axis, you can watch a group of sunspots travel from one edge of the disk to the other over a period of about 10 days. A few long-lived spots will come around for a second period of visibility.

Sunspot observations are often best around mid-morning. At this time of day, you can see the Sun without a lot of atmosphere in the way and before it has heated your surroundings too much. To count sunspots visually through your telescope, use a solar aperture filter and a magnification that shows the whole disk. Note the number and positions of the sunspot groups, then switch to an eyepiece giving 65 power, the agreed-upon standard. Revisit each group and count the individual spots. The total estimated number of sunspots, R, is given by the formula: $R = 10g + s$, where g is the number of groups, and s is the total number of spots seen. Summaries of these numbers are regularly published in astronomy magazines.

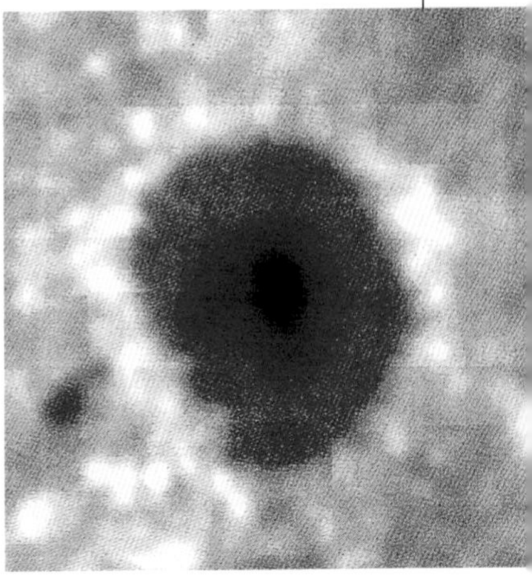

HOT AND COOL *This infrared image shows the temperature of a sunspot's umbra and penumbra. The hottest areas are white and the coolest areas are black.*

SKETCHING SUNSPOTS

Sketching sunspots is easiest using the projection method. Decide on a standard size of sketch—6 inches (150 mm) is common—and prepare blank forms ahead of time with a circle for the Sun's diameter and places to note date and time, weather, seeing conditions, and the equipment you used. Tape the blank onto the projection screen, adjust it so the Sun just fills the circle, and with an HB or softer pencil, lightly trace the locations of the sunspots.

By making regular sketches, you can produce a record of the changing face of the Sun.

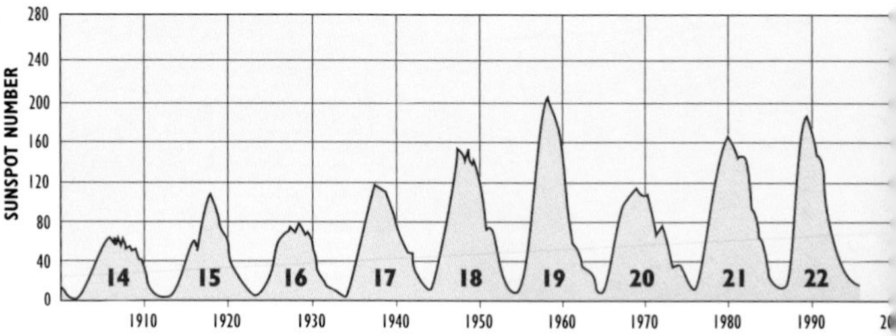

SUNSPOT NUMBER
280
240
200
160
120
80
40
0
14
15
16
17
18
19
20
21
22
1910
1920
1930
1940
1950
1960
1970
1980
1990

SUNSPOT ACTIVITY *follows a clear 11-year cycle (above). An infrared image taken in 1981 (right) shows the Sun near the peak of a sunspot cycle, with the spots positioned in two bands around 10 degrees north and south latitude.*

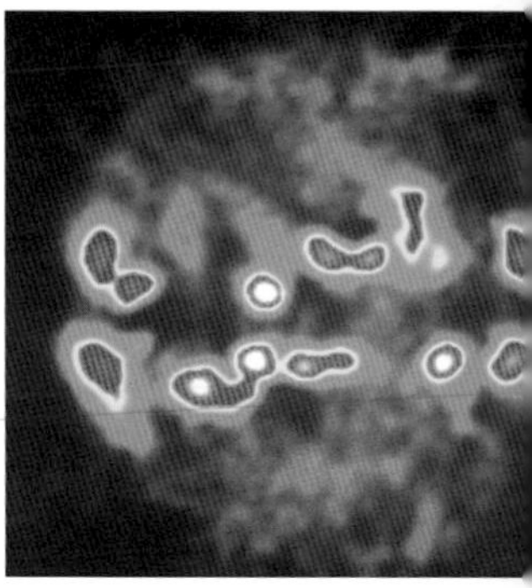

THE SUNSPOT CYCLE

In the nineteenth century, German amateur astronomer Heinrich Schwabe was systematically scanning the Sun's surface, hoping to catch in transit a planet then thought to be orbiting inside Mercury To avoid mistakes, he tabulated all the sunspots he saw. Schwabe never found the planet—which does not, in fact, exist—but he did make the discovery that the number of sunspots rose and fell in a pattern lasting about a decade

Today, after observations over many years, scientists have determined that sunspot follow an 11-year cycle, durin which the number of spots goes from a minimum to a maximum. Solar astronomers reckon Cycle 0 as being the cycle that peaked in 1750. On this reckoning, Cycle 22 peaked in mid-1989 and ende in late 1996, and Cycle 23 is expected to peak around 2001 Each cycle has an asymmetric shape, rising to maximum activity in about five years anc declining over about six years

At the start of a new cycle, the first spots appear around latitudes 35 degrees north and south. As the cycle progresses, sunspots become more numerous and appear at latitudes

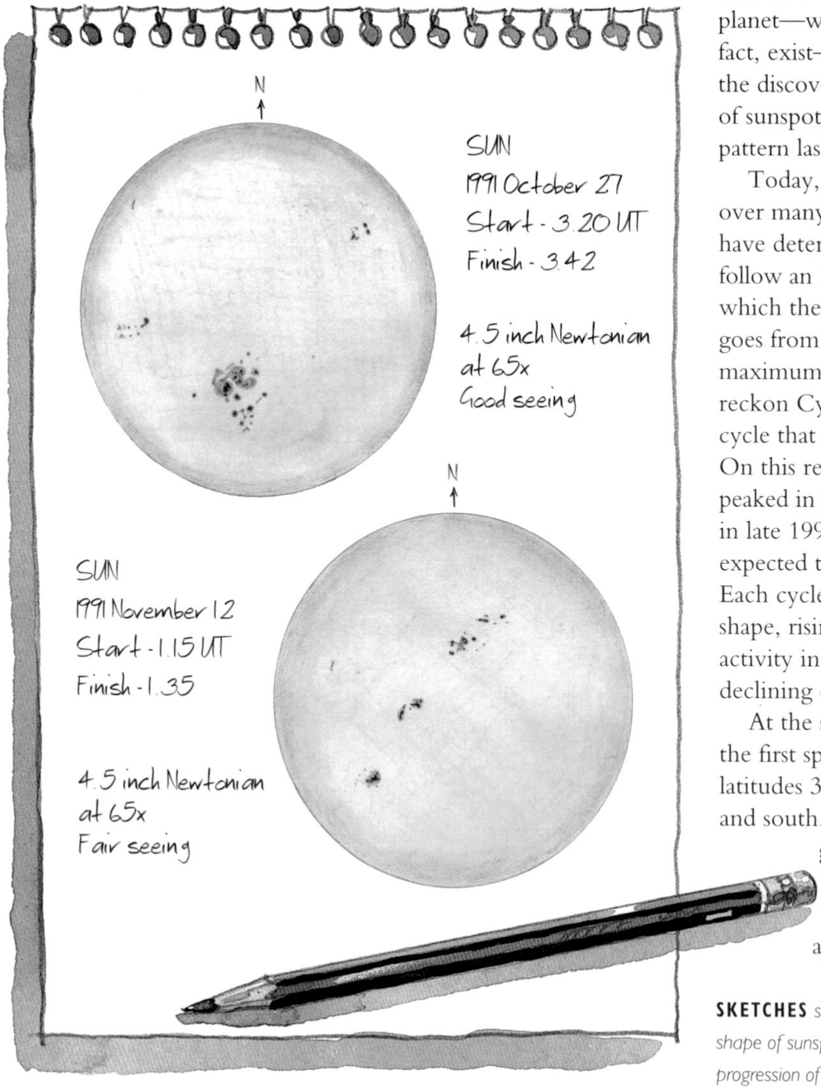

N
SUN
1991 October 27
Start - 3.20 UT
Finish - 3.42
4.5 inch Newtonian
at 65x
Good seeing
N
SUN
1991 November 12
Start - 1.15 UT
Finish - 1.35
4.5 inch Newtonian
at 65x
Fair seeing

SKETCHES *showing the location and shape of sunspots can record the progression of solar activity over time.*

SUNNY RADIO SIGNALS

Because most cosmic radio signals are very weak, radio astronomy usually demands large, sophisticated equipment. The Sun, however, is by far the strongest radio source in the sky and this presents opportunities to the amateur with an interest in electronics.

Solar flares induce changes in the Earth's ionosphere—the layer of ionized particles in the upper atmosphere. The ionosphere usually reflects very low frequency (VLF) signals (5 to 50 kilohertz), permitting long-distance radio communication. Disruption of the ionosphere by a solar flare, however, can cause communications fade-out. You can monitor this effect with a radio receiver, antenna, and recorder that cost less than a medium-size reflector telescope.

The Sun itself can be detected with a short-wave receiver for 1 to 30 megahertz frequencies and a simple antenna. A radio-frequency preamplifier will improve the sensitivity of the system. You can even collect the Sun's signals using a backyard satellite dish, with minor modifications. The Small Radio Telescope (right) at Haystack Observatory, Massachusetts, uses a 9 foot (2.8 m) satellite dish to receive signals of 1 to 4 gigahertz frequencies.

To find out more about radio astronomy, check our Resources Directory (p. 274).

The adventure of the sun is the great natural drama by which we live, and not to have joy in it ... is to close a dull door on nature's sustaining and poetic spirit.

The Outermost House,
HENRY BESTON (1888–1968),
American naturalist and writer

earer the Sun's equator. By he end of the cycle, sunspots ppear around 7 degrees north nd south latitude. As one ycle is tailing off, the first spots f the next cycle are already ppearing at high latitudes.

Sunspots are highly magetized, and astronomers ave discovered that there is ctually a 22-year magnetic ycle superimposed on the rdinary 11-year cycle. When ney measure the magnetic elds of sunspots during any 1-year cycle, they find that ne spots leading each group cross the disk show opposite agnetic polarities in the two emispheres. Then, in the ext 11-year cycle, all leading oots switch polarity.

DOES THE SUN VARY?

Compared to some stars, which can vary in apparent brightness by factors of a hundred or more, the Sun seems like constancy itself. On the other hand, there are those 11-year sunspot cycles. Furthermore, astronomical history shows that sunspots can virtually disappear for decades on end, with notable climatic consequences.

For example, the British amateur astronomer E. Walter Maunder drew attention to the fact that from 1645 until 1715 sunspots virtually disappeared for reasons still unknown. Astronomers label this period the Maunder Minimum, and note that it coincides with a period that climatologists now call the Little Ice Age, when winters were more severe than normal, and summers were shorter and cooler.

Scientists are only starting to explore the connections between activity on the Sun and the weather on Earth. Our climate system is highly complex, and there is little established theory to explain just how the Sun might affect it, but it is already clear that there is a definite link.

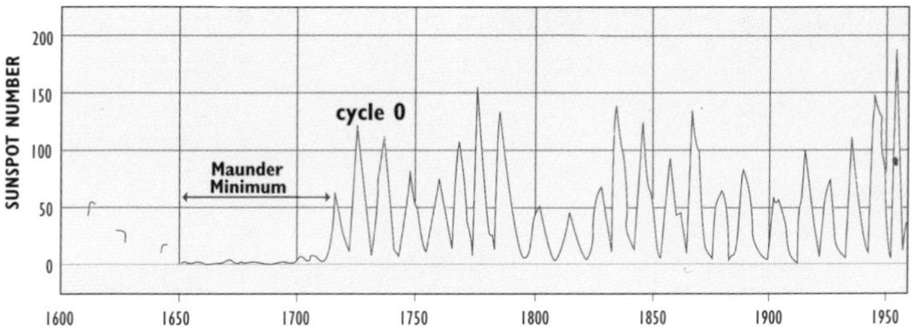
SUNSPOT NUMBER
200
150
100
50
0
cycle 0
Maunder Minimum
1600
1650
1700
1750
1800
1850
1900
1950

HE LITTLE ICE AGE, *a period of ery cold weather, coincided with the launder Minimum (right). London's ver Thames froze over and was the e of frequent frost fairs (above right).*

Solar Eclipses

Watching the Moon creep across the Sun, blot out its brilliance, and then slowly reveal it once more is an experience that fills us with awe.

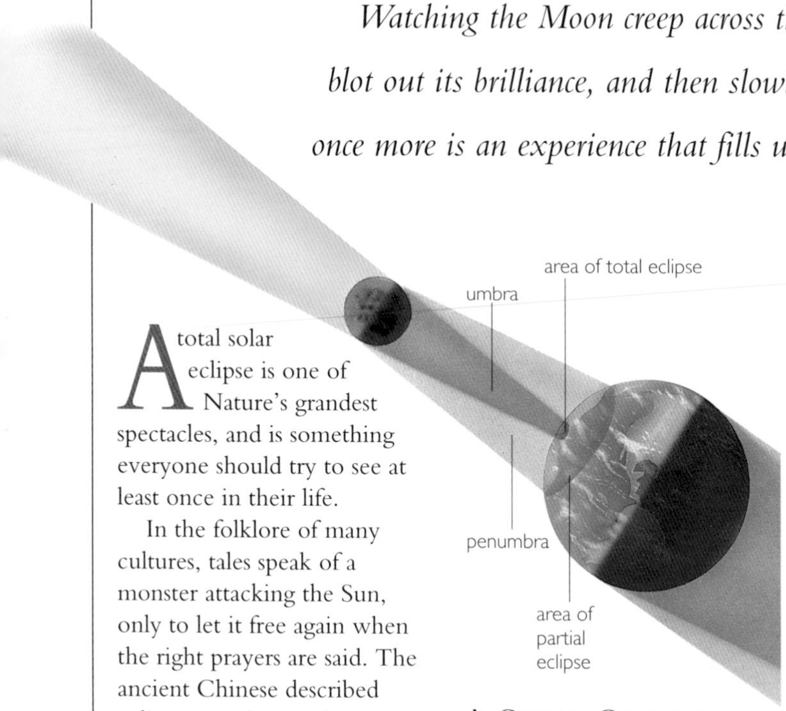
area of total eclipse
umbra
penumbra
area of partial eclipse

THE LUNAR UMBRA *traces a narrov path on Earth. Within the penumbra, people see only part of the Sun eclipse*

A total solar eclipse is one of Nature's grandest spectacles, and is something everyone should try to see at least once in their life.

In the folklore of many cultures, tales speak of a monster attacking the Sun, only to let it free again when the right prayers are said. The ancient Chinese described eclipses as a dragon devouring the Sun—in fact, their word for eclipse came from their word meaning "eat."

Eclipses have even stopped wars. Herodotus writes of an eclipse on 28 May 585 BC. The Medes and Lydians of Asia Minor had been fighting for more than five years and were in battle that day, when the Sun grew dark unexpectedly. At the frightening sight, both sides immediately sought peace. Though neither side knew it, the eclipse had been predicted by Thales of Miletus, the first such prediction on record.

A Great Coincidence

A solar eclipse occurs whenever a New Moon passes directly between the Sun and the Earth. We do not see an eclipse at every New Moon because the Moon's orbit inclines about 5 degrees to the ecliptic. In most months, the Moon passes above or below the Sun instead of across it.

Solar eclipses can be total, partial, or annular, depending on how much Sun is covered by the Moon. Total eclipses occur when the Moon covers the entire disk of the Sun. A total eclipse may last just a second or two; the longest possible total eclipse is only about 7½ minutes.

It is pure accident that the apparent sizes of the Moon an the Sun are roughly the same. The Moon is about 400 times smaller than the Sun—but it also happens to be about 400 times closer. In fact, since the Moon formed, its orbit ha been moving very slowly outward. From our point of view it is steadily shrinking. Eventually, it will orbit too far fror Earth to ever completely cove the Sun, and total solar eclipse will be a thing of the past.

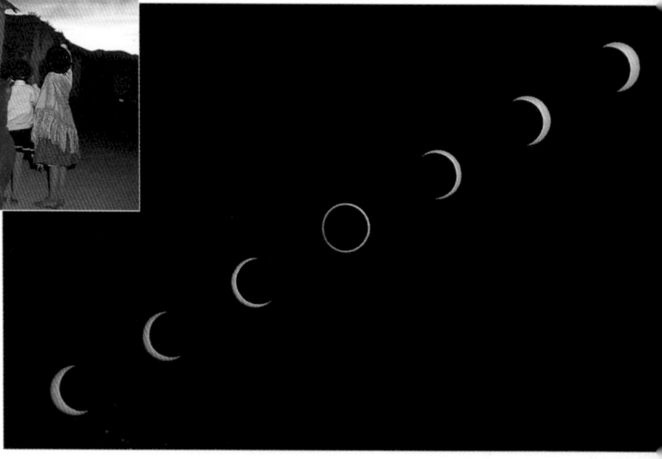

TWO GIRLS AND A LLAMA *watch a solar eclipse in Chile in 1994 (above). This multiple-exposure shot captured the stages of an annular eclipse (right).*

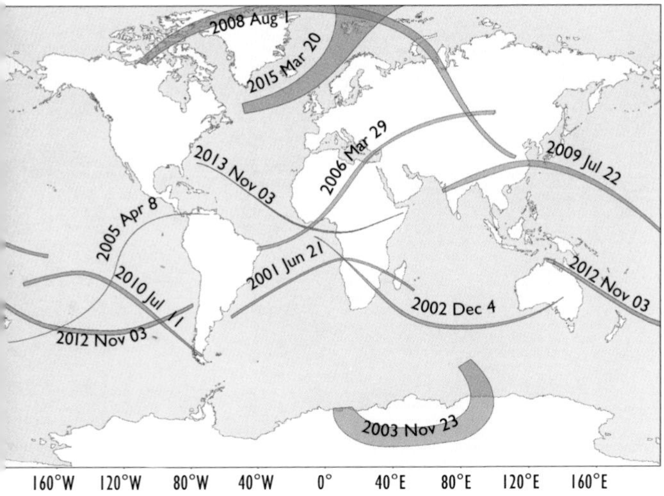
2008 Aug 1
2015 Mar 20
2006 Mar 29
2009 Jul 22
2013 Nov 03
2005 Apr 8
2010 Jul 11
2001 Jun 21
2012 Nov 03
2002 Dec 4
2012 Nov 03
2003 Nov 23
160°W 120°W 80°W 40°W 0° 40°E 80°E 120°E 160°E

ECLIPSE TOURS *The world map (left) shows the paths of total solar eclipses until the year 2015. Use it to plan your eclipse travels. This group (below) visited India in 1995 to observe a solar eclipse.*

Because Earth is rotating ıd the Moon is moving, the ınar shadow, or umbra, traces curving path on Earth. The mbra barely reaches Earth, so ıe path is relatively narrow—here the Moon is directly verhead, the path will be at ıost about 170 miles (270 km) ride. This means that total clipses are selective, occurring nly rarely at any particular lace, although in most years ıere are one or two eclipses omewhere on Earth.

In recent decades, a new ranch of the travel industry as developed, offering solar-clipse expeditions to remote estinations. Tour-company dvertisements start to appear ı astronomy magazines about year before the event.

ING OF SOLAR FIRE

'otal eclipses are rare, but artial eclipses occur more equently. While enjoyable) view with the right equip-ıent, partial eclipses lack the verwhelming drama of a total clipse and few people will avel very far to see one.

A possible exception is the nnular eclipse, a special type f partial eclipse. Annular clipses happen because the Moon's orbit around Earth—nd Earth's orbit around the un—are not circular but lliptical. Thus the apparent sizes of the Sun and Moon vary by a small amount. On occasion, the Moon may appear too small (or the Sun too large) for a total eclipse to occur, even though other circumstances are right for it.

The result is that at maximum eclipse the Moon does not cover the Sun completely, and a ring (or annulus) of the Sun's uneclipsed surface surrounds the black circle of the Moon. Annular eclipses will not show you the Sun's corona (unless the eclipse is about 99.99 percent total), but they can be quite spectacular all the same, especially if the Sun is sitting near the horizon.

AN ECLIPSE FOR EINSTEIN

Total solar eclipses are dramatic, but few have been awaited more eagerly than the one in Brazil on 29 May 1919. At this eclipse, a British expedition made the first attempt to confirm (or refute) Albert Einstein's general theory of relativity, published in 1916.

A key prediction of the theory was that a ray of light grazing the edge of the Sun would be bent by the Sun's gravity by a small, but detectable, amount. A total eclipse offered the only chance of "turning off" the Sun's light so that astronomers could measure the positions of background stars and find the deflection—if it existed.

The 1919 eclipse was the first for which major expeditions could be arranged following World War I (1914–18). Teams were sent to two separate sites in case of clouds, with one expedition headed by Britain's foremost astrophysicist, Arthur Stanley Eddington.

After the eclipse, when the astronomers measured their photos, the deflection of starlight came out exactly as Einstein had predicted. What made the eclipse notable (and even poignant) at the time was that it was British astronomers confirming a theory developed in the depths of wartime by a German physicist. Einstein (right) suddenly found his name blazoned across every newspaper. Literally overnight, he became the most famous scientist in the world.

Turning Off the Sun

As with any observation of the Sun, you must take special precautions before viewing a solar eclipse. However, it is not dangerous to be outdoors, as some over-cautious warnings may lead you to believe. You can watch the progress of the eclipse through a telescope rigged for solar observation (see p. 77) or you can hand-hold a solar filter. When the Moon has fully covered the Sun, you can look directly with the naked eye, but you must turn away as soon as the Sun begins to reappear.

There are many ways to record an eclipse. To photograph the partial stages, use the same techniques as you would for the uneclipsed Sun (see p. 77). During totality, other kinds of photographs become possible (see Box).

Videotaping an eclipse provides a record of the whole event—including all the enthusiastic (and often funny) comments from bystanders. You will need a camcorder, a solar filter, a tripod or a telescope mount, and adequate power (plenty of batteries or a line-current adapter). You can remove the filter during totality, but be ready to replace it when the Sun starts to reappear at the end. The Earth's rotation will carry the Sun and Moon two solar diameters every four minutes. You will either have to track them with the camcorder on an equatorial mount (see p. 52), or zoom back enough that they stay within the frame.

AFTER FIRST CONTACT, *the Moon slowly creeps across the Sun (above). Moments before second contact, the last rays of sunlight create Baily's beads (right).*

A HAND-HELD FILTER *protected this Parisian's eyes during the 1912 eclipse.*

Four Contacts

A total solar eclipse has four stages: first contact, second contact (or start of totality), third contact (or end of totality), and fourth contact. In your eclipse-trip planning, check astronomy magazines and almanacs for the exact time of each stage.

The eclipse begins at first contact, the moment when the lunar disk first touches the solar one. First contact can be detected only through a telescope prepared for solar viewing.

Over the next hour, the partial phase unfolds, as the Moon steadily creeps across the Sun. At first, there is little apparent change. But then you will notice a difference in the quality of the light. Shadows grow sharper, while the light becomes more like that on an overcast day. The air grows cooler, and insects, birds, and other animals begin to act as if it is nightfall. If you are standing near a tree that is in leaf, look down. The small gaps between the leaves act like pinhole lenses, throwing hundreds of images of overlapping crescent Suns onto the ground.

The diminished light may tempt you to look at the Sun with the naked eye, but the Sun is still bright enough to damage your vision.

A few minutes before second contact—the start of totality—changes begin to occur in rapid succession. The sky grows quite dark, and the air feels distinctly cool. A breeze usually picks

TOTALITY *reveals the Sun's corona—outer atmosphere—with streamers at reach out a solar diameter or more.*

p, perhaps blowing dust cross delicate equipment. In e gloom, observers may mble over camera settings d mislay eyepieces and filters.

Then, the last rays of sunght stream through valleys n the edge of the Moon, a henomenon called Baily's eads, after an English solar bserver, Francis Baily, ho first described them, in 836. The last of the beads ay linger a moment, creatg what is known as the amond-ring effect. Then omes a flash of pinkish light om the chromosphere, the rst layer above the Sun's rface—and totality begins.

SOLAR FLOWER

The sky darkens to deep twilight, and planets and stars can be seen above the glow that circles the horizon. It is now safe to remove solar filters, but they should be kept handy. Around the black disk of the Moon is a soft fringe of pearl-colored light. This is the Sun's corona, its outer atmosphere. The corona is much too dim to be seen except at totality.

Now it is time to turn a telescope on the edge of the Moon, where ruddy tendrils stand out. These are solar prominences—huge eruptions of gas suspended by magnetic force above the Sun's surface. They appear to extend up from behind the Moon. The prominences on one edge of the Moon slowly disappear while those on the other side emerge from behind the lunar limb. This is a reminder that the Moon is moving on and totality is near its end.

When only a few instants of the total phase are left, reach for your solar filter. Abruptly, there will be a bright burst of light from the Moon's eastern edge, then another, and daylight floods back. This is third contact, and the end of totality. The corona disappears into the rapidly brightening sky.

For many observers, the end of totality means the end of the eclipse, despite the hour or so that remains until fourth contact—the exit of the Moon from the Sun's disk. People pack up their equipment, and talk excitedly about the sights just seen, reliving the brief moments of totality. Precious rolls of film are rewound, and telescopes are packed away. The grand show has ended—that is, until the next eclipse!

PHOTOGRAPHING A TOTAL ECLIPSE

To capture the feel of a total eclipse, some photographers use a camera with a wide-angle lens perched on a tripod and aimed where the Moon will be at mid-totality. The eclipsed Sun will be small in the frame, but the photo will catch the main ingredients of the scene: the Sun and Moon, the brighter planets and stars, the horizon glow, and the blurry silhouettes of observers.

Other photographers concentrate on the corona. With a camera piggybacked on a telescope (see p. 64) and a 1,000 mm or 2,000 mm lens, you could take a long exposure, showing its maximum extent. Or you could make a series of exposures starting with short ones and running progressively longer to capture the corona in stages, from the brightest parts near the lunar limb to the outer streamers. Another possibility is to photograph the eclipsed Sun through a telescope, perhaps using eyepiece projection for extra magnification (see p. 66). This will capture the pinkish prominences that reach above the lunar disk.

With any photo, the correct exposure time depends on the film speed and f-ratio. For ISO 100 film, try the following exposures.

- outer corona 1/2 second at f/8
- inner corona 1/15 second at f/8
- prominences 1/125 second at f/8
- diamond ring 1/500 second at f/8
- scenic shots 1/2 to 8 seconds at f/2.8

Attaching a camera to a telescope is just one method of eclipse photography.

Lunar Eclipses

You need not travel far, nor wait for many years, to behold the eerie sight of the Moon as it moves into the Earth's shadow and turns dusky red.

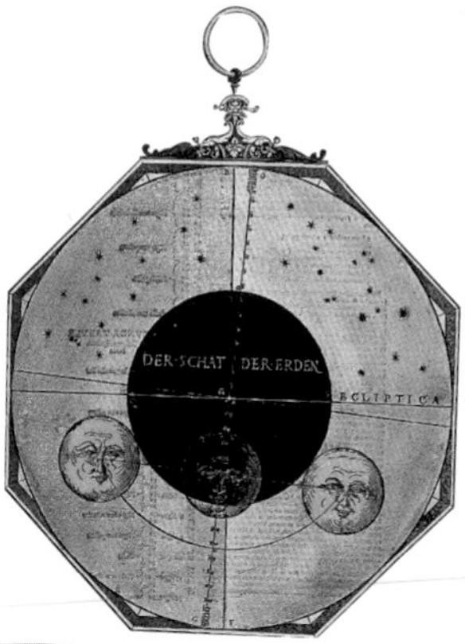
DER SCHAT DER ERDEN
ECLIPTICA

ECLIPSE CALCULATOR *This 1540 woodcut illustrates a method of calculating eclipses of the Sun and Moon.*

During most months, the tilt of the Moon's orbit in relation to the Earth's orbit around the Sun ensures that the Full Moon passes north or south of the Earth's shadow. But at roughly six-month intervals, the Full Moon lies near (or inside) the shadowy cone of darkness that extends far into space directly from Earth. It is during these "eclipse seasons" that lunar and solar eclipses take place. Lunar eclipses can be partial or total, depending on how deeply the Moon goes into the shadow.

The shadow of our planet points at the stars like a finger of night. It starts off as wide as the Earth itself, but shrinks by the time it crosses the Moon's orbit. Although it is still almost three times as wide as the Moon, the shadow is a small area in astronomical terms.

Assuming skies are clear, anyone who is on the night side of Earth while a lunar eclipse is happening can see it. Observers often find lunar eclipses a good deal more relaxing than solar ones because the event usually lasts hours instead of the minutes that measure solar eclipses.

Eclipse Photos

As a lunar eclipse progresses, viewers with binoculars and telescopes watch the Moon closely, timing when the shadow reaches individual craters and then estimating how dark the eclipse is. Other observers get a camera, put it on a tripod, and take scenic shots of the dark, coppery Moon in the sky.

Some photographers piggyback the camera on an equatorially mounted telescope. With the telescope tracking the stars, they take a multiple exposure that captures the Moon in stages as it moves through Earth's shadow.

Another popular multiple-exposure photograph that is even easier to take is to frame the Moon in the sky next to a tree or building in the foreground. Then, without moving the camera (or changing its aim), you take a series of identical exposures, precisely 5 or 10 minutes apart. The result will be a striking scenic view with a number of Moons marching across the

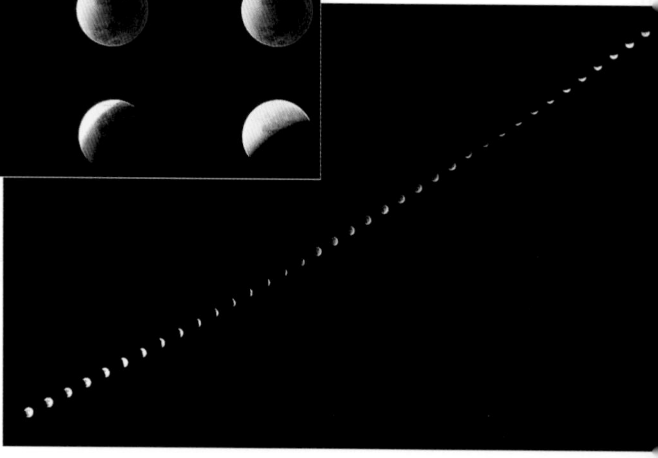

THE PHASES *of a lunar eclipse can be photographed in close-up (left), or they can be captured as they occur in the sky in a single scenic image (below).*

What is there in thee, moon,
that thou shouldst move
My heart so potently?

Endymion, John Keats (1795–1821), English poet

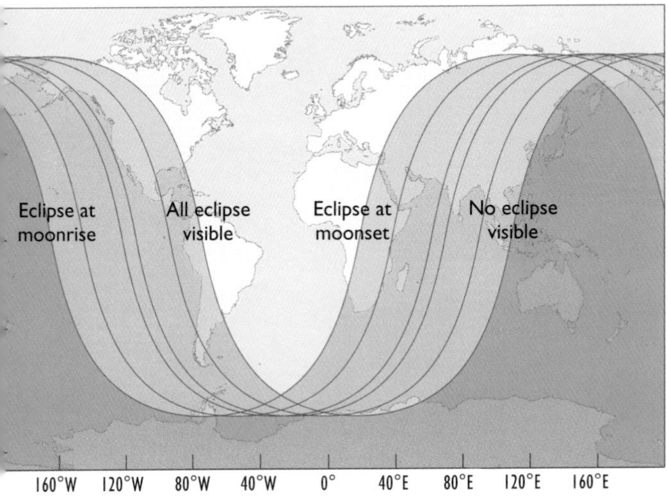
Eclipse at moonrise
All eclipse visible
Eclipse at moonset
No eclipse visible
160°W
120°W
80°W
40°W
0°
40°E
80°E
120°E
160°E

ECLIPSE 2003 *This map for 9 November 2003 shows how much of the world can experience a total lunar eclipse.*

ky, changing color and brightness as the eclipse progresses.

With ISO 200 film, you an capture the partial phases vith a 1/60 second exposure t f/8. During totality, use seconds at f/4. If the eclipse eems especially dark, you hould double (or even uadruple) the exposure. Because it is so hard to guess he Moon's brightness ahead f time, you should plan to ake a range of exposures. Make one shot at the "normal" xposure, followed by one t twice that exposure, and third at four times the ormal exposure. At least ne of the frames should ome out just right.

As the Moon dwindles in rightness during the eclipse, eautiful star fields all over the ky become more prominent. This effect is strongest for bservers at dark sites well way from light pollution.

MOON COLORS

The spookiest thing about lunar eclipse is the eerie eddish color that covers the Moon. If Earth had no atmoshere, lunar eclipses would imply turn the shadowed part f the Moon jet-black. But Earth's air acts like a simple ens or prism. It bends part of he Sun's light into the shadow nd stains this light a deep, ruddy copper color. In essence, what illuminates the Moon during a lunar eclipse is the light of every sunset and sunrise happening on Earth.

Experienced eclipse gazers try to guess how dark the shadow will be. With some eclipses, you see a pale amber Moon. But especially in years following big volcanic eruptions, the shadow can become so dark that the Moon virtually disappears, even when seen through a telescope. In 1883, the volcano Krakatau near Java exploded, sending several cubic miles of pulverized rock high into the atmosphere. At the next lunar eclipse several weeks later, the Moon all but vanished. Similarly, dark eclipses have followed other big eruptions, such as that of Mount Agung on Bali, which produced two "black Moon" eclipses in 1963 and 1964. More recently, a very dark eclipse followed the eruption of the Mexican volcano El Chichón in 1982.

DANJON SCALE

To compare lunar eclipses, an observer can note the degree of darkness using a five-step scale, named for the French astronomer André Danjon, who devised it.

L = 4 Strikingly bright copper-red or orange eclipse; bluish tint where the inner shadow (umbra) meets the outer shadow (penumbra).
L = 3 Brick-red eclipse; umbra has light gray or yellow rim.
L = 2 Deep red or rust-colored eclipse; umbra usually has a very dark center and relatively bright outer edge.
L = 1 Dark eclipse; gray or brownish coloration; surface features hard to make out.
L = 0 Very dark eclipse; Moon nearly invisible.

Make your estimate at maximum eclipse. You can use decimal steps between the classes: 3.5 or 2.3, for example.

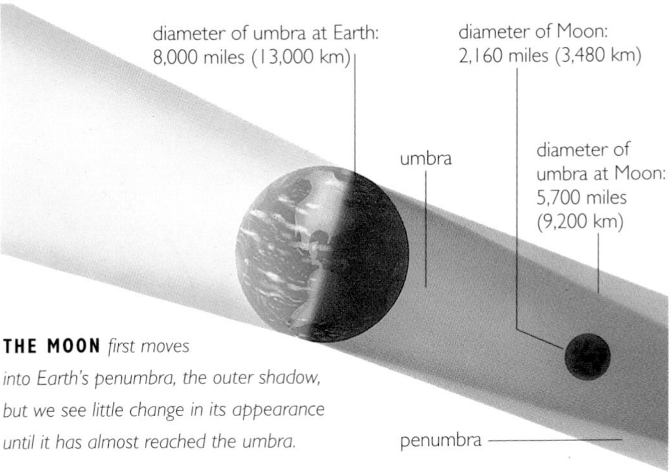
diameter of umbra at Earth: 8,000 miles (13,000 km)
diameter of Moon: 2,160 miles (3,480 km)
umbra
diameter of umbra at Moon: 5,700 miles (9,200 km)
penumbra

THE MOON *first moves into Earth's penumbra, the outer shadow, but we see little change in its appearance until it has almost reached the umbra.*

OBSERVING *the* MOON

For as long as people have thought about the world around them, they have looked up at the Moon and pondered its varying ways.

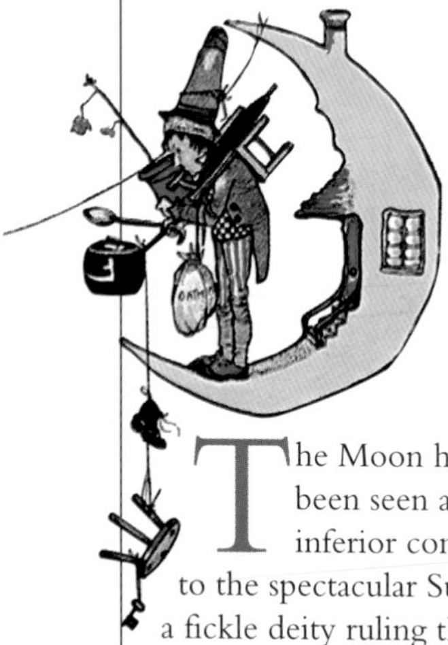

The Moon has often been seen as an inferior companion to the spectacular Sun or as a fickle deity ruling the night sky. Its phases heralded many important cultural and religious events. In today's urban culture, lunar folklore holds a smaller place than it once did, but some still refer to people as being "moonstruck."

The light of the Moon aided hunters and farmers from the very beginning, and nearly every culture recognized the Moon's influence on the ebb and flow of the tides. The Moon also provided humankind with its first calendar—the word "month" even comes from the word "moon."

The Moon is the first celestial object most new telescope owners look at—and with good reason. No matter what size instrument you use, the sight is spellbinding. Craters, rays, and mountains all parade in testimony to the Moon's violent geological past.

To the naked eye, the Moon shows two kinds of area—one dark and the other light. Ancient astronomers believed the dark areas were oceans like those on Earth and named them using the Latin word for sea: mare, pronounced MAH-ray (plural maria, MAH-ree-uh). These terms persist, despite our knowledge that the "seas" are really smooth sheets of congealed lava that oozed from beneath the Moon's surface after meteorite impacts. The lighter areas, known as highlands or terrae, are pieces of ancient lunar crust, battered by innumerable meteorites.

MOON FACTS

Distance from Earth 238,856 miles (384,401 km)
Mass 0.012 x Earth's mass
Radius 0.272 x Earth's radius
Apparent size 31 arcminutes
Apparent magnitude –12.7 at Full Moon
Rotation 27.3 Earth days
Orbit (relative to stars) 27.3 Earth days
Orbit (seen from Earth) 29.5 Earth days

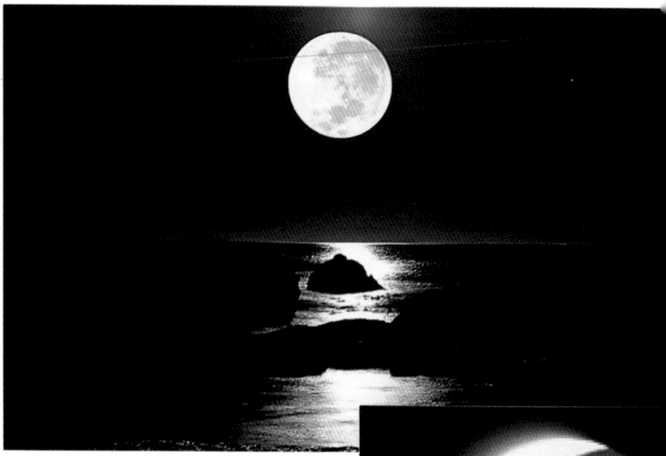

THE FULL MOON *rising over water (above) and a crescent Moon bathed in earthshine (right) make fine photographic subjects. Many people claim they can see the Man in the Moon (top).*

MOON-ROVING "Tides" raised in the rocky body of the Moon by Earth's gravity have caused the Moon to rotate on its axis in exactly the same time as it takes to orbit Earth. This is why the same side is always turned toward us.

As it orbits, the Moon shows a varying amount of its illuminated face because the angle between the Sun, Earth and itself changes. These variations are called phases.

While Full Moon has its attractions, the best time to explore the Moon's cratered surface with a telescope is when sunlight strikes it at a shallow angle, creating long

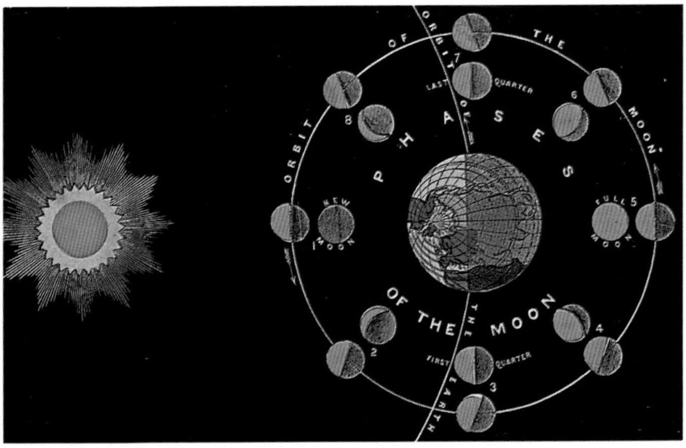
ORBIT OF THE MOON
PHASES OF THE MOON
THE EARTH
NEW MOON
FULL MOON
LAST QUARTER
FIRST QUARTER

THE LUNAR PHASES, *as explained in his 1840 diagram, are: (1) New Moon; (2) Waxing Crescent; (3) First Quarter; (4) Waxing Gibbous; (5) Full Moon; (6) Waning Gibbous; (7) Last Quarter; and (8) Waning Crescent.*

hadows and throwing the eatures into sharp relief. This happens twice a month or about a week at a time, entered on the dates of the First and Last Quarter phases.

Start with the lowest magnification your telescope has and slowly increase the power as one feature or another catches your eye. If the view grows blurry, drop back in power. At first, the variety of features may be bewildering, but compare your view with one of the eight Moon maps on pages 96 to 103. Identify any large dark maria you can see, and use them as a guide to take you farther. You will see the most details near the terminator—the dividing line between the illuminated and dark portions of the Moon.

A Two-week Day

Any feature of the Moon will change appearance during the two weeks that sunlight falls on it—the lunar day for that area. A fascinating exercise is to follow the play of sunlight across a big crater such as Copernicus (see Map 6, p. 101).

A day or two after First Quarter, as the Sun rises over Copernicus, the peaks of the rim turn into a ring of light. As dawn creeps inside, the light reveals terraces on the crater's walls and its central peaks—and finally a flat floor of lava.

At Full Moon, sunlight pours down on the crater, flattening the view of the terrain but revealing rays—streaks of pulverized rock that shot from the newborn crater in the first moments after the meteorite impact.

As the Sun descends from lunar noon, shadows return and Copernicus takes on form and shape once more.

At lunar sunset, the crater is completely hidden by shadows. Then, one fine evening you see a young Moon low in the west and can start preparing to watch the show again.

Sketchy Details

Anyone trying to sketch the Moon's face soon gives up hope of capturing all its details. Instead, try mapping just one small area throughout its phases. You will see how even small lighting changes can reveal crater terraces—or make a rille pop into view on a mare.

Pick a region or a feature, and settle down with your telescope. Use a magnification that captures the whole feature in the field of view. Lightly sketch the main parts of the feature using a sharp HB pencil, and then shade in details with a 2B pencil, smudging them with your finger or touching up with an eraser.

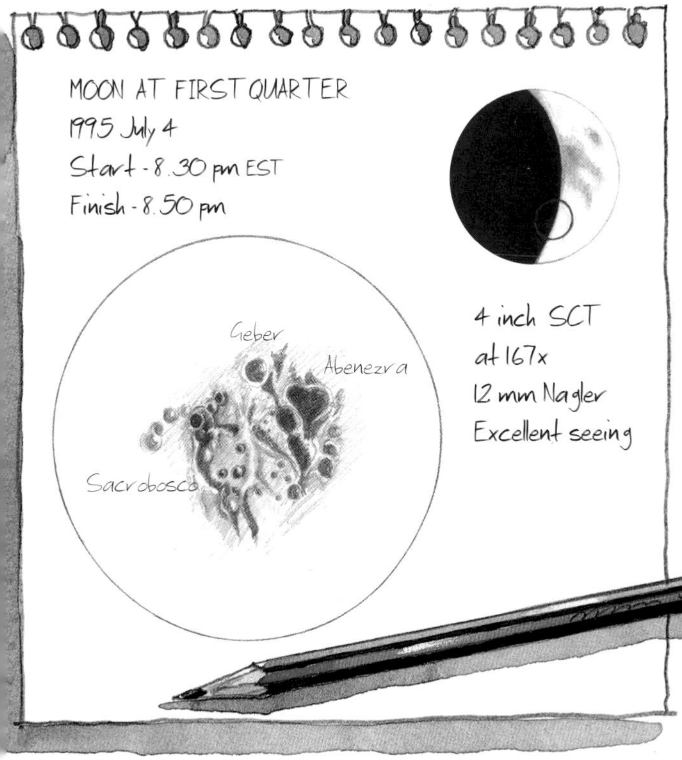
MOON AT FIRST QUARTER
1995 July 4
Start - 8.30 pm EST
Finish - 8.50 pm
4 inch SCT
at 167x
12 mm Nagler
Excellent seeing
Geber
Abenezra
Sacrobosco

WHEN YOU FINISH *a sketch, note details such as the date, the telescope used, and the seeing conditions.*

Reading *the* Moon's History

One of the biggest lunar questions has been—where did such a body come from?

WALLACE AND GROMIT *visit the Moon in the animated film* A Grand Day Out. *They discover the Moon is made of cheese and slice off pieces for their picnic.*

Some scientists dub the current theory for the Moon's origin "The Big Whack." It emerged in 1984 after more than 10 years' study of rock and soil samples taken from the six Apollo landings (1969–72) and it still offers the most complete account of the birth of the Moon. According to this theory, the proto-Earth formed alone. By the time Earth had reached nearly its present size, it had melted throughout and separated into a core and mantle.

At this point, roughly 4.5 billion years ago, Earth was struck a glancing blow by another nearly formed planet about the size of Mars. This impact sped up Earth's rotation, probably tipped its axis, and heated its surface intensely. The force of the collision vaporized the impactor and sent a great spray of superhot vapor into orbit around Earth. This vapor quickly cooled into a ring of debris and in only a few tens of thousands of years, the fragments coalesced into the Moon.

But the newborn Moon looked very different from the one you might gaze at tonight. It loomed much larger in the sky because it orbited less than half as far away. A thin, brittle crust covered a global ocean of molten rock. Bright glowing wounds opened with every meteorite impact—and such impacts were constant at first, as the new Moon swept up the remaining swirling debris.

A Quiet World

Millions of years passed. The lunar crust cooled, thickened, and stiffened. Collisions continued at a great rate, but because the crust was tougher, impacts began to excavate basins and craters, and to pile up debris around them.

About a billion years after the Moon coalesced, the last few large impacts blasted a handful of basins on the Moon's Earth-facing side. Pools of dark lava flooded the basins, filling them and spilling over to make the face of the Man in the Moon. Although impacts continued to occur, the rate dropped off dramatically.

Finally, the bombardment all but ceased. For the last three billion years, the Moon

APOLLO ASTRONAUTS *left behind footprints (above) that will remain for millions of years because the Moon has virtually no erosion. During the Apollo 17 mission, Harrison Schmitt (left) spent 22 hours collecting rock samples.*

ROUNDED HILLS *flank the valley where Apollo 17 landed (left). This lunar rock (below) came back with Apollo 16.*

as been a fairly quiet place. t present, it is struck only arely. The crater Copernicus s 800 million years old, and he most recent major impact ormed the crater Tycho bout 109 million years ago. ompared to Earth, with its onstant erosion, mountain uilding, and volcanic activity, he Moon is a dead world ndeed, although unusual vents are sometimes reported y skywatchers (see p. 94).

MPACTS, CRATERS, ND BASINS

A telescope shows an amazing rray of lunar features. Most re the result of impacts by neteorites, asteroids, and omets. When a piece of rock, ce, or anything hits the Moon t several miles per second, it lows up in a cloud of vapor, lasting a scar on the Moon's urface. Small impacts make idy, bowl-shaped craters, uch as Aristarchus, which is nly 25 miles (40 km) across see Map 5, p. 100). These mall craters are surrounded y aprons of ejected rock ragments, seen as bright areas round the younger craters.

Bigger impacts make larger raters with a more complex tructure. A good example is opernicus, 57 miles (92 km) across, with central peaks formed from pieces of deep lunar crust thrust upward by the shock of the impact. Landslides have terraced the inside of the crater's walls, and hardened pools of impact-melted rock are littered across its floor (see Map 6, p. 101).

Craters that are larger still tend to resemble walled plains. They have flat or convex floors and their central peaks may form concentric rings. These big craters rarely have rays, however, because rays disappear in about a billion years through the action of tiny meteorites striking the surface—and all of the largest craters are older than that.

The biggest craters are called basins. Filled by Mare Imbrium (see Map 6, p. 101), the Imbrium basin spans 720 miles (1,160 km)—about the size of Texas—and was created by a small asteroid about 3.9 billion years ago. The Apennine Mountains form the southeastern part of the Imbrium basin's rim.

LIQUID ROCK

Lunar geologists describe the maria as vast lava flows, many of which fill large shallow basins excavated by impacts. After a basin forms, millions of years may pass before lava flows into it. In the meantime, impacts can form craters on the basin floor. Look at Archimedes in Mare Imbrium—its walls are intact but its floor is flooded by lava. When lava floods a basin, it flows easily and for long distances, being about as runny as motor oil at room temperature. Mare lavas are made up of several layers that may form a total thickness of less than 2 miles (3 km)—paper-thin in relation to their large extent.

Full Moon is a good time to look for differences in hue that reveal separate episodes of flooding. In places, the channels that fed the flows are visible as thin, winding grooves, known as rilles, that meander and then disappear. The Apollo 15 astronauts landed near the Hadley Rille, a sinuous rille at the foot of the Apennines (see Map 6, p. 101).

Where the basin floor buckles under the weight of the lava, the mare surface pushes up in what is known as a wrinkle ridge. Where mare lava bends over a rise, the brittle lava cracks in straightish lines to form long troughs, called grabens. Several sets of grabens are visible on the eastern shore of Mare Humorum, from about 11 days after New Moon (see Map 7, p. 102).

MAPPING *the* MOON

Since the invention of the telescope, many lunar observers have devoted their efforts to mapping the seemingl endless geological detail of the Moon.

In 1610, Galileo Galilei was the first person to use a telescope to survey the Moon in detail through a full cycle of phases. His sketches were crude, but they were soon improved on by other astronomers. These early maps, with their plains, mountains, and craters, are still amazing to see, especially since they were done with unwieldy telescopes much inferior to today's amateur scopes.

Many observers hoped to record changes in the Moon's features, but despaired of ever capturing by hand all the detail visible in their eyepieces. About a century ago, photography stepped in to offer a solution, and Moon charts today are usually based on photos, even if the final map is drawn or painted.

The era of lunar mapping has not yet ended. The Space Age brought new maps of the unseen far side of the Moon, long a source of speculation, as well as much more detailed maps of the near side. But it was only in 1996 that scientists were finally able to map the Moon's global topography precisely, thanks to the laser altimeter on the Clementine spacecraft. The altimeter fired a laser beam at the ground, timing its return very accurately. With both Clementine's orbit and the speed of light being known, elevations of th landscape could be determinec

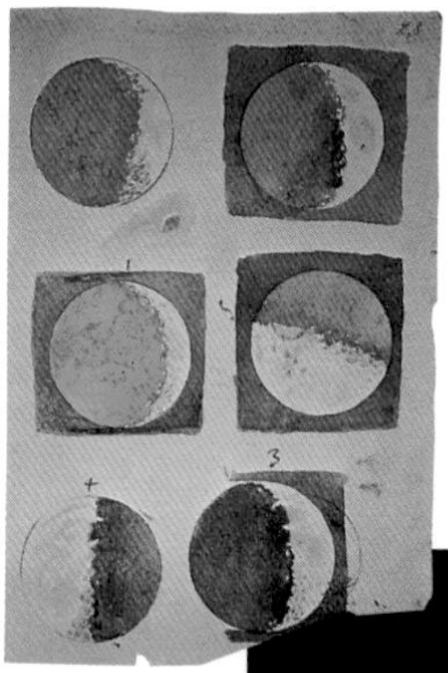

MAPS OF THE MOON *range from Galileo's 1616 sketches (left) to Clementine's 1994 topographic map (top), which shows highlands as pink and red, and lowlands as blue and purple.*

IS THE MOON ACTIVE?

Searches for activity on the Moon began with Galileo almost 400 years ago and have continued ever since. Observers have claimed to see mists obscuring craterlets on Plato's floor, a glow on the central peak of Alphonsus and the "disappearance" of the Linné crater, to name just a few examples. Are these transient lunar phenomena (TLPs) the result of meteorite strikes or gaseous escapes fron the Moon's interior—or are they merely tricks of the light and an over-eager imagination

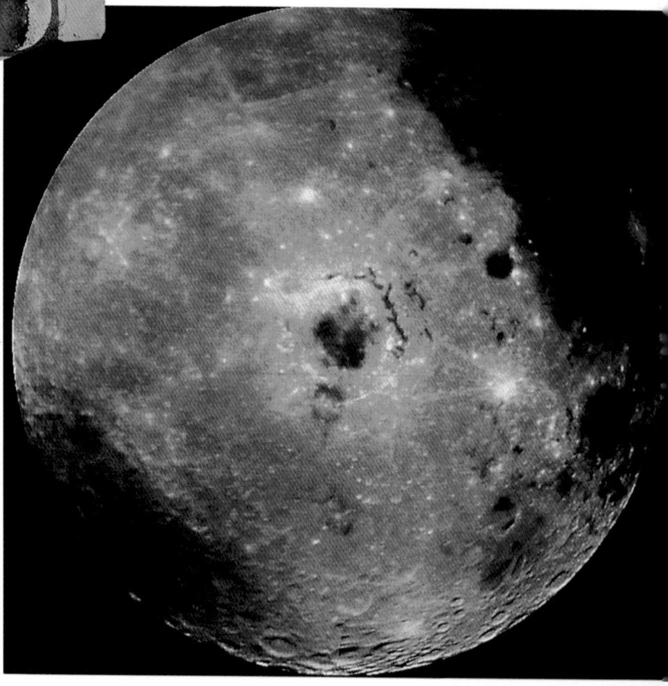

THE LUNAR FAR SIDE, *which is partly visible in the left half of this image, shows almost no areas of mare. This may be because its crust is too thick for lava to break through the surface easily.*

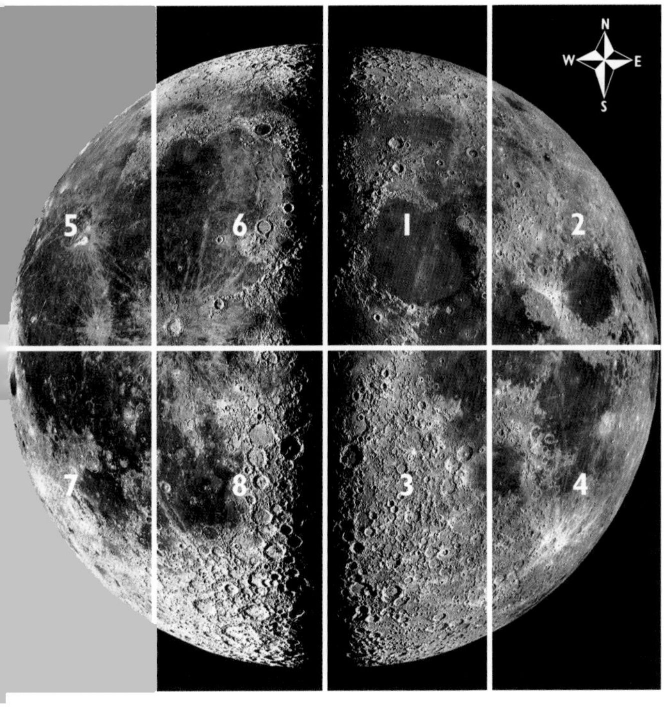
N
W
E
S
5
6
1
2
7
8
3
4

THE NEAR SIDE *is divided into eight Moon Maps on pages 96 to 103. Maps 1 to 4 show the Moon at First Quarter, and Maps 5 to 8 show it at Last Quarter.*

Many TLP reports grew om the long-standing belief iat the Moon's features were vholly or largely volcanic in rigin. Since volcanoes are lways erupting somewhere on arth, it was natural to expect milar activity on the Moon.

But scientists now know iat lunar features, especially raters, are overwhelmingly ie result of impacts. The Aoon's real volcanic activity iainly produced the maria, nd research has shown that ven the youngest mare has een cold, hard rock for at least billion years. Instruments left y Apollo astronauts reported iat the Moon's interior is seis- iically quiet, indicating that iere is no volcanic activity.

While the Moon's geo- ɔgical engine may have shut own ages ago, it is no less nteresting a place. In fact, s surface records something f great value: a history of ie Solar System's earliest ays. The record of this period n Earth has all but vanished, ianks to erosion. But in the Aoon, nature has conveniently laced a veritable geo-museum a our cosmic backyard.

Navigating the Moon

The eight Moon maps on the following pages portray the Moon with north up. This is how it appears to naked-eye observers in the Northern Hemisphere. Depending on what type of telescope you use, your telescope image may be oriented differently (see p. 50). Also bear in mind that the Moon appears south up in the Southern Hemisphere.

East and west on these maps match the directions used for Earth, rather than those usually used for the sky. This convention was chosen by the International Astronomical Union, which agreed with NASA's astronauts that for people on the Moon, the Sun should rise in the east and set in the west.

You may sometimes see slightly more or less of the Moon than our maps show because the tilted, elliptical orbit of the Moon causes "librations," which give us occasional glimpses of extra territory around its limb edges.

It is a most beautiful and delightful sight to behold the body of the Moon.

Galileo Galilei (1564–1642), Italian astronomer

ICE ON THE MOON

If rocky asteroids and meteorites can hit the Moon, so can comets. But a comet would have a different effect from an asteroid because comets are extremely rich in water. The impact of a large comet would, like any asteroid, create a big crater. But it would also give the Moon a tenuous and temporary atmosphere loaded with water vapor. The Sun's heat would drive most of the water vapor off into space after a few weeks, but any that drifted near the lunar poles could possibly condense into ice, especially in those extremely cold areas of the Moon that remain untouched by sunlight. The craters in these regions could retain ice for millions of years.

Recently, scientists using the Clementine spacecraft believe they may have found evidence for ice just under the surface near the Moon's south pole (right). If confirmed, ice could provide resources for a new generation of lunar explorers.

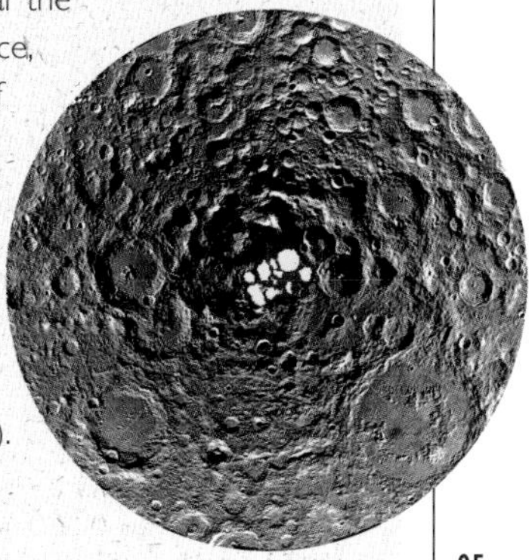

FEATURES OF MOON MAP 1

- ● 1 W. Bond
- ● 2 Barrow
- ● 3 Meton
- ▼ 4 Vallis Alpes
- ● 5 Aristoteles
- ● 6 Mitchell
- ■ 7 Mare Frigoris
- ● 8 Gärtner
- ● 9 De La Rue
- ● 10 Cassini
- ● 11 Alexander
- ● 12 Eudoxus
- ■ 13 Lacus Mortis
- ● 14 Bürg
- ● 15 Mason
- ● 16 Grove
- ● 17 Hercules
- ● 18 Atlas
- ■ 19 Lacus Somniorum
- ● 20 Daniell
- ● 21 Posidonius
- ● 22 Chacornac
- ■ 23 Mare Serenitatis
- ▲ 24 Montes Caucasus
- ● 25 Aristillus
- ● 26 Aurolycus
- ● 27 Linné
- ▲ 28 Mons Hadley
- ● 29 Bessel
- ● 30 Le Monnier
- ▲ 31 Serpentine Ridge
- ● 32 Plinius
- ▲ 33 Promontorium Archerusia
- ▲ 34 Montes Haemus
- ▲ 35 Montes Apenninus
- ■ 36 Mare Vaporum
- ● 37 Manilius
- ● 38 Julius Caesar
- ● 39 Ross
- ● 40 Jansen
- ■ 41 Mare Tranquillitatis

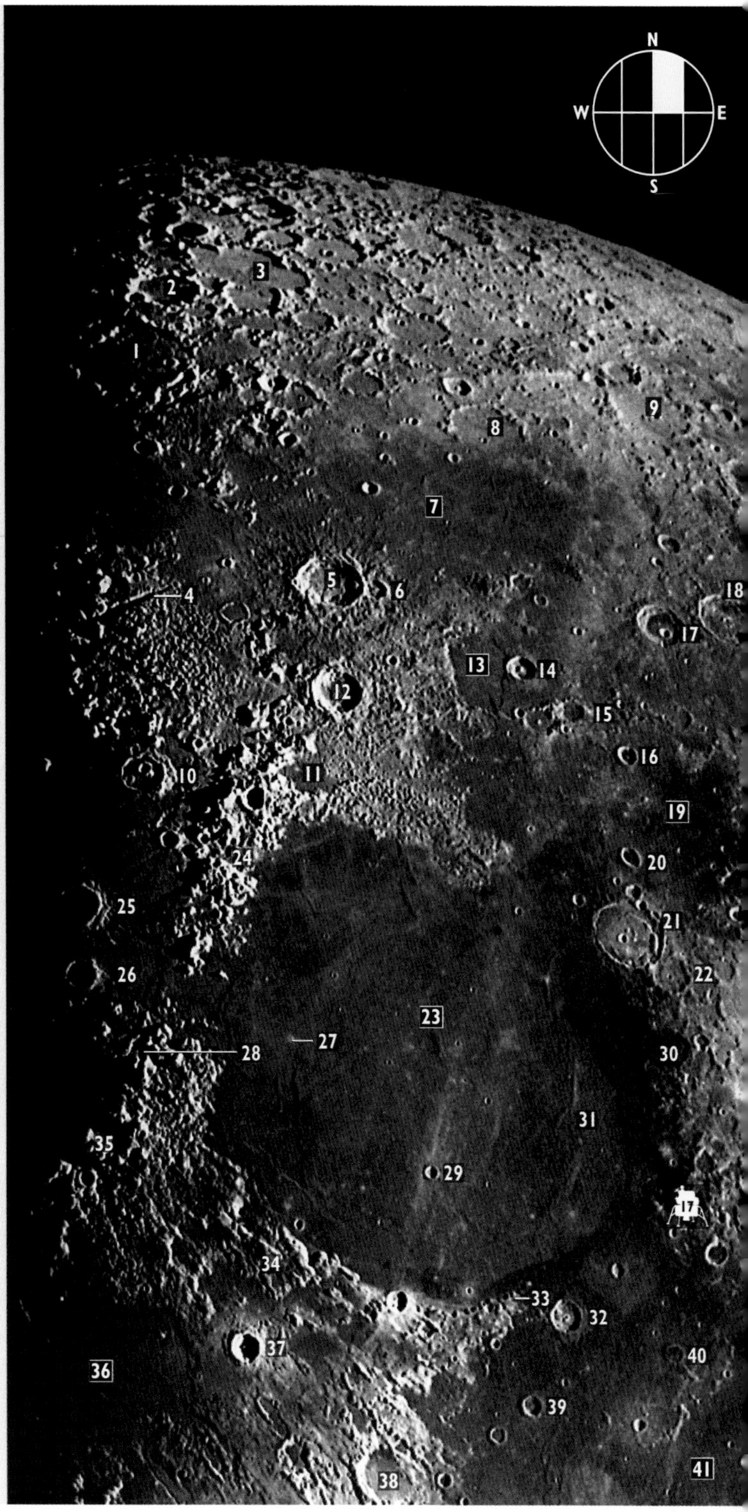
N
W
E
S
1
2
3
4
5
6
7
8
9
10
11
12
13
14
15
16
17
18
19
20
21
22
23
24
25
26
27
28
29
30
31
32
33
34
35
36
37
38
39
40
41

Mare Serenitatis (23)**, the Sea of Serenity, is a sheet of lava 380 miles (610 km) across that marks the Man in the Moon's left eye. It shows best toward First Quarter. The distinct tints in the lava suggest several flooding episodes. In the mare's east lies **Serpentine Ridge (31)**, a superb wrinkle ridge formed when the basin subsided and squeezed the lava. On the northeastern shore lies the flooded crater **Posidonius (21)**, with a curved ridge crossing its floor. The bright streak near **Bessel (29)** is a ray from Tycho, a young crater some 1,400 miles (2,200 km) away (see Map 8). The western rim of the mare, formed by the mountain range **Montes Apenninus (35)** and **Caucasus (24)**, is shared with the Imbrium basin (see Map 6). North of Serenitatis lie several large craters and the eastern reaches of **Mare Frigoris (7)**, the Sea of Cold

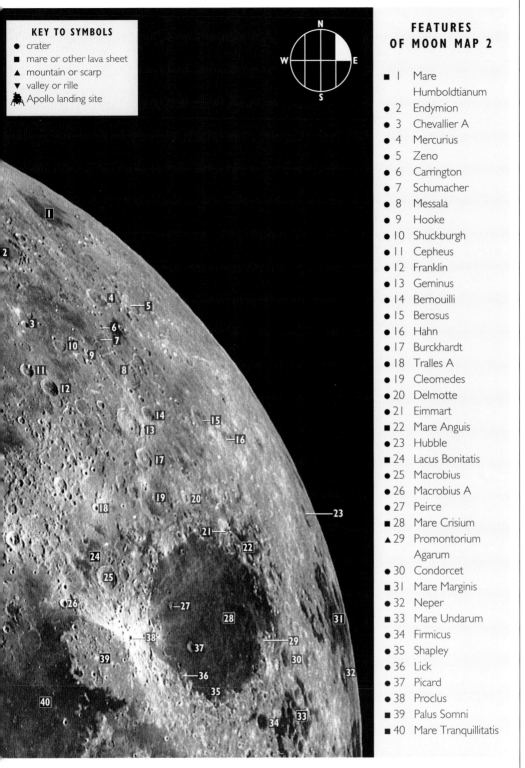

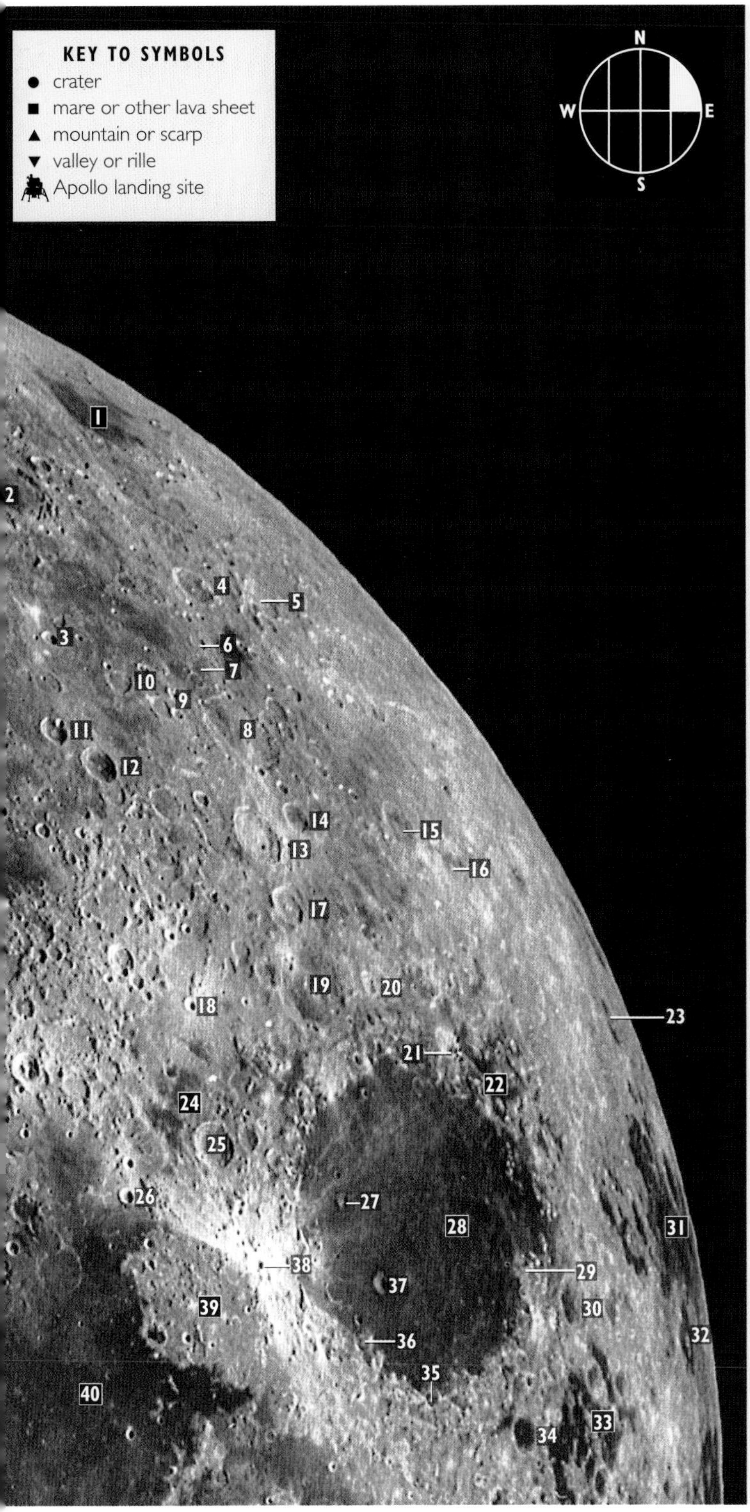
KEY TO SYMBOLS
● crater
■ mare or other lava sheet
▲ mountain or scarp
▼ valley or rille
Apollo landing site
N
W
E
S
1
2
3
4
5
6
7
10
9
11
8
12
14
15
13
16
17
19
20
18
23
21
22
24
25
26
27
28
31
38
29
37
39
30
32
36
35
40
34
33

FEATURES OF MOON MAP 2

- ■ 1 Mare Humboldtianum
- ● 2 Endymion
- ● 3 Chevallier A
- ● 4 Mercurius
- ● 5 Zeno
- ● 6 Carrington
- ● 7 Schumacher
- ● 8 Messala
- ● 9 Hooke
- ● 10 Shuckburgh
- ● 11 Cepheus
- ● 12 Franklin
- ● 13 Geminus
- ● 14 Bernouilli
- ● 15 Berosus
- ● 16 Hahn
- ● 17 Burckhardt
- ● 18 Tralles A
- ● 19 Cleomedes
- ● 20 Delmotte
- ● 21 Eimmart
- ■ 22 Mare Anguis
- ● 23 Hubble
- ■ 24 Lacus Bonitatis
- ● 25 Macrobius
- ● 26 Macrobius A
- ● 27 Peirce
- ■ 28 Mare Crisium
- ▲ 29 Promontorium Agarum
- ● 30 Condorcet
- ■ 31 Mare Marginis
- ● 32 Neper
- ■ 33 Mare Undarum
- ● 34 Firmicus
- ● 35 Shapley
- ● 36 Lick
- ● 37 Picard
- ● 38 Proclus
- ■ 39 Palus Somni
- ■ 40 Mare Tranquillitatis

The northeastern limb of the Moon is best seen when the Moon s about five days old, or about wo weeks later, just after Full Moon. The dominant feature s **Mare Crisium (27)**, the Sea of Crises, some 340 miles 550 km) across and visible o the naked eye. From Earth, it looks elongated north and south, but it is actually round, with an extension on the eastern side. Just to the west of Crisium lies the small, bright-rayed crater **Proclus (38)**, 18 miles (29 km) across. The irregular ray pattern indicates that the meteorite was traveling northeast at a shallow angle when it struck. North of Proclus, **Macrobius (25)** shows a dark floor, as does **Cleomedes (19)**, which sports a rille on its northern floor. Scientists think the lava that flooded the Crisium basin also underlies these two craters and other parts of the region where it shows up as dark patches.

FEATURES OF MOON MAP 3

- ● 1 Triesnecker
- ▼ 2 Rima Hyginus
- ▼ 3 Rima Ariadaeus
- ● 4 Arago
- ■ 5 Mare Tranquillitatis
- ● 6 Ritter
- ● 7 Sabine
- ● 8 Delambre
- ● 9 Horrocks
- ● 10 Taylor
- ● 11 Torricelli
- ● 12 Hipparchus
- ● 13 Halley
- ● 14 Mädler
- ● 15 Theophilus
- ● 16 Albategnius
- ● 17 Cyrillus
- ● 18 Abulfeda
- ■ 19 Mare Nectaris
- ● 20 Almanon
- ● 21 Catharina
- ● 22 Geber
- ● 23 Fracastorius
- ● 24 Abenezra
- ▲ 25 Rupes Altai
- ● 26 Sacrobosco
- ● 27 Apianus
- ● 28 Werner
- ● 29 Piccolomini
- ● 30 Aliacensis
- ● 31 Walter
- ● 32 Stiborius
- ● 33 Stöfler
- ● 34 Maurolycus
- ● 35 Janssen
- ● 36 Breislak
- ● 37 Heraclitus
- ● 38 Pitiscus
- ● 39 Baco
- ● 40 Vlacq
- ● 41 Hommel
- ● 42 Lilius

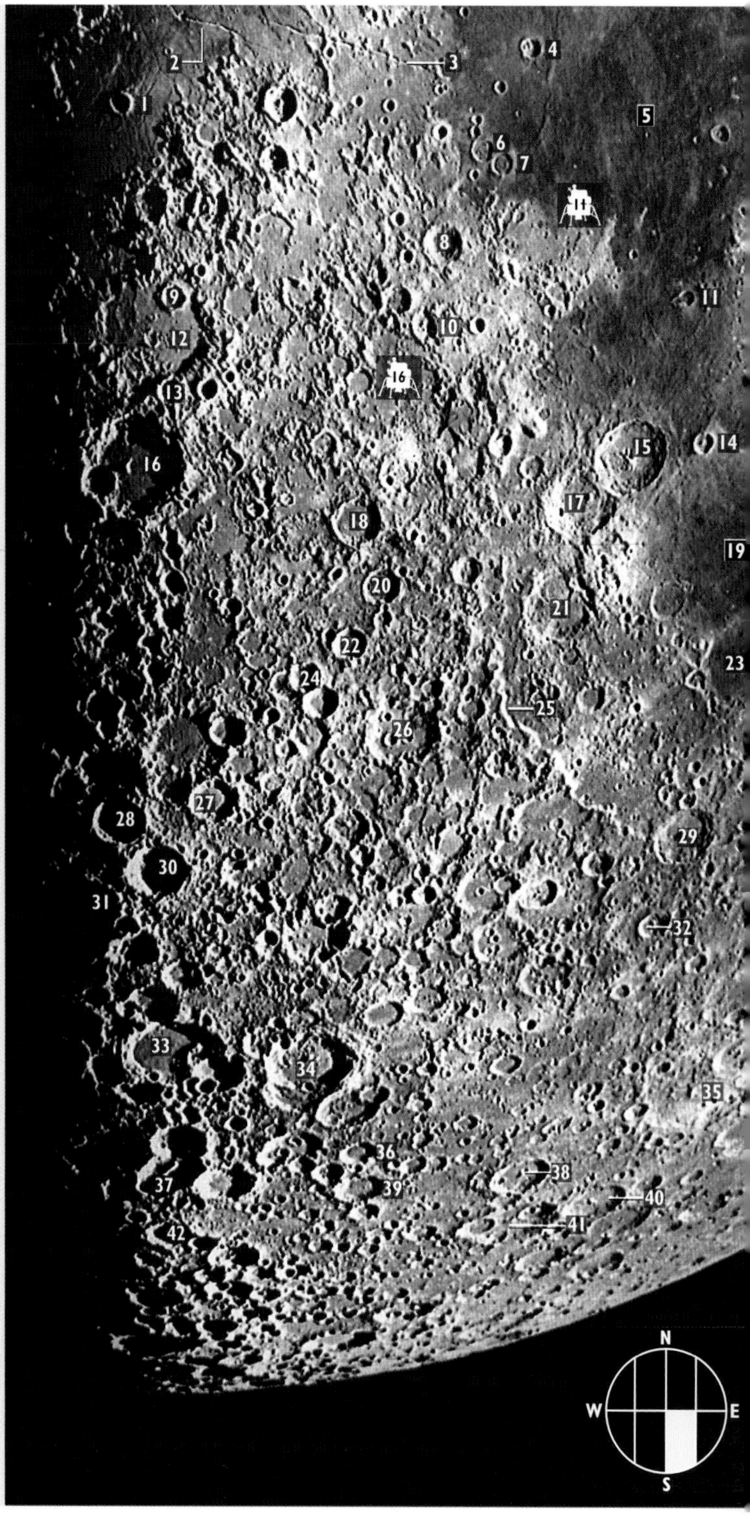
1
2
3
4
5
6
7
8
9
10
11
12
13
14
15
16
17
18
19
20
21
22
23
24
25
26
27
28
29
30
31
32
33
34
35
36
37
38
39
40
41
42
N
W
E
S

While Mare Nectaris, the Sea of Nectar, is shared with Map 4, its western features lie here in Map 3. **Mare Nectaris (19)**, 220 miles (350 km) across, is a thin sheet of lava that fills just the inner part of the much larger Nectaris basin, 540 miles (860 km) in diameter. On the southern shore of the mare sits the crater **Fracastorius (23)**, with one wall melted away. **Theophilus (15)** is the youngest and best preserved crater of a trio lying west of the mare. North from Theophilus, **Mare Tranquillitatis (5)**, the Sea of Tranquillity, contains the landing site of Apollo 11, which carried the first men to the Moon. **Rupes Altai (25)**, th Altai Scarp, is a curved, east-facing cliff that marks part of the Nectaris basin's rim. Run-ning southwest from the scarp are the southern highlands, an ancient terrain where craters of all sizes interrupt each othe in a saturated jumble.

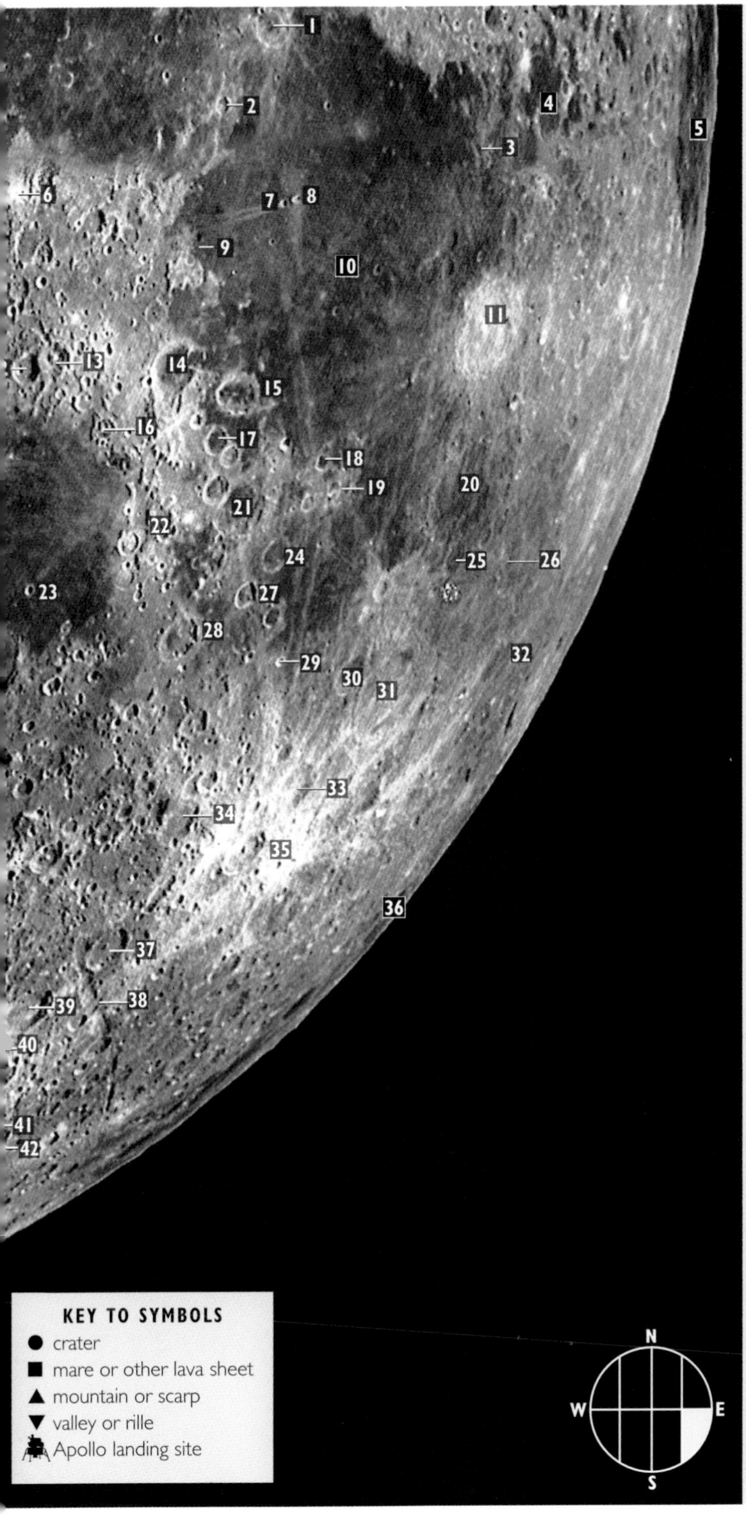
KEY TO SYMBOLS
● crater
■ mare or other lava sheet
▲ mountain or scarp
▼ valley or rille
Apollo landing site
N
W
E
S

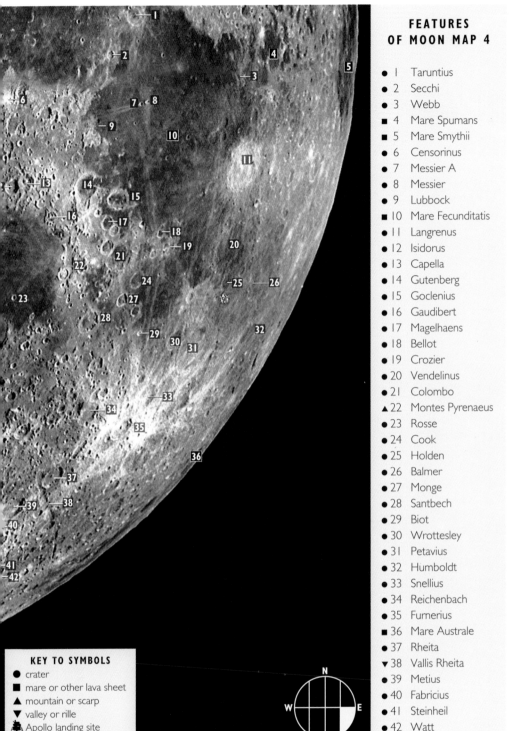

FEATURES OF MOON MAP 4

- ● 1 Taruntius
- ● 2 Secchi
- ● 3 Webb
- ■ 4 Mare Spumans
- ■ 5 Mare Smythii
- ● 6 Censorinus
- ● 7 Messier A
- ● 8 Messier
- ● 9 Lubbock
- ■ 10 Mare Fecunditatis
- ● 11 Langrenus
- ● 12 Isidorus
- ● 13 Capella
- ● 14 Gutenberg
- ● 15 Goclenius
- ● 16 Gaudibert
- ● 17 Magelhaens
- ● 18 Bellot
- ● 19 Crozier
- ● 20 Vendelinus
- ● 21 Colombo
- ▲ 22 Montes Pyrenaeus
- ● 23 Rosse
- ● 24 Cook
- ● 25 Holden
- ● 26 Balmer
- ● 27 Monge
- ● 28 Santbech
- ● 29 Biot
- ● 30 Wrottesley
- ● 31 Petavius
- ● 32 Humboldt
- ● 33 Snellius
- ● 34 Reichenbach
- ● 35 Furnerius
- ■ 36 Mare Australe
- ● 37 Rheita
- ▼ 38 Vallis Rheita
- ● 39 Metius
- ● 40 Fabricius
- ● 41 Steinheil
- ● 42 Watt

The region on the Moon's southeastern limb contains few show-stopper" features, but is easy to explore using lare **Fecunditatis (10)**, the ea of Fertility, as home base. 'he two best times to investi- ate this region are about five ays after New Moon and about two days after Full Moon. Mare Fecunditatis itself is roughly circular and fills an ancient impact basin 430 miles (690 km) in diameter. Lying in its center are two tiny craters, **Messier (8)** and **Messier A (7)**, with two rays extending west from the latter. **Langrenus (11)** is a large, pale crater with a small central peak. To its south lies a similar crater, **Petavius (31)**, with terraced walls and rilles on its lava-filled floor. Around the medium-size crater **Furnerius (35)** lies a bright patch of rays coming from a diminutive but very fresh crater called Furnerius A.

FEATURES OF MOON MAP 5

- ● 1 Oenopides
- ● 2 Markov
- ● 3 Repsold
- ■ 4 Sinus Roris
- ● 5 Sharp
- ● 6 Louville
- ● 7 Lavoisier
- ▲ 8 Mons Rümker
- ● 9 Mairan
- ● 10 Naumann
- ● 11 Lichtenberg
- ● 12 Gruithuisen
- ● 13 Heis
- ● 14 Nielsen
- ● 15 Russell
- ● 16 Briggs
- ● 17 Krieger
- ● 18 Delisle
- ● 19 Diophantus
- ▼ 20 Vallis Schröteri
- ● 21 Schiaparelli
- ● 22 Eddington
- ● 23 Seleucus
- ● 24 Herodotus
- ● 25 Aristarchus
- ● 26 Brayley
- ● 27 Euler
- ● 28 Krafft
- ■ 29 Oceanus Procellarum
- ● 30 Cardanus
- ● 31 Bessarion
- ● 32 T. Mayer
- ● 33 Marius
- ★ 34 Reiner Gamma
- ● 35 Reiner
- ● 36 Milichius
- ● 37 Kepler
- ● 38 Hevelius
- ● 39 Suess
- ● 40 Encke
- ● 41 Hortensus

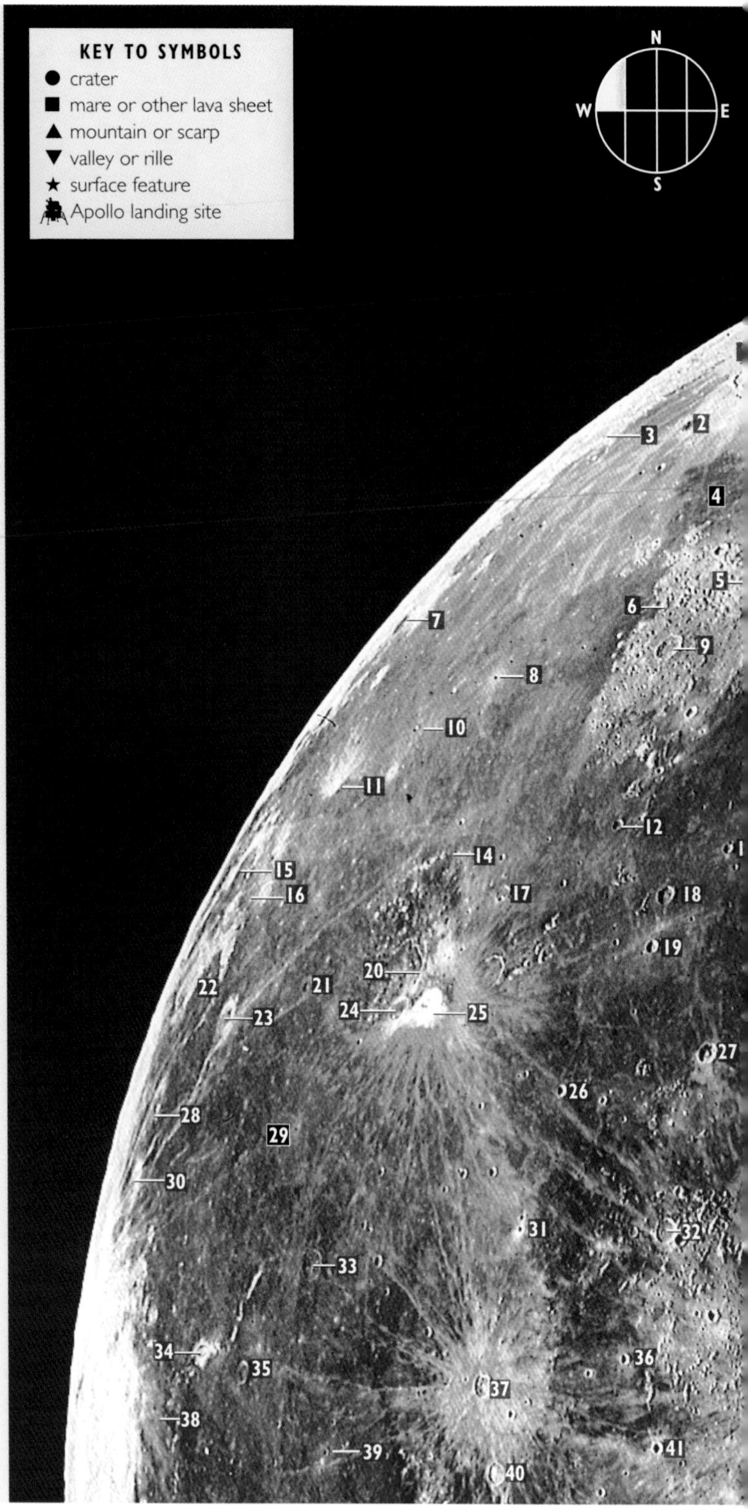
KEY TO SYMBOLS
● crater
■ mare or other lava sheet
▲ mountain or scarp
▼ valley or rille
★ surface feature
Apollo landing site
N
W
E
S

Compared to other parts of the Moon, this section looks almost empty. Most of the area is covered by a vast sheet of lava—**Oceanus Procellarum (29)**, the Ocean of Storms. Sitting atop a large plateau that saw some of the Moon's most violent activity is one of its most interesting craters—**Aristarchus (25)**, just 25 miles (40 km) across yet visible to the eye from Earth thanks to its dazzling brightness. Next to it is the crater **Herodotus (24)**, the source for Schröter's Valley, or **Vallis Schröteri (20)**, a winding sinuous rille carved by flowing lava.

Another crater to explore her is **Kepler (37)**, which looks like a miniature Copernicus (marked on Map 6). See also the mysterious feature **Reine Gamma (34)**, a swirl of ligh markings on the mare surface Compared to other material on the Moon, this feature has a strong magnetic field.

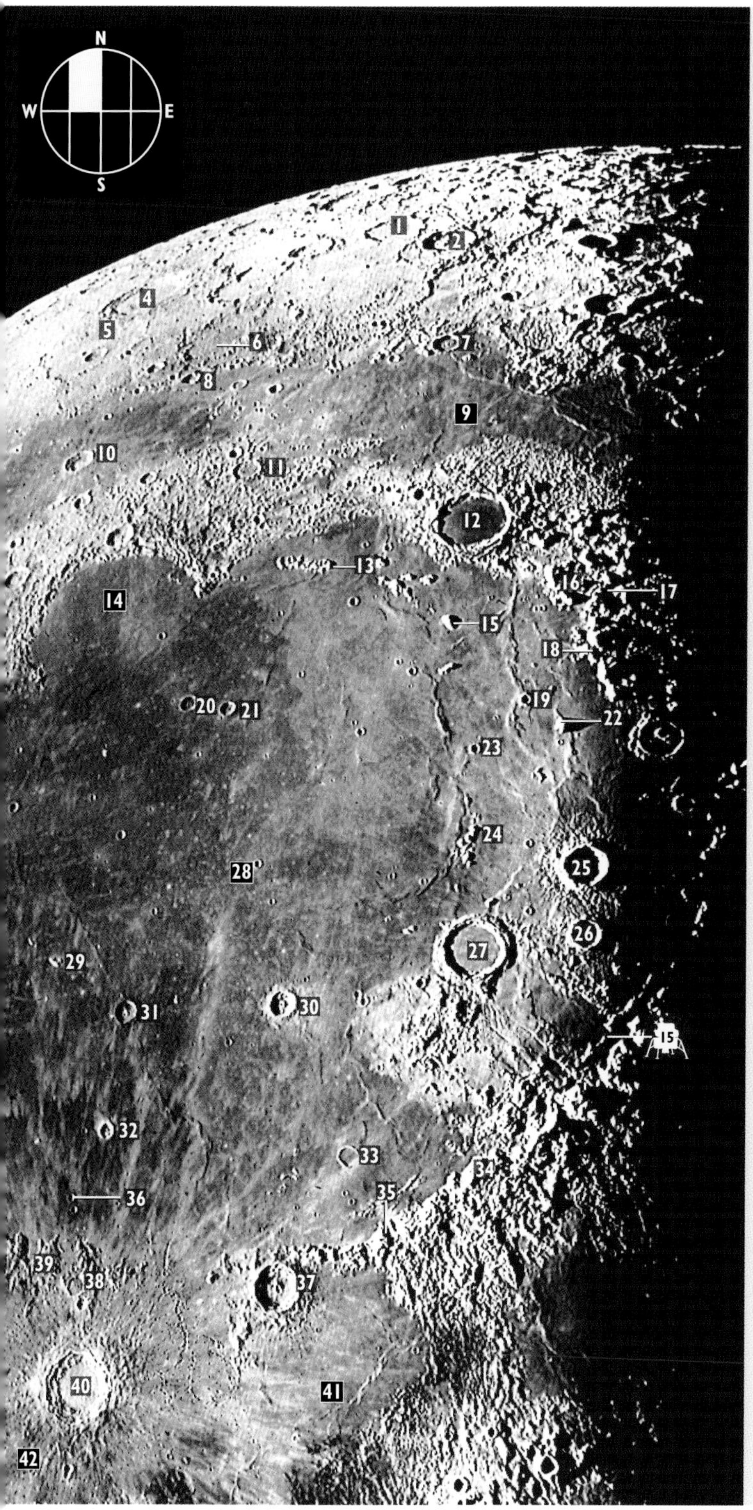
N
W
E
S
1
2
3
4
5
6
7
8
9
10
11
12
13
14
15
16
17
18
19
20
21
22
23
24
25
26
27
28
29
30
31
32
33
34
35
36
37
38
39
40
41
42

FEATURES OF MOON MAP 6

- ● 1 Anaximenes
- ● 2 Philolaus
- ● 3 Goldschmidt
- ● 4 Pythagoras
- ● 5 Babbage
- ● 6 J. Herschel
- ● 7 Fontenelle
- ● 8 Horrebow
- ■ 9 Mare Frigoris
- ● 10 Harpalus
- ● 11 La Condamine
- ● 12 Plato
- ▲ 13 Montes Recti
- ■ 14 Sinus Iridum
- ▲ 15 Mons Pico
- ▲ 16 Montes Alpes
- ▼ 17 Vallis Alpes
- ▲ 18 Mont Blanc
- ● 19 Piazzi Smyth
- ● 20 Helicon
- ● 21 Le Verrier
- ▲ 22 Mons Piton
- ● 23 Kirch
- ▲ 24 Montes Spitzbergen
- ● 25 Aristillus
- ● 26 Autolycus
- ● 27 Archimedes
- ■ 28 Mare Imbrium
- ▲ 29 Mons La Hire
- ● 30 Timocharis
- ● 31 Lambert
- ● 32 Pytheas
- ● 33 Wallace
- ▲ 34 Montes Apenninus
- ▲ 35 Mons Wolff
- ● 36 Draper
- ● 37 Eratosthenes
- ● 38 Gay-Lussac
- ▲ 39 Montes Carpatus
- ● 40 Copernicus
- ■ 41 Sinus Aestuum
- ■ 42 Mare Insularum

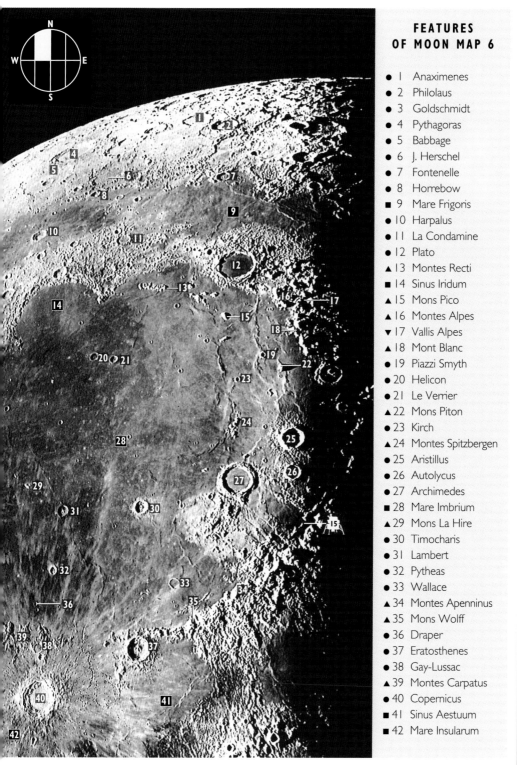

As the Moon enters its gibbous phase, Mare Imbrium and the pennine Mountains emerge om their lunar night. **Mare mbrium (28)**, the Sea of ains, fills the Imbrium basin, depression created in a gantic impact that affected ost of the Moon's near side. Marking the basin's rim are **Montes Apenninus (34)**, the Apennines, plus **Montes Alpes (16)**, the Alps, and **Montes Carpatus (39)**, the Carpathian Mountains. Craters that formed in the time between the basin impact and the lava eruptions that filled it are **Archimedes (27)**, **Plato (12)**, and the Bay of Rainbows, **Sinus Iridum (14)**. More recent features are **Eratosthenes (37)** and the splendid **Copernicus (40)**, which formed some 800 million years ago. Copernicus has a large ejecta apron—bright streaks of rock flung out by the impact—and a complex internal structure (see p. 93).

FEATURES OF MOON MAP 7

- ● 1 Riccioli
- ■ 2 Grimaldi
- ● 3 Hermann
- ■ 4 Oceanus Procellarum
- ● 5 Flamsteed
- ● 6 Lansberg A
- ● 7 Wichmann
- ● 8 Lansberg
- ● 9 Reinhold
- ● 10 Euclides
- ▲ 11 Montes Riphaeus
- ■ 12 Mare Cognitum
- ■ 13 Mare Orientale
- ■ 14 Lacus Aestatis
- ● 15 Crüger
- ● 16 Sirsalis
- ● 17 Hansteen
- ● 18 Billy
- ● 19 Zupus
- ● 20 Gassendi
- ● 21 Lubiniezky
- ● 22 Mersenius
- ● 23 Liebig
- ■ 24 Mare Humorum
- ● 25 Hippalus
- ● 26 König
- ● 27 Palmieri
- ● 28 Doppelmayer
- ● 29 Campanus
- ● 30 Lagrange
- ● 31 Piazzi
- ■ 32 Lacus Excellentiae
- ■ 33 Palus Epidemiarum
- ● 34 Schickard
- ● 35 Phocylides
- ● 36 Hainzel
- ● 37 Schiller

1
2
3
4
5
6
7
8
9
10
11
12
13
14
15
16
17
18
19
20
21
22
23
24
25
26
27
28
29
30
31
32
33
34
35
36
N
W
E
S
KEY TO SYMBOLS
● crater
■ mare or other lava sheet
▲ mountain or scarp
▼ valley or rille
Apollo landing site

In the last days before Full Moon, the Sun reaches the area of the southwestern limb. The main landmark here is the 220 mile (350 km) expanse of **Mare Humorum (24)**, the Sea of Moisture, which you can see with the naked eye. Two craters stand on its shores, both partly breached by lava: **Gassendi (20)** in the north and **Doppelmayer (28)**, the more engulfed crater, in the south. Between Mare Humorum and the limb lies **Schickard (34)**, a large subdued crater with patches of dark lava covering about half its floor. Near Schickard, scientists have found terrain they call cryptomare—dark patches of lava that seem to have oozed up from beneath craters and hills during a late phase of flooding. Perhaps also from this period is **Grimaldi (2)**, an isolated patch of mare that fills a 140 mile (225 km) basin near the western limb.

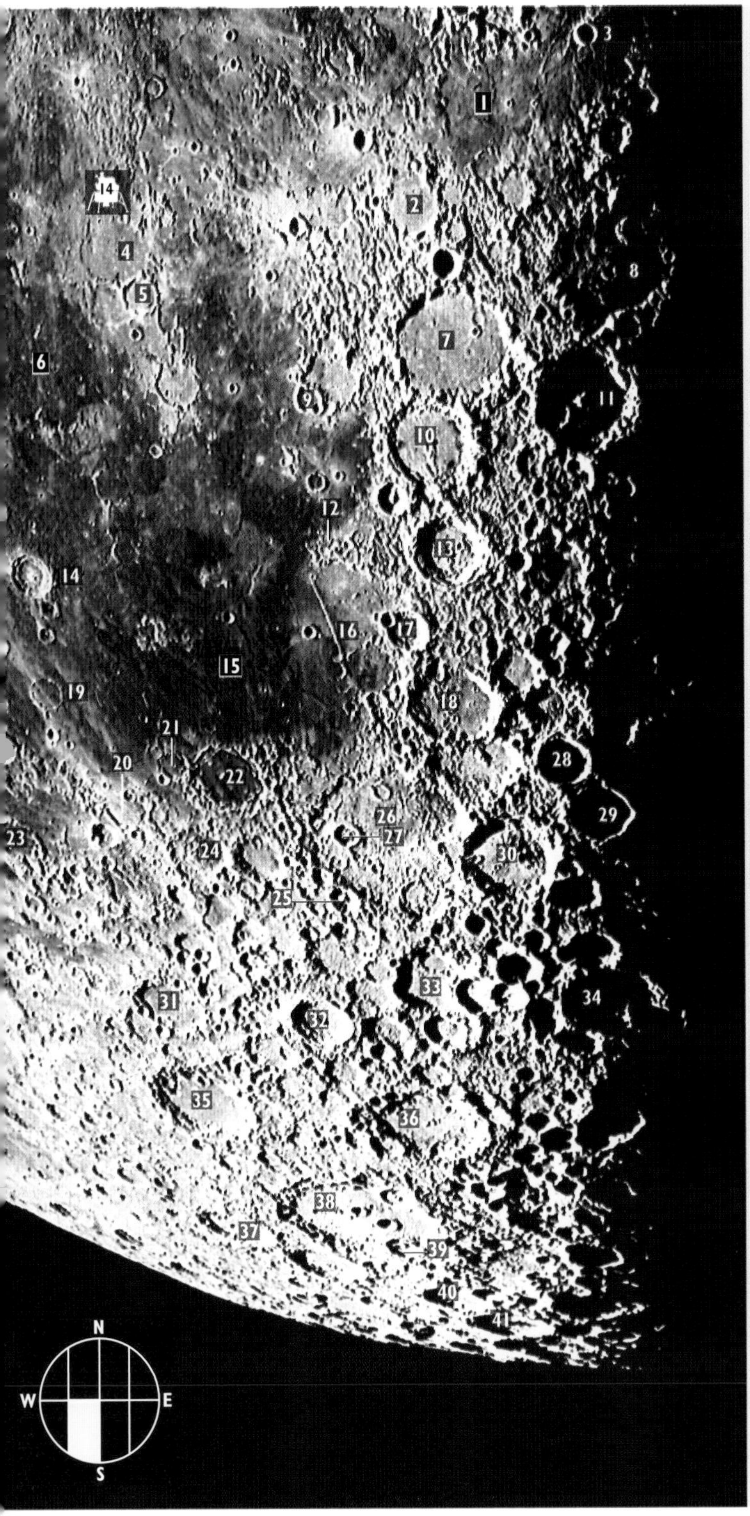
N
W
E
S

FEATURES OF MOON MAP 8

- ■ 1 Sinus Medii
- ● 2 Flammarion
- ● 3 Triesnecker
- ● 4 Fra Mauro
- ● 5 Parry
- ■ 6 Mare Cognitum
- ● 7 Ptolemaeus
- ● 8 Hipparchus
- ● 9 Davy
- ● 10 Alphonsus
- ● 11 Albategnius
- ▲ 12 Promontorium Taenarium
- ● 13 Arzachel
- ● 14 Bullialdus
- ■ 15 Mare Nubium
- ▲ 16 Rupes Recta
- ● 17 Thebit
- ● 18 Purbach
- ● 19 Kies
- ● 20 Weiss
- ● 21 Hesiodus
- ● 22 Pitatus
- ● 23 Capuanus
- ● 24 Wurzelbauer
- ● 25 Ball
- ● 26 Deslandres
- ● 27 Hell
- ● 28 Werner
- ● 29 Aliacensis
- ● 30 Walter
- ● 31 Wilhelm
- ● 32 Tycho
- ● 33 Orontius
- ● 34 Stöfler
- ● 35 Longomontanus
- ● 36 Maginus
- ● 37 Scheiner
- ● 38 Clavius
- ● 39 Rutherfurd
- ● 40 Gruemberger
- ● 41 Moretus

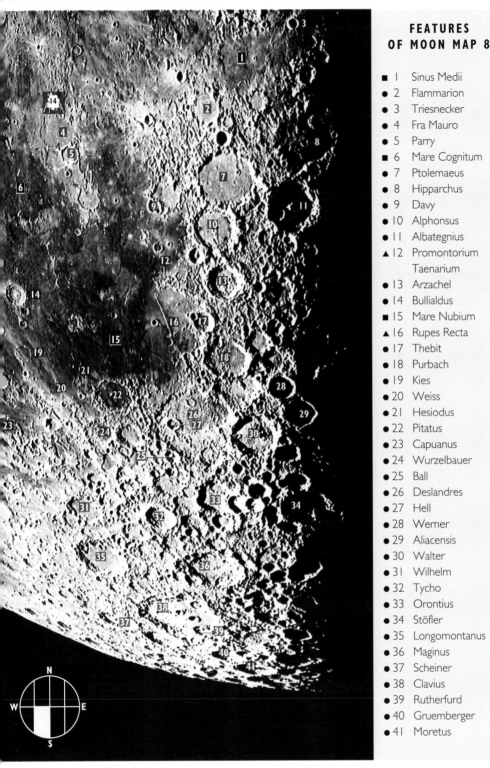

In the evenings that follow First Quarter, the Sun rises over some of the Moon's inest sights. In the south lie wo notable craters, giant **Clavius (38)**, which spans 140 miles (225 km), and maller **Tycho (32)**. Catch Clavius when the Sun first trikes, revealing a convex floor with a curving chain of craterlets. Tycho, only 109 million years old, lies at the hub of a splash pattern of rays (best seen at Full Moon) that stretches for more than 1,000 miles (1,600 km). Look for a trio of craters—**Arzachel (13)**, **Alphonsus (10)**, and **Ptolemaeus (7)**—aligned north-south. **Rupes Recta (16)**, the Straight Wall, is a fault scarp that changes from a dark line to a bright one toward lunar noon. **Deslandres (26)**, to the south, is a ruined crater 146 miles (235 km) across. Nearby **Pitatus (22)** has several rilles on its floor and partly flooded wall.

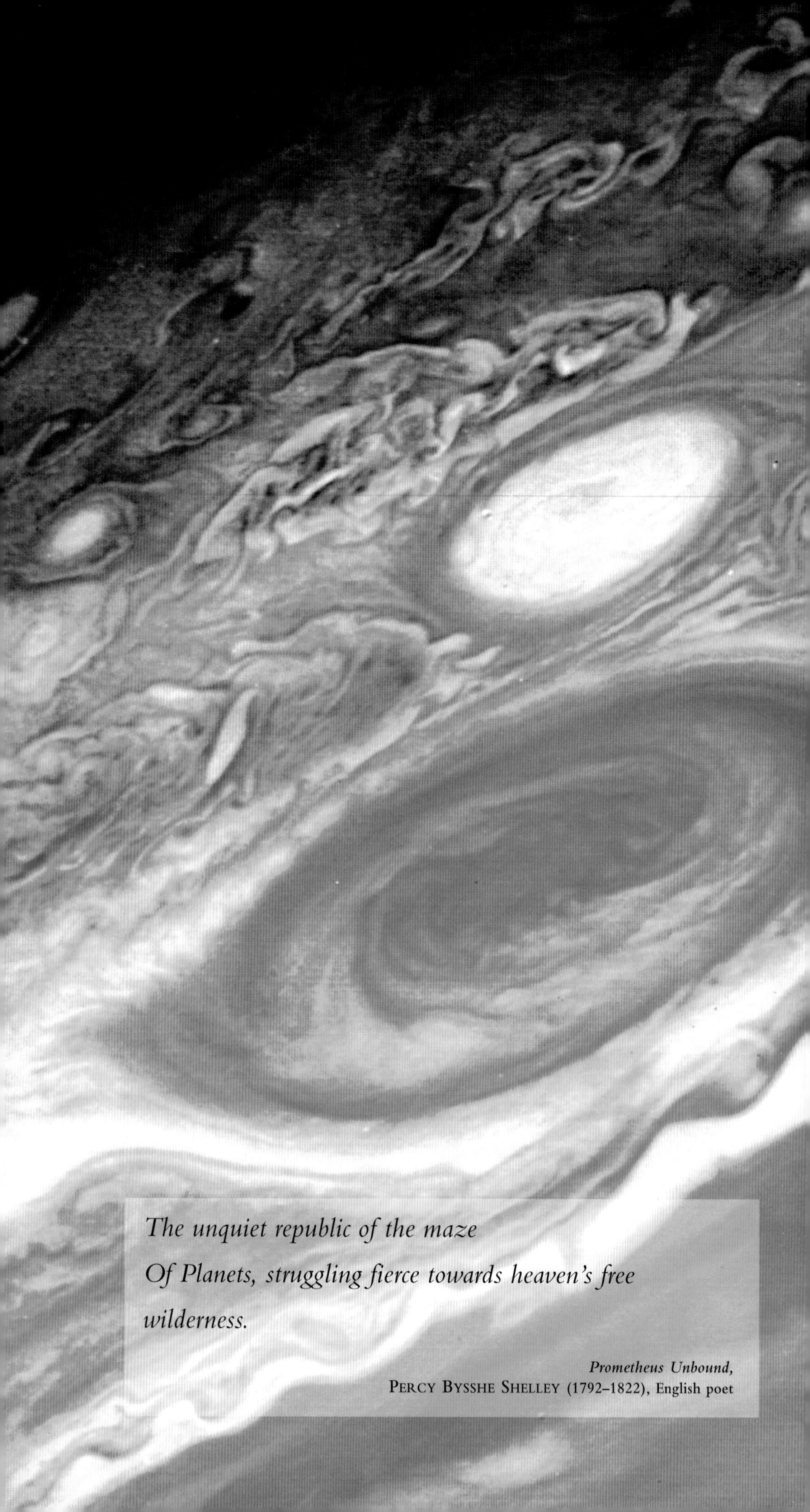

The unquiet republic of the maze
Of Planets, struggling fierce towards heaven's free
wilderness.

Prometheus Unbound,
PERCY BYSSHE SHELLEY (1792–1822), English poet

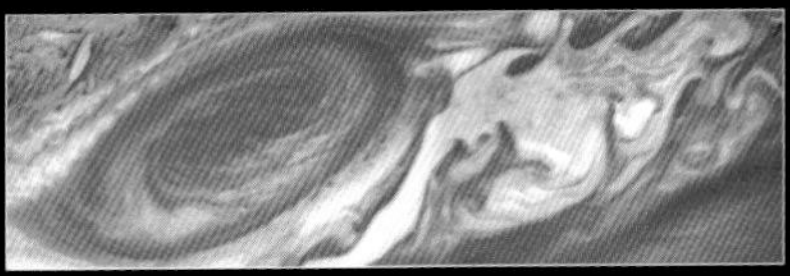

Chapter Four

The Planets

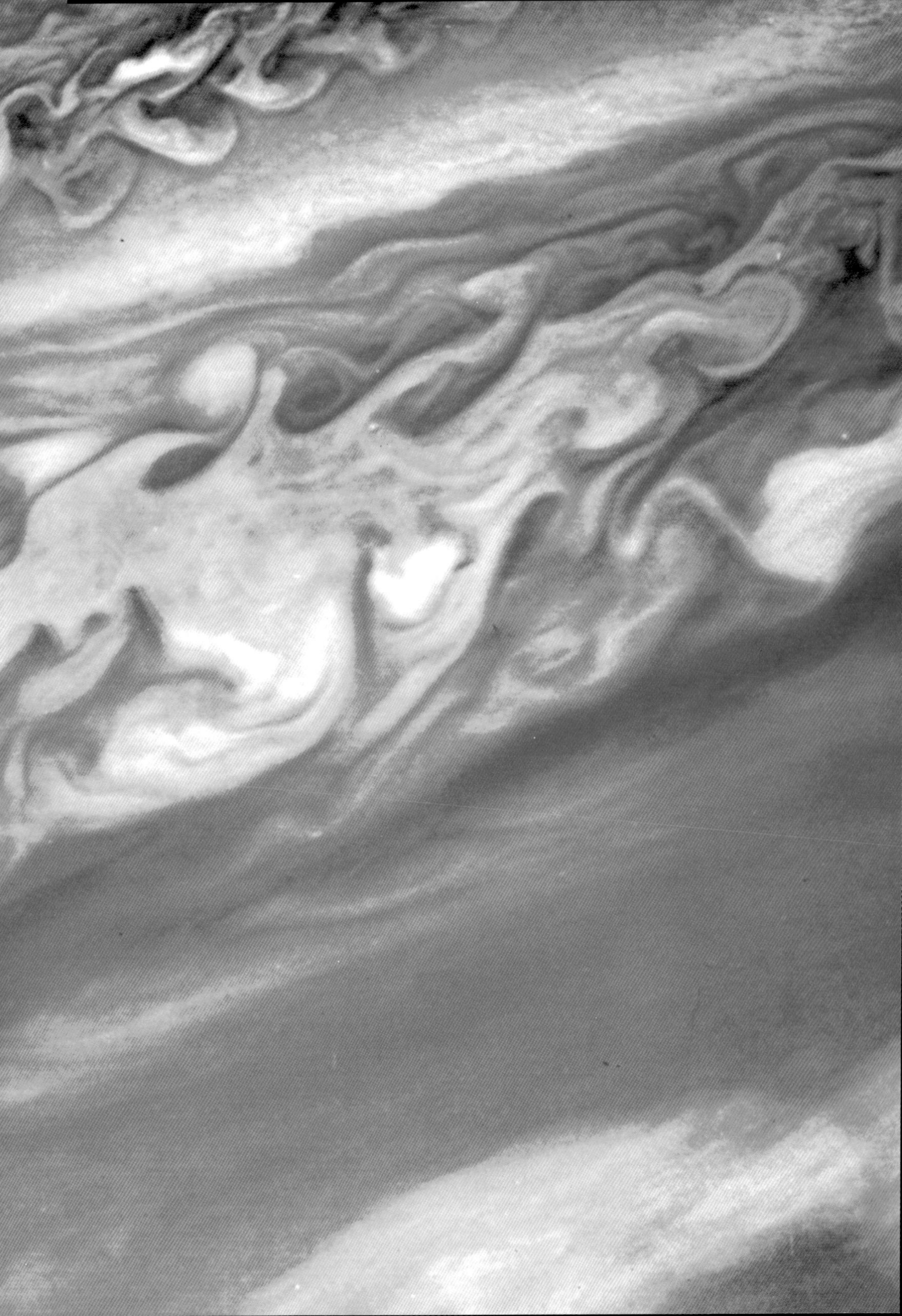

Observing Planets

Telescopes may provide less spectacular views than spacecraft do, but they let you check out Mars, Saturn, or Jupiter on any clear night of the year.

AN ORRERY *(left) is a mechanical model of the Solar System. The first such device was made for Charles Boyle, the Earl of Orrery (1676–1731).*

The Space Age has revolutionized our understanding of the planets, and future spacecraft missions will no doubt continue to dramatically alter the picture. Often as not, however, it is an amateur astronomer who alerts scientists that a new dust storm has begun on Mars, or that the placid equatorial zone of Saturn has been disturbed by a huge spot.

A telescope is also the closest thing to a spaceship most observers are ever likely to have. No one has yet stood on the sands of Mars, but you can peer into your eyepiece and see hazy clouds lit by afternoon sunlight over the Red Planet's volcanoes.

Rocks and Gas

You can sort planets into two categories—two ways. The first pair of categories refers to the planets' physical nature. Earth is a terrestrial planet, so too are Mars, Venus, and Mercury. Jupiter, on the other hand, is a gas giant, along with Saturn, Uranus, and Neptune. The terrestrial planets are small and rocky with relatively thin atmospheres. Gas giants are at least a dozen times more massive and consist of deep, heavy atmospheres that surround small, rocky cores. Pluto is an oddball that does not fit either category well. Scientists are beginning to think it most resembles an enormous comet nucleus. But its nature remains in question, partly because it has never been visited by spacecraft.

Interior Planets

The other way to sort planets is into interior and exterior ones. The two interior planets are Mercury and Venus; they circle the Sun inside Earth's orbit. Exterior planets are those from Mars outward; they stay outside of Earth's orbit.

This difference matters because it controls how and where you look for the planet. An interior planet always stays near the Sun in the sky; you will never see one at midnight, for example. As Mercury or Venus orbits the Sun, you will see it first in the evening sky, moving day by day out from the Sun's glare. After a few weeks, it reaches a point of maximum separation from the Sun called greatest eastern elongation, usually the time of best visibility.

The planet passes between Earth and the Sun at a point called inferior conjunction, when it briefly disappears into the solar glare. Greatest western elongation follows in the predawn, again bringing good visibility. The planet completes its orbit on the far side of the Sun, and disappears once more. This ends one apparition, or viewing season, and begins the next.

INSIDE AND OUT *Interior planets such as Venus reach best visibility at the points of greatest elongation, while exterior planets such as Mars are best seen at opposition. All planets disappear into the Sun's glare at conjunction.*

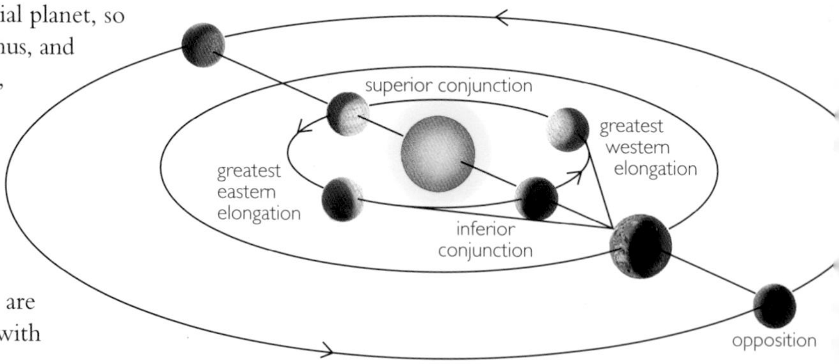
superior conjunction
greatest western elongation
greatest eastern elongation
inferior conjunction
opposition

HE NINE LANETS *of the ɔlar System orbit ɪe Sun in ellipses, a fact scovered by Johannes Kepler the seventeenth century.*

꞉XTERIOR PLANETS

xterior planets have a simpler ɔutine. In all cases, Earth is rbiting faster than they are, ɔ we overtake them. We first otice the planet rising just head of the Sun, low in the redawn sky. As weeks pass, rises earlier each night.

Many weeks before oposition, the planet's motion gainst the background stars ops, and it then appears to ɪove backward (westward). 'his retrograde motion is ɪused by Earth's greater ɔeed—in effect, the outer lanet behaves like a car aveling in the adjoining lane you pass it on the highway.

At opposition, the planet ·aches a point where it ses as the Sun sets. This is ɪe time of best visibility. he planet then slips into ɪe evening sky, where it sets ter the Sun. After several 'eeks, it stops moving retro-ade and recommences ɪstward motion. Eventually, ɪe planet disappears into the ɪn's glare and reaches conɪnction. Conjunction is llowed by the predawn start f a new apparition.

'he Planets in their

ations list'ning stood,

Vhile the bright Pomp

꞉cended jubilant …

radise Lost,

HN MILTON (1608–74),

ɪglish poet

FINDING PLANETS

Planets orbit close to the ecliptic, the plane of Earth's orbit. This makes them easier to find—a line marking the ecliptic is usually drawn on sky charts.

Unlike constellations, however, whose seasons of visibility can be predicted easily, planets are on the move. Astronomy magazines provide regular monthly guides and publish feature articles on upcoming events. Various annual handbooks and almanacs do likewise on a yearly basis. Planetarium-type computer programs can show you what is visible on any particular night.

OBSERVING TIPS

Planets are generally bright, so a dark observing site is less critical than it is for deep-sky objects. And since planets seldom stray more than a few degrees from the ecliptic, the site does not need to provide access to the entire sky.

Planet watching demands no special type of telescope. Steadier and sharper views let you see more with less effort, so the most desirable items are a sturdy equatorial mounting and high-quality optics.

Because atmospheric seeing is rarely steady, most planet observing is done using eyepieces that yield about 200x magnification or less. But planet watchers should have on hand at least one eyepiece that provides 300x or more. You may not be able to use it very often, but as all planets have apparent sizes smaller than many of the craters on the Moon, large magnifications are called for at times. Many amateurs also use a set of filters—either glass eyepiece filters (see p. 57) or the less expensive gelatin ones available from camera stores.

For sketching, a clipboard with a dim red light attached will be handy. Purchase a set of circle and ellipse templates from an art store, along with pencils in several grades (2H to 3B), a stub stick, and a white eraser. Photography with film or CCDs can be challenging, but is another satisfying way to record the planets.

Observing planets is a skill like any other—you get better with practice. Beginners who take their first look at Jupiter, for instance, are often disappointed. "Is that all there is to see?" they think, when their telescope shows a tiny oval disk with just two faint, dusky bands crossing it. Likewise, reports of colors seen in astronomical objects often strike newcomers as exaggerated.

When the sight in the telescope eyepiece is unfamiliar, we need to educate our brain to recognize and understand the image it is receiving. The best way to see more detail is to spend lots of time with your telescope exploring the sky.

MERCURY

Few skywatchers can boast of spotting Mercury on more than isolated occasions. How can a planet that becomes as bright as the brightest stars be so elusive?

MERCURIUS, *sculpted here by l'Antico (1460–1528), was the messenger of the gods in Roman mythology.*

Of all the planets that are visible to the naked eye, Mercury is the least often observed. The reason lies in its orbit.

Because Mercury is the closest planet to the Sun, its year lasts just 88 days, and it never strays more than 28 degrees from the blaze of light surrounding the Sun. Observers usually glimpse it just before sunrise or soon after sunset, and they can never see it in a fully dark sky.

For all this, antiquity's astronomers knew of Mercury and carefully logged its appearances. In Mesopotamia, where clear desert twilights and low horizons made it easier to see, this come-and-go object was named Nabu; he was a scribe and a messenger of the gods. The planet's swift apparitions also prompted the Greeks to name it after a heavenly messenger—Hermes. The Romans translated the name into Latin as Mercurius.

FIRE AND ICE

In 1974 and 1975, the Mariner 10 spacecraft photographed about half of Mercury. It revealed an ancient Moon-like face with craters and scarps, but without the Moon's dark lava sheets. Mercury has an unusually large iron core, perhaps the result of a catastrophic collision early in its history that removed part of the planet's mantle. Mercury also has a magnetic field, but with only 1 percent the strength of Earth's. Although the planet's gravity is too weak to retain any significant atmosphere, there is a tenuous layer of helium and hydrogen (probably captured from the solar wind), as well as sodium (possibly from the surface rocks).

With a diameter of only 3,024 miles (4,878 km), Mercury is the smallest planet apart from Pluto. It also has the second-most eccentric orbit, again after Pluto's. Mercury's temperatures range more than any other planet. In daytime, the equatorial regions can reach about 800 degrees Fahrenheit (430° C), while at night, the surface temperature can fall to −300 degrees Fahrenheit (−185° C). Mercury may have subsurface ice in permanently shadowed craters at the poles. If so, the ice probably condensed from water vapor released in comet impacts.

OBSERVING MERCURY

From the Northern Hemisphere, Mercury is best seen at an evening apparition during March or April, and

VISITING MERCURY *An artist's impression (above) shows Mariner 10 flying by Mercury. The spacecraft's cameras sent digital images to Earth, including this close view of grabens—depressions caused by fracturing (right).*

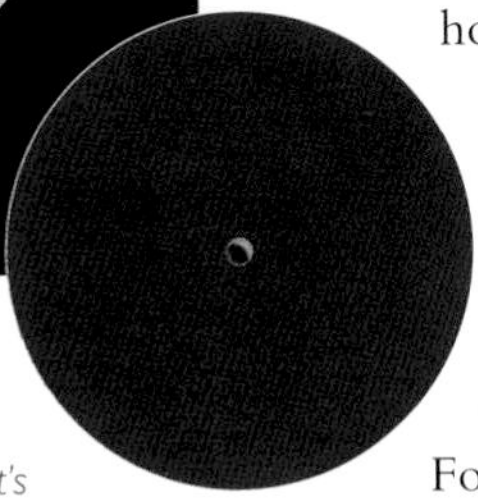

SMALL TELESCOPE *ill show you Mercury's hases (right), but the planet's atered surface can be seen only in lariner 10 photographs (above).*

t a morning one in Sep- ember or October. (In the outhern Hemisphere, the est evening apparitions come n September or October, and norning ones in March or April.) At these seasons, Mer- ury's orbit tilts most steeply o the horizon, so it reaches naximum separation from the un and best visibility.

At a good evening appari- ion, you can follow Mercury or about two weeks before he date of greatest elongation, nd a week after; for morning pparitions, it is a week before nd two weeks after.

Mercury's apparent size aries according to its phases. When Mercury first appears in he evening sky, it is coming round the far side of its orbit oward us. Seen in a telescope, t presents a warm-gray disk that looks like a tiny gibbous Moon devoid of features. At greatest eastern elongation, it appears half lit. As it approaches inferior conjunction, it grows in size while its illuminated portion shrinks to a crescent. After passing between Earth and the Sun, it repeats the phases in reverse order, finally reaching superior conjunction and starting the cycle anew.

Mercury typically presents a small disk that shimmers from the poor seeing near the horizon. Occasionally, with good seeing, you can use 200x magnification or more, yielding a view like that of the naked-eye Moon.

For better seeing, try viewing in twilight, when Mercury is higher in the sky, and there is less contrast between sky and planet. This strategy works best for a morning apparition—you locate the planet low in a relatively dark sky and track it as it rises and the sky brightens.

If your telescope has an equatorial mounting, you can observe Mercury in full day- light. On a clear day, find the coordinates of both the Sun and Mercury in an almanac, magazine, or computer pro- gram. Work out the difference in their positions, then center the Sun (with proper precau- tions; see p. 77), and step off the distance to Mercury. Scan with a low-power eyepiece to locate the planet.

You must not expect to see at sight … Seeing is in some respects an art which must be learned.

WILLIAM HERSCHEL (1738–1822), English astronomer

CRESCENT MOON *is about to hide, occult, Mercury (right). Occultations re often closely followed by amateurs.*

TRAVERSING THE SUN

At rare intervals, observers can see Mercury cross in front of the Sun. The next two such transits of Mercury will occur on 7 May 2003 and 8 November 2006, and both will last about five hours. When transit day arrives, rig your telescope for viewing the Sun (see p. 77) and look for Mercury as a tiny dot crawling across its face. It will resemble a small sunspot, but will move detectably after several minutes.

THE FACTS BOXES *(such as Mercury Facts, below) list statistics for each planet, mostly in relation to Earth. The distance between Earth and the Sun is known as an astronomical unit—or 1 AU.*

MERCURY FACTS

Distance from Sun 0.39 AU
Mass 0.06 x Earth's mass
Radius 0.38 x Earth's radius
Apparent size 5 to 13 arcsecs
Apparent magnitude –2 to +3
Rotation (relative to stars) 58.7 Earth days
Orbit 88.0 Earth days

VENUS

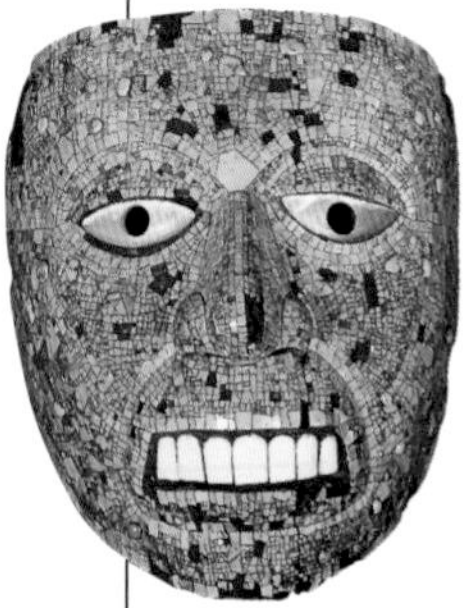

Venus is a planet of paradox. No other planet comes closer to us—or appears larger or brighter—but your telescope will show you nothing of its surface, thanks to a veil of globe-girdling clouds.

Venus was well known to the ancients. The Greeks believed it was two distinct objects: Phosphoros in the morning sky and Hesperos in the evening.

The Maya saw Venus as the god Kukulkán, a figure who symbolized the death and rebirth of the universe. They based a complex 418-year calendar on the observation that in the space of eight years, Venus makes five complete apparitions. To the Aztecs, several centuries later, Venus was Quetzalcoatl, the feathered serpent who symbolized the power of life emerging from earth, water, and sky.

THE EVENING STAR

Venus circles the Sun once every 225 days. But because Earth is also moving, it takes 584 days before Venus reappears in the same part of our sky. This makes the typical Venus apparition a far more leisurely business than Mercury's frantic scurry. It begins when Venus emerges from the Sun's glow, low in the evening twilight. At this point, it looks small (about 10 arcseconds across) and round in phase.

Each night, the planet climbs higher, moving farther from the Sun. At greatest eastern elongation, it stands highest above the western horizon. It is in half-lit phase, but it has grown to about 25 arcseconds. Now at magnitude –4.5, the planet will be brighter than anything in the sky except the Sun or Moon. Although it does not move visibly, Venus resembles an airplane with landing lights on, and has caused many a UFO report.

After greatest elongation, Venus loses altitude but grows even brighter as it approaches Earth. The point of greatest brilliancy occurs about 4 week[s] before inferior conjunction.

Venus is then bright enough to cast a shadow. In a dark location, hold a sheet of white paper facing Venus. Le[t] your eyes dark-adapt, then move your hand just in front of the paper. The shadow wi[ll] be sharp-edged compared to shadows cast by the Sun or

VENUS AND THE MOON *Venus is sometimes the brightest point of light in the sky apart from the Moon (below). An occultation of Venus by a crescent Moon (right). A mask of Quetzalcoatl (top).*

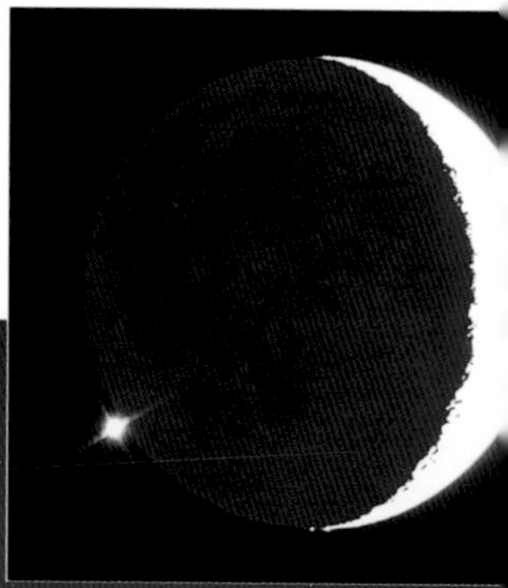

Ishtar Terra
Maxwell Montes
Maat Mons
Aphrodite Terra
Ovda Regio

NEAR AND FAR *From Earth, Venus appears largest in its crescent phase (below left). Up close in 1979, Pioneer saw cloud details (below right). Magellan used radar to map the surface through the clouds in 1993 (far left).*

ıoon. This is because Venus a point-source of light, ather than a broad disk.

A week before inferior con-ınction, Venus is in crescent hase, but has grown to nearly 0 arcseconds across—just within the resolving power of uman eyesight. Not every-ne's vision is acute enough ɔ see the crescent, however, nd binoculars can help.

At inferior conjunction, 'enus overtakes Earth, leaves he evening sky, and shifts into he morning. Thereafter, it ises before the Sun, repeating he stages of its evening per-ɔrmance, but in reverse order.

Superior conjunction, when 'enus lies on the other side f the Sun from us, marks the lose of one apparition—and he start of the next.

ᴛELESCOPIC VENUS

ronically, the very brightness hat makes Venus so easy to ind works against the telescope ser. Seen in your scope's yepiece in a dark sky, Venus s too dazzling an object for good or even comfortable viewing. Experienced Venus-watchers observe at twilight—or in full daylight by offsetting the telescope from the Sun (see p. 109).

Unlike the other terrestrial planets, Venus hides its surface features under a veil of clouds. The clouds, roughly 30 miles (50 km) above the surface, are made of sulfuric acid. They are almost as reflective as snow, which is why Venus looks so bright. The atmosphere below the clouds is generally clear, but Earth-bound observers usually see just the opaque and featureless shroud.

Although atmospheric features are hard to detect, for more than a century visual observers have noted faint, temporary streaks—probably real cloud features. Likewise, the terminator sometimes shows an irregular edge instead of a smooth curve—the jags into the dark part of the disk could well be clouds at higher elevations.

A famous atmospheric feature is the "ashen light," an apparent brightening of the unlit portion of Venus occurring a month or two before and after inferior conjunction. Most observers say the light has a warm color to it. The ashen light has not been verified by spacecraft measurements, but scientists think it could be a kind of airglow, similar to auroras on Earth.

To see cloud effects, consider using filters. The best gelatin filters to use are blue or violet ones (Kodak's Wratten 38A or 47). Step up to the pricier glass ones if the results seem worthwhile.

PENCIL AND CAMERA

The toughest part of sketching Venus is to avoid exaggerating its features, which are all of low contrast. Prepare sketch blanks ahead of time, using a standard scale, such as 1 mm per arcsecond, and checking the size of Venus in an almanac, magazine, or software.

It is fairly easy to take photos of a crescent disk, but the cloud features of Venus call for advanced techniques (see p. 66). For best results, photograph in UV light by using a UV *transmitting* filter, such as a Wratten 18A. A reflector telescope with a long focal length will give a suitably large image, but avoid Schmidt-Cassegrains, which have glass corrector plates that absorb UV light. Use eyepiece projection to enlarge the image, but make sure the lens elements are made of quartz or fluorite.

VENUS FACTS

Distance from Sun	0.72 AU
Mass	0.81 × Earth's mass
Radius	0.95 × Earth's radius
Apparent size	10 to 64 arcsecs
Apparent magnitude	–4.0 to –4.6
Rotation (relative to stars)	243 Earth days, retrograde
Orbit	225 Earth days

THE LAVA PLANET

To peel away the clouds and map Venus's geological features, scientists used an imaging radar on a spacecraft named Magellan. Building on work by US and Soviet probes and Earth-based radar, Magellan inventoried a museum of volcanic features, completing its global survey in 1994.

Sixty percent of Venus's surface is a lava plain. The rest falls mainly into two continents: an equatorial one named Aphrodite Terra, and a northern one called Ishtar Terra. These stand a few miles higher than the plains. The remainder of the surface consists of individual mountains and broad-based volcanic peaks.

On the plain, Magellan revealed long channels that mark past lava flows, and found fissures and vents from which vast sheets of lava once poured, often in repeated eruptions. Scientists identified pancake-like lava domes, and found strange oval features called coronae that have no counterpart on Mercury, Earth, or Mars. The coronae may have formed when a rising blob of molten rock pushed up the surface, cooled, and subsided, leaving a distorted ring around a central depression.

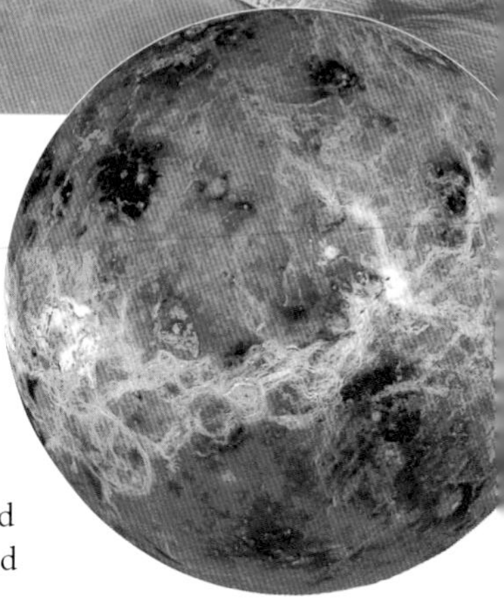

MAGELLAN DATA *was used to create a global mosaic of Venus (right), and a computerized image of the intensely deformed highland of Ovda Regio (above).*

CONTINENTAL DIVIDE

The continents preserve some of Venus's oldest terrain—the tesserae. These regions, which include Ovda Regio (shown above), have undergone a phenomenal amount of folding and faulting. They look like islands of an older surface that escaped the volcanic flooding that covered most of Venus.

Venus may have had a plate-tectonics cycle. Portions of both Aphrodite and Ishtar resemble the regions on Earth where plate tectonics have been active, building new crust and destroying the old.

Maxwell Montes, a rugged range in Ishtar, contains the highest point on Venus and stands more than 7 miles (11 km) above the global lava plain. It is much too massive to be supported as a single block of crust. Only plumes of molten rock rising from the mantle could sustain it.

TRANSITS OF VENUS

Every 100 years or so, observers on Earth can watch Venus at inferior conjunction pass across the face of the Sun. These rare events are called transits of Venus, and they occur in a pair, 8 years apart. The last two transits were in 1874 and 1882; the next two take place on 8 June 2004 and 6 June 2012. In 2004, Venus crosses the southern part of the Sun, and in 2012, the northern part.

In earlier times, transits of Venus let astronomers measure the distance from Earth to Venus, and, by extension, the scale of the Solar System. After the telescope was invented, a handful of astronomers made individual efforts for the 1639 transit. But for the 1761 and 1769 events, the British and French sent out expeditions all over the globe, among them the famous exploring voyage of James Cook, which went to Tahiti for the 1769 transit.

Better distances for the Solar System did emerge from these efforts, but the most notable finding was about Venus itself. Observing the 1761 transit, the Russian scientist Mikhail Lomonosov discovered that Venus has an atmosphere. He noted the halo it produced around the black dot of the planet as it slipped onto the solar disk and off again.

The portable observatory used by James Cook.

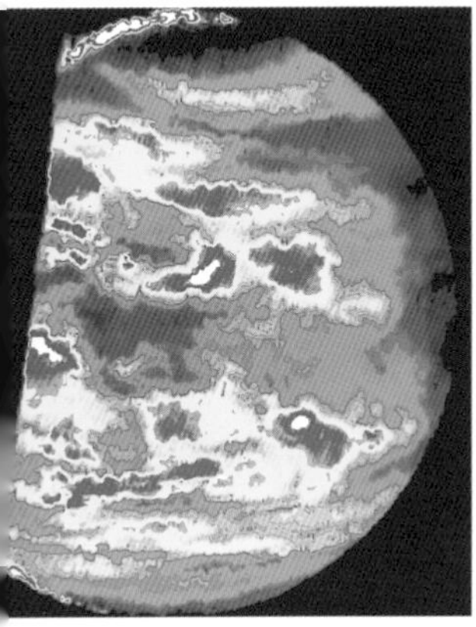

IAVING AN IMPACT

ike every planet, Venus has een struck repeatedly by neteorites. The craters un-overed in Magellan's images how a spectrum of types, nany familiar from other noons and planets. The largest s a flooded basin called Mead, which is 175 miles (280 km) vide. Most craters show signs f post-impact lava floods, lmost as if they tapped a ource of molten material elow the surface.

Venus's atmosphere is ense enough to filter out maller meteorites, and all its raters appear to be bigger han 3 miles (5 km) in diam-ter. Magellan did, however, nd dark splotches where the urface seems to have been ulverized. The culprit may ave been a shock wave—an erial hammerblow from a neteorite that broke up before triking the ground.

EPAVE THE PLANET

bout 600 million years ago, omething reset the geo-clock n Venus's surface, and obody knows why or how. mpact craters are by far the nost common landform in the olar System, but Venus has ss than a thousand of them. iven the rate at which new npacts occur, this means that e planet has somehow erased l traces of the first 85 percent f its history. There are various eories to explain this.

THIS INFRARED IMAGE *of Venus's lower cloud deck shows areas of thinnest clouds as white and red. High-speed winds cause the streaky appearance.*

Venus may have a thin, soft crust that cannot preserve ancient terrain. Or episodes of plate tectonics might have erased old features, but these periods alternate with times (like now) when it is quiescent. Or maybe Venus has spasms of volcanism. Or perhaps we are seeing Venus soon after its geo-engine shut down. Or some combination of the above. The fact is, scientists do not yet know.

Some scientists think that Venus is still volcanically active. The amount of sulfur dioxide—a common volcanic gas—in the atmosphere has changed over the last 15 years, perhaps indicating a recent eruption followed by a settling-out of the gas. Also, the tallest mountains are too high to be held up by Venus's crust alone, so either they are recently built or they are supported by hot rock still rising from below.

VOLCANIC VENUS *The giant volcano Maat Mons in a computer-generated view (below), and a lava dome, with the informal name of "Tick Mons," seen from above (right).*

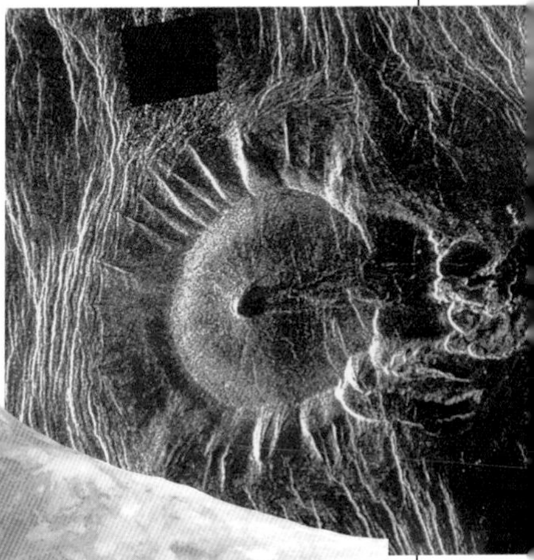

GREENHOUSE EFFECT

Spacecraft have also studied conditions above the ground. The atmosphere of Venus is 90 times heavier than Earth's. And with temperatures of about 890 degrees Fahrenheit (480° C), Venus has the Solar System's hottest surface. Because Venus is closer to the Sun than Earth is, the sunlight it receives is twice as strong. After filtering through the clouds, the light is colored orange and is about as bright as on an overcast day on Earth.

When sunlight warms the surface rocks, they try to radiate the heat back into space at infrared wavelengths. But the heavy carbon-dioxide atmosphere is largely opaque to infrared light and acts like a thick blanket to keep the heat in. As a result, it is as hot at the poles of Venus as at the equator, and the night side is no cooler than the day side.

EARTH

Compared to the other planets in our Solar System, Earth is astonishingly active. Its surface is continually recycled and its atmosphere sustains a stunning variety of life.

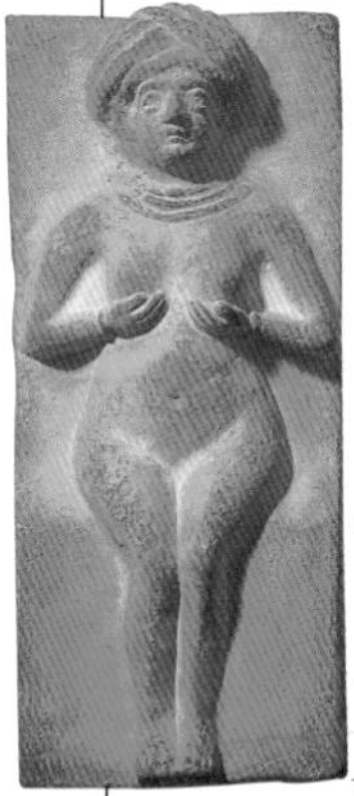

A FEMALE FIGURE *has often symbolized Earth, as in this 2000 BC bas-relief of the Sumerian earth goddess.*

Earth is the largest of the four rocky, or terrestrial, planets. While radically different from one another, Mercury, Venus, Earth, and Mars are easily distinguished from the gas-giant planets—Jupiter, Saturn, Uranus, and Neptune—by their size, density, and chemical composition.

Terrestrial planets are composed mainly of dense, rocky materials, such as basalt, and metals, such as iron. Gas-giant planets are more like the Sun in composition: they have enormous atmospheres, mainly of hydrogen and helium, wrapped around rocky centers. The terrestrial planets have far greater densities than the gas planets, but are low in mass—Earth has only one-fifteenth the mass of the smallest gas giant, Uranus.

EARTH'S BIRTH

Some 4.6 billion years ago, the Solar System formed out of a cold cloud of interstellar gas and dust. The densest part of the cloud coalesced into the proto-Sun, while the remainder flattened into a far-flung disk called the solar nebula. Particles in the nebula collided and stuck, gradually accreting to form the planets.

Earth's characteristic properties were instilled at birth. It formed hot, and was heated further by the decay of radioactive elements in its interior. The heat turned Earth molten and caused the heaviest materials, such as nickel and iron, to sink to the center, while lighter elements, such as silicon, rose to the surface. Even the impact of the small planet that probably led to the birth of the Moon (see p. 92) did not disrupt this arrangement for very long.

PEELING THE LAYERS

Although Earth's interior is nearby in astronomical terms—its center lies less than 4,000 miles (6,400 km) below our feet—it remains frustratingly out of reach. To look inside our planet, scientists rely on earthquakes, studying how seismic waves rebound and refract as they travel through the interior. In simple terms, the picture they give of Earth resembles an onion made of rock.

At the center is the core, which extends upward to a point 1,800 miles (2,900 km) below the surface. The core is about 90 percent nickel and iron, and has two parts. The inner core is more than hot enough to be liquid—about

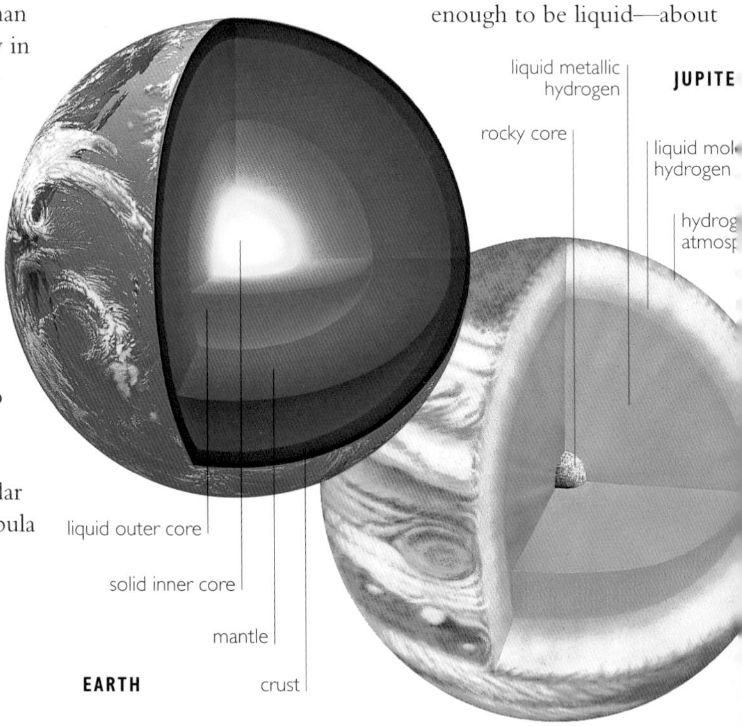

liquid outer core
solid inner core
mantle
crust
EARTH
liquid metallic hydrogen
rocky core
JUPITE
liquid mol
hydrogen
hydrog
atmos

CROSS SECTIONS *of Earth and Jupiter (not to scale) show the contrasting structures of terrestrial and gas-giant planets.*

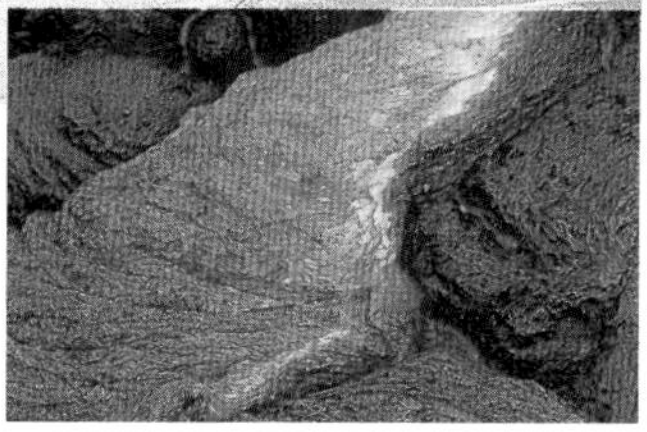

IGNEOUS ROCKS *are forged by heat from the Earth's interior. The hot, molten rock (right) is often forced to the surface, resulting in volcanic eruptions (above).*

9,000 degrees Fahrenheit (5,000° C)—but it is kept solid by the enormous pressure, which is nearly four million times that at the surface.

In the outer core, the pressure is lower, temperatures are about 900 degrees Fahrenheit (500° C) cooler, and the iron remains liquid. This liquid nickel-iron flows in turbulent currents that give rise to the Earth's magnetic field. The field reaches up through the planet, past the surface where it controls our magnetic compasses, and out into space where it affects the flow of charged solar particles blowing past Earth (see p. 79).

Above the core is a layer called the mantle, composed of high-density iron and magnesium-silicate minerals. While not liquid like the outer core, the rocks of the mantle are hot enough to flow, and the energy that they carry powers all the tectonic activity we see at the surface.

The rocks that make up Earth's surface form a thin but distinct layer called the crust. We live on the continental type of crust. It is made of lightweight granitic and carbonate rocks and is between 20 and 40 miles (30 and 65 km) thick. A second kind of crust, oceanic, is made of heavier and denser basalt. It is only 3 miles (5 km) thick.

Most oceanic crust is covered by miles of sea water, so, until recently, geologists mainly studied the continental crust. However, there are a few places, such as Iceland, where oceanic crust is exposed.

Types of Rock

Crustal rocks and minerals appear in a bewildering array of types, since chemical elements can combine in myriad ways. But in broad terms, geologists recognize three kinds of rock.

Rocks that form in a molten state are called igneous. These include basalt, which erupts as molten lava, and granite, which forms deep within the Earth and is exposed by erosion or geological activity.

Sedimentary rocks are formed by the slow accumulation of layers of particles, such as sand, mud, or organic debris. Shale, sandstone, and limestone are common sedimentary rocks.

Metamorphic rocks are those that have been altered by heat, pressure, or both. Metamorphism makes a rock harder and denser, but traces of the rock's original nature are usually preserved. The process can turn shale into slate, sandstone into quartzite, limestone into marble, and so on.

Earth is a geologically active planet, and these three types of rock are involved in a continual process of recycling, driven by weathering, heat, and pressure. Although Earth formed some 4.6 billion years ago, the oldest known rocks are about 3.9 billion years old, and these are rare exceptions; most of the surface is only about 100 million years old. This explains why we do not find the number of craters here that we do on the Moon or Mars. While Earth experienced just as many impacts, most craters have been eroded by the weather or erased by geological activity.

SEDIMENTARY ROCKS, *such as sandstone (below left and right), often form in distinct layers and may be eroded into dramatic shapes.*

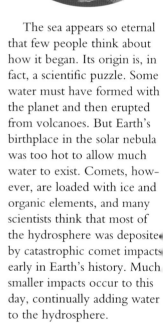

LAND *covers only 30 percent of Earth's surface (below). The collision of tectonic plates created landforms such as the Himalayas seen here from space (left).*

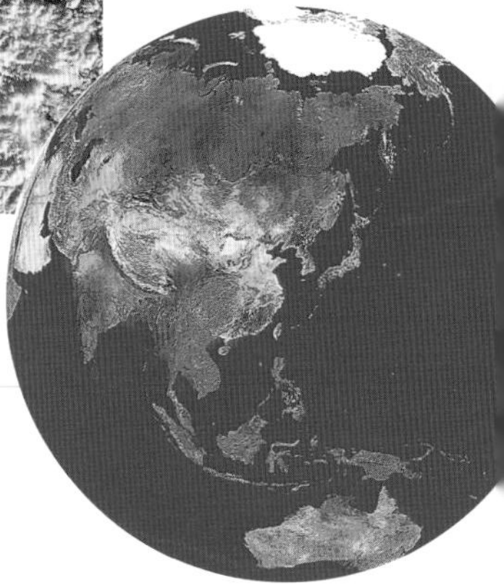

Continents Adrift

Earth differs from other planets in that its crust is made up of slabs of thin rock called plates. Driven by currents in the upper mantle, the plates, which number more than a dozen, carry the continents on their backs like rafts.

Meteorologist Alfred Wegener argued for continental drift (as he called it) in the early 1900s. However, the idea was not generally accepted until the 1960s, when geophysicists found that new oceanic crust was constantly being formed in mid-ocean ridges on the seabed. The theory of how new crust moves away from the mid-ocean ridges and disappears at the edges of continents is called plate tectonics, and it has revolutionized geology.

As the edges of crustal plates collide, they rift, overlap, and fold, recycling their rocks and remaking the surface of the planet. The Himalaya Mountains, for example, mark where the Indian plate is being forced under the edge of the Asian plate. The Rift Valley of East Africa, on the other hand, marks where a continent is rifting apart, to eventually form a new ocean.

A Watery Planet

Conditions on Earth's surface depend on the interaction of two spheres: the atmosphere and the hydrosphere.

The hydrosphere is the water on, or near, the surface. This water is found in oceans, rivers, and lakes; it is found below the surface as groundwater; it is locked up in frozen polar caps and glaciers; and it exists as vapor carried in the atmosphere. Earth is the only planet where temperatures allow surface water to exist in solid, liquid, and gaseous states.

Oceans make up 97 percent of the hydrosphere and contain enough water to cover the planet 2 miles (3 km) deep. The only reason we are not underwater is that the continental crust (some 30 percent of Earth's surface) has an average elevation high enough to keep it above sea level.

The sea appears so eternal that few people think about how it began. Its origin is, in fact, a scientific puzzle. Some water must have formed with the planet and then erupted from volcanoes. But Earth's birthplace in the solar nebula was too hot to allow much water to exist. Comets, however, are loaded with ice and organic elements, and many scientists think that most of the hydrosphere was deposited by catastrophic comet impacts early in Earth's history. Much smaller impacts occur to this day, continually adding water to the hydrosphere.

An Ocean of Air

Above the Earth's surface lies the atmosphere, an ocean of air that sustains life, drives our weather, and aids the breakdown of rocks into their constituent minerals.

The composition of the atmosphere is constantly evolving. If you went back three billion years in time,

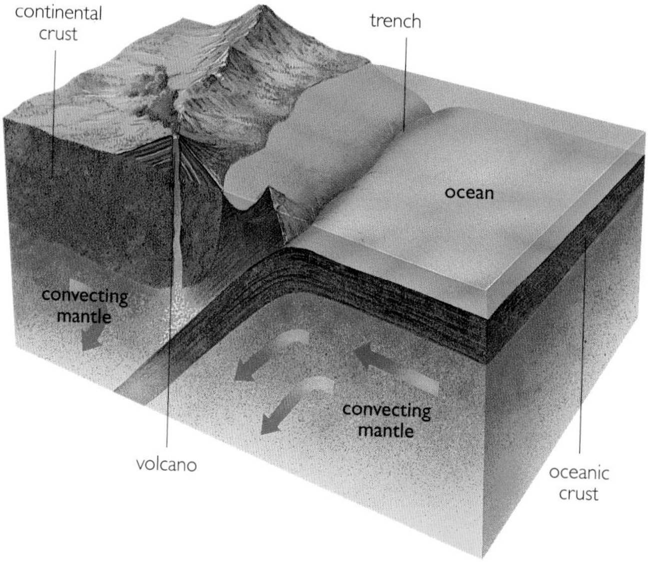
continental crust
trench
ocean
convecting mantle
convecting mantle
volcano
oceanic crust

A DYNAMIC EARTH *When an oceanic plate collides with a continental plate, the dense ocean crust is drawn under and melted in a process known as subduction.*

ARTH'S ATMOSPHERE *creates ramatic weather, such as this hurricane iewed from space (right), and forms a ocoon in which life thrives, as in this eeming tropical rain forest (below right).*

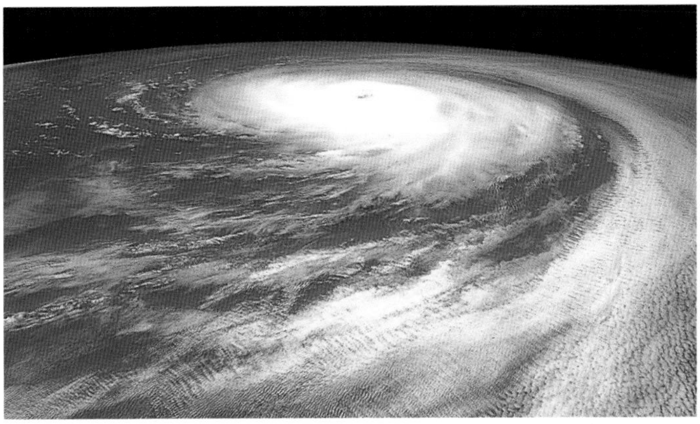

ou would be unable to reathe. At that time, the tmosphere contained almost o oxygen—it was mostly arbon dioxide and nitrogen. Today, Earth's air is mostly oxygen and nitrogen.

Oxygen began to accu-nulate in the atmosphere only vhen primitive life forms (first acteria and then plants) devel-oped photosynthesis, excreting oxygen as a waste product of he process. Not only does oxygen let us breathe, it also eacts with sunlight to form ozone, which protects us from armful ultraviolet rays.

Carbon dioxide acts as a "greenhouse gas," allowing unlight to reach the surface, out preventing heat from escaping. In this way, it helps warm our atmosphere, just as an abundance of it turned Venus into an inferno. Today, the burning of fossil fuels is releasing carbon dioxide into the air, subtly changing the constitution of the atmosphere and contributing to a global increase in temperatures.

EARTH'S WEATHER

The interaction between the Sun's heat and Earth's atmos-phere creates our weather. Other vital ingredients are the 23.5 degree tilt of Earth's axis, and the combination of land and water on Earth's surface. Because of its tilt, the Earth is heated unevenly, and this is accentuated by the different heat-absorbing characteristics of land and water. In its quest for equilibrium, the atmosphere directs warm air to cold places, and cold air to warm, creating our basic system of weather patterns.

Today, Earth has a benign, stable climate with a narrow range of temperatures. In the past, however, there have been periods of rapid and dramatic climate change. Small variations in Earth's orbit around the Sun or in the tilt of its axis could be enough to drastically alter our climate, creating condi-tions far less conducive to life than those that currently exist.

THE ORIGIN OF LIFE ON EARTH

No one has yet figured out exactly how Earth passed from a state of having all the chemical building blocks necessary for life to a state where life actually existed. But the change, while enormously important, may not have been all that much of a leap.

The oldest rocks in the world are about 3.9 billion years old, and ones showing fossil traces of life are barely younger. This suggests that either life conveniently arose when these rocks were formed (rather unlikely), or it arose even earlier. In any case, the date is important because at that time the Solar System was still a pretty hostile place. Large, devastating impacts by asteroids and comets were frequent, and Earth's surface looked more like the Moon's southern highlands than like any landscape on Earth today.

From what we know about today's Earth, life is outstandingly tenacious, adaptable, and resourceful. Given this tenacity, it is possible that life actually started on Earth several times, but was wiped out by impacts on each attempt but the last.

What does this say for the chances of finding life else-where, either in the Solar System or outside it? Actually, the chances look fairly good. Astrophysics shows that chemistry is chemistry the universe over, so nature ought to be able to duplicate in another place the pathway that led to life on Earth. But whether life can persist elsewhere, and for how long, we can only guess.

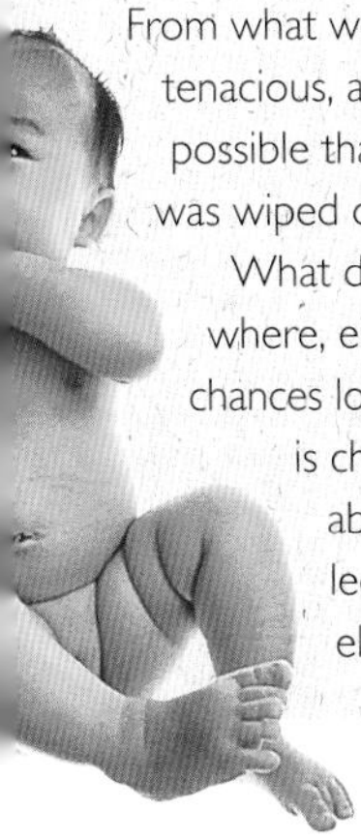

EARTH FACTS

Distance from Sun 92,752,000 miles (149,600,000 km) = 1 AU
Mass 6×10^{21} tons
Radius 3,954 miles (6,378 km)
Rotation (relative to stars) 23 hours 56 minutes
Orbit 365.24 days

MARS

The Red Planet is Earth's closest cousin, and a recent discovery has raised the tantalizing possibility that life once existed on Mars.

Mars, with its ocher color, relatively swift motion among the stars, and changeable brightness, has long been associated with war, disaster, death, and pestilence. To many ancient people, who feared that changes in the sky portended changes for the worse on Earth, such an association was natural. Our name for the planet is that of the Roman war god, who took on the same attributes as his Greek forerunner, Ares.

THE MARTIAN CYCLE

Depending on when you look at it, Mars's apparent size can vary greatly—from 4 to 25 arcseconds. A typical apparition begins with Mars low in the east in the glow of morning twilight. As weeks pass, the planet grows. It moves eastward against the stars and does not rise out of the dawn quite as quickly as they do. Earth, orbiting faster than Mars, begins to overtake it, and Mars's apparent eastward motion slows and stops. Mars then spends about 11 weeks moving westward against the background stars before halting once more and resuming eastward motion. In the middle of the westward (or retrograde) part, Mars reaches opposition and best visibility around midnight, and is at its maximum apparent size for the apparition.

Finally, over the following months, Mars slips into the early evening sky and eventually disappears in the twilight glow. After remaining hidden in the Sun's glare for several weeks while it passes conjunction, Mars reemerges at dawn to begin its next apparition.

Mars comes to opposition about every 780 days, or 2 years 7 weeks (see Table, p. 120). Because it has an eccentric orbit, Mars's distance from the Earth and the Sun at opposition varies, so its apparent size and brightness also change. The most favorable opposition is when Mars lies closest to both the Earth and the Sun. Such oppositions are called perihelic, and they occu when opposition is in July, August, and early September—about once every 17 years. At these times, Mars is just 35 mil lion miles (56 million km) fror Earth, with an apparent size of about 25 arcseconds and a magnitude of −2.6. At aphelic oppositions (in January, February, and early March), Mars is 63 million miles (101 million km) from Earth. Its disk i only about 14 arcseconds wid and it shines at magnitude −1.(more than four times dimmer

OBSERVING MARS

On first viewing Mars throug a telescope, beginners are ofte struck by how small it appear The planet is only half the siz of Earth and, even at favorabl oppositions, it appears no larger than a lunar crater. Firs views typically show an ocher disk, a few faint markings, an perhaps a whitish polar cap.

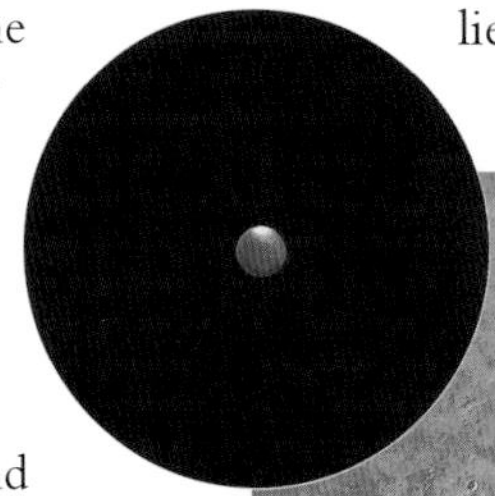

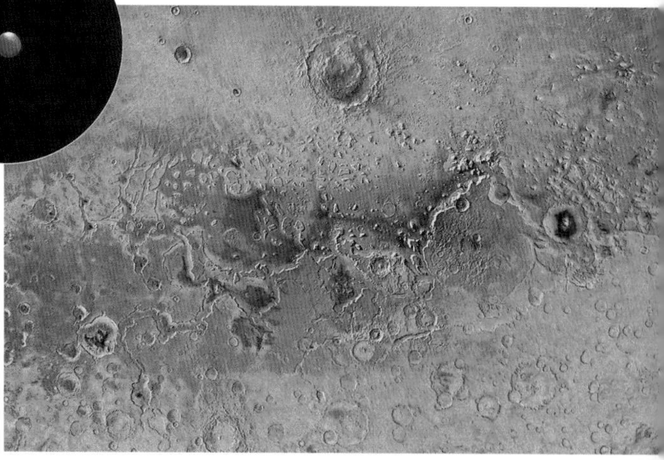

MARS, GOD OF WAR *(top), in a Diego Velázquez (1599–1660) painting. A telescope view of the real Mars (above) reveals only a small disk, but Viking images show a cratered surface (right).*

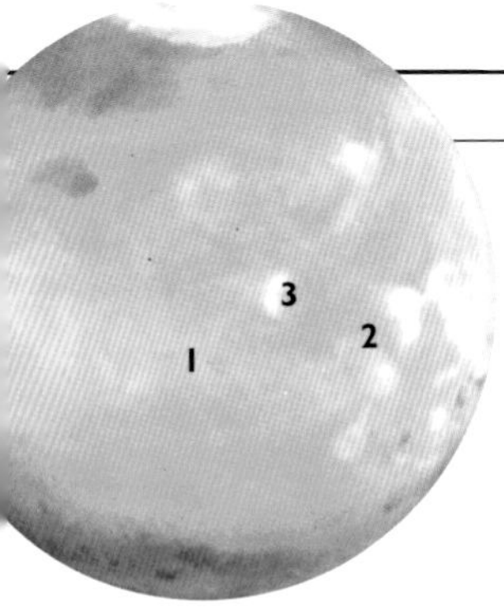
3
2
1

IREE HUBBLE VIEWS *of Mars in 95. At 160 degrees longitude, features lude: (1) the Amazonis region; (2) the arsis region; and (3) Olympus Mons.*

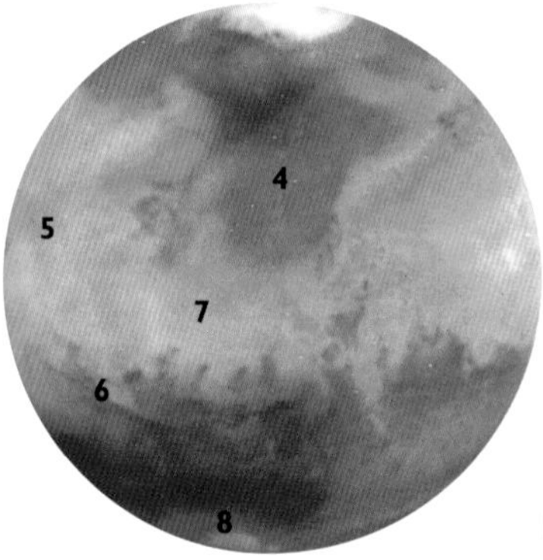
4
5
7
6
8

AT 60 DEGREES LONGITUDE, *the main features are: (4) Mare Acidalium; (5) Ascraeus Mons; (6) Valles Marineris; (7) Chryse basin; and (8) Argyre basin.*

AT 270 DEGREES LONGITUDE, *look for: (9) Syrtis Major; (10) the Hellas basin; and (11) the Elysium basin.*

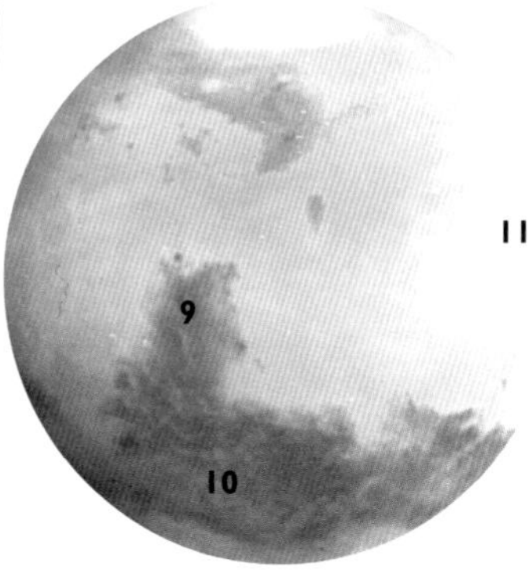
11
9
10

he most powerful telescopes 1 Earth give views of Mars ıly about as detailed as ıked-eye views of the Moon. his helps explain why Martian udies progressed little before e Space Age.

Early observers of Mars saw ırk markings and thought ey were either old seabeds areas of vegetation. Today, ese are known to be lava ɔws and boulder fields.

In 1877, the Italian astrono- er Giovanni Schiaparelli ported seeing a network of raight lines he called *canali*, eaning channels. This was idely translated as canals, ggesting these features were tificial and fueling popular eculations about life on Mars. s it turned out, the canals ere merely optical illusions.

In 1964, the first close-up ews of Mars were taken by ASA's Mariner 4 spacecraft. he features in these photos at most shocked scientists ıd the public alike were the aters—very few people had pected to see such Moon- e features. The fact that a loon-like Mars was such a ock reveals the depth of ost people's assumptions that lars was just a "little Earth."

ETTING EQUIPPED

nall telescopes can show you mething of Mars, but to see tails you will need at least a inch (125 mm) refractor or 8 inch (200 mm) reflector, and long focal ratios (f/8 to f/10 or more) are preferable. Good seeing is also essential.

With Mars, unlike most planets, colored filters will really help your observations. High-quality glass filters that screw into your eyepiece are available from telescope-accessory dealers. Many observers, however, prefer to start out with less-expensive gelatin filters, such as Kodak's Wratten filters, which can be bought from camera stores.

A useful set would include a blue (Wratten 44a) or blue-violet (W47); a green (W58) or a yellow (W12 or 15); an orange (W21 or 23a); and a red (W25). To use these filters, simply hold them up to the eyepiece and look for features that appear enhanced.

THE WAYWARD RED PLANET

The Red Planet and its motions challenged every astronomer from antiquity onward. By the late 1500s, the Danish astronomer Tycho Brahe (1546–1601) had amassed decades of observations of Mars and the other planets. His observations were by far the most accurate to date, yet he failed to weld them into a coherent system of the universe. Then Tycho hired a young Austrian mathematician-astronomer named Johannes Kepler (1571–1630) and put him to work on the orbit of Mars. When Tycho died shortly afterward, Kepler inherited his position and, more importantly, his observations.

Kepler (right) battled with Mars for much of a decade, finally emerging with the first two of his three laws of planetary motion. These include the conclusion that planets orbit in ellipses, with the Sun at one focus. (Every prior astronomer had insisted on combinations of circles.)

Kepler's work paved the way for Isaac Newton's *Principia* (1687), the cornerstone of all modern physical science.

A Variable Planet

With the Moon, a look through any telescope tends to instantly impress. Mars, on the other hand, is a planet with delicate, elusive features, and the experienced eye will see much more than the beginner's. Mars has a great many features, but you will not catch them all on your first night's viewing. Experience is crucial, and the key to getting the most out of your observations of Mars is to view the planet as many times as you can.

Backyard observers who are patient and continue to watch Mars over many hours will see darkish markings move across the disk, gradually slipping out of sight as the Martian day passes by. Over the months leading up to opposition, you can see Mars's features at ever-better resolution as its globe grows and your eye becomes more practiced.

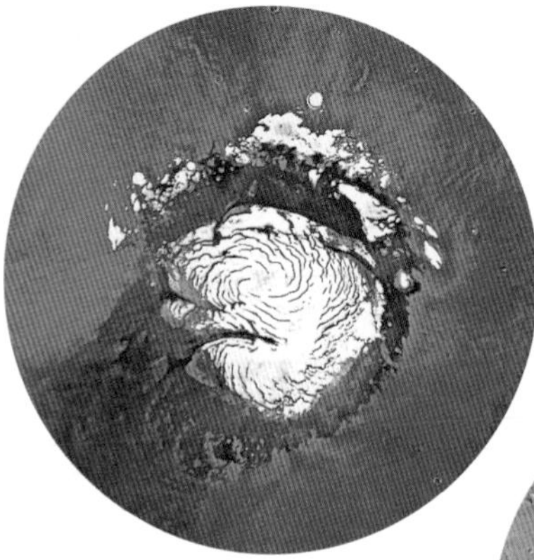

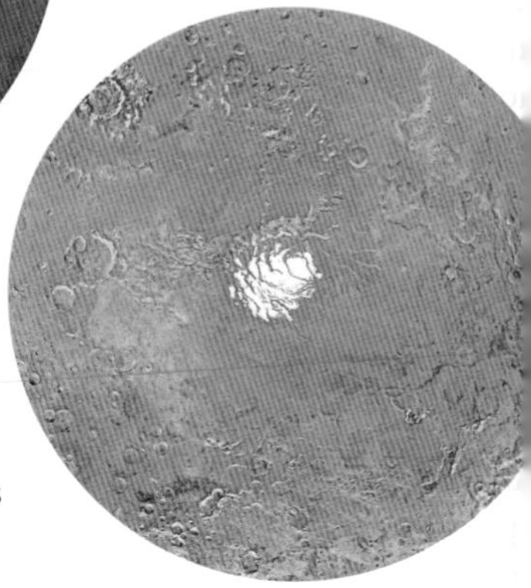

THE ICE CAPS *of the northern (left) and southern (below) poles. While only a tiny vestige of the southern cap remains in summer, more of the northern cap survives the cooler northern summer.*

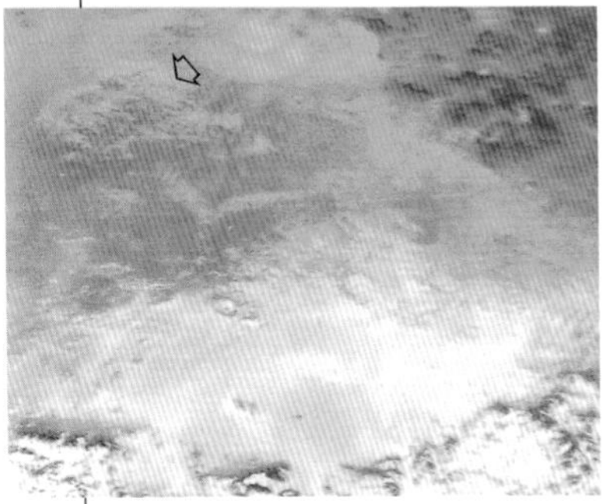

LOCAL DUST STORM *In this Viking 2 picture of Mars's surface, an arrow indicates a bright, turbulent dust cloud 190 miles (300 km) across.*

Martian Seasons

If you observe Mars from one opposition to another, you can see the Martian weather and seasons change. Mars has a tilt similar to Earth's and experiences four seasons, each lasting about twice as long as ours because Mars's orbit is that much larger.

Martian seasons do not correspond exactly with Earth's, because Mars's polar axis points in the direction of the star Deneb, whereas Earth's points to Polaris. A handy rule of thumb to remember is that the season on Mars at each opposition is one season in advance of Earth's at that time.

For Mars's northern hemisphere, spring and summer generally feature a clear atmosphere with little dust in the air. Whitish clouds, however, can appear near the sunrise line and over high elevations. Frost deposits turn the large impact basins of Argyre, Hellas, and Elysium into bright patches. You will find that filters excel at revealing details such as these, especially when alternated with unfiltered views.

The northern polar cap vanishes under a hood of clou during its fall and winter. The plummeting temperatures renew the ice cap by condensing carbon dioxide from the atmosphere. The carbon-dioxide ice comes and goes with the seasons, but a permanent water-ice cap in the north remains intact throughout the summer.

UPCOMING OPPOSITIONS OF MARS

Date	Apparent size in arcseconds	Distance (x 1,000,000)	
		in miles	in km
24 April 1999	16.2	54.2	87.4
13 June 2001	20.5	42.4	68.2
28 August 2003	25.2	34.7	55.8
7 November 2005	19.8	43.7	70.3
24 December 2007	15.8	55.8	88.7
29 January 2010	14.0	61.7	99.3
3 March 2012	14.0	62.7	100.8
8 April 2014	15.1	57.7	92.9
22 May 2016	18.4	47.3	76.1

If you are bored with these weary calculations, take pity on me who had to go through at least 70 repetitions of them.

Johannes Kepler (1571–1630), Austrian mathematician-astronomer, writing about his observations of Mars

Because the southern-
misphere summer occurs
perihelion—when Mars
s closest to the Sun—
nlight is 44 percent
onger at this time. This
akes southern summers
otter, and winters colder,
an northern ones.
The southern polar cap
mostly carbon-dioxide ice.
uring summer, it shrinks
a tiny remnant (with a
re of water ice), but does
ot vanish completely.
bout every 17 years, when
position coincides with
rihelion, the southern polar
p is tilted toward Earth,
owing clearly as it shrinks
nder the Sun's strong rays.

UST STORMS

ast lava flows and boulder
lds, such as Syrtis Major,
ually appear as dark mark-
gs on the surface. However,
eir visibility varies because
inds carry dust across them,
oscuring them in whole or in
rt. Blowing dust can also hide smaller markings or make them difficult to identify.

Sometimes you will see no markings at all for days or weeks on end, thanks to one of the global dust storms to which Mars is prone. These storms occur most often during the southern hemisphere's spring and summer. The last global dust storm was in 1982, but localized ones occurred in 1988, 1990, and 1992, and observers should always be alert to their onset.

CAPTURING MARS

Sketching Mars is relatively easy. Many observers use standard-size blank forms—for example, 2 inches (50 mm) to the Martian diameter—but these are too large for times when the Martian disk is small. They will tempt you to put down details you cannot really see. A better course is to use an image scale of 2 mm per arcsecond. Circles work well for much of the time, but three to four months before and after opposition, Mars's disk looks distinctly gibbous—much like the Moon three or four days before Full—and you should modify your sketch blank to reflect this shape.

Photographing Mars, on the other hand, is a lot more difficult. The small planet demands a long focal length, and this means you need exposures of several seconds. "Windows" of good seeing seldom last this long, so capturing a sharp image on film depends mostly on luck. Many Mars photographers now use CCD cameras. These are just as subject to seeing requirements, but they let you quickly take many short exposures, at least some of which will catch Mars during moments of crisp seeing.

MARS FACTS

Distance from Sun 1.52 AU
Mass 0.11 × Earth's mass
Radius 0.53 × Earth's radius
Apparent size 4 to 25 arcsecs
Apparent magnitude +1.8 to –2.6
Rotation (relative to stars) 24.6 hours
Orbit 687 Earth days

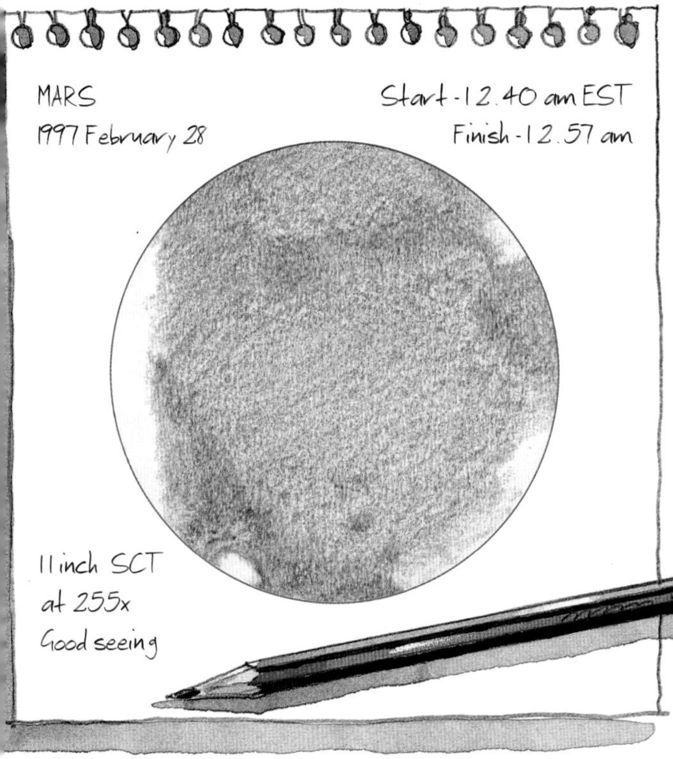
MARS
1997 February 28
Start - 12.40 am EST
Finish - 12.57 am
11 inch SCT
at 255x
Good seeing

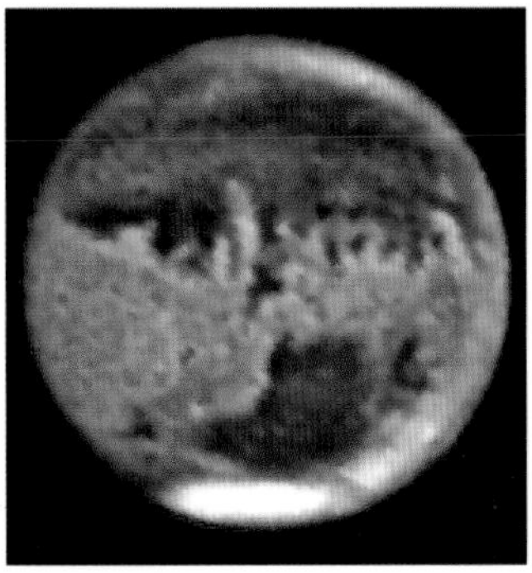

RECORDING MARS *Observers wanting to keep a permanent record of Mars can try sketching the planet (left). Or they might photograph it—CCD cameras can produce images such as this (above).*

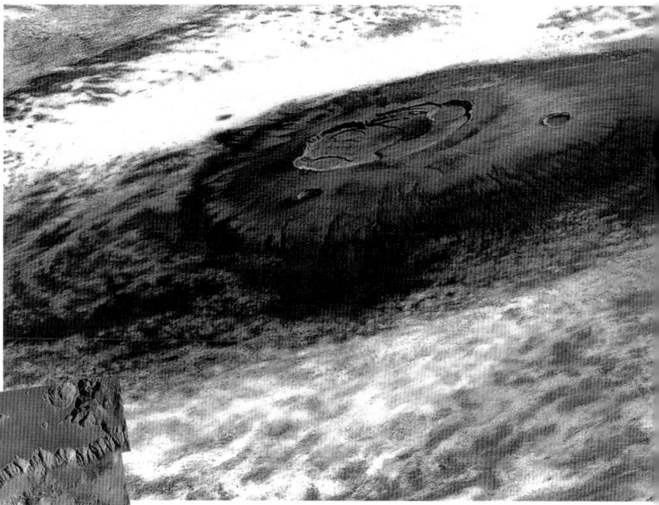

SURFACE FEATURES *of Mars include Candor Chasma (left)—part of the giant Valles Marineris—and Olympus Mons (above).*

Red-rock Geology

Of all the Sun's other planets, Mars is the most Earth-like. Its polar axis tilts 25 degrees to its orbit (Earth's tilts 23½ degrees), the Martian day lasts just 41 minutes longer than ours, and it has clouds, seasons, and polar ice caps. Despite these similarities, Mars is a frigid desert world, geologically inactive and hostile to life as we know it. Its atmosphere is too thin to allow liquid water at the planet's surface and offers no barrier to ultraviolet light from the Sun.

Mars has two distinct types of terrain, each occupying about half of the planet: in the south are the older highlands with many craters, while in the north, a relatively uncratered plain lies a few miles lower. How half of Mars could lie so much lower than the rest of the planet mystifies scientists. What causes the difference and where did all that crustal rock go?

The southern highlands have geological puzzles of their own. They contain the largest volcanic region on Mars—Tharsis—with its four immense volcanoes. Tharsis is about the size of North America and appears to have been built by episodes of crustal uplift followed by intense volcanism, which deposited lava on top of the uplift. Tharsis is topped by Olympus Mons, which is 16 miles (25 km) high, and 300 miles (500 km) wide.

From the number of craters on the highlands, scientists have calculated that this terrain is at least three billion years old. However, some portions, such as the slopes of Olympus Mons, appear quite young in geological terms. They have no craters and may be only a few million years old.

Radiating east from Thars is a huge crack called Valles Marineris—2½ miles (4 km) deep and long enough to reac across the United States from coast to coast. It seems to ha started as a tectonic fault, then as the fault tapped sources of groundwater, the walls collapsed and eroded, opening u the valley. Parts of it are now 60 miles (100 km) wide.

Other regions flanking Tharsis show evidence of actu rivers. The largest channels drain from the highlands across the crustal boundary and onto the northern lowlands in the region of Chryse, where the Viking 1 spacecraft landed. (This site was chosen in the hope of detecting life that might have fed on the water.) Scientists believe that these flows were catastrophic and that they occurred when internal heat or meteorite impacts released groundwater in sudden flood Although the volume of wate in these flows would have been enough to fill Earth's Amazon River 100 times, scientists think the flows were brief, because Mars probably did not have enough water tc sustain them continuously.

THE NORTHERN PLAINS *as record by the Viking 1 lander. The photo was taken at about noon, local Mars time.*

WETTER PAST

unning water, lakes, possibly ain, a warmer climate, and a hicker atmosphere—ancient Mars might well have been emarkably Earth-like. So what happened? Two things re the keys to understanding resent-day Mars.

First, being a small planet, Mars's geological engine seems have run down. If it has not opped entirely, it is nowhere ear as active as Earth's. It also ems that Mars has never had plate-tectonics cycle—as one ientist put it, Mars appears be a one-plate planet. A gid crust could explain why harsis is such a large region. n a planet with active plate ctonics, such as Earth, the rust above an erupting hot ot would keep moving and volcano would never grow big as Olympus Mons.

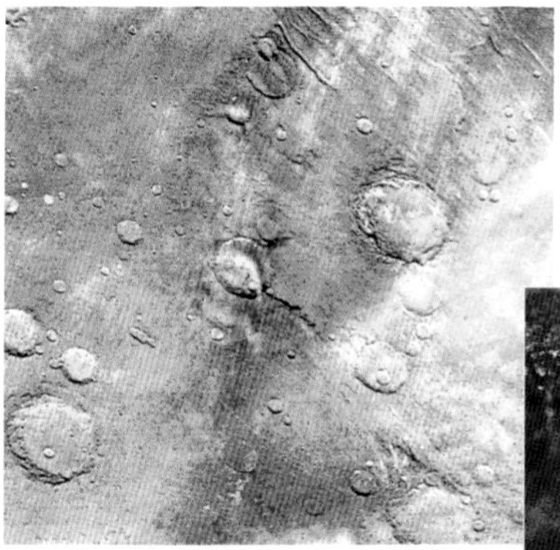

A THIN LAYER OF FROST *on the surface of Mars (above), composed of water and carbon-dioxide ice.*

MINERAL *globules (below) in a Martian meteorite may have been formed by primitive organisms.*

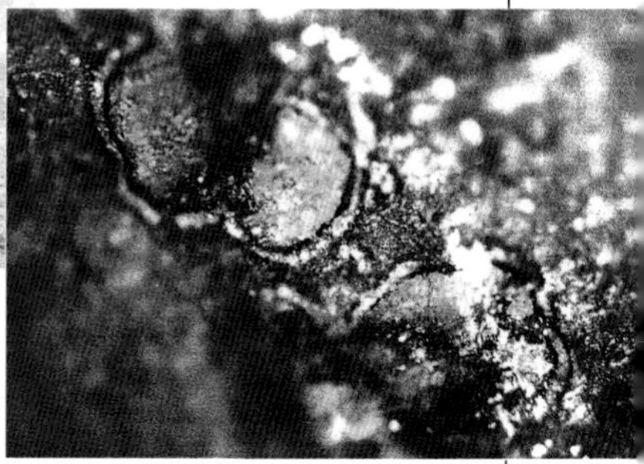

The second key to present-day Mars is its lack of a large satellite like our Moon. Mars has only two small moons, Phobos and Deimos, which are probably asteroids captured from the nearby main belt. Both are mere pebbles compared to Mars. Lacking the stabilizing influence of a large moon, Mars's axial tilt could change abruptly (in geological terms), throwing its climate from warm, wet conditions into a global ice age and back again, perhaps several times over in the course of a billion years.

LIFE ON MARS?

Possible similarities between Earth and ancient Mars beg the question: if life emerged here, why not on Mars?

Maybe it did. A recently discovered meteorite that came from Mars may contain traces of life. This meteorite, designated ALH84001, is believed to have been blasted off Mars by a comet or asteroid impact, after which it wandered through space before crashing to Earth. ALH84001 contains gas bubbles whose chemistry matches the Martian atmosphere that was sampled by the Viking landers.

What excited scientists was the presence of curious structures, thought to be the fossilized traces of microscopic, bacteria-like organisms. Whether or not these features are actually Martian fossils will probably not be clear until more samples are collected directly from Mars. But even if the features in this particular meteorite fail the test, the question of ancient life on Mars remains a tantalizing possibility.

WAR OF THE WORLDS

On the evening of 31 October 1938, radio producer Orson Welles broadcast a version of the H. G. Wells novel, *The War of the Worlds*. The novel tells of a Martian attack on London. It was published in 1897, at a time of enormous public interest in Mars, occasioned by recent close oppositions.

In preparing his radio play for an American audience 40 years later, Orson Welles had a stroke of genius and changed the locale and timing of the Martian attack. Instead of London, the Martians landed near Princeton, New Jersey. And instead of 1897, the show led listeners to believe that Martians were invading that very night.

The result was a broadcast production so realistic that many who heard it took it for an actual news flash from the scene of an extraterrestrial invasion. The radio show tapped directly into a mother lode of latent paranoia, brought to hypersensitivity by the gathering clouds of war over Europe. Panic ensued in some places, and America's pre-war jitters edged up a notch. Sixty years later, it is easy to smile at those who fell for the ploy. But would we respond any more skeptically today if some TV network staged a modern version of the hoax?

An illustration from H. G. Wells's science-fiction novel, The War of the Worlds.

JUPITER

Rightfully called the "king of worlds," the planet Jupiter is more massive than all the rest of the planetary system put together.

To the naked eye, Jupiter displays a brilliant white gleam that is unmistakable, especially in dark skies. In its slow course around the ecliptic, Jupiter paces out a long zodiacal "year"—it takes 12 years to orbit the Sun, spending about a year in each constellation of the zodiac. Modern skywatchers can often recall the constellation that Jupiter was passing through the first time they saw it. For many ancient cultures, the numerical coincidence seems to have indelibly marked this planet as the celestial symbol for the leader of the gods.

BABYLONIAN BELIEFS

Jupiter was the chief of the Roman pantheon, while to the ancient Greeks he was Zeus. More than a thousand years earlier, this planet was associated with Marduk, the most important figure in Mesopotamian cosmology and the patron god of the city-state of Babylon.

As the story goes, Marduk took on Tiamat, the goddess of chaos, and her 11 monsters. With scheming and titanic effort, Marduk defeated them one by one, and split Tiamat's body in two, thus dividing heaven from Earth. Marduk came to symbolize the rule of heavenly order over the universe. In this role, he placed the wandering star Jupiter specifically in charge of the night sky.

ZEUS *was the Greek version of Jupiter, ruler of the gods. He took the form of an eagle to carry Ganymede (the namesake of a Jovian moon) to Mount Olympus.*

BORDER PLANET

At Jupiter, the Solar System changes character. All the inner worlds are small, rocky, and in the case of asteroids, fragmentary. From Jupiter out to Neptune, the worlds are large with thick, gaseous atmospheres, and each commands a miniature solar system of moons and rings.

Jupiter is almost entirely made up of hydrogen and helium, so what you see is the cloudy top of a deep atmosphere. (See diagram of Jupiter, p. 114.) Traces of methane, ammonia, water, phosphine, and other hydrocarbons combine with solar ultraviolet light to produce the color and banding seen in a telescope.

The cloud tops lie at a level where the pressure is almost the same as at Earth's surface, while above them an atmosphere of hydrogen thins out into space. Descending below the clouds, the pressure increases until the hydrogen starts to behave like a liquid. About 12,000 miles (20,000 km) down, the hydrogen becomes a liquid metal. At Jupiter's center lies a rocky-iron core, perhaps 15 times the mass of Earth and having a temperature of 34,000 degrees Fahrenheit (19,000° C).

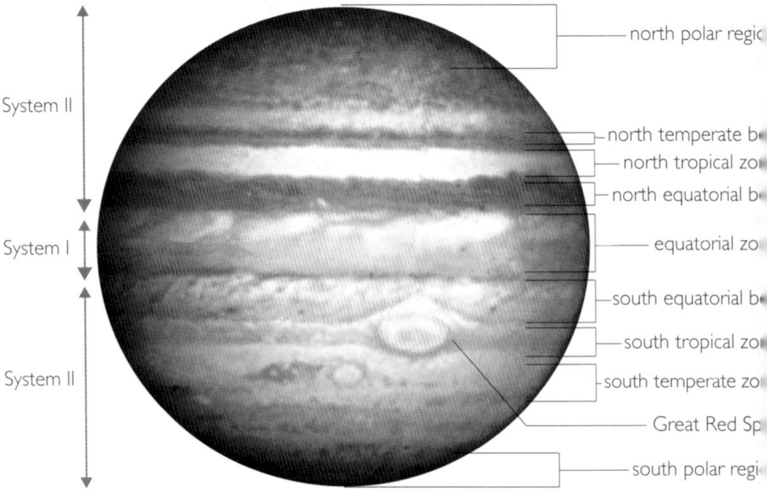
System II
System I
System II

THE BRIGHT ZONES *and dark belts of Jupiter divide into two rotation "systems."*

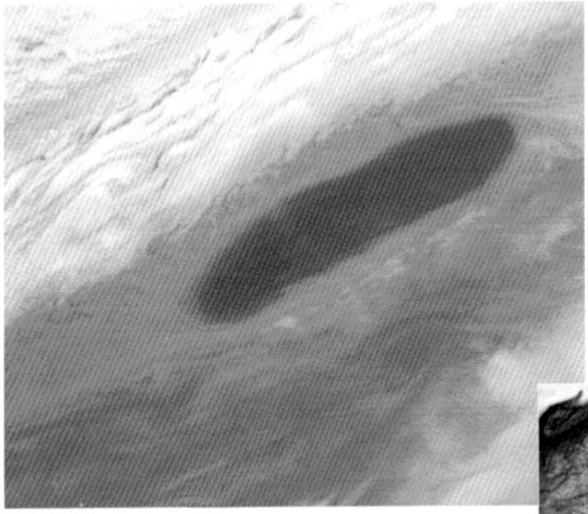

A TELESCOPE VIEW *of Jupiter (bottom) is impressive, but spacecraft images have revealed details in the Great Red Spot (below) and a brown oval (left).*

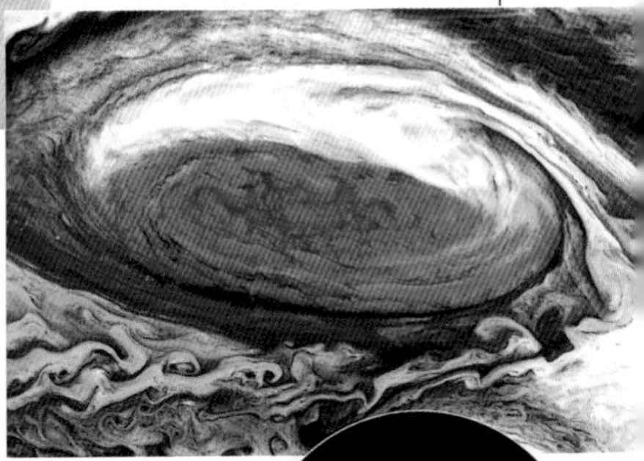

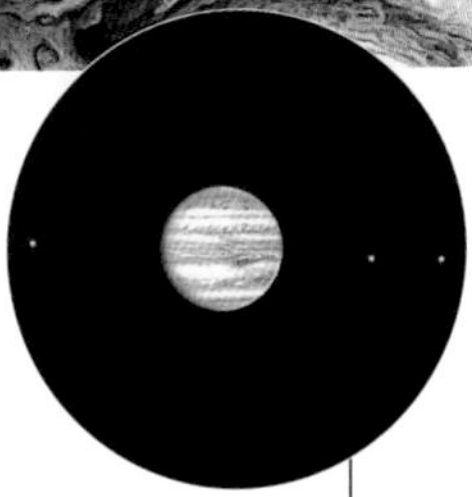

OUND AND ROUND

piter is still cooling off from formation, and the heat els phenomena ranging from e planet's immense magnetic ·ld to convection currents in clouds. The light-colored ›nes tend to be high, cool ouds in regions of updrafts, hile the darker belts mark armer areas of downdrafts. piter's rapid rotation—less an 10 hours—smears the oud features into east-west ·ipes paralleling the equator. his rapid rotation also causes piter's marked oval shape—e planet is 7 percent wider its equator than at its poles.

Lacking a solid surface, piter displays differential tation, like the Sun does. he period for System I (the juatorial zone) is 9 hours minutes, while System II he higher latitudes) rotates bit more slowly and has a ·riod of 9 hours 56 minutes.

BSERVING JUPITER

piter is one of the planet atcher's greatest delights. offers a wealth of ever-ıanging detail that can be en in almost any telescope. also rewards patience, since oservers who take the time become familiar with this anet benefit the most.

As with any planet, taking lvantage of nights of good ·eing is the key to enjoyable observing, and the view gets better the longer you look. If your telescope has a motorized equatorial mounting, it is worth taking the trouble to align it so that the drive tracks correctly (see p. 53).

What can you expect to see? With a 2.4 inch (60 mm) telescope and 50x to 100x magnification, you should be able to see two broad dusky belts paralleling the equator. Through a 4 inch (100 mm) scope at 100x or more, you can detect several more bands extending into higher latitudes. And with 6 inch (150 mm) or larger scopes, use 150x to 300x and the view will be overwhelming—details will cover most of Jupiter's disk and quickly pass in review as the planet rotates.

The bright zones and dark belts display subtle shades of brown, tan, yellow, orange, and blue-gray. The north and south equatorial belts are the most constant, but all the bands vary in strength and change position slightly. The edges of the bands become irregular, developing projections and indentations as storms stir up the clouds, and winds reach speeds of up to 250 miles per hour (400 km/h).

STORM SYSTEMS

The most famous storm system on Jupiter is the Great Red Spot, which appears to be a vast high-pressure system, about twice the size of Earth. Its size has varied over the decades, and it tends to become redder as solar activity increases (see p. 80). When the Great Red Spot is faded, look for the Red Spot Hollow, an indentation that surrounds it in the south tropical zone.

Jupiter has other storms, too: three white ovals reside in the south temperate belt, where they were first noticed in the late 1930s. A lasting characteristic of the white ovals is that they slowly drift eastward relative to the Red Spot. Smaller white spots appear now and then as well.

Additional cloud features to look for are festoons—dark, linear features linking two belts across an intervening zone—and condensations—dark spots or short lines that usually appear within the belts.

To enhance atmospheric details, experienced Jovian observers view through gelatin or glass filters. The most helpful is light blue (Kodak Wratten 80A or 82A), which can enhance the edges between belts and zones. A yellow or orange (W12 or 21) filter brings out festoons and polar details that are bluish in color.

JUPITER FACTS

Distance from Sun 5.20 AU
Mass 318 × Earth's mass
Radius 11.2 × Earth's radius
Apparent size 33 to 50 arcsecs
Apparent magnitude –1.2 to –2.5
Rotation (relative to stars)
9.84 hours at System I;
9.93 hours at System II
Orbit 11.86 Earth years

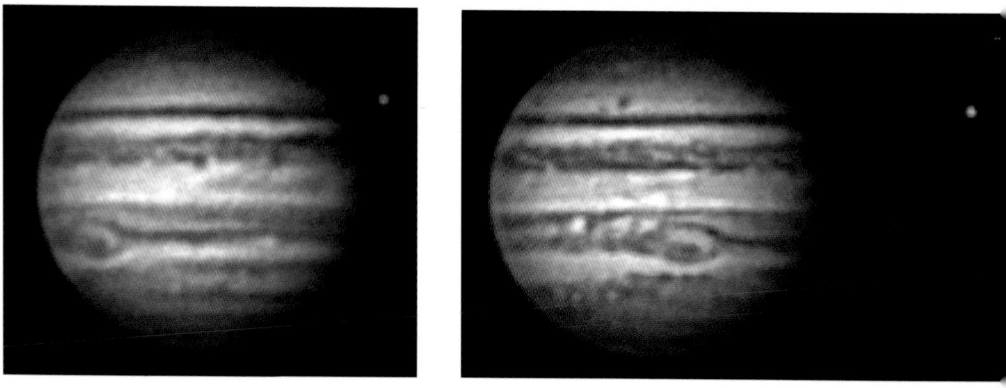

THESE CCD IMAGES *of Jupiter and Ganymede were taken 90 minutes apa They clearly show the rapid rotation o the planet and motion of the moon.*

SKETCHING TOOLS AND TECHNIQUES

Sketching Jupiter is a rewarding and useful way to train your eye to see more, and capturing the details on paper will give you an outstanding familiarity with the planet's features. You have to work quickly, however, spending no more than 10 minutes on each sketch because Jupiter's rotation is fast enough to alter the view on time-scales longer than this.

Two approaches to sketching have been popular with Jupiter observers, one showing the whole Jovian disk, the other being the "strip sketch."

The full-disk drawing uses a prepared blank outline, usually on a scale of 2 inches (50 mm) to the planet's diameter. This is about 1 mm per arcsecond, and some observers standardize on that, varying the diameter of the blank as Jupiter's apparent size changes. (The planet's size ranges from 33 arcseconds near conjunction to 50 arcseconds at perihelic oppositions such as those of 1998 and 1999.) Whether you vary the blank's size or not, be sure that it properly shows the oval disk. A 70 degree ellipse template from an art-supply store fits the shape perfectly.

The strip sketch focuses o a range of latitude, such as th zone containing the Great Red Spot and its two flankin belts. Because they do not attempt to capture all of Jupit at once, strip sketches are often easier and considerably less nerve-wracking for beginners. By focusing on an area of interest, you can spen your 10 minutes per sketch more carefully.

Besides the Jupiter drawin the page should include spac to note the date and time, th instrument and magnification used, the weather and seeing conditions, and any other factors that seem relevant.

JOVIAN PHOTOS

For a planet whose brightnes averages magnitude –2.5 (exceeded only by Venus), Jupiter can be surprisingly difficult to photograph well. Its small size demands the use of eyepiece projection (see p. 67) and exposures of severa seconds, during which seeing conditions may blur fine detail. Jupiter's limb is significant darker than the center of its disk.

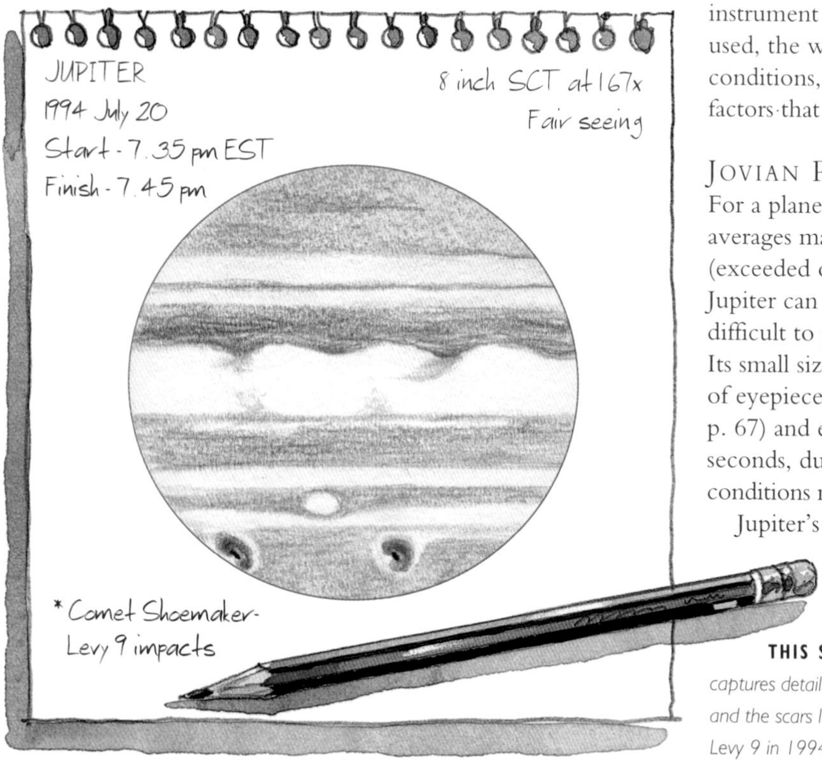
JUPITER
1994 July 20
Start - 7.35 pm EST
Finish - 7.45 pm
8 inch SCT at 167x
Fair seeing
* Comet Shoemaker-
Levy 9 impacts

THIS SKETCH OF JUPITER *captures details of the bands and belts, and the scars left by Comet Shoemake Levy 9 in 1994 (see also p. 146).*

While this is not obvious to the eye, it is readily detected on film. In images processed for high contrast (to reveal belt and zone details), the limb regions can darken and merge with the sky background.

Astrophotographers resort to a few darkroom tricks to get around this difficulty. One way is to expose a print for the limb regions and burn in (give more exposure to) the center of the disk—or inversely, hold back the limb by using dodging (reducing the exposure). Another trick is to sandwich the negative with a slightly out-of-focus positive image. This "unsharp mask" can reveal wonderful details.

While many beautiful color photos of Jupiter have been taken, most planet photographers prefer to use black-and-white film, usually Kodak's Technical Pan. It offers fine grain, good contrast control, and costs less than color. With an f-ratio of about f/60, try exposures of 1, 2, and 4 seconds (see also p. 61).

Many astrophotographers are now switching to CCDs, despite the high initial costs (see p. 68). The main reason is that electronic imaging affords greater flexibility in processing. Software offers unsharp masking and many other enhancements, all at the click of a mouse.

The fairest thing we can experience is the mysterious. It ... stands at the cradle of true art and true science.

The World As I See It,
ALBERT EINSTEIN (1879–1955), German-born US physicist

PARTICLES AND FIELDS

The Jovian world is more than what we see. If our eyes could see Jupiter's magnetosphere, it would be a highly impressive sight, with an apparent diameter equal to four times that of the Moon. The magnetosphere is a vast, complex structure full of fleeting charged particles. In fact, the density of radiation near Jupiter is high enough to kill an unprotected human in just a few minutes. This radiation also causes spacecraft designers major headaches, since delicate electronics must be shielded from it.

Jupiter is one of the most intense radio sources in the sky. It produces emissions at millimetric, decimetric, decametric, and kilometric wavelengths. The decametric emissions, at frequencies between 0.6 and 39.0 megahertz, are the best known and first discovered. Within this band, Jupiter emits bursts of radiation triggered by interactions with its moon Io. These can last from a few seconds to several minutes.

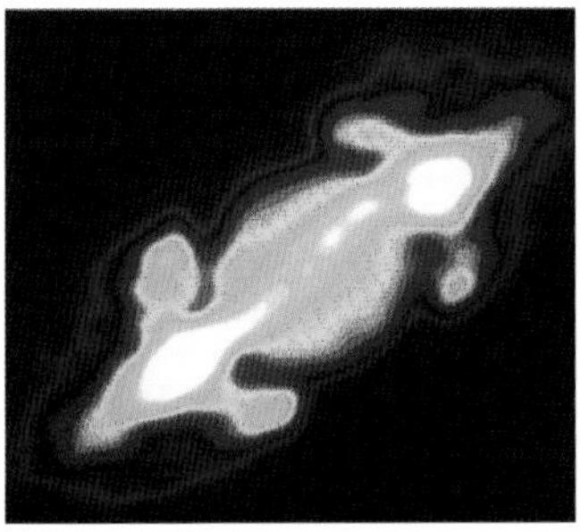

THIS RADIO IMAGE *of Jupiter shows the planet in the center, with bright radiation belts on either side of it.*

OCCULTATIONS OF JUPITER

The Moon occasionally covers up, or occults, a planet or star, producing a sight not to be missed. Occultations of Jupiter (below) are particularly captivating because the planet is relatively large in the telescope's field of view.

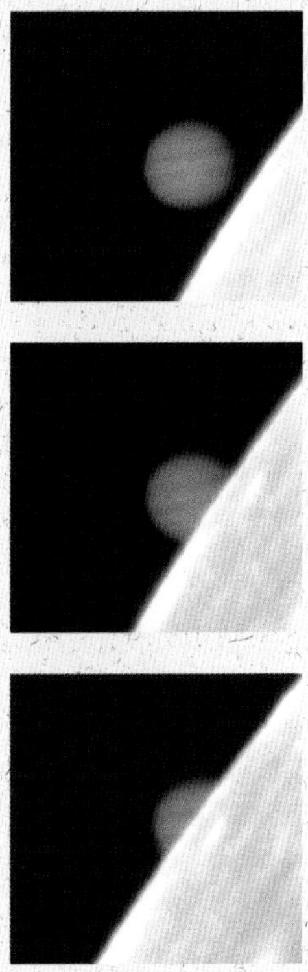

Consult astronomy magazines, almanacs, or sky-charting programs to find out when occultations are going to happen, and whether they will be visible from your location. Amateur groups will often arrange parties to watch these events—check with your local club.

Set up your telescope at least 30 minutes before the scheduled disappearance. If you have an equatorial mounting, observing will be easier if it is aligned on the pole (see p. 53). Use a low-power eyepiece to center the telescope on the planet.

As the Moon approaches, get ready. It always seems to take forever to draw close, then the actual event runs its course quickly. After disappearing, the planet will emerge from behind the Moon's limb—it may be an hour later or it may be only a few minutes. The occultation predictions will tell you where on the limb to look for it.

A VOYAGER I VIEW *of Jupiter with its moons Io (far left) and Europa (left).*

Miniature Solar System

When Galileo discovered Jupiter's four large moons—Io, Europa, Ganymede, and Callisto—he was excited to see that Jupiter mimicked the Copernican model of the Solar System.

Being 5th to 6th magnitude in brightness, the four Galilean moons are conspicuous in any telescope. You can even duplicate Galileo's discovery with a pair of binoculars. Observe Jupiter at the same time each night, and plot the positions of the planet and the minute points of light lying beside it. After a week or two, a pattern will emerge, and you may experience some of the emotion that gripped Galileo.

From time to time, one or another of the Galilean moons will transit Jupiter's disk, where the dark shadow it casts will be more conspicuous than the moon itself. For observers with scopes of 10 inches (250 mm) or more, transits are a good time to look for the color of the moon because Jupiter provides a light background. Io has a yellowish cast, Europa looks grayish white, Ganymede tan-gray, and Callisto bluish gray.

These moons orbit in the plane of Jupiter's equator, which is presented nearly edge-on to us. Every six years, when Earth passes through this plane, we witness a series of mutual eclipses and occultations among the moons. Consult astronomy magazines and almanacs for dates of these events (see p. 276).

Jupiter has at least 12 other moons. Unfortunately, all of them are small, with the brightest, Himalia, reaching only 15th magnitude.

Four Worlds

Each of the Galilean moons has a unique character, but none is more bizarre than Io. This world, larger than our own Moon, is in a state of

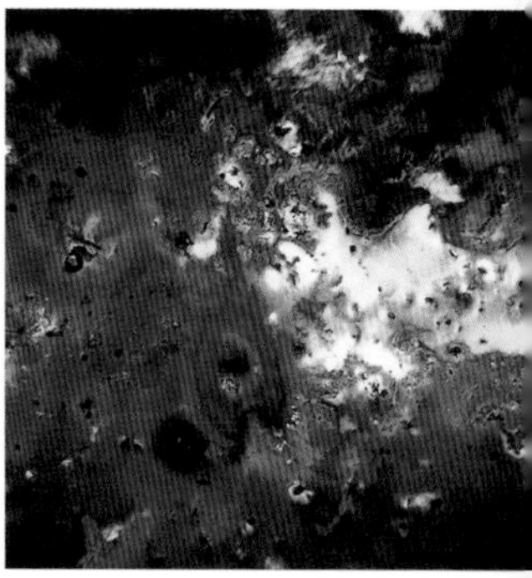

PELE, *the heart-shaped feature just left of center in this Voyager I image, was the first active volcano discovered on Io.*

THE GALILEAN MOONS AND THE SPEED OF LIGHT

On 7 January 1610, Galileo Galilei (1564–1642), professor of mathematics at the University of Padua, aimed a new device of some glass lenses mounted in a piece of organ pipe—he called it a "perspicillum"—toward the southeastern sky. There, to the upper right of the gibbous Moon, was Jupiter, a bright dot of light.

With his telescope yielding about 20x magnification, Galileo (below) noted three little stars near Jupiter, one to the west and two to the east of it. The next night all three lay to the west of Jupiter. Two nights later, he saw only two stars, both to the east. This went on for two weeks before Galileo realized he was seeing a total of four worlds orbiting Jupiter. No one had ever seen these bodies before. Today, we know them as the Galilean moons.

By the 1670s, astronomers had determined the periods of revolution for the Galilean moons to within seconds of their modern values. But the predictions for when a moon would enter or leave Jupiter's shadow were often up to several minutes in error. These events seemed to occur earlier when Earth was nearer the Jovian system and later when it was farther away from it. In 1675, Danish astronomer Ole Römer realized these errors arose from the fact that the speed of light was not infinitely quick. As a result of this study involving the Galilean moons, the first scientific determination for the speed of light was made.

SATELLITE SURFACES *Ganymede, Europa, and Callisto all display water-ice surfaces, but Ganymede (below) and Callisto (right) are heavily cratered, while Europa (below right) has few craters.*

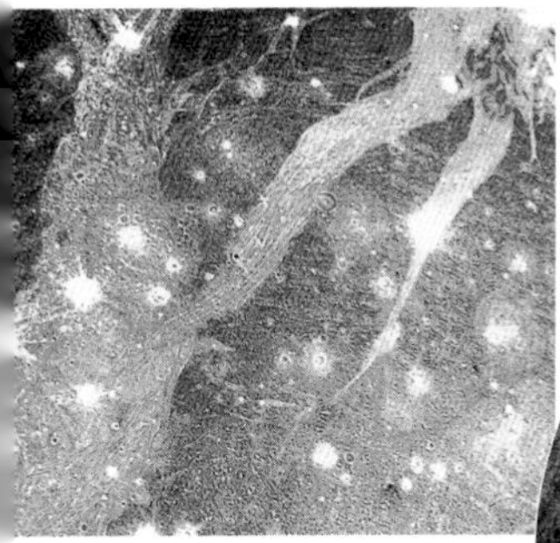

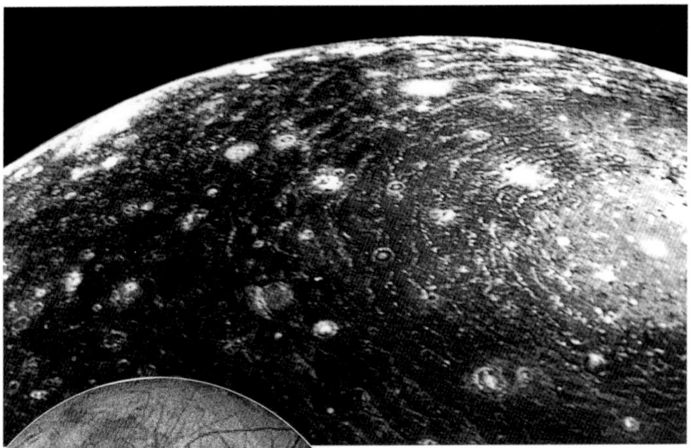

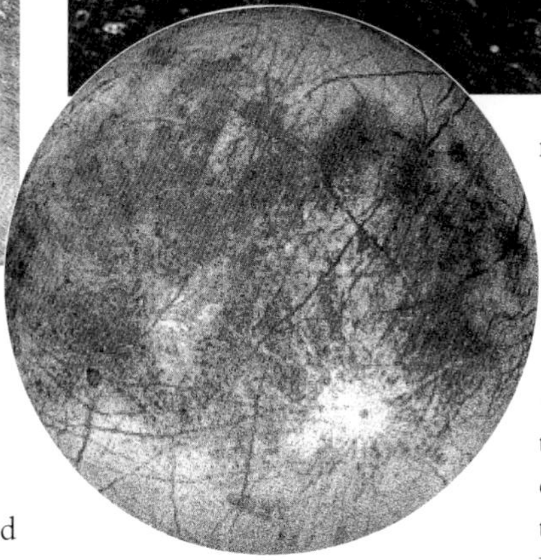

continual volcanic eruption, thanks to gravitational tugs from Jupiter, Europa, and Ganymede. Io's volcanoes have covered it with a multicolored tapestry of sulfur deposits.

Europa, the next moon out, is a little smaller. Its surface is also young, but instead of lava flows, it is covered with ice floes floating on a deep ocean of liquid water.

The third Galilean moon, Ganymede, is bigger than Mercury. Its surface tells of a history with tectonic activity and impact cratering.

The fourth large moon is Callisto, about the size of Ganymede. It appears to be a sphere of ice and rock that has evolved little since it formed. The surface is pitted by countless impact craters, testimony to its long geological dormancy.

Jove's Ringlet

Many scientists were surprised when Voyager 1 revealed a ring around Jupiter. The most tenuous of the gas-giants' ring systems, it is about 3,700 miles (6,000 km) wide and less than 20 miles (30 km) thick, and lies some 35,000 miles (56,000 km) above the cloud tops. It can be imaged from Earth only at near-infrared wavelengths.

Solar System Pot-stirrer

With its great mass, Jupiter exerts a major influence on the orbital mechanics of the asteroids and comets that pass near it. In the early Solar System, it prevented any planet from forming in the region that is now the asteroid belt by winnowing the population of large planetesimals in that vicinity. (Planetesimals are the small bodies that coalesced to form the planets.)

Today's asteroid belt has gaps in it where few objects orbit. Any that wander into these areas are soon removed by perturbations—changes in their orbit caused by the gravity of Jupiter.

Scientists are also starting to hold Jupiter largely responsible for the creation of the Oort Cloud of comets (see p. 142). The theory is that as comet nuclei move inward from the Kuiper Belt through perturbations with Neptune, they can end up near Jupiter. At that point, Jupiter's gravity may bend their orbits into elongated paths that approach the Sun and Earth. More often, though, it will fling them outward, to the Kuiper Belt again, into the Oort Cloud, or perhaps even out of the Solar System altogether.

Jupiter can also destroy comets, as the world saw in July 1994 with Comet Shoemaker-Levy 9. This hapless object passed close enough to Jupiter that tidal effects broke the comet apart. Jupiter then sent the comet on one last looping orbit around itself. When the comet returned to Jupiter, its 21 pieces crashed into the planet's southern hemisphere, leaving Earth-size dark splotches that could be seen even in small telescopes.

THIS INFRARED IMAGE *shows the plume (at bottom left) from the collision of fragment G of Comet Shoemaker-Levy 9 with Jupiter on 18 July 1994.*

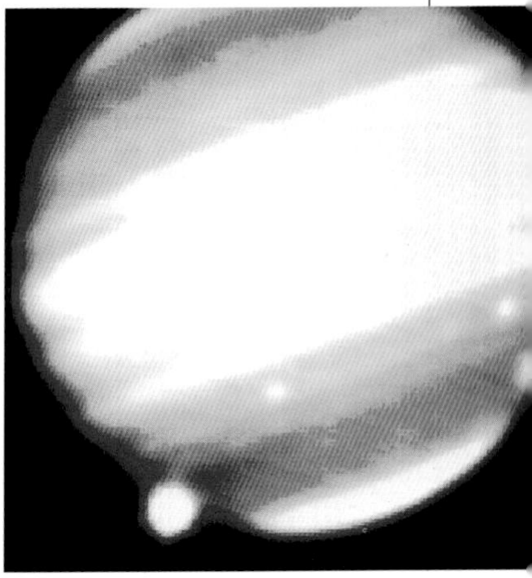

SATURN

While Saturn and its rings have captivated observers for hundreds of years, the full wonder of this planet and its satellites is only just being realized.

THE EARLY FRUITS *of the Earth offered to Saturn, god of the harvest, from a fresco by Giorgio Vasari (1511–74).*

Saturn is the one object, apart from the Moon, that always gets a gasp of delight from first-time telescope viewers. Once known as *the* Ringed Planet, we now know that Jupiter, Uranus, and Neptune also have rings. However, Saturn's rings are the only ones you can see in any backyard telescope, and once seen, that tiny image is truly unforgettable.

DISCOVERING SATURN

Saturn held few attractions for the ancients, who knew it as the slowest moving planet in the sky. Our name for the planet comes from Roman mythology, but the Greeks called it Kronos after Zeus's father, the overthrown ruler of the universe and a weary old man. Earlier still, Mesopotamian astronomers had called it "the old sheep" or even "the eldest old sheep."

The modern era for Saturn began in 1610, when Galileo Galilei turned his telescope on the planet. He thought it was a triple-bodied object. Other observers reported that it had "handles" or "ears," and fierce debate ensued.

Finally, in 1659, Christiaan Huygens (who discovered Titan, Saturn's largest satellite) announced that Saturn was circled by a broad, flat ring inclined to the ecliptic. This explained why the rings seem to disappear at about 15-year intervals. Twice each Saturnian year, Earth passes through the plane of the rings and they become too thin to be seen. The last such passage occurred in 1995–96, and the next will take place in 2009.

THE SPACECRAFT ERA

In 1980 and 1981, the space-probes Voyagers 1 and 2 flew past the planet. Their close-up images of Saturn, its multitudinous rings, and the diverse family of moons revolutionized our view of the planet. The spacecraft revealed new details in Saturn's cloud belts, showed the rings were ribbons of particles moving in complex mathematical resonances, and mapped the intricate geologies of most of the larger satellites.

In 2004, the Cassini space-craft is scheduled to go into orbit around Saturn and spend four years studying the Saturn-ian system. It will also place a lander on the moon Titan.

THE VIEW FROM HERE

Although Saturn orbits nearly 10 times farther from the Sun than Earth does, it is relatively bright, shining at about 1st magnitude. That makes it easy to locate. Moreover, until it enters the star-rich region of Taurus, around the year 2000, Saturn will be among the dim

SATURN AND ITS MOONS *This image was composed from photographs taken by the Voyager spacecraft.*

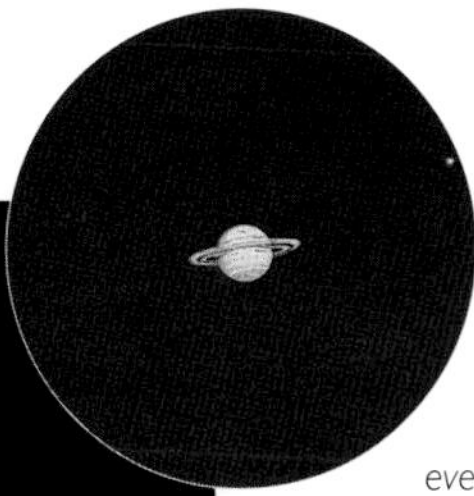

A STORM SYSTEM *near Saturn's equator, revealed in a Hubble image (far left). The planet's rings are obvious even in a small telescope (left).*

onstellations of Pisces, Cetus, nd Aries, where few bright ars compete with it.

Being a gas-giant planet, aturn has a structure much ke Jupiter's—mostly gas and quid with a small, dense core. 'he visible surface is the top f an atmosphere containing ydrogen, helium, and other ompounds such as methane, nd the features we see are onstantly shifting structures in ie planet's cloud tops. Since orbits farther from the Sun ıan Jupiter does, Saturn's nvironment is colder. This ieans it has less "weather" nd displays fewer features.

To viewers with 3 inch 75 mm) scopes or smaller, aturn's disk will appear ›atureless and creamy white. ut in larger scopes, especially uring good seeing, a pair f dusky bands paralleling ie planet's equator becomes isible at about 20 degrees orth and south latitude. These re called the north and south quatorial Belts.

Viewing in twilight helps eveal these and other features. ı full darkness, use a yellow lter to improve visibility. It elps to remember that belts e dark while zones are light.

The features contained etween the two Equatorial elts form what astronomers call System I, and they rotate once every 10 hours 14 minutes. System II includes everything else and has a rotation period of 10 hours 38 minutes.

At rare intervals, a white cloud breaks out in Saturn's Equatorial Zone. This occurred in 1990, when a small, bright patch appeared. Scientists think it was produced by a gigantic bubble of ammonia gas that rose from the depths and, upon reaching the chilly cloud tops, froze into white crystals and spread around the planet. Such eruptions seem to occur about every 30 years, but the cause remains unknown.

Another event occurs when Saturn lies 90 degrees away from the Sun in the sky—about 90 days before and after opposition. At this time, we can peer a little past one limb of Saturn and detect its shadow falling on the rings. The effect, though small, provides a visual hint of the planet's three-dimensional reality.

CASSINI AND HIS DISCOVERIES

Giovanni Domenico Cassini (1625–1712) is famous for discovering the gap between the A and B rings of Saturn. He also proposed that the rings were made up of individual particles rather than being a solid body as many thought at the time.

Born in Italy, Cassini (below) taught at Bologna and determined the rotation period of Mars to within two minutes of its correct value. He later did the same with Jupiter. In 1669, Louis XIV of France invited him to take over the Paris Observatory. Cassini stayed in France for the rest of his life and became a French citizen.

His most important work came in 1672 when he used observations of Mars to determine the size of the planets' orbits. He derived a value for the Earth–Sun distance that was only about 7 percent too small—the most accurate estimate at that time.

In 1671, Cassini discovered the first of four new moons of Saturn. This moon now bears the name Iapetus. The following year he found another, Rhea, and on one night in 1684 he discovered two more moons, Dione and Tethys.

In 1675, he observed that the "ring" of Saturn contained a dark gap. To this day, it is known as the Cassini Division.

THE UNDERSIDE *of Saturn's rings, as seen by Voyager 1. Thousands of "ringlets" make up each of the rings.*

Lord of the Rings

If the Voyager spacecraft showed too many rings to number, the view from Earth is much simpler—we see only three. The outermost is called the A ring and it is 9,000 miles (14,500 km) wide. It is separated from the B ring by the dark gap of the Cassini Division, 2,600 miles (4,200 km) across. The B ring is both the brightest and, at 16,000 miles (26,000 km) across, the widest. On its inner edge is the gauzy C (or crepe) ring, 10,500 miles (17,500 km) wide.

While the B ring looks solid enough to walk on, in reality it, like the other rings, is a loose collection of particles orbiting in complex ways. This was dramatically illustrated in 1989, when Saturn passed in front of the star 28 Sagittarii. All over the world, professional and amateur astronomers watched the star flicker as it appeared to traverse the rings. The occultation revealed even more structure in the rings than Voyager had shown.

The Cassini Division is hard to see when the rings are nearly edge-on, as in the 1990s, but it becomes easier to see as the tilt increases. Voyager showed that the division is not empty, but simply contains less material than the rings. The division occurs because the gravity of the moon Mimas perturbs ring particles that orbit there, selectively removing them.

The C ring is also hard to view when the rings are relatively closed. When the rings are open, it is easy to spot the C ring in front of Saturn's disk.

Saturn's rings are probably transitory. They originated when one or more moons strayed too close to Saturn to withstand its tidal force and broke apart. Many scientists now think the rings will last only a few tens of millions of years before mutual collisions rob the particles of energy and they spiral into Saturn.

Worlds of Ice

Saturn has some 18 known moons, many being mixtures of ice and rock. Most, however, lie beyond the reach of amateur-size telescopes. The largest is 8th magnitude Titan, visible even in a 3 inch (75 mm) scope. Titan circles Saturn every 16 days, and

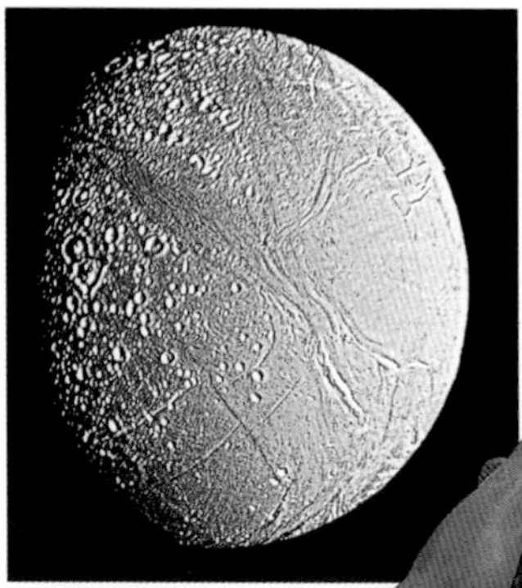

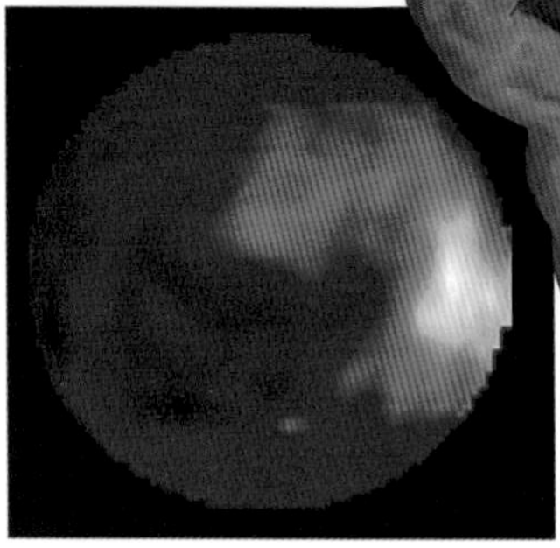

Oh Titan, warmed by a
hydrogen blanket,
ice ribbed volcanoes
jet ammonia
dredged out of a glacial heart

The Planets,
Diane Ackerman (b. 1948),
American poet

SATELLITES *A Voyager image of Enceladus (top). A near-infrared Hubble image of Titan shows light and dark surface features (above). Titan is named after the giants of Greek mythology (right).*

hen lying due east or west
Saturn, is about 4½ ring-
ameters away from the center
the planet. This moon is
trange world indeed. At
nost 3,200 miles (5,100 km)
ross, it is bigger than the
anet Mercury. An opaque,
noggy atmosphere of nitro-
n and methane blankets its
rface, which may feature
kes of hydrocarbons.

Inside Titan's orbit, other
oons to look for are magni-
de 9.9 Rhea, magnitude 10.6
ione, and magnitude 10.4
ethys. Careful observation
ay also catch the magnitude
.9 Enceladus.

Be sure not to overlook
ldball Iapetus. Because its
ading side is coated with
rk dust, Iapetus is more than
vice as bright when it lies
est of Saturn than when it
s east of it—ranging from
)th down to 12th magnitude.
petus circles outside the or-
t of Titan every 79 days at a
stance of 13 ring-diameters.

APTURING SATURN

aturn's low contrast and
mplex aspect make drawing
a major challenge. Many
oservers concentrate on
ortions—the Equatorial
one or one side of the rings.

SATURN FACTS

Distance from Sun 9.54 AU
Mass 95.2 x Earth's mass
Radius 9.4 x Earth's radius
Apparent size 15 to 21 arcsecs
Apparent magnitude 0.6 to 1.5
Rotation (relative to stars) 10.2 hours
Orbit 29.5 Earth years

CCD TECHNOLOGY *allows amateurs to capture splendid color images of distant objects such as Saturn (right).*

There is no time to dawdle—Saturn's quick rotation means that 15 minutes is long enough to work on a sketch.

Conventional photography of Saturn requires fairly advanced techniques. Saturn's small apparent size means you must use eyepiece projection on an equatorially mounted telescope and track the sky during the exposure (see p. 67). You also need the good luck to capture a few seconds of good seeing. The result, however, can be an outstanding record of a beautiful object.

Many astrophotographers have switched to CCD cameras (see p. 68), despite the greater expense. CCDs allow you to take many images in succession, which improves your chances of catching moments of good seeing.

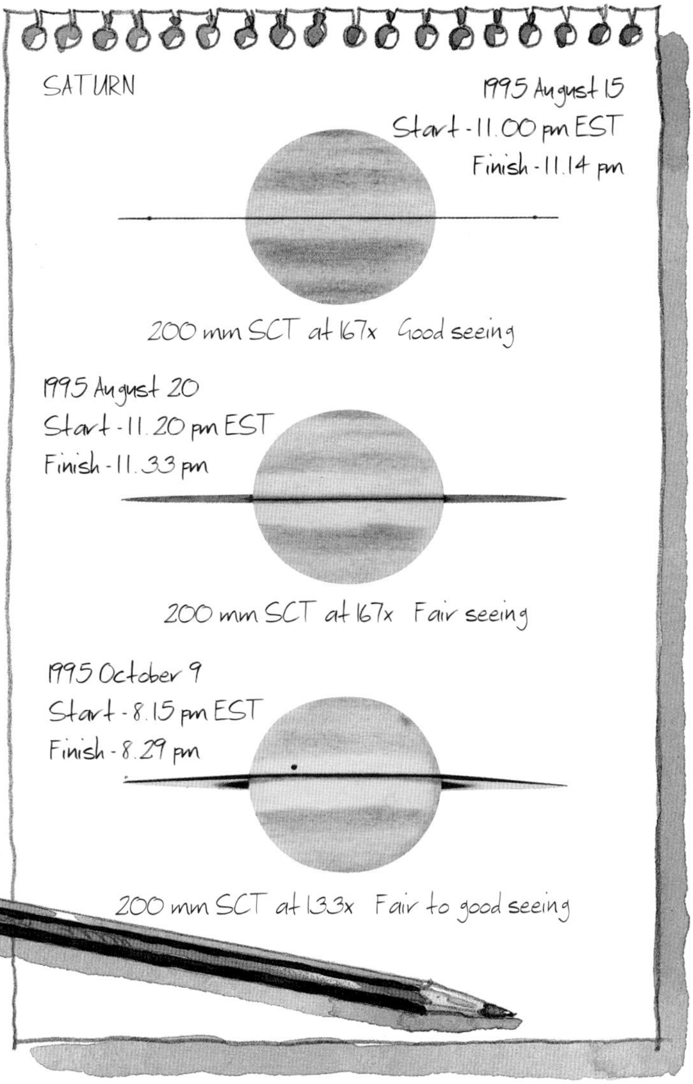
SATURN
1995 August 15
Start - 11.00 pm EST
Finish - 11.14 pm
200 mm SCT at 167x Good seeing
1995 August 20
Start - 11.20 pm EST
Finish - 11.33 pm
200 mm SCT at 167x Fair seeing
1995 October 9
Start - 8.15 pm EST
Finish - 8.29 pm
200 mm SCT at 133x Fair to good seeing

KETCHES *made at different mes show the varying tilt of Saturn's ngs. Sketching can often record details at would be blurred in photographs.*

URANUS

This distant, blue-green planet is something of a mystery. Even after Voyager 2's flyby in 1986, there remains great deal for future astronomers to discover.

ARIEL *(far left) with Prospero in a scene from Shakespeare's play* The Tempest. *All of Uranus's moons are named for Shakespearean characters.*

Uranus's brightness comes just within reach of the naked eye, and it was mistaken for a star many times before it was finally discovered.

The first recorded sighting was made in 1690 by England's Astronomer Royal, John Flamsteed, who cataloged it as 34 Tauri. Another observer, Pierre Lemonnier, logged Uranus as a star a total of 12 times—6 of those over one 9-day period in 1769. These early records later helped to establish the planet's orbit.

When William Herschel discovered Uranus in 1781, he initially thought it was "either a nebulous star or perhaps a comet." He soon confirmed, however, that the 6th magnitude object was a planet, circling the Sun every 84 years. It was named Uranus after the Greco-Roman god who personified the universe and was the father of Saturn.

Being so distant, little was learned about Uranus until well into the twentieth century. Much of our current understanding came from the Voyager 2 flyby in 1986.

URANUS FACTS

Distance from Sun 19.2 AU
Mass 14.5 x Earth's mass
Radius 4.0 x Earth's radius
Apparent size 3 to 4 arcsecs
Apparent magnitude 5.5 to 5.9
Rotation (relative to stars) 17.2 hours
Orbit 84.0 Earth years

A Big Blue World

The planet's density, 1.3 times that of water, groups Uranus with the other gas giants, such as Jupiter. Its atmosphere is mostly hydrogen and helium, with some methane.

It is believed that Uranus consists of two, or possibly three, layers. At the surface, the mix is gaseous, but increasing temperature and pressure make it behave as a liquid about a third of the way to the center. The remainder of the planet is a hot, slushy mixture of water, methane, and ammonia, together with rocky components. It may also have a nickel-iron core. Unlike the other gas-giant planets, Uranus does not seem to have any internal heat source.

In a telescope, Uranus looks pale blue-green because of the methane gas in its atmosphere. Its disk almost always appears featureless, although past observers have noted faint dark bands paralleling its equator.

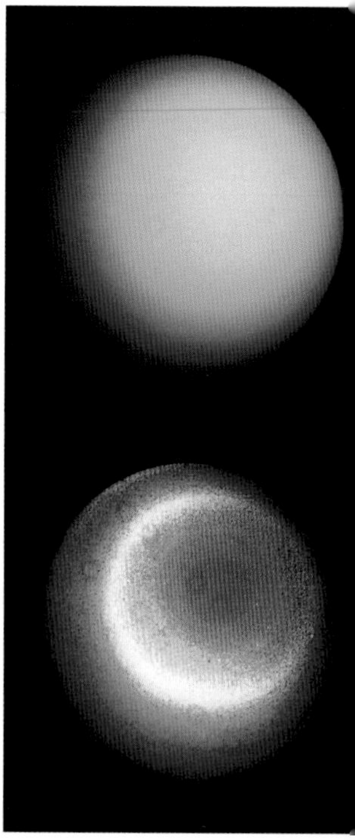

URANUS *appears blue and featureless in true color (top), and reveals hardly any more detail even with strong enhancement in false colors (above).*

Orbiting Sideways

Uranus orbits almost "on its side" at a tilt of 98 degrees, possibly as a result of a collision with an Earth-size object early in its history. Its moons and rings may be leftover fragments from this impacting body. The tilt gives the planet peculiar seasons, with one pole or the other pointing toward the Sun for several decades. Such long cycles of sunlight and darkness heat the polar regions more than the equator. The resulting energy nbalance would xplain the 450 nile per hour 720 km/h) winds etected during ne Voyager 2 flyby.

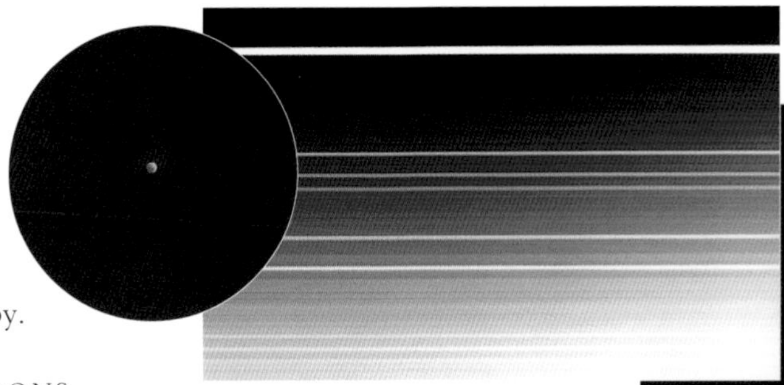

URANUS'S RING SYSTEM, *as shown in a false-color image (above) and in a montage with Miranda in the foreground (right). A telescope view of Uranus shows only a tiny disk (top left).*

MASHING MOONS

'he first two of Uranus's five rge moons, Titania and)beron, were found by Villiam Herschel in 1787. The nird and fourth moons, Ariel nd Umbriel, were not dis-overed until 1851. Miranda, ne fifth, held out until 1948. 'hese major moons certainly ose a challenge to observers: ne brightest is Titania, at mag-itude 13.7, and the dimmest, Miranda, is 16th magnitude.

Due to their faintness, little as known of these moons ntil Voyager 2's visit. Its ameras revealed Oberon and Jmbriel to be heavily cratered vorlds of ice and rock, with o sign of geological activity. riel and Titania, on the ther hand, appear to have ndergone some kind of icy olcanism that has "relaxed" neir surfaces, erasing craters n many areas and producing ong fault valleys.

The surprise was Miranda. : displays a geology so com-lex that, more than a decade fter the Voyager visit, sci-ntists are still arguing about vhat causes its dominant feature—the grooved terrain known as coronae. One theory is that Miranda is a second-generation satellite, reassembled out of a shattered moon or moons. Its surface shows three enormous coronae, whose roughly oval patterns may mark places where large fragments sank into Miranda's interior.

Voyager 2 also showed that Uranus has 10 more moons, all of them small, dark mixtures of ice and rock. Like the larger moons, these have been named after Shakespearean characters, including Ophelia, Portia, Juliet, Desdemona, and Puck.

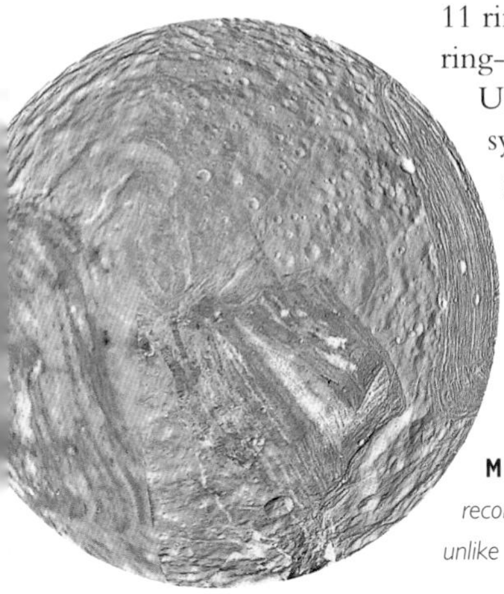

MIRANDA, *seen in this computer reconstruction, displays a surface unlike any other in the Solar System.*

SLIM RINGS

In March 1977, astronomers watched Uranus slide in front of a 9th magnitude star and discovered that Uranus has rings. The Voyager 2 flyby confirmed that Uranus has 11 rings, as well as several ring-arcs, or partial rings.

Unlike Saturn's grand ring system, Uranus's rings are gauzily thin and difficult to detect from Earth. Most of the rings are not circular, being tugged out of shape by "shepherd" moonlets that confine their widths. The ring particles are as dark as coal and range from boulder- to house-sized fragments.

WATCH THIS SPACE

At the time of the Voyager 2 flyby, Uranus displayed an all-but-blank appearance. This disappointed many scientists, who had hoped to find atmospheric features like those of Jupiter. Any features the planet had were hidden deep in the hydrogen-methane haze.

However, the Uranus that Voyager saw in 1986 may not tell us much about the planet we will see in the future. In 1985, the planet's south pole was pointing at the Sun. Past telescopic observations suggest that cloud markings appear on Uranus in the years around its equinoxes, when the Sun is over the planet's equator.

Uranus will reach its next equinox in 2007, but even if it does develop new features, they will not be easy to see in smaller instruments. But there is always the chance that an amateur astronomer with a big scope and a sharp eye—or a CCD camera—will get lucky.

NEPTUNE

Although it is a dim sight from Earth, this deep-blue planet with its bright cloud bands is one of the most varied and interesting in the Sun's family.

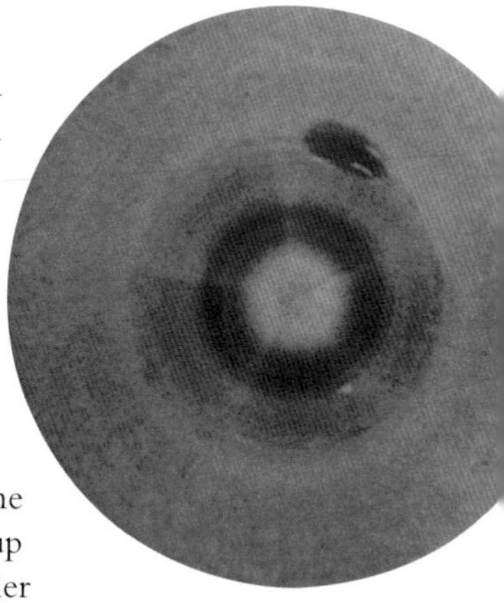

NEPTUNE, *the Roman god of water (left), is an appropriate namesake for the planet. A polar view (right).*

Neptune is four times the size of Earth, but because it orbits 30 times farther from the Sun, observers see only a small, 8th magnitude disk.

DISCOVERY TRIUMPH

The discovery of Neptune was a great triumph of mathematical astronomy. By the 1840s, astronomers had known for years that Uranus was not moving as predictions said it should, and that there must be another planet affecting its orbit. Acting independently, two mathematician-astronomers, Urbain Le Verrier in France and John Couch Adams in England, took the discrepancies and tried to determine where this planet might be, assuming likely values for distance, mass, and so on.

Their research produced nearly the same result, but the British were slow to follow up Adams's prediction. Le Verrier had more luck, and on 23 September 1846, following Le Verrier's directions, Johann Galle and Heinrich d'Arrest at the Berlin Observatory found the planet on the first night of searching—in fact, within half an hour. In the end, after much embittered finger-pointing among British astronomers, Adams received joint credit with Le Verrier for the discovery.

VOYAGING TO NEPTUN

Because of the planet's distanc from Earth, information abou Neptune accumulated slowly William Lassell saw Triton, the largest moon, within a month of the planet's discovery, but the second moor Nereid, was discovered only in 1949, by Gerard Kuiper. I the 1970s and 1980s, astronomers received strong hints of Neptune ring system, but cou not make sense of the data.

In August 1989, Voyager flew by the planet, and the modern era for Neptune bega Voyager found a big, blue ball, with many markings and cloud bands—a pleasant chang from the bland Uranus the spacecraft had visited thre years earlier. Voyager's instruments collected dat about Neptune's composition. There may be

A CHANGING FACE *Neptune's Great Dark Spot (far left) has disappeared since Voyage 2 flew by in 1989. Urbain Le Verrier (lef*

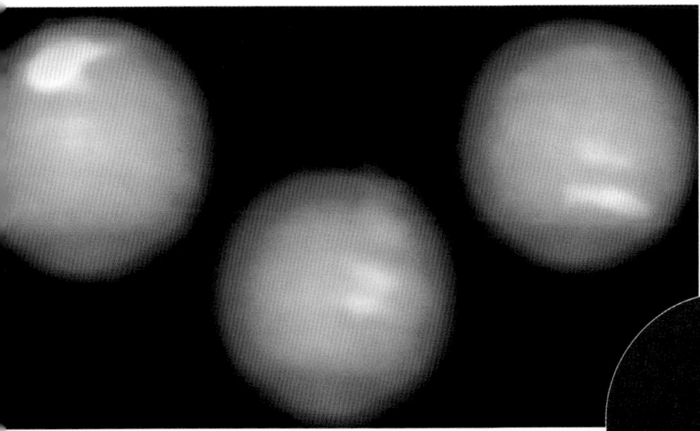

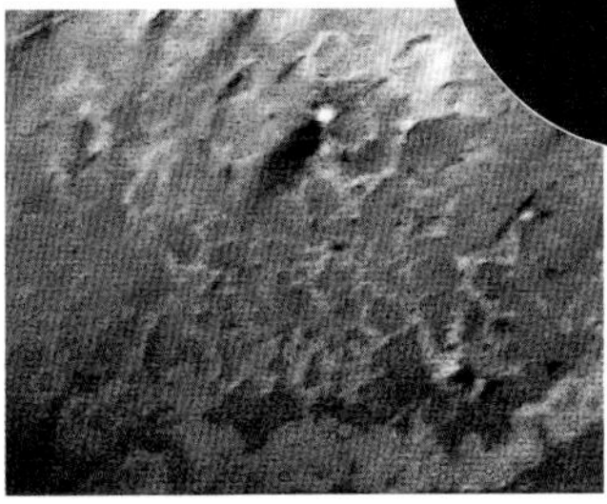

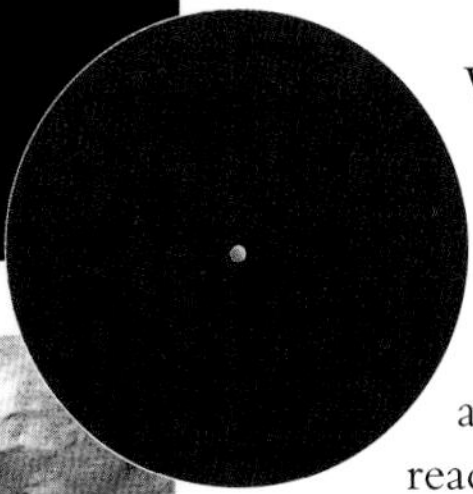

NEW CLOUDS *on Neptune, seen by the Hubble Space Telescope in 1995 (left), are invisible in the amateur telescope view (below). A false-color Voyager 2 image of Triton's surface shows possible nitrogen geysers (below left).*

small rocky core, but the ulk of the planet is probably deep ocean of water. This ıen merges into an atmoshere of hydrogen and helium. ⁄Iethane in the top of the tmosphere gives the planet its ronounced blue color. The ıterior bubbles up more than wice as much heat as the urface receives from the Sun.

Neptune spins on its axis very 16 hours 7 minutes. In s equatorial zone, winds :ream westward at 900 miles 1,500 km) per hour, powering uge storm systems.

Voyager froze Neptune in me with a snapshot, but the Iubble Space Telescope has nce tracked its changing eatures. The spacecraft saw everal storms, notably the Great Dark Spot (the size of arth), which drifted toward he equator and vanished ometime after the flyby.)ther noted markings, such s the Scooter, the second)ark Spot, and various white treaks, have also disappeared r changed greatly, while ew features have emerged. longer view is needed to ılly understand the planet, o observations continue.

;EYSERS AND ICEFIELDS

⁄oyager 2 confirmed the existnce of five thin rings around Ieptune and added six satellites) the two already known. It lso took a close look at the rgest satellite, Triton.

Two-thirds the size of our Moon, Triton displayed a varied face to the cameras—plains, impact craters, a strange cantaloupe terrain of pits and depressions crossed by ridges, and a southern hemisphere with a thin, pinkish ice cap of nitrogen. Triton also has plumes—geysers of nitrogen gas shooting 5 miles (8 km) straight up into a thin atmosphere of nitrogen and some methane. Winds then blow the gas plumes about 60 miles (100 km) downrange.

Triton is the only major satellite with a retrograde orbit. Scientists think it was captured by Neptune, maybe after drifting in from the Kuiper Belt (see p. 138). However, it has a much rockier composition than the comet-like objects orbiting in the belt. Perhaps it changed during capture, which possibly involved collisions with other, now-vanished moons.

Planetary scientists believe Triton may be providing us with a preview of Pluto, which is probably another Kuiper Belt refugee and has not yet been visited by any spacecraft.

VIEWING THE BLUE WORLD

Seeing Neptune is not particularly difficult—almost any telescope can reach 8th magnitude. Knowing exactly where to look, however, can certainly be a challenge for beginners. Astronomy magazines regularly publish finder charts for Neptune, and any sky-charting software will give its current position. Until about 2012, the planet can be seen in Capricornus, a relatively star-poor region, making it easier to identify.

When you locate Neptune, chances are you will recognize it. At moderate power (about 70x), the planet shows a tiny disk and it appears distinctly blue-gray, because of the methane in its atmosphere. At higher power (150x or more), you can see the 2.3 arcsecond disk more clearly.

If you are hunting for moons, Triton is the only possibility. At 13th magnitude, it looks just like a star and is difficult to find. It calls for a telescope of 8 inches (200 mm) aperture or larger.

NEPTUNE FACTS

Distance from Sun 30.1 AU
Mass 17.1 x Earth's mass
Radius 3.88 x Earth's radius
Apparent size 2.3 arcsecs
Apparent magnitude 7.9
Rotation (relative to stars) 16.1 hours
Orbit 165 Earth years

PLUTO

There is an undeniable thrill in seeing this distant member of the Sun's family, and Solar System buffs cannot call their observations complete without it.

Percival Lowell, best known for his theories about Martian life, began the search for Pluto in 1905. An accomplished mathematician, he believed that residual irregularities in the motion of Uranus could be explained by the gravitational pull of a planet orbiting beyond Neptune. Lowell's approach had good precedent: discrepancies in Uranus's motion had led directly to the discovery of Neptune in 1846.

However, in the case of Pluto, Lowell was off track—we now know that the real Pluto is much too small to have any such effect. But Lowell's obsession with "Planet X," as he called it, did eventually pay off. After Lowell's death in 1916, the observatory he founded in Flagstaff, Arizona, dropped the search for more than a decade. Then in 1929, Lowell Observatory hired Clyde

Pluto, the grisly god,

who never spares,

Who feels no mercy, and

who hears no prayers.

The Iliad (Book IX),
HOMER (c. 8th century BC),
Greek epic poet

Tombaugh, a 26-year-old amateur astronomer from Kansas. He started making a photographic survey of the ecliptic—and, working systematically, he found the planet within a year. It was named Pluto for the Roman god of the underworld.

VISITOR FROM THE KUIPER BELT

Pluto is an odd object. Its 248-year orbit is the most eccentric and inclined of all the planets, and, for part of it, Pluto travels closer to the Sun than Neptune does.

Astronomers think Pluto is an object that wandered in from the Kuiper Belt, a region of the Solar System beyond the zone of the planets. The belt begins at roughly 35 astronomical units and may extend to about 1,000 astronomical units. The bodies in the Kuiper Belt are icy planetesimals—comets without tails. These never accreted into larger objects like Neptune because they orbit too slowly and there are not enough of them to make planet-building collisions likely.

THE BEST VIEW *so far of Pluto and Charon (left), from the Hubble in 1994. Pluto a ruler of the underworld (far le Clyde Tombaugh (below).*

NEWTONIAN TELESCOPE

A MOON FOR PLUTO

In 1978, James Christy discovered that Pluto has a moon. He named it Charon, after the mythological boatman who ferried the souls of the dead across the River Styx to Hades. When astronomers determined Charon's orbit, they were startled to note that between 1985 and 1990, observers on Earth could see the Pluto-Charon system essentially edge-on, and watch the two bodies eclipse and occult one anothe every 3.2 days.

By taking measurements during these eclipses, scientists worked out that Pluto has a diameter of 1,425 miles

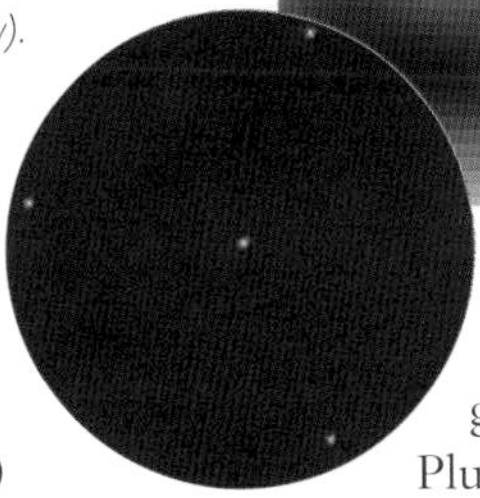

HE FAINT OBJECT CAMERA *on the Hubble Space Telescope produced a map f Pluto (right). From the ground, Pluto oks like a faint star (below).*

2,300 km), and Charon is fully half ts size: 760 miles 1,220 km). They re separated by 2,100 miles (19,500 m), making almost a "double planet." Scientists also stimate the masses of Pluto nd Charon together amount o less than 1/400 Earth's mass.

A Layer of Gas

A second stroke of luck ccurred in June 1988, when Pluto passed in front of a 2th magnitude star. As hey watched the star blink ut, astronomers discovered hat Pluto has a thin, cold tmosphere consisting mainly f nitrogen and methane. The data also hinted that he atmosphere might have azes. But the strangest thing s that Pluto's atmosphere is oth dynamic and transient.

Pluto's weak gravity cannot old the atmosphere forever, nd it is escaping into space nuch like a comet's gases do.

The atmosphere is also episodic: it is gaseous only when Pluto is around perihelion, the peak of its orbital "summer." Pluto reached perihelion in 1989. As the planet retreats toward aphelion (the coldest part of its orbit), it is cooling, and scientists expect the entire atmosphere will condense into snow or frost by the 2010s. And it will remain frozen until Pluto again draws near the Sun, in the twenty-second century.

Seeing Pluto

Pluto is a tough challenge for any planet watcher. At about 14th magnitude, it is a very faint object. You need a telescope of at least 8 inches (200 mm) to find it—but a 10 inch (250 mm) or larger scope makes the job a lot easier. With an apparent diameter of less than 0.1 arcsecond, Pluto displays no visible disk and looks no different than a star.

So how do you know when you have found it? Basically, you locate the correct field of view, map (by hand or with a camera) every star you see down to 15th magnitude, and recheck the field a night or two later. The star that has moved in the interval is Pluto.

Astronomy magazines publish finder charts for Pluto, which are a great help for getting your telescope into the right area of the sky. Computer-driven telescopes (see p. 58) can make the job even easier. But actual identification still calls for observations over several nights to spot the moving dot.

And do it within the next few years or so. Pluto already lies north of Antares in Scorpius and is approaching the Milky Way. Once there, the crowded star fields will make identifying this distant world much more difficult.

PLUTO FACTS

Distance from Sun 39.5 AU
Mass 0.002 × Earth's mass
Radius 0.18 × Earth's radius
Apparent size 0.08 arcsec
Apparent magnitude 13.7
Rotation (relative to stars) 6.39 Earth days
Orbit 248 Earth years

SPACECRAFT *are yet to visit Pluto and its moon Charon, but NASA is planning the Pluto-Kuiper Express. The small craft (left) would take 7 to 10 years to arrive.*

Asteroids

While these rocky fragments pose a challenge to the amateur astronomer, they have much to tell us about the history of our Solar System.

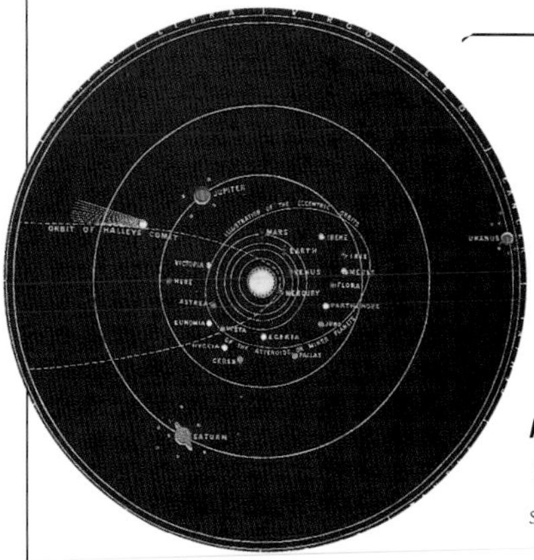
JUPITER
ORBIT OF HALLEYS COMET
MARS
EARTH
MERCURY
FLORA
VICTORIA
EUNOMIA
VESTA
EGERIA
HYGEIA
CERES
PALLAS
JUNO
URANUS
SATURN

A CHART *of the Solar System from 1857, showing, among other things, some of the known asteroids.*

Their very name means "star-like," and it tells the telescope user not to expect details like those seen on the Moon or Jupiter. Asteroids always look just like points of light; only the Hubble Space Telescope or radar observations can image a few of their surfaces in any detail from Earth.

For all that, asteroids are fascinating objects, being remnants of a much larger population of small bodies that formed and evolved under the gravitational control of Jupiter (see p. 129). Most known asteroids are still found within a main belt, between the orbits of Mars and Jupiter. Collisions within this belt may hurl asteroid fragments into orbits that cross Earth's orbit, and these fragments are the main source of meteorites.

Missing Planet?

Eighteenth-century astronomers saw a pattern in the spacing of the planets, which led them to look for something orbiting between Mars and Jupiter. Several observatories joined the search and, in 1801, the first asteroid, 1 Ceres, was discovered.

Since that time, more than 35,000 asteroids have been identified. No one knows how many asteroids there are in all, although we have probably found every object larger than 60 miles (100 km) across and about half of the 6 mile (10 km) objects. Estimates suggest there may be a million asteroids larger than 1/2 mile (1 km) in diameter.

The largest asteroid, at barely 600 miles (1,000 km) across, is Ceres, and the next biggest, Pallas and Vesta, are about half that size. Planetary scientists studying the main belt are beginning to see its current inhabitants as the remains of a much larger population. They believe that collisions—with each other and with the planets—have steadily reduced larger asteroids to smaller ones, and smaller ones to dust and fragments.

Types of Asteroid

Scientists classify asteroids into two main kinds, based on their reflectivity. The darker kind reflect 5 percent or less of the sunlight falling on them, and resemble dark stony meteorites (see p. 156). The brighter kind reflect about 20 percent of sunlight and look like light colored stony meteorites. A rare third type resemble iron meteorites and may be the shattered metallic cores of ancient proto-asteroids.

Hunting Asteroids

Most asteroids are too faint for amateur instruments to detect. Still, there are several hundred within reach of a 3 inch (75 mm) telescope. For most observers, the question is: where do I look?

Astronomical almanacs publish positions, called ephemerides, for the brightest asteroids, and more can be

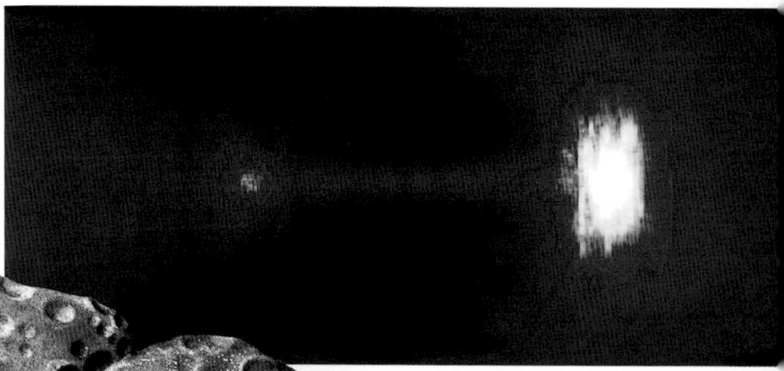

FRAGMENTS *This computer-simulated view of the whole sky (above) shows the dusty fragments from asteroids that lead, at left, and trail, at right, Earth in its orbit. The binary asteroid Toutatis (left).*

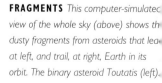

STEROID *243 Ida and its moon* *actyl (below) have been color-enhanced* *ere to reveal mineralogical details. A* *omputer model of near-Earth asteroid* *769 Castalia (right), "a contact-binary."*

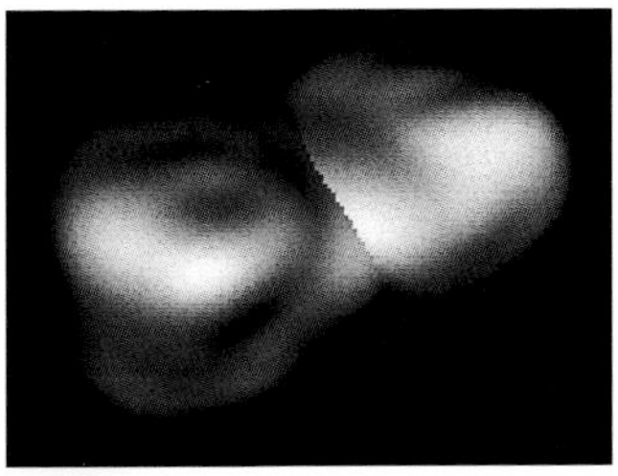

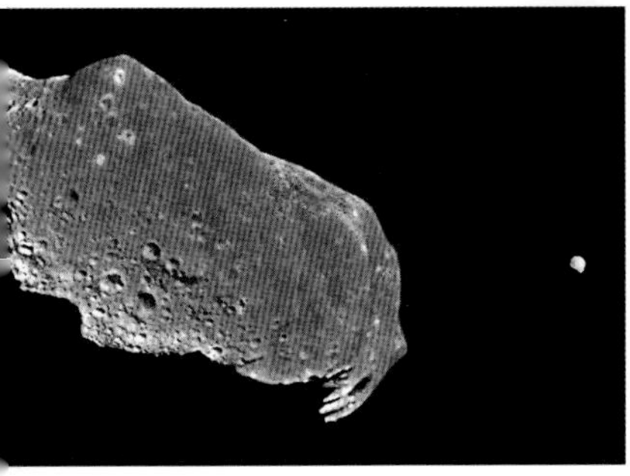

ound in computer software, stronomy magazines, and nnual astronomical hand- ooks (see p. 276).

When you have positions or an asteroid, plot them on a tar chart and observe the field. tart with one of the brighter steroids—these just reach aked-eye visibility. When ou locate the right area, com- are your sky chart with the tar field using your lowest- ower eyepiece. If you are ucky, the asteroid will be easy o identify as the "star" not lotted on the chart.

If the asteroid is not ob- ious, you will have to identify he most likely area for it to e in and plot all the star-like bjects you see, going down ne magnitude fainter than hat predicted for the asteroid. Wait several hours, or come ack the next night, and re- bserve the field. Your steroid is the star that moved.

From time to time, an steroid will pass in front f a star, briefly blocking its ght. Such events, called ccultations, are extremely aluable in determining an steroid's physical size and hape. Often it is the only vay to gain this information. Amateurs join professionals n expeditions to locations where they hope to see the event. As it is difficult to know precisely where the best view of the event will be visible, the more observers there are, the better. Anyone interested in helping with this kind of observing should contact their national amateur organization (see p. 277).

Finding New Ones

A handful of programs at professional observatories are devoted to the search for asteroids. One approach is to photograph a portion of sky, wait a day or two, and re-photograph it. By comparing the two images, any object that moved between the exposures—a definite asteroid candidate—can be identified.

A second method uses CCD detectors to image a portion of sky through a telescope. As the chip records the field of view, it electronically subtracts a reference image for that area containing just the stars. Anything not in the reference image will stand out clearly.

Neither of these approaches is perfect. The film technique imposes delays, but covers a much wider field of view. The CCD method produces results at the telescope in real time (or close to it), but it covers only a tiny portion of sky in each exposure.

JOHN, PAUL, GEORGE, AND RINGO

Or as they are officially designated, minor planets 4147 Lennon, 4148 McCartney, 4149 Harrison, and 4150 Starr.

Unlike other astronomical objects, asteroids can carry the names of living people or commonplace objects. Whoever discovers an asteroid is allowed to suggest a name for it, and the name is usually accepted by the ruling body, a committee of the International Astronomical Union. The name is combined with a number reflecting the asteroid's order of discovery: 1 Ceres was the first asteroid found.

With more than 35,000 asteroids identified, fewer than half have names. The names of the first few hundred asteroids tend to be of females from classical mythology (16 Psyche, 34 Circe). Other religions and mythologies were explored (77 Freia, 1170 Siva), then wives and girlfriends followed (607 Jenny, 1434 Margot).

As the number of asteroids grew, the list broadened to include 2825 Crosby and 4305 Clapton, 3656 Hemingway and 4474 Proust, 4511 Rembrandt and 6677 Renoir, and even 6000 United Nations.

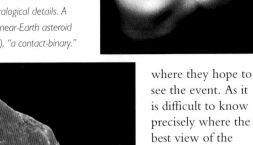

COMETS

Once interpreted as omens, comets remain a source of wonder and excitement. In recent years, our skies have played host to several spectacular examples.

HAROLD

THE BAYEUX TAPESTRY *shows King Harold I being told of Halley's Comet before the Battle of Hastings in 1066.*

After a long time during which there were very few bright comets, the 1990s will be remembered as the comet decade. Comet Levy in 1990, Comet Shoemaker-Levy 9 in 1994, Comet Hyakutake in 1996, and Comet Hale-Bopp in 1997 captivated skywatchers around the world.

ANCIENT COMETS

Comets are among the oldest objects in the Solar System, almost unchanged since they were formed billions of years ago. The primordial Solar System was a large, flattened cloud that spun slowly, its center slowly building to become the Sun, its outskirts condensing to become the planets. Wandering through this cloud were the comets. As Jupiter and the other giant planets grew, their gravity slung many of these comets into a large sphere, now called the Oort Cloud and located well beyond Pluto's orbit. Others survive in the Kuiper Belt, just beyond Pluto and more closely confined to the plane of the Solar System.

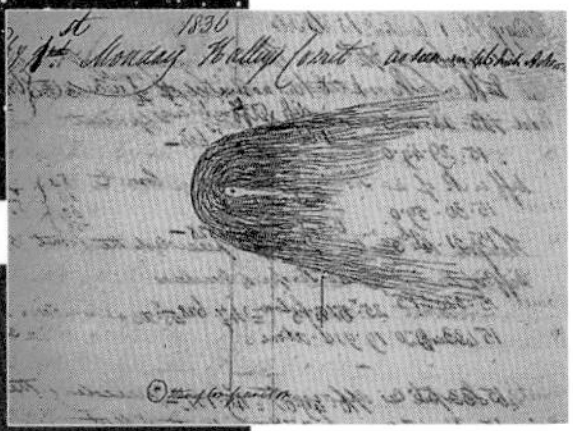

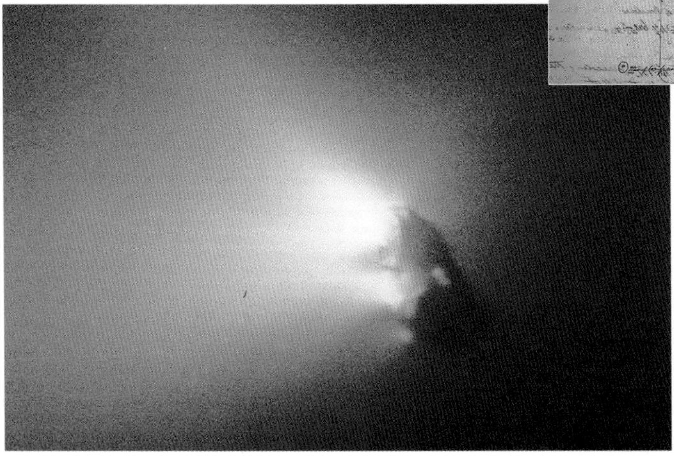

As far as a light-year away from the Sun, the Oort Cloud contains billions of comets, their ices forever frozen unless disturbed by the gravitational tug of a passing star or by the Solar System's passage through the galaxy. A comet may then leave the cloud, either ejected from the Solar System altogether or following a new orbit that will head it toward the Sun. After 4½ billion years in deep space, the comet begins to boil off its gases, or sublimate, as it nears the Sun. A comet newly discovered may be in just this state, warming up for the first time and releasing ancient gases.

WHAT IS A COMET?

Imagine a snowball several miles wide, loosely packed with rocky or muddy material. Far from the Sun, a comet consists of little else. Within the snowball are ices of water, cyanogen, and other materials as well as what scientists call CHON particles—organic material containing carbon, hydrogen, oxygen, and nitrogen. As the snowball, or

HALLEY'S COMET *sketched in 1836 (above), and photographed from Earth in 1986 (above left). Also in 1986, the Giotto spacecraft revealed the comet's nucleus, which is surprisingly dark (left).*

HE STRUCTURE OF A COMET *is illustration shows a typical comet, th its tiny nucleus inside a gaseous ma, trailed by two distinct tails.*

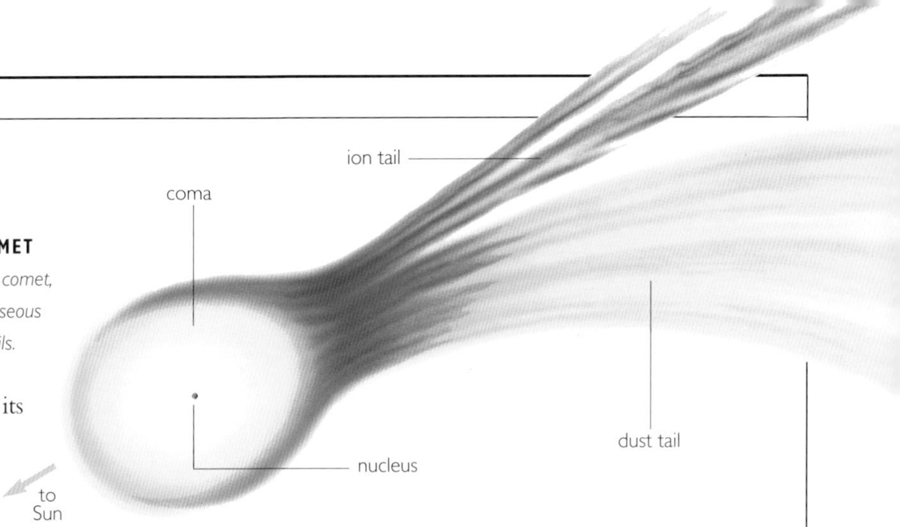
ion tail
coma
dust tail
nucleus
to Sun

ucleus, nears the Sun, its es begin to sublimate, leasing gas and dust. his material forms large head, or coma, ound the nucleus, and then reams behind the coma to rm a tail. The coma of a rge comet might be thou-nds of miles across, its tail ns of millions of miles long. n active comet might bleed way only the topmost few et of material during its onths near the Sun. Even ough the comet appears be burning away, it has nough stored ice and dust last for hundreds of volutions about the Sun.

OMET TAILS

here are two main types of omet tail. Ionized gas forms n ion tail, which appears uish, while dust particles rm a yellowish dust tail. The ust tail curves out from the omet under the pressure of nlight. The ion tail reacts the charged particles of e solar wind and is pushed rectly away from the Sun. his means that both tails ream behind the comet when approaching the Sun, but when the comet moves away from the Sun, the tails lead.

In addition to these tails, small jets of dust and gas may erupt. Such eruptions are triggered as the nucleus rotates and sensitive areas heat up quickly as the Sun rises over them. Seen through a tele-scope, jets are a striking sight.

HOW COMETS TRAVEL

Like the planets and their moons, comets travel about the Sun in paths known as orbits. The orbits of the major planets are almost circular, but the orbit of a comet such as Halley's is a much more elongated ellipse. A comet coming in from the Oort Cloud has an even more ex-tended orbit: it speeds round the Sun and then heads out for a trip that either never ends or will not bring it back for millions of years. On rare occasions, a comet travels in on an elliptical orbit, but then encounters a large planet such as Jupiter, whose gravity puts it into a new, hyperbolic orbit that ejects it from the Solar System forever.

Comets that travel in elliptical orbits and return to the vicinity of the Sun within a period of 200 years are called short-period comets. The most famous of these is Halley's, whose return every 76 years allows each generation on Earth a chance to view it. Most short-period comets return much more often, with 6 years being the average. With a period of 3.3 years, Encke's Comet has the shortest period.

Comets Hyakutake and Hale-Bopp, the great comets of 1996 and 1997, are good examples of long-period comets. Returning for its first visit in 9,000 years, Comet Hyakutake passed close to Earth in March 1996. A year later, Comet Hale-Bopp rounded the Sun on its first visit in more than 4,000 years. An approach to Jupiter has shortened its orbit, so that its next visit will occur in only 2,400 years. It is possible that repeated encounters with Jupiter could shorten the period even more. In a hundred thousand years, Hale-Bopp might return as often as Halley's does now!

COMET HYAKUTAKE *The comet tail stretched across the sky during its closest approach to Earth, in March 1996.*

WHAT IF YOU THINK YOU HAVE DISCOVERED A COMET?

1 Check to see whether the object has a tail.
2 Move the object around the field of view to make sure that it is not a reflection in your eyepiece from a nearby planet or bright star.
3 If the object is faint, use high power to verify that the object is not a faint star, or group of stars, that might have appeared diffuse at low power.
4 Check to see if the object is marked in a star atlas or catalog.
5 Draw a simple sketch, showing the object's position in relation to nearby field stars.
6 Look at it a half hour later to see whether it has moved.
7 Make sure that your object is not one of the known comets in the sky. The positions of the brighter comets are published in major astronomy magazines and other sources (see p. 276).
8 Have your sighting confirmed by an experienced observer. Tell that person the position, suspected nature, and direction of motion of the object you have found.
9 If you still think that you have discovered a new comet, it is now time to notify the Central Bureau for Astronomical Telegrams in Cambridge, Massachusetts (see p. 276).
10 If you have actually found a new comet, it may well be named after you.

I think I've found a squashed comet!

CAROLYN SHOEMAKER, American astronomer, on discovering Comet Shoemaker-Levy 9, 25 March 1993

HOW TO SEARCH FOR COMETS

A comet can appear anywhere in the sky. In 1995, Alan Hale and Tom Bopp found Comet Hale-Bopp when they both happened to be looking at Messier 70, a globular star cluster in Sagittarius. Most comets, however, brighten only when they are near the Sun, and appear either in the western sky after dusk or in the eastern sky before dawn.

Through a telescope, most comets look like fuzzy stars or galaxies. The best way to search for them is to move your telescope slowly—about one field every few seconds—in either an up-and-down or a left-to-right motion. Many comet-like objects will be visible, but a star atlas will usually identify them as galaxies, star clusters, or nebulas.

OBSERVING COMETS

Although some 25 comets are found each year, most of these are much too faint to appear in small telescopes. However, several comets of 11th magnitude or brighter appear annually, usually when they are close to the Sun. Almost always discovered by amateur astronomers, you can often see these comets in small scopes when the sky is dark and moonless. If the comet is brighter than 8th magnitude and the coma is condensed, it should be visible through 6 inch (150 mm) scopes. Dark skies are essential: a comet's ghostly light is easily swamped by a bright sky.

What makes comets so interesting to observe is their rapidly changing nature. A single comet can change radically in appearance as it swings past Earth, its geometr shifting so that we see differe parts of the tail from night to night. It is also common t see the comet's coma change in shape and structure from week to week.

Occasionally, the ion tail c an active comet may appear t separate itself from the head. This "disconnection event" occurs when the comet's mag netic field reacts to changes i the solar wind. A new tail ca be rebuilt in as little as a half hour, but the disconnection may recur every few days.

Less frequently, a comet will suddenly brighten by one or several magnitudes. This can be brought on by the onset of jet activity or by the ejection of a fragment from the nucleus. As Comet West rounded the Sun in 1976, the stresses of the Sun's gravity broke its nucleus into four pieces. This released great amounts of dust and gas,

COMET WEST *(left), rounding the Sun in 1976, and Comet Ikeya-Seki (below), photographed in 1965*

OMET HALE-BOPP *(below)*
ppeared in the northern skies in 1997.
false-color view of a comet (right).

:ausing the comet to brighten rom magnitude 0 up to −2, vhich is as bright as Jupiter lthough more spread out.

RECORDING YOUR OBSERVATIONS

Estimating brightness is one of the most common activi-ies for a comet watcher, out it can also be one of the trickiest. Among experienced observers, a popular pro-cedure is the Sidgwick, or 'in-out," method.

First, fix in your mind the "average" brightness of the comet's coma. Unfortunately this "average" tends to vary among observers. Then choose a comparison star, and take your telescope out of focus until the star reaches the size of the in-focus coma. Compare the star's surface brightness with the memorized average brightness of the coma.

Repeat this procedure and try to find a star that matches the coma's brightness. You will usually find that the brightness is somewhere between the magnitudes of nearby stars.

Comets vary greatly in appearance. Observers often record how condensed the coma appears. A standard scale ranges from 0 (diffuse image, no condensation) to 9 (a bright, star-like image).

If the comet has a visible tail, you may like to note its length. A tail equal in length to the Moon's diameter is 30 arc-minutes long. Also record the direction in which the tail is pointing. As the comet cruises past Earth, the orientation of the tail can change rapidly over a few days.

CAPTURING A COMET ON FILM

Although photographing a comet is a challenging task, the results can be spectacular. Photographers have to be aware of two things: first, that the star field moves in response to the Earth's rotation; second, that the comet moves slowly in the field of stars. When taking long exposures, either through a telescope or with your camera piggybacked on a telescope, you will need to guide it (see pp. 64–67). You can guide your telescope on the comet if the coma is sharply defined. More often, however, the comet will be diffuse, without a solid point to focus on. In that case, center the guiding telescope on a star. If the exposure lasts for less than 10 minutes, it is unlikely that the comet will show perceptible motion among the stars.

CHARLES MESSIER: THE COMET FERRET

Parisian Charles Messier (1730–1817) found his first comet in 1759, and proceeded to locate new ones almost every year thereafter. He was so successful that King Louis XV called him the "Comet Ferret." Messier (below) worked at the Marine Observatory in Paris, supported by a pension from his friend, Jean Baptiste de Saron, President of the Paris Parliament and an expert in comet-orbit calculation.

The French Revolution forced Messier to leave Paris, but in September 1793 he found a comet in the constellation of Ophiuchus. By this time, de Saron had been accused as an enemy of reform and was in prison. Despite this, he used Messier's positions to calculate an orbit for the comet from his prison cell, predicting that it would move closer to the Sun, then swing away and reappear in the morning sky.

Messier confirmed de Saron's predictions on 29 December, and smuggled the news to the prisoner. On 20 April, just three months before the end of the Reign of Terror, de Saron was guillotined. Although Messier survived, he was left virtually penniless.

Comet *and* Asteroid Impacts

Comets and asteroids are fascinating not just for what they are, but also for what they can do.

MIMAS, *one of Saturn's moons, bears the scar of a devastating collision with a comet or asteroid billions of years ago.*

In July 1994, 21 fragments of Comet Shoemaker-Levy 9 collided with Jupiter, one about every six hours. The impacts produced energy greater than the combined force of every nuclear weapon on Earth, providing a graphic demonstration of what could happen if such an object were to strike our planet.

The Jupiter collisions brought comet impacts to the forefront of research by planetary astronomers and geologists, and offered a revised view of the evolution of life on Earth. Instead of a smooth, gradually changing process, it now appears that life on Earth progressed erratically. Long periods of relative stability were punctuated by periods of sudden change, possibly produced by comet and asteroid impacts.

We live in a peaceful time in Solar System history. Our sky is almost unchanging, with bright comets appearing only about once a decade. However, it was not always this way—the youthful Solar System was a violent place and the Earth was so hot that all its organic materials were vaporized away. As Earth cooled down, its sky was crammed with bright comets, and major impacts took place often.

The Origin of Life

It is possible that the combination of impacting comets and slowly falling cometary dust provided the building blocks for life on Earth—carbon, hydrogen, oxygen, and nitrogen. Observations of Halley's Comet in 1986 determined the presence of these materials in almost identical amounts as exist on Earth. It is possible that comets also brought our water supply. Astronomers have calculated that Comet Hale-Bopp, the great comet of 1997, carried with it one trillion tons of water, an amount 50 percent greater than all the water in North America's Great Lakes.

Mass Extinction

If comets can bring life, they can also take it away. Much evidence supports the theory that a comet or asteroid struck the Caribbean basin 65 million years ago. In this scenario, the first result was an incredible earthquake measuring 12 on the Richter scale. Millions of tons of dust surged upward to form a gigantic cloud. The excavated material rushed out with such force that it quickly circled the Earth. For more than an hour, the surface of the Earth was bombarded with this debris, sending temperatures soaring

FRAGMENTS OF COMET *In February 1994, the Hubble Space Telescope captured Shoemaker-Levy 9 (below) on a collision course with Jupiter. It later recorded eight of the impact sites (right).*

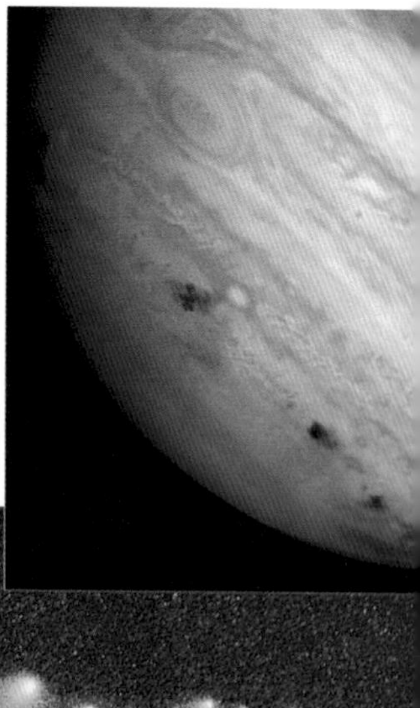

IMPACTS, GREAT AND SMALL *A 100,000x magnification of the crater left on a space satellite by a loose paint flake (left). Meteor Crater, Arizona (below).*

nd setting off a global fire-torm. Larger, slow-moving lebris landed close to the nain crater, in the Gulf of Aexico, creating miles-high sunamis that devastated the oasts of Mexico and Florida.

Soon the whole planet was hrouded in a cloud of dust nd soot, and for more than a nonth there was no sunlight nywhere on Earth. The huge mount of nitric oxide in the ir created rain dense with ulfuric acid. As the air finally leared, the temperatures rose gain and Earth suffered a evere greenhouse effect lasting or centuries. It is widely be-ieved that this impact, and the onsequent climatic changes, aused the mass extinction hat claimed 80 percent of all pecies, including the dino-aurs. This extinction, in turn, ed to the rise of mammals.

THE RISK OF IMPACT

Asteroids that remain in the nain belt pose no threat to us. Iowever, we know of some 200 asteroids with orbits that ross the orbit of Earth, and there are probably 10 times that number we do not know about, as well as an undetermined number of comets. Any of these objects could collide with our planet.

The threat of an impact is real. Earth already bears the scars of previous collisions. Meteor Crater is a 1 mile (1.6 km) diameter crater in northern Arizona, made 50,000 years ago by a piece of nickel-iron just 100 feet (30 m) wide. As recently as 1908, an object—possibly a 100,000 ton asteroid—collided with Earth, exploding at an altitude of about 6 miles (10 km) with the force of a 20 to 30 megaton nuclear bomb. The explosion occurred largely over uninhabited Siberian forest, but had the asteroid landed in a densely populated area, the impact would have been catastrophic.

SEARCHING FOR THE NEXT ONE

We do not know when Earth will be hit again. Statistically, a 6 mile (10 km) diameter comet—large enough to cause a mass extinction—should strike once every 100 million years. Smaller 1/2 mile (1 km) diameter objects could strike every 100,000 years, and would certainly still be devastating.

Astronomers are hoping to get funding to systematically survey the sky for impact hazards, plotting the orbit of every asteroid or comet that could inflict global damage. The numbers are daunting—there are millions of objects to be discovered and sifted through to identify the 2,000 or so most dangerous ones.

Next time a comet "stalks the sky," we should enjoy its beauty and mystery. And as we gaze, we should be mindful of the awesome potential for disaster should such an object strike the Earth. But also remember: were it not for such an impact, 65 million years ago, we quite possibly would not be here today.

HE EXTINCTION *of the dinosaurs nay have been caused by a comet or arge asteroid colliding with Earth.*

Stars scribble in our eyes the frosty sagas,
The gleaming cantos of unvanquished space.

HART CRANE (1899–1932),
American poet

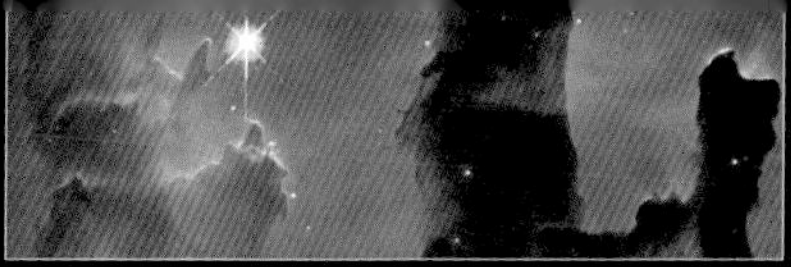

Chapter Five
Other Lights *in the* Sky

TYPES *of* OBSERVING

Beyond the Sun, Moon, and planets, the universe presents a treasure-trove of near and far celestial wonders.

Glittering star clusters, wispy nebulas, and misty galaxies abound in the dark regions between the stars. Intriguing star patterns, double and multiple stars, and the billowy star clouds of the Milky Way add to the night sky's boundless riches. Closer to home, Earth's upper atmosphere also puts on a show—with meteors, auroras, and meteorological phenomena such as noctilucent clouds, sun dogs, sun haloes, and the green flash.

DECIDING WHAT TO OBSERVE

Where to begin? What to observe? The answers depend on whether you have access to a pair of binoculars or a telescope, the quality of your observing site, and your geographical location.

Even without a telescope or star chart, you can venture outside on any clear evening and watch the stars gradually emerge from the deepening twilight. About an hour after dusk, you may see the sudden bright flash of a meteor or spot an artificial satellite moving steadily among the stars.

Looking beyond our Solar System out into deep space, we see myriad star clusters, nebulas, and galaxies. Some of these deep-sky objects are large and bright enough to see with the naked eye. With even a modest pair of binoculars, the star clouds and clusters of the Milky Way are incomparable.

However, most deep-sky objects require at least a 4 to 6 inch (100 to 150 mm) telescope and a dark observing site far from city lights.

A telescope will also help you track down double and variable stars, even from the semi-darkness of the suburbs.

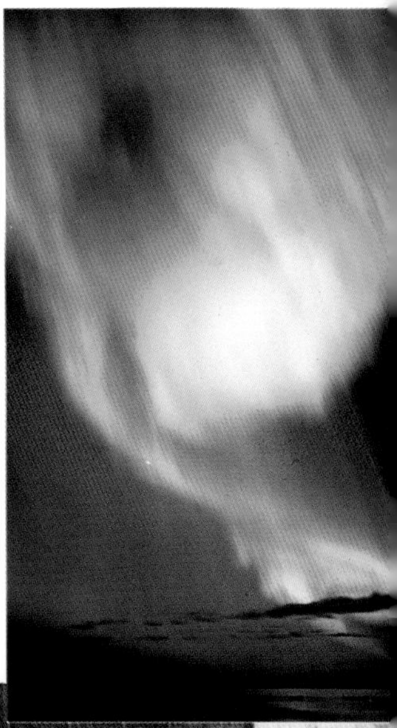

THE SPECTACLES *of a meteor (top) and an aurora (right). The more distant universe fascinated the amateurs at a 1917 meeting of the American Association of Variable Star Observers (below).*

DEEP-SKY WONDERS *include M66, a spiral galaxy in Leo (right); the globular cluster Omega Centauri (below); and the diffuse nebulas of NGC 6559 and IC 1274–75 in Sagittarius (below right).*

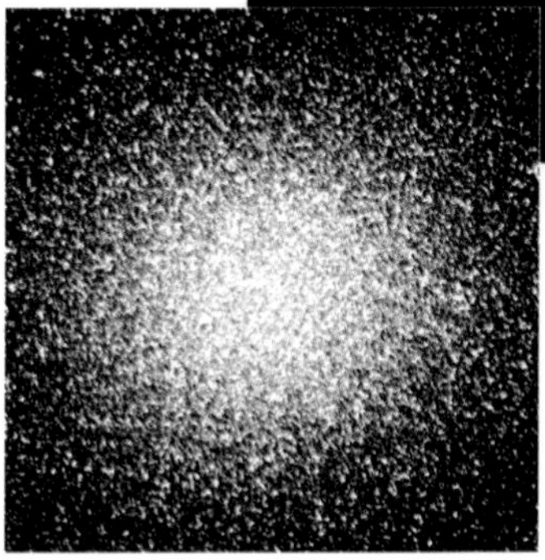

No one can see everything that happens in the sky from one place on Earth, and your geographical location will determine what you can and cannot observe. The Magellanic Clouds, for example, cannot be seen from much of the Northern Hemisphere, while most Southern Hemisphere observers are unable to see the Big Dipper. Sky phenomena such as auroras and noctilucent clouds are more frequently observed at high latitudes, but the zodiacal light is more obvious in the middle latitudes.

The Observing Specialist

No matter where you live, it does not take many nights out under the stars to realize that there is a lot of universe and that we can only see a very small swath of it at a time. A good way to structure your observing is to follow the star-hops described in Chapter 6. These encourage you to explore one patch of sky in detail.

As you become more experienced, you may find yourself returning time and again to a particular aspect of skywatching—counting meteors, say, or searching for double stars. Before you know it, you are on the road toward becoming a serious, specialized observer.

Specialized observing can be an extremely rewarding pursuit, particularly if you submit your observations to an organization that archives these for scientific reference. (A list of these organizations is provided in the Resources Directory, see p. 277.) Because there are a limited number of observatories in the world, and time on the big telescopes is always oversubscribed, professional astronomers do not have the opportunity to oversee every aspect of the sky. But amateur astronomers are free to react quickly to discoveries, and can make a significant contribution to many of the fields described in this chapter. For instance, they are often the first to report a supernova in a distant galaxy or to plot the brightness fluctuations of an erratic variable star.

The technological developments of the twentieth century, which initially threatened to displace amateurs, have actually opened up new areas of study. The accessibility of photographic equipment and, more recently, CCD technology has enabled amateurs to support professionals in specialized fields. Amateurs often capture useful pre-discovery images of novae, supernovae, and variable stars. Using a device called a photometer, they can measure variations in the intensity of a variable star's light. With a micrometer attached to a telescope, amateurs can measure the separation of double stars and contribute to the field of astrometry, which determines the precise positions of celestial bodies.

Even if you do not have the time or desire to become a "serious" amateur astronomer, there are many personal rewards in studying the near and far universe—not the least of which is a greater appreciation of the boundless beauty of the sky above us.

NATURAL SKYLIGHT

The ghostly glow of auroras and the gentle radiance of the zodiacal light never fail to enchant skywatchers.

The phenomena of auroras and the zodiacal light demonstrate that Earth is a planet within a system of planets. An aurora is the reaction of Earth's upper atmosphere to particles streaming outward from the Sun, while the zodiacal glow is created when sunlight reflects off countless specks of interplanetary dust.

CURTAINS OF LIGHT

The compelling sight of the aurora borealis, or northern lights, has inspired, frightened, and otherwise transfixed people living in Arctic regions for centuries. To the Inuit of North America, the undulating gossamer curtains of light were the play of unborn children, or the torchlight held by the dead to aid the living in winter. Some say that if you whistle gently, the lights will respond by drawing nearer. In Scandinavia, it was once common to use auroras for weather forecasting, and auroras are still called "wind lights" and "weather lights."

Auroras are spawned around the Earth's magnetic poles by activity on the Sun (see p. 80). Generally, the more active the Sun, the more prominent the aurora. Solar activity increases and decreases over about 11 years, with most sunspots and solar flares appearing toward the middle of the cycle. The flares produce tremendous outflows of charged particles that intensify the solar wind.

PHOTO TIPS *Longer exposures will pick up a greater range of colors (left), while short exposures can freeze rippling curtains of light (above). A photo taken from the space shuttle shows the full extent of an aurora australis (below).*

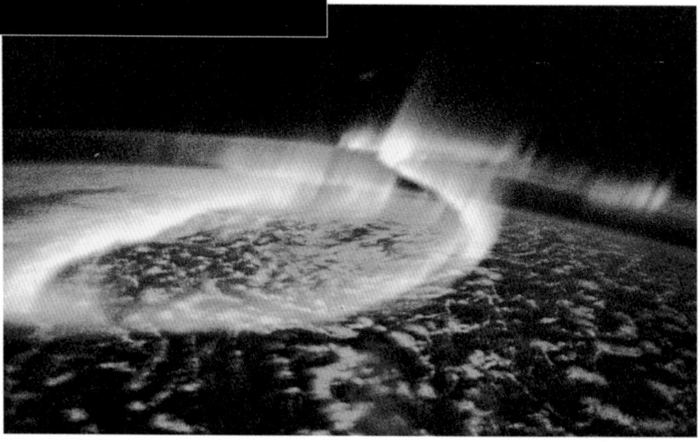

When the particles reach Earth, they excite, or ionize, neutral oxygen and nitrogen molecules in Earth's upper atmosphere. This ionization produces the eerie colored glow that we see as an aurora

In both the Northern and Southern hemispheres, most auroral activity is confined to high latitudes—the region above 65 degrees. Auroras ar commonly seen in Alaska, Canada, and Scandinavian countries, but in the Souther Hemisphere—where the phenomena is called the aurora australis, or southern lights—few inhabited locations are far enough south to see it often. During periods o high solar activity, auroras ca extend down to latitudes of 40 degrees or less, and during extreme activity, they may even reach the tropics. One major outburst of solar activit in March 1989 produced auroras that were seen from the Caribbean Sea.

Auroras are usually visible for about an hour, but during peaks in solar activity, they can last all night. For Norther

PINK AND GREEN *auroras appear when the solar wind ionizes oxygen molecules in our atmosphere. When nitrogen molecules dominate, blue auroras occur.*

lemisphere observers, an ırora begins as a dome f reddish or greenish light the north. The glow ventually shapes itself into a istinctive arc that gradually reeps southward. During its eak, green, red, and blue reamers of light may extend ı the zenith and beyond.

Spectacular photographs of ıroras can be made using a ipod-mounted camera, fast lm, and a wide lens. Simply ım the camera toward the ırora and expose for 10 to 0 seconds (see also p. 63). Color photographs will often eveal subtle colors that the ye does not detect.

HE ZODIACAL LIGHT

)f all the astronomical henomena visible to the aked eye, the zodiacal light one of the most delicate. It ppears as a diaphanous wedge f light, no brighter than the Milky Way, extending up om the horizon well after ınset. The glow is faint and ou can easily see dim stars ırough it. In fact, even its ıost "brilliant" display is often ıistaken for sunset glow r uplighting from a nearby ity or town.

The zodiacal light occurs when sunlight scatters off the ountless dust particles that orbit the Sun in a great disk. Like the planets, the dust particles are confined to the ecliptic—the plane of the Solar System. The constellations through which the ecliptic passes are collectively known as the zodiac, and since this is where the light is concentrated, it is called the zodiacal light.

Because the zodiacal light is aligned along the ecliptic, it reaches a higher altitude when the ecliptic is steeply inclined to the horizon. In middle-latitude locations, it is best seen in the west after nightfall during the local spring, and in the east before dawn during the local fall. In the tropics, where the ecliptic is more often nearly perpendicular to the horizon, the zodiacal light can be seen throughout the year in both the evening and morning sky.

To observe the zodiacal light, find a dark-sky location away from city lights. Give your eyes about 20 minutes to grow accustomed to the dark. Scan along the ecliptic from west to east, looking for a slight difference in brightness between the darker sky and the border of the zodiacal light. If it is not immediately visible, try using averted vision and look for the zodiacal light out of the corner of your eye.

AN ELUSIVE LIGHT

A manifestation of the zodiacal light appears in another part of the sky, where it has a different name—the gegenschein, or counterglow. A very faint elliptical glow some 10 degrees across, this phenomenon appears opposite the Sun in what is called the antisolar region of the sky.

The best time to locate the gegenschein is around midnight in early fall or early spring. This is when it is projected into a relatively star-poor region of the sky, away from the Milky Way. Dark, clear moonless nights are requisite to see this most elusive of sky glows. The technique of averted vision will also help you see the gegenschein.

THE ZODIACAL LIGHT *appears here as a delicate wedge of light extending up from the western horizon after sunset.*

Sunlit Clouds

By day and by night, the interaction of sunlight and moisture in our atmosphere creates delightful sights in the sky.

A MOON HALO *is depicted in this 1872 painting (left). Noctilucent clouds sometimes show a crossed pattern known as herring structure (below).*

As sunlight passes through the ice crystals and water droplets suspended in the air above us, it creates an endless variety of arcs, rings, dapples of color, and splashes of light that can be seen throughout the year.

Most of these phenomena occur within Earth's lower atmosphere, which means that they are actually meteorological in nature. Noctilucent clouds, however, can claim both meteorological and astronomical ties.

Noctilucent clouds appear at night—usually an hour or so after sunset, when the Sun is more than 10 degrees below the horizon. This phenomenon is known to appear around midsummer in both the Northern and Southern hemispheres, but it can only be seen between latitudes of 45 and 60 degrees, with most sightings occurring in Canada and northern Europe.

These silvery-blue clouds form at the edge of space at an altitude of 50 miles (80 km), which is almost five times higher than most of Earth's weather-making systems. At such heights, the temperature can dip to −150 degrees Fahrenheit (−100° C). The scant amount of water vapor present at this altitude condenses onto meteoric dust and charged particles in the atmosphere, forming the clouds.

Sightings of noctilucent clouds have doubled since the 1950s. A possible explanation is that industrial pollution has increased the amount of particles at high altitudes around which the clouds can form.

Over the Rainbow

Perhaps the most familiar of all atmospheric sights is the rainbow. A rainbow begins when sunlight enters a raindrop. Most of the light passes straight through the drop, but some of it is refracted, or bent, into the colors of the spectrum. It then reflects off the back of the raindrop and is refracted again as it exits. The result of this refraction and reflection through millions of raindrops is a multicolored ring around a point opposite the Sun. From the ground, we see a beautiful arch, which is part of the ring.

Sun Companions

Another common sight in daytime skies is the halo, sometimes called "a ring around the Sun." Haloes are caused by the refraction of sunlight through hexagonal ice crystals in cirrus clouds. As the sunlight passes through one of the six sides of the tumbling crystal, it is bent through a 22 degree angle, producing a halo with a radius of 22 degrees. Occasionally, when sunlight passes through the ends of hexagonal prisms, or through cubic ice crystals, it is bent through a greater angle, producing a second, fainter, 46 degree halo.

When the Sun is low in the sky, and when high cirrus clouds are present, look for

THE SUN *and Earth's atmosphere conspire in different ways to produce delicate sun haloes and sun dogs (left), and the spectacular green flash (below).*

ʼo bright spots of sunlight on :her side of the Sun. These ots, which will be tinged red the inside and blue on the tside, are commonly called n dogs or mock suns. The chnical name is parhelia.

Sun dogs are produced by nlight refracting through xagonal ice crystals that are iented with their bases paral- to the horizon. Sun dogs metimes touch the outside ge of a 22 degree halo.

The safest way to observe n dogs and haloes is to block ur view of the Sun with a arby structure such as a tree nb, chimney, or flagpole.

Yet another ice-related enomenon, sun pillars, can seen around sunrise or sun- :. When low-angle sunlight flects off ice crystals in cirrus ouds, a spike—or pillar—of ht may be projected up from e just-set or pre-dawn Sun.

At night, moonlight can eate moon haloes, moon gs or mock moons, and oon pillars, but these effects fainter and less common an the Sun phenomena.

ne can enjoy a rainbow ithout necessarily forgetting e forces that made it.

een *Victoria's Jubilee,*
RK TWAIN (1835–1910),
erican writer and humorist

A Flash of Green

One of the most spectacular solar-terrestrial events, as well as one of the rarest, is the sudden flash of green light seen just as the Sun rises or sets. The green flash, as it is called, is the result of Earth's atmosphere bending different colors of light by various amounts, especially near the horizon. You effectively have red, green, and blue suns that set at very slightly different times. The blue sun should be last, but the air usually blurs it out, leaving our final glimpse of the Sun as a flash of green.

To see the green flash, you will need a well-defined, unobstructed horizon and clear, haze-free conditions. An ocean panorama is ideal, which is why the green flash is most commonly seen from ships and from coastal plains and mountains.

Even when setting, the Sun can damage your eyesight, so only ever look at it for a few brief moments at a time.

CREPUSCULAR RAYS

Sunsets are often depicted in paintings with majestic beams streaming up radially into the sky from the already set Sun. Exaggerated as the artist may make them appear, sunbeams can often be seen in clear skies 10 to 15 minutes after sunset (or before sunrise). This phenomenon is called crepuscular radiation (from the Latin word *crepusculum*, meaning "twilight").

Crepuscular rays are caused by a cloud below the horizon partially blocking the Sun. The shafts of sunlight that get past the cloud illuminate dust particles in the air above the horizon, creating spectacular rays that occasionally reach to the opposite point in the sky (below). As the cloud moves and the Sun's angle changes, the rays, too, shift across the sky. The phenomenon may last for several minutes.

METEORS

Whether you are wishing upon a shooting star or counting the hourly rate of the Leonid shower, meteors are sure to capture your attention.

GIANT ROCKS *This enormous meteorite was found in Oregon, USA.*

The brief flash of light we call a meteor or, more popularly, a shooting or falling star, was once interpreted as a dead person's soul on its way to heaven, or even as a warning of death. The scientific explanation, however, is that a piece of space grit known as a meteoroid is burning up in Earth's upper atmosphere.

SPEEDY METEORS

When meteoroids enter our atmosphere, they travel at enormous velocities. Meteoroids that overtake Earth travel at about 6 miles per second (10 km/s). Those that plunge head-on into our atmosphere, however, can reach speeds of 45 miles per second (75 km/s), more than 225 times the speed of sound.

The surface of the meteoroid, rapidly heated by friction to more than 2,000 degrees Fahrenheit (1,100° C), begins to vaporize, or ablate. The meteoroid's particles and the particles in the atmosphere collide violently and are stripped of electrons, creating the meteor's intense luminosity, which can be seen more than 90 miles (150 km) away.

Tiny meteoroids last less than a second before burning up completely. Gravel-size meteoroids produce brighter and longer displays. Meteors as bright as the planet Venus are known as fireballs. These can cast shadows and last several seconds, leaving curling trails of glowing gas in their wake. Long-duration fireballs that explode or appear to fragment are called bolides.

The majority of meteoroids range in size from dust grains to small pebbles. Occasionally, larger meteoroids survive their fiery entry into Earth's atmosphere and reach the ground intact. When they do, they are called meteorites.

SPORADICS, SHOWERS, AND STORMS

On any moonless night, you should be able to see four or five meteors per hour. These unexpected shooting stars—or sporadic meteors—can appear anywhere in the sky. Their rate tends to increase from local midnight until dawn, when the observer's location is carried toward Earth's leading orbital edge and placed in the path of more interplanetary particles.

You never really know how many sporadics you will see, but some meteors can be anticipated. On certain nights each year, the rate of meteor

ROCKS FROM SPACE

When a meteoroid reaches the ground as a meteorite, it can help complete our picture of the universe. Many meteorites are pieces of matter unchanged since the earliest days of the Solar System, while others are bits of Mars or the Moon blasted into orbit by comet or asteroid impacts. Meteorites are divided into three main types according to their composition: stony meteorites; iron meteorites; and stony iron meteorites.

Stony meteorites (above) are angular gray or brown stones, made up of stony minerals. They tend to be fairly smooth and often have a glassy black fusion crust, formed as the rock ablated in the atmosphere.

Iron meteorites (below) contain 90 percent iron and 10 percent nickel, making them heavy for their size. Their surfaces can be black or rusty brown and are often pitted with thumbprint-shaped depressions.

Stony iron meteorites bridge the stony and iron classifications. They have the coarsest surface, with crystals of olivine in a network of nickel-iron.

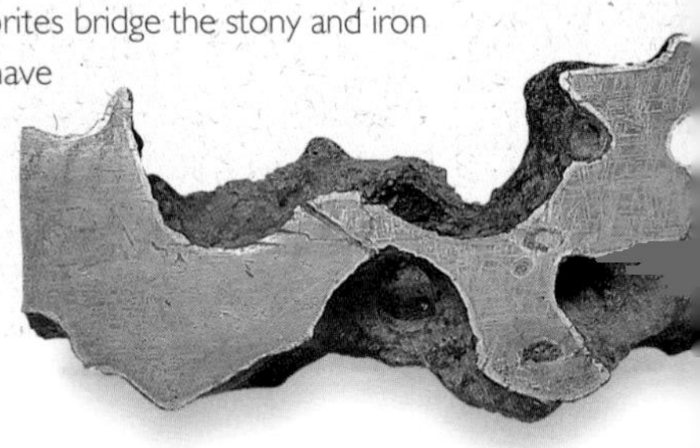

A SHOOTING STAR, *or meteor, provides a burst of movement in an otherwise still starry sky (left). Very bright meteors are known as fireballs (below).*

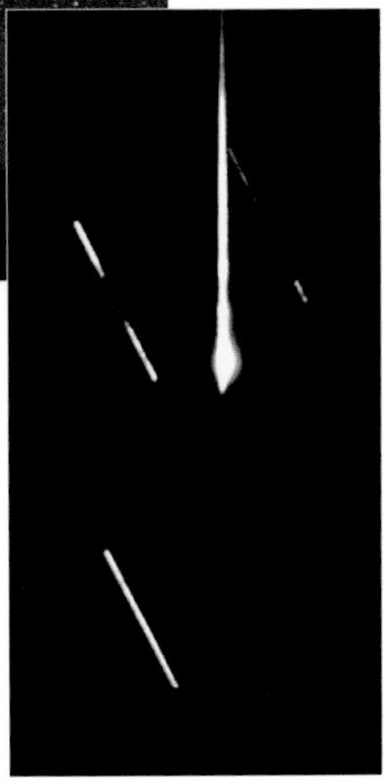

ot only increases, but the ıeteors appear to radiate from particular region of the sky. hese predictable displays are ılled meteor showers.

Compared with the 4 or meteors per hour you can xpect to see on any dark, ıoonless night, the 20 to 50 ıeteors per hour of a meteor ıower might be considered downpour. Even a shower, ıough, appears like a sprinkle ext to the meteor storm. A ɔbust meteor storm can pro-uce thousands of shooting ars per hour—sometimes 0 to 20 per second. Meteor orms, however, are rare vents (see Box, p. 159).

ΛETEOR STREAMS

Λeteor showers and storms ccur when Earth passes ırough a meteoroid stream, vhich is a trail of dust and ebris that orbits the Sun.

Most meteoroid streams are ıe detritus of comets, although ɔme streams have been linked ɔ asteroids. When a comet merges from the deepfreeze f space and moves through ıe inner Solar System, its rozen surface is warmed by ıe Sun. As the comet's icy rust vaporizes, billions of ɔns of loose material spew from its surface, and a concentrated path of debris forms in its wake. When Earth sweeps through this path, the meteoroids plunge into our atmosphere, where they burn up as meteors.

From the ground, it appears as if the meteors are streaming in from a particular region in the sky. That region is known as the radiant and its location gives the shower its name. For example, the meteor shower that appears to emanate from the constellation Gemini every 14 December is called the Geminids, which means "children of Gemini."

HOW MANY METEORS?

A meteor shower's activity is gauged by its zenithal hourly rate, or ZHR. This value, often quoted in the press and astronomy publications, has sometimes been the source of mis-understanding and disappointment.

The ZHR is an ideal value. It is, by definition, the number of meteors a single observer could possibly see during a shower's peak with the radiant directly overhead in a dark, clear sky.

Most observers, however, will not see as many meteors as the ZHR suggests. Obviously, if peak activity occurs during the day or when the radiant is below the horizon, few or no meteors will be visible.

Also, the more ambient light—artificial or natural—at your location, the fewer meteors you will see. City observers will see only the brightest meteors, and a bright Moon will wash out fainter meteors, even in the country.

Luck is another factor. If you happen to be looking in the wrong direction of the sky, a fast, short-lived meteor can easily sneak past you.

HE IMPACT *of a large meteorite eated Wolfe Creek Crater, in Western ustralia. It is more than half a mile ide. Fortunately, most meteorites are nall enough to fit into your pocket.*

THE LEONID METEORS *show up as yellow streaks in this false-color photograph. The blue trails are stars.*

Preparing for a Meteor Shower

You do not need a telescope or binoculars to observe a meteor shower. The naked eye, in fact, makes the best instrument because of the large amount of sky it can take in at any one time. However, binoculars can help you locate fainter meteors. Scan in and around the radiant zone and be on the lookout for fast flashers zipping among the stars. You may even be able to see a meteor break up in flight and leave behind a "smoke" train.

Meteor watching often involves long sessions, so it is important to be comfortable. A deck chair or recliner is an easy way to keep your head inclined at the right angle. Position yourself so you face the direction of the meteor's radiant, but do not ignore other parts of the sky. Meteors in a particular shower will sometimes appear elsewhere in the sky, even on the opposite side of the sky from the radiant. This is why it can be especially worthwhile to monitor a meteor shower in a group, with each observer facing a different direction.

A BOLIDE *depicted in an 1870 painting. Bolides are rare, but you are most likely to see one during a meteor shower.*

When to Observe

While there is no need for expensive equipment, you do need to choose your observing time carefully. The table below gives the dates of peak activity for major showers. For the best observing times on those dates, however, you will need to check local newspapers, astronomy magazines, or the Internet. Because of Earth's rotation and orbital motion, the exact time of a shower's peak varies slightly each year, usually by about six hours. Also, the shower's radiant ma[y] not be visible in your area at the time of maximum activit[y].

Say you live in Chicago and would like to observe th[e] Perseid shower. The Perseids peak on 12 August. In 1997, the peak occurred at 18 hour[s] Universal Time (UT), which converts to 1 pm in Chicago—obviously not a good time to look for meteors. You could observe either before dawn (some 9 to 10 hours before the peak), or in the evening (8 to 9 hours after the peak). An early morning session would probably be th[e] better choice because the daw[n] side of Earth faces into its orbital path.

You also need to check when the shower's radiant will be above the horizon, and wher[e] in the sky it will be. (A simple planisphere or the sky charts in *The Natur[e] Company Guides: Skywatching* can help you determin[e] this.) The radiant of the Perseids is in

ANNUAL METEOR SHOWERS

Shower	Active period	Peak	Rate per hour	Parent comet
Quadrantids	1–5 Jan	3 Jan	40–100	
Lyrids	16–25 Apr	22 Apr	15–20	Comet Thatcher
Eta Aquarids	19 Apr–28 May	4 May	20–50	Comet Halley
Delta Aquarids	8 July–20 Sept	29 July	20	
Perseids	17 July–24 Aug	12 Aug	50–100	Comet Swift-Tuttle
Orionids	10 Sept–26 Oct	22 Oct	25	Comet Halley
Taurids	15 Sept–26 Nov	3 Nov	12–15	Comet Encke
Leonids	14–21 Nov	17 Nov	10–15	Comet Tempel-Tuttle
Geminids	7–17 Dec	14 Dec	50–80	Asteroid 3200 Phaeton
Ursids	17–26 Dec	22 Dec	10–20	Comet Tuttle

he constellation Perseus. For Chicago observers during August 1997, Perseus rises in he northeast at about 10 pm nd stays visible until dawn.

THE METEOR LOG

ome meteor watchers go n to pursue serious meteor bserving. Should you decide o do this, your observations ould contribute important nformation about the me- eoroids in a particular stream.

A log of each night's bservations should note the ollowing general details:

the observing session's date, tart and stop time, and time- ut for breaks;

the longitude and latitude f your observing site;

sky conditions;

the limiting naked-eye agnitude (the magnitude f the faintest star visible o your eye); and

the direction of the area f sky you are observing, or, etter yet, the right ascension nd declination of the center f your viewing area.

Your observation records ould include the number f meteors that you see per our, and the magnitude nd direction of each meteor. ou can estimate magnitudes y comparing the meteor's rightness with that of the djacent stars. To plot the irection a meteor takes, mark track line on a star map owing where the meteor ppeared and disappeared.

In addition, describe each eteor, noting its duration nd the presence of smoke ains or fragmentation. If ossible, include an estimate f the meteor's speed. The andard way to do this is to stimate how far it would ave traveled had it persisted or one second (most meteors last a fraction of a second). For example, if a meteor traveled 6 degrees in about half a second, its speed would be noted as 12 degrees per second.

This can be a daunting amount of detail to write down, especially when the meteors are frequent, so some observers use a tape recorder during the session and transcribe the information later.

If you are interested in contributing to meteor astronomy, you can approach the International Meteor Organization for additional information. Contact details are provided in the Resources Directory (see p. 277).

As I flit through you hastily,
soon to fall and be gone,
what is this chant,
What am I myself but one
of your meteors?

Year of Meteors,
WALT WHITMAN (1819–92),
American poet

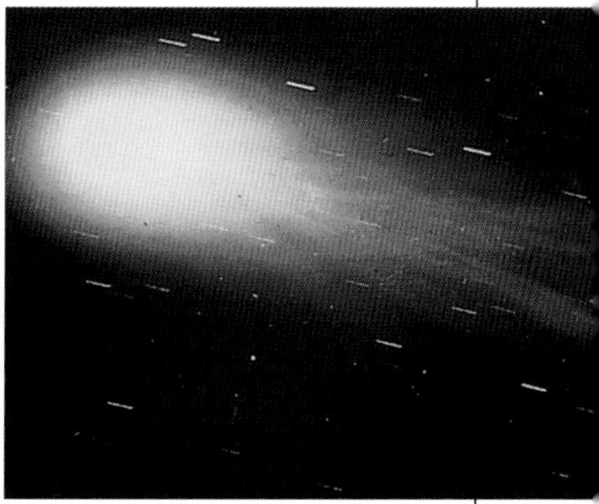

COMETS *shed tons of debris that can form meteor streams. Halley's Comet (right) has been linked to both the Orionid and the Eta Aquarid showers.*

THE LEONID METEOR STORM

One of the most famous meteor storms on record occurred on 17 November 1833, during the Leonid shower. All along North America's east coast—including Niagara Falls, New York (below)—stunned observers saw hundreds of meteors per minute. They described the meteors as falling like snowflakes or heavy rain. Estimates of the hourly rate range from 50,000 to 200,000 meteors.

Every 33 years, the Earth passes through an especially dense portion of the Leonid stream, resulting in a brief storm. Following the 1833 event, thousands of meteors again rained down in 1866. The Leonid stream can be perturbed by Jupiter's gravity, which may explain the poor performances in 1899 and 1932. The storm returned in full force, however, on 16 November 1966, when an estimated 150,000 meteors were seen a few hours before dawn. And in 1999 the Leonids "roared" again, although changes in the meteor stream reduced the numbers greatly compared to 1966.

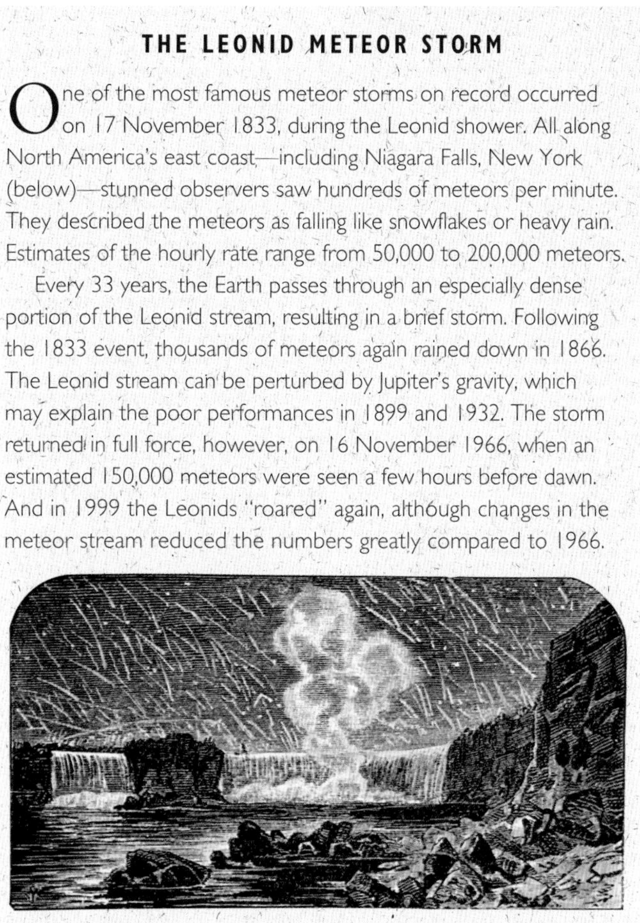

ARTIFICIAL SATELLITES

When stargazing, you will occasionally notice a "star" moving steadily across the sky—one of the hundreds of artificial satellites orbiting our planet.

SKYLAB, *the first US space station, could be seen orbiting Earth from 1973 until 1979, when it broke apart as it reentered the atmosphere over Australia.*

Artificial satellites are the wallflowers of the sky. They are reserved, silent, and usually pass overhead without attracting a single glance. Yet, when they are seen, usually inadvertently, they can garner as much attention as a shooting star. At star parties, someone will often point up into the sky and shout, "Satellite!"

There are hundreds of satellites and spacecraft orbiting Earth today. They are used for everything from telecommunications, meteorology, and navigation, to Earth-resources monitoring, geophysics, and astronomy. Not all satellites are visible to the naked eye, but many are, and you can usually spot at least one on any night.

Satellites are placed into one of three types of orbit: equatorial, polar, or geostationary (see p. 22). Equatorial satellites travel from west to east. Because they are usually placed at low altitudes, they tend to look like bright stars and take just a few minutes to move across the sky. Polar-orbiting satellites are placed in high-altitude orbits and travel north to south or south to north. Because of their greater altitude, they look like faint stars and seem to move more slowly than equatorial satellites do. Geostationary satellites are placed at extremely high altitudes and remain fixed over the same point on Earth. They can be glimpsed only in telescopes, and even then they are no brighter than a 14th or 15th magnitude star.

When a satellite or a piece of space debris reenters Earth's atmosphere, it may flare up so brightly that it casts a shadow. Such objects are often mistaken for fireballs (see p. 156), which, in a sense, they are, except that they are artificial.

SATELLITE SPOTTING

For people living from low t
middle latitudes, satellites wi
appear anywhere in the sky a
hour or two after sunset or
before sunrise, when objects
great altitude are still bathed
sunlight. Observers living at
high latitudes can expect to
see satellites for longer perioc

A satellite's visibility depends on a number of factors
• Altitude: Large satellites in low orbit can be brighter tha 1st magnitude stars; those in high orbit are fainter.
• Elevation: The light from a satellite that appears low in tl sky travels through more laye of air to reach the observer, and thus appears fainter than would if it were overhead.

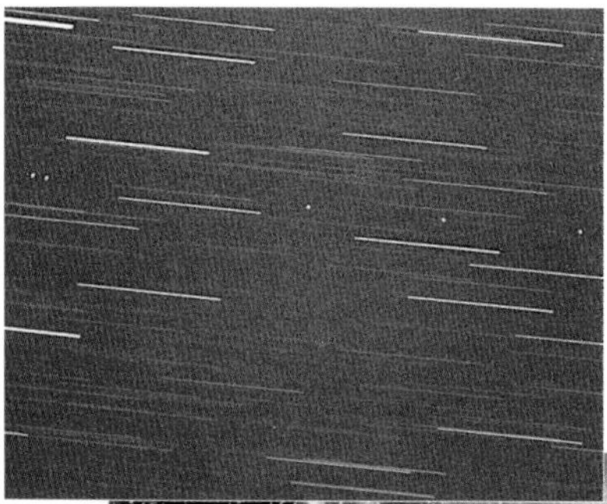

SATELLITE TRACKS *The dots amor the star trails (left) are five geostation communications satellites. A polar-orbiti satellite left a trail across this photo of the Vela supernova remnant (below).*

/isual Satellite Observer's Home Page
rom left to right the above images are: A Soyuz manned capsule. Satellite captured on
ng exposure photograph. The external tank caught just after separation during the STS
uttle mission.
mirror of these pages with predictions for the UK, Australia, South Afric

KEEP UP TO DATE *by visiting the Satellite Observer's web site (left). It provides advance warning of events such as the space shuttle's reentry (below).*

Observer-Sun-satellite angle: satellite is brightest when it fully lit by the Sun; that is, hen it is seen opposite the un, in the east after sunset or the west at sunrise.
Size: The larger the satellite, e easier it will be to see.
Satellite surface: A polished rface reflects more light than dull one does; a faceted surce makes a tumbling satellite ppear to "twinkle."

UCKY SIGHTINGS ND PREDICTIONS

atellite observing can be njoyed on many levels. Most ywatchers simply enjoy keeping an eye out for satellites hile they are doing other inds of observing. For this, l you need is a lucky glance the right direction at the ght time. Satellites can even e rather intrusive, leaving eir trails on long-exposure strophotographs, or startling eep-sky observers by racing rough their field of view.

Interested observers can btain predictions for when tificial satellites will appear their area. Favorite targets clude the space station Mir, e space shuttle, and the Iubble Space Telescope. redictions are supplied by roups such as the Belgian Vorking Group Satellites, r via web sites such as the isual Satellite Observer's Iome Page (see pp. 276–77 or contact details).

A few dedicated amateurs obtain software packages that enable them to make their own satellite predictions, and then diligently monitor satellite appearances, positions, and brightnesses. These observations can help scientists to accurately track satellite orbits and study how they are affected by changing conditions in the atmosphere and fluctuations in Earth's gravitational field.

To determine a satellite's position, you note exactly when and where the satellite crosses an imaginary line between two stars that are close to each other in the sky. This can be done by eye, but if you wish to submit your records, you will need to use a telescope or take photographs. For more information on contributing positional measurements of satellites, contact the Royal Greenwich Observatory, Satellite Laser Ranging Team (see p. 277 for contact details).

UFOS DO EXIST!

Do Unidentified Flying Objects exist? The answer is, yes—at least until they are identified.

A common UFO report is of an object that rapidly changes color, moves anomalously, and, sometimes, follows the observer. Disappointing as it may be to those who hope for close encounters with intelligent aliens, the two culprits usually responsible for such displays are the planet Venus and the brightest star in the night sky, Sirius.

At its most brilliant, Venus creates the illusion of being closer than it is, and may appear to trail after the observer. Sirius, being one of the brightest point sources in the sky, twinkles violently, especially when it is near the horizon. Twinkling, or scintillation, can involve rapid color changes and is caused by turbulence in the air.

Other common UFOs are fireballs, planetary conjunctions, glinting or flashing satellites, and reentering space debris. None, however, can be attributed to alien spacecraft.

The enormous popularity of the television series The X-Files *attests to the public's fascination with UFO sightings.*

Stars

People have always sought to understand the stars. The journey begins when you go outside on a clear night and look up.

In ancient times, people saw stars as little more than ornamental lights affixed to the vault of the heavens. They arranged stars into sky pictures, or constellations, and gave names to the bright stars.

Ancient astronomers could hardly have guessed that stars generate light and energy deep in their cores by fusing hydrogen gas to form helium. Nor that these enormous spheres of hot gas are actually scattered throughout space.

Star Bright

Astronomers designate a star's brightness in visible light by its apparent magnitude. The faintest star you can see with the naked eye is about magnitude 6. Most stars brighter than this are numbered from 5 to 0, with each difference equivalent to a jump in brightness of 2.512 times. Apart from the Sun, four stars are even brighter than magnitude 0, and have negative magnitudes: Alpha Centauri (–0.01), Arcturus (–0.04), Canopus (–0.7), and Sirius (–1.5).

Apparent magnitude does not tell us how luminous a star truly is, because the stars lie at various distances from Earth. A star's true luminosity is defined in terms of absolute magnitude, which is how bright it would appear at a standard distance of 32.6 light-years. The Sun's nearness yields an apparent magnitude of –26.7, but its absolute magnitude is a humble 4.8.

DIFFERENT COLORS *of stars are captured in this photo of the Milky Way around the Southern Cross (left). A stained-glass window in France depicts the creation of the stars (far left).*

Star Colors

A star's luminosity, color, and surface temperature are correlated. Most red stars are cool and dim, and most white stars are hot and bright. But there are some important exceptions.

The Hertzsprung-Russell, or H-R, diagram plots star color against brightness. Its major feature is the "main sequence," the narrow diagonal band where a star spends most of its life. Ninety percent of all stars fall within the main sequence, with brighter, blue stars at the upper left of the band, and fainter, red stars at the lower right. The Sun, a medium-bright star, lies between these two extremes.

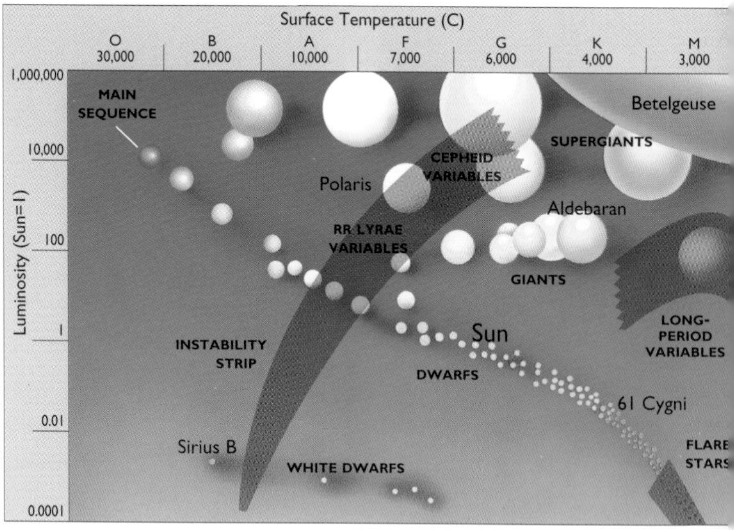

Surface Temperature (C)
O 30,000
B 20,000
A 10,000
F 7,000
G 6,000
K 4,000
M 3,000
Luminosity (Sun=1)
1,000,000
10,000
100
1
0.01
0.0001
MAIN SEQUENCE
Betelgeuse
SUPERGIANTS
CEPHEID VARIABLES
Polaris
Aldebaran
RR LYRAE VARIABLES
GIANTS
LONG-PERIOD VARIABLES
INSTABILITY STRIP
Sun
DWARFS
61 Cygni
Sirius B
WHITE DWARFS
FLARE STARS

IN THE H-R DIAGRAM, *most stars are dwarfs lying within the main sequence. Stars in the gray zones vary in brightness.*

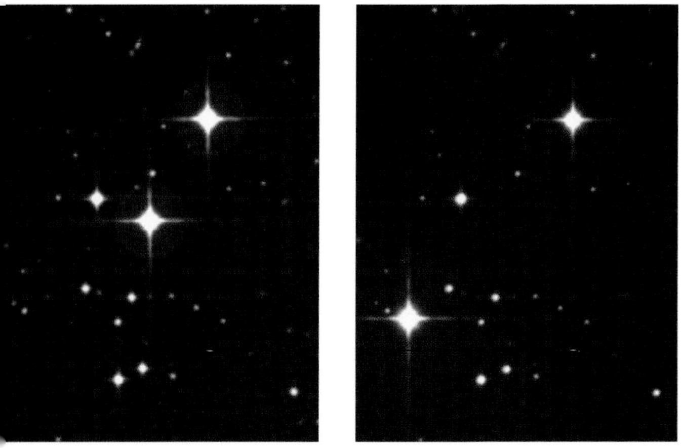

THE PROPER-MOTION SHIFT *of Kapteyn's Star is obvious in two photos taken in 1975 (far left) and 1990 (left) from the Palomar Sky Surveys.*

There are other stars on the diagram that have left the main sequence and are nearing the end of their lives. Cool, red giant and supergiant stars, in the upper right corner, have a large surface area that makes them very luminous. Hot, white dwarfs, in the lower left corner, are faint because they are only about as big as Earth.

Based on their locations across the H-R diagram, that is, their color and temperature, stars are divided into seven main spectral types. The hot, blue to white stars are type O, B, or A; moderate, yellow to orange stars are F, G, or K; and cool, red stars are type M.

Distances

The stars lie at mind-boggling distances. The nearest stars are a little more than 4 light-years away; those making up the dusty Milky Way around Auriga lie at an average distance of 2,500 light-years.

Distances to the nearest stars are obtained by measuring their annual parallax. This is the slight shift of a nearby star against the more distant background stars that occurs as Earth orbits the Sun. The nearer the star, the greater its displacement. With stars that are much farther away, astronomers estimate the star's distance by calculating the difference between its absolute and apparent magnitudes.

Star Motions

Compared to their distances from Earth, the space stars traverse is so minuscule that they appear immobile. Stars do move, though—the Sun, for example, has a space velocity of about 12 miles per second (20 km/s) in the direction of the star Vega.

This positional shift is called proper motion. For most stars, an appreciable difference can be discerned only after several thousand years. Some stars, however, have unusually large proper motions. Kapteyn's Star in the constellation of Pictor moves 8.9 arcseconds per year. In 200 years, it will have moved the width of the Moon from its current position in the sky.

Observing Stars

To appreciate the aesthetics of a starry sky, all you need are your eyes and perhaps a pair of binoculars. But a telescope helps you study each star up close. You may notice that some stars appear to have a faint companion star next to them. These are double stars (see p. 164), and thousands can be seen with a small telescope. Other stars appear to vary in brightness from one night to the next. These variable stars (see p. 166) have periods ranging from minutes to years. On the H-R diagram, many kinds of variables fall within a transitional region called the "instability strip."

ANNIE JUMP CANNON

The monumental task of classifying the spectra of stars was undertaken in the early 1900s by Annie Jump Cannon (1863–1941) at the Harvard College Observatory, Massachusetts. Cannon (below) worked as a "computer," which meant someone who tirelessly examined hundreds of thousands of photographic plates of stellar spectra. The data was eventually compiled into the *Henry Draper Catalog*, which contains the spectral classification of 225,300 stars.

Cannon refined the patterns seen in the endless array of spectra into a single sequence. Her elegant classification system arranged the stars by color, rather than by the strength of hydrogen lines, as the first classification scheme did. Cannon's sequence went from the hottest, blue stars of type O, through the Sun-like G stars, and ended with the cool, red M stars. Almost a century later, her system is still in use.

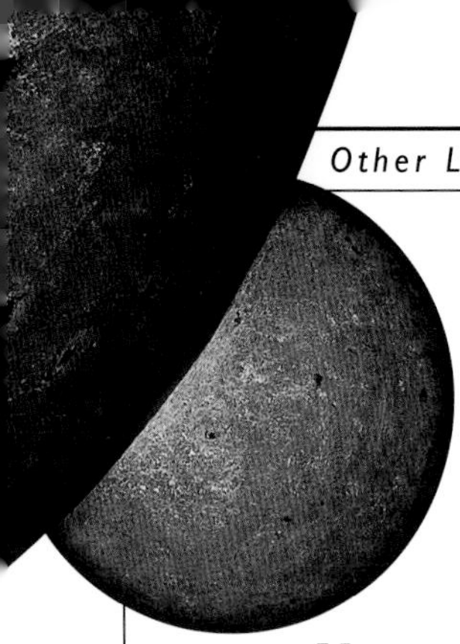

Observing Double Stars

You can enjoy the different colors of a double, meet the challenge of splitting a close pair, or precisely measure separations.

Astronomers estimate that more than half of all the stars in our galaxy have one or more companion stars. These are true binary- or multiple-star systems in orbit about a common center of gravity. From Earth, we also see a few "optical doubles"—chance alignments of two stars that actually lie at different distances.

Seeing Double

You do not need a large telescope to observe double stars. A 3 inch (75 mm) refractor can produce excellent results, since its narrow field renders high-contrast, pinpoint star images. A 6 inch (150 mm) scope, though, is better for resolving close doubles.

You can observe double stars from the suburbs and even during a bright Moon. Because stars are point sources, ambient light generally does not overwhelm them.

Double-star observing is enjoyable on many levels.

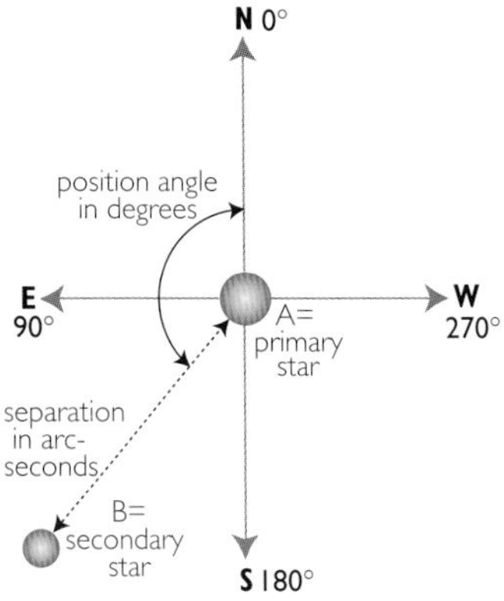
N 0°
position angle in degrees
E 90°
W 270°
A= primary star
separation in arc-seconds
B= secondary star
S 180°

THE POSITION ANGLE *of a secondary star is measured eastward from a line extending north of the primary star. This system has a PA of 135 degrees.*

THE COLORS *and separations of three multiple-star systems (below). Double blue giants (top).*

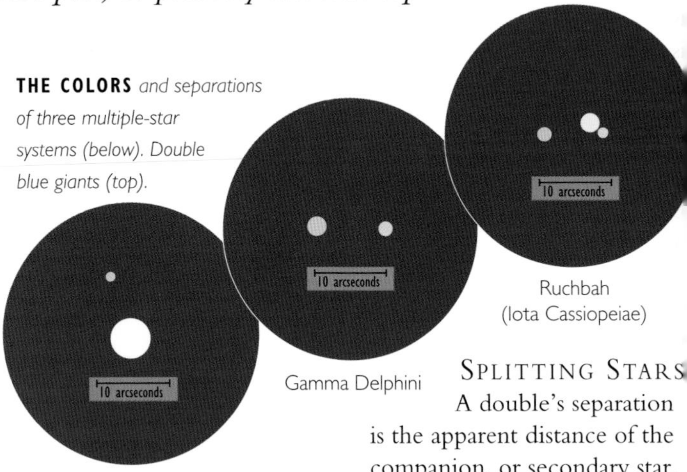
10 arcseconds
Ruchbah (Iota Cassiopeiae)
10 arcseconds
Gamma Delphini
10 arcseconds
Rigel (Beta Orionis)

Some observers delight in aiming their telescopes at random stars and looking for companions, or they might try to "split" close pairs.

Others take a more serious bent—measuring, sketching, and photographing double stars, and then contributing their observations to a group such as the Double Star Section of the Webb Society (see p. 277). Such organizations particularly need help monitoring long-period binaries; with separations of at least 0.5 arcseconds, these are the binary type most suited to small telescopes. There also exists a great opportunity for amateurs with CCD cameras to make a significant contribution. There are thousands of wide faint pairs that have been unobserved for years.

Splitting Stars

A double's separation is the apparent distance of the companion, or secondary star, from the brighter star, the primary. Albireo (Beta Cygni), a beautiful double with colors described as gold and sapphire, has a separation of 34 arcseconds. This is twice as wide as Saturn's apparent disk size, so it is an easy object in a 3 inch (75 mm) telescope. Separations of less than about 5 arcseconds can be challenging in telescopes smaller than 6 inches (150 mm), especially when the seeing conditions are poor, the stars are of unequal brightness, or the stars are faint.

The separation of double stars can be measured using a filar micrometer, a complex device that attaches to the telescope's eyepiece.

Position Angle

The position of the secondary star with respect to the primary is called the position angle, or PA. The primary star is designated component A, and the secondary, component B. Other components, if any, are listed as C, D, and so forth.

You can estimate the ›osition angle by eye, although filar micrometer allows you › make precise measurements.

Position angles help astrono- ›ers keep track of the orbital ›otion of the secondary over ›he years. When its position ›ngle is increasing, the motion › said to be direct. When it › decreasing, it is said to be › retrograde.

›AGNITUDE AND ›OLOR DIFFERENCES

›ouble stars of different ›agnitudes and small separa- ›ions can be both a joy and challenge to observe. Tau^4 ›ridani's 5.7 arcsecond ›eparation would not usually ›e a problem for, say, a 6 inch 150 mm) telescope to split. ›ut in this case, the primary's ›agnitude of 4.0 can over- ›helm the magnitude 10.0 ›econdary, particularly in poor ›eeing. Still, it is not impos- ›ible, and seeing a bright star ›ext to a faint one with just hairbreadth between them ›an be breathtaking.

Many doubles display ›ramatically different colors, ›hich add to the overall splen- ›or of the system. One of the ›nest doubles in the sky, Eta ›assiopeiae, has yellow and ›ed stars 12.7 arcseconds apart.

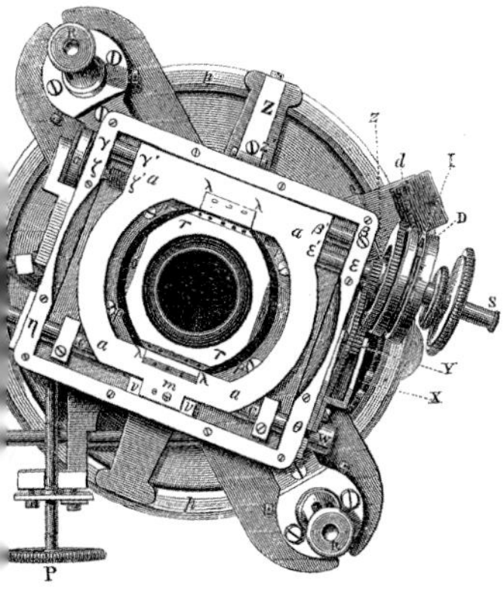

› **FILAR MICROMETER** *from the ›ineteenth century, used to precisely ›easure the separations of double stars.*

OTHER TYPES

Some multiple-star systems contain more than two stars. One famous example is the Trapezium, a nexus of four prominent stars at the heart of the Great Nebula in Orion.

When the orbit of the fainter member of a binary system takes it in front of the brighter member, the system fluctuates in brightness. Such a system, called an eclipsing binary, is of interest to variable-star observers (see p. 167).

Some binary stars are so close together that they cannot be visually separated. However, a spectroscope (see p. 16) will detect tiny, periodic shifts in the motion of the brighter star. These shifts are caused by the pull of the invisible companion's gravity as it orbits. The A component of the visual double star Mizar in Ursa Major is an example of a spectroscopic binary.

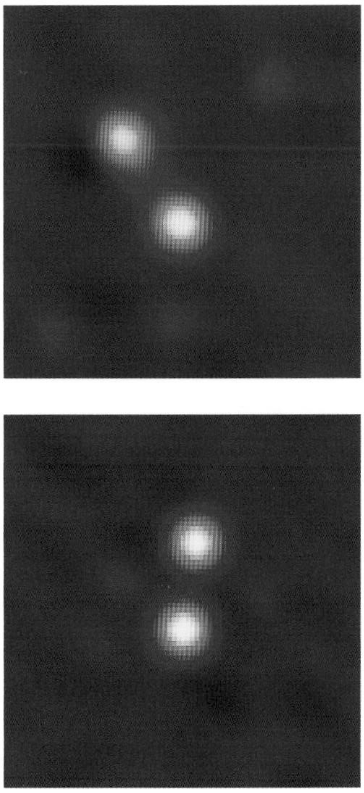

THE MOTION *of the double star Capella is shown in these images taken two weeks apart with a special imaging technique. The stars are too close to separate in a conventional telescope.*

DOUBLE DEDICATION

The first double star ever recorded—Mizar in the Big Dipper—was discovered accidentally in 1650 by the Italian astronomer Giovanni Battista Riccioli. Subsequent discoveries by other astronomers were also accidental. By 1779, enough observations had been compiled to inspire the indefatigable William Herschel (1738–1822) to begin a systematic search for these stellar curiosities. Two years later, he had discovered more than 800 new double stars, assessing each pair with a filar micrometer, a device that allowed him to precisely measure the separation and orientation of the components. Later measurements of these stars by Herschel and others revealed that some of them were, in fact, orbiting each other.

The American astronomer S. W. Burnham (1838–1921) kicked off a new age of double-star discoveries in 1873 when he published a list of 81 new pairs he had found with his 6 inch (150 mm) refractor. Over the next four decades, this tireless, sharp-eyed observer discovered an additional 1,340 double stars using telescopes of various sizes. In 1906, his observations were collected in the *General Catalogue of Double Stars.*

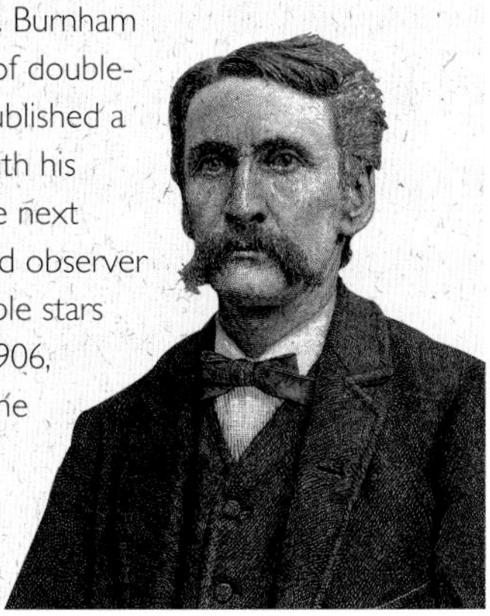

The amateur astronomer S. W. Burnham.

Observing Variable Stars

Not all stars shine with a steady, invariable light. Many fluctuate in brightness over periods ranging from minutes to year.

Astronomers have listed some 30,000 variable stars of many different types and brightnesses. By observing these stars, amateurs can significantly contribute to the science of astronomy.

To begin observing variable stars, you need a telescope or binoculars, a list of variables noting their magnitude ranges and periods, and star charts showing their locations.

How Bright?

After locating a variable star, many observers estimate its apparent magnitude. There are various methods, but most use an interpolative method, comparing a variable's brightness with that of nearby stars.

The American Association of Variable Star Observers (AAVSO) provides finder charts for individual variables that show the magnitudes of surrounding stars. Say the variable is not as bright as comparison star A, but is brighter than comparison star B. From your AAVSO chart, you know that star A is magnitude 8.5 and star B is 9.0, so you might estimate the variable's magnitude as 8.7.

The British Astronomical Society's method also uses comparison stars, but you do not need to know the magnitudes of the comparison stars until after the observing session.

Observing a star that is too bright for your telescope may result in an inaccurate magnitude estimate. A 4 inch (100 mm) telescope should not be used on stars brighter than 7th magnitude. The limit for 6 to 8 inch (150 to 200 mm) telescopes is magnitude 8.5. You can cut down the aperture of your telescope by making an aperture mask.

A VARIABLE-STAR SAMPLER						
Name	**RA**	**Dec.**	**Max.**	**Min.**	**Period (days)**	**Type**
Lambda Tauri	04h 01m	+12° 29'	3.3	3.8	3.95	EclB
R Doradus	04h 37m	–62° 05'	4.8	6.6	338	SemR
Zeta Geminorum	07h 04m	+20° 34'	3.6	4.2	10.20	Ceph
R Leonis	09h 47m	+11° 26'	4.4	11.3	312	Mira
RR Lyrae	19h 26m	+42° 47'	7.1	8.1	0.57	RRLyr
RT Cygni	19h 43m	+48° 46'	6.0	13.1	190	Mira

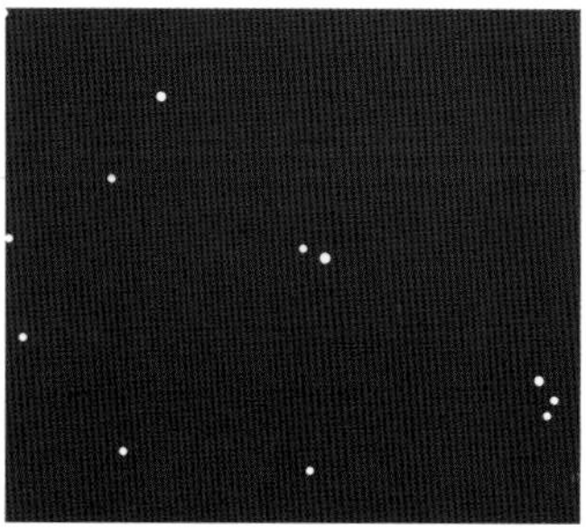

MIRA, *the brightest long-period pulsat ing variable, may go from magnitude 9 minimum (above left) to magnitude 3 maximum (above). The AAVSO logo (top*

Some observers photograp variable stars. This provides a permanent record and allows them to estimate the magnitude away from the telescope

For precise measurements of a variable's brightness, ama teurs can use a photoelectric photometer, a device that attaches to the telescope's focu CCD technology can also be used for photometry.

Groups such as the AAVS or the Variable Star Section of the British Astronomical Association welcome visual, photographic, and photometri observations from amateurs, and can provide further advic about variable-star observing.

Variable Terms

The brightest magnitude a variable star reaches is called its maximum, while the faintes magnitude is its minimum. The difference between the star's maximum and minimun is its amplitude, and the time

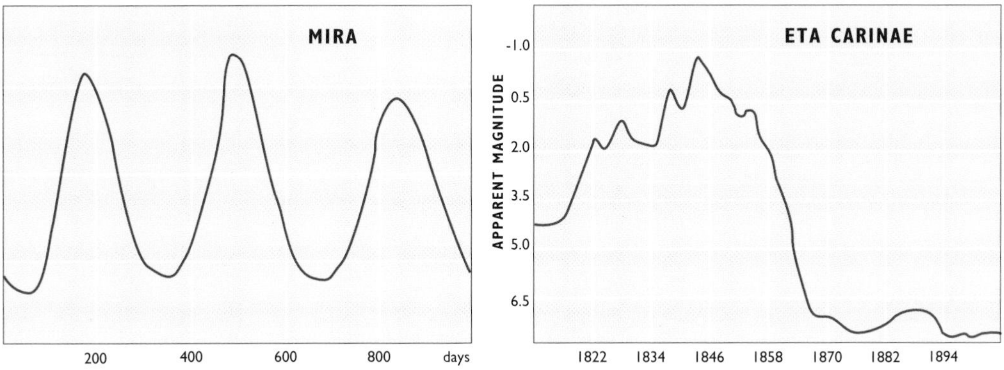
MIRA
200
400
600
800
days
ETA CARINAE
APPARENT MAGNITUDE
-1.0
0.5
2.0
3.5
5.0
6.5
1822
1834
1846
1858
1870
1882
1894

THE LIGHT CURVE *of a pulsating variable such as Mira follows a regular pattern (above left). The cataclysmic star Eta Carinae has an erratic curve (above).*

:tween successive maxima or ninima is called the period. A ght curve is a plot of the star's nanging brightness over time.

Many variables have com- non names or Bayer Greek tters, but others are known y their official variable-star esignation—for example, . Leonis or X Aquarii. The tters R to Z are used for the rst variables discovered in a onstellation. Additional vari- bles are assigned RR to RZ, S to SZ, and so on, up to ZZ. he sequence then progresses om AA to AZ, BB to BZ, nd concludes with QZ. Since nere are only 334 available tter combinations, the 335th ariable in a constellation is beled V335, and so on.

CLASSES OF VARIABLES

here are three broad classes f variable stars: pulsating, ruptive, and eclipsing.

Pulsating variables brighten nd fade as their outer layers nythmically contract and xpand. They are further clas- fied according to the pattern f their periods.

A well-known pulsating ariable, Delta Cephei, is the rototype for the Cepheid ariables. These have regular eriods ranging from one to everal days. The longer a Cepheid's period, the greater its absolute magnitude. This relationship is so reliable that astronomers use Cepheids as "standard candles" to calculate the distances to nearby galaxies.

Long-period stars, such as Mira, are red giants that pulsate like Cepheids do, but their cycles are less regular and their periods are longer, ranging from 80 days to 5 years.

Other pulsating variables forming their own subtypes include RR Lyrae and RV Tauri. Semi-regular and irregular variables are also kinds of pulsating variables.

• **Cataclysmic,** or eruptive, variables exhibit sudden and large brightness outbursts. Many are novae, binary systems that erupt once in a cycle that can last thousands of years (see p. 168). Dwarf novae, such as U Geminorum, erupt more frequently, with cycles lasting weeks. Supernovae are often classed as cataclysmic, but their outbursts happen just once and destroy the star.

Sometimes included in the cataclysmic category are the peculiar R Coronae Borealis stars. At irregular intervals, these rapidly fade by as much as 6 to 8 magnitudes, possibly because of an outbreak of car- bon soot in their atmosphere.

• **Eclipsing** variables, such as Beta Lyrae, are binary systems in which one star eclipses the other during each orbital period. From Earth, we see this as a periodic decrease in light output, followed by a return to normal brightness.

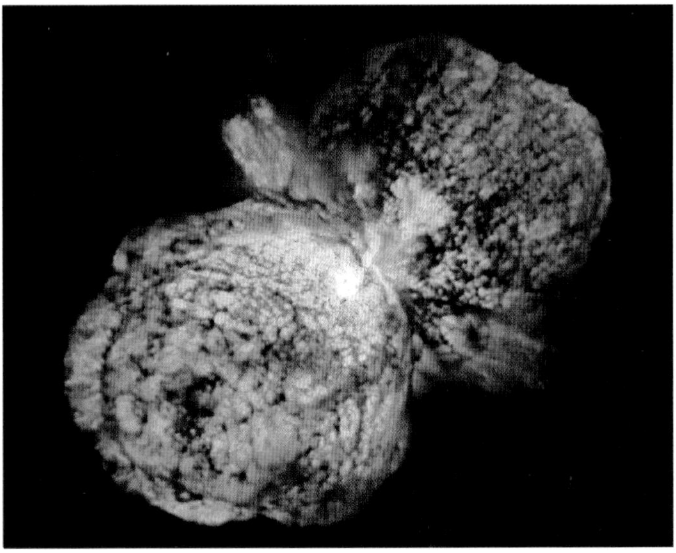

TA CARINAE, *a strange cataclysmic ıriable, suffered a violent outburst 50 years ago that produced two polar bes on either side of an equatorial disk.*

NOVAE *and* SUPERNOVAE

Novae and supernovae are produced by different processes with dramatically different finales: the violent eruption of a nova may recur, while a supernova signals complete stellar destruction.

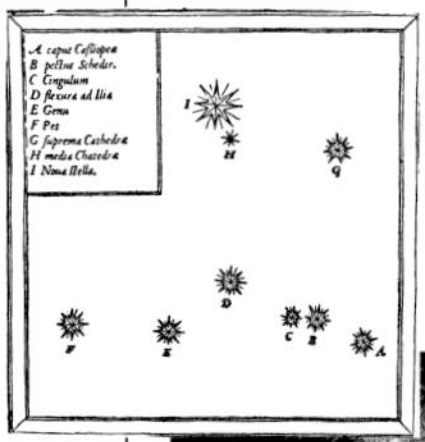

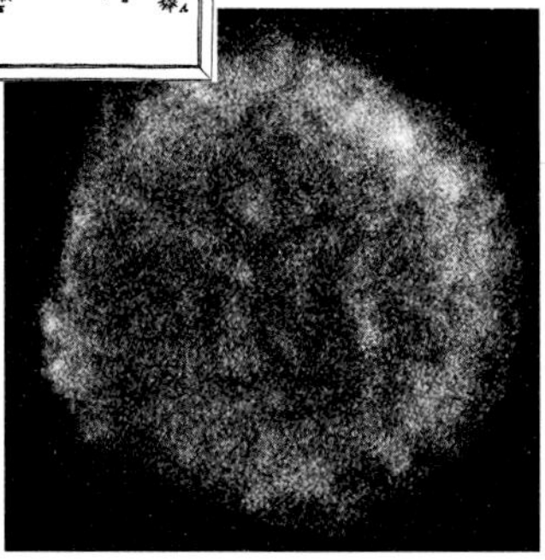

THEN AND NOW *In 1572, Tycho Brahe sketched a supernova in Cassiopeia (top left). All that remains of it today is Tycho's supernova remnant, seen here at X-ray wavelengths (left).*

The search for novae and supernovae is a fascinating and worthwhile amateur pursuit. These extraordinary exploding stars are of great interest to astronomers studying stellar evolution, nucleosynthesis (the creation of elements via nuclear reactions in stars), and cosmology.

NOVAE

Novae are a class of cataclysmic variable star. They occur in binary systems made up of a white dwarf and a red giant. Hydrogen gas from the giant is drawn by gravity onto the surface of the white dwarf, and after hundreds or thousands of years, enough material accumulates to trigger a thermonuclear detonation. Over a period of a few days, the star brightens by perhaps 10 magnitudes and remains bright for several days before slowly fading again. The explosion may be repeated when sufficient new material builds up.

Famous novae include Nova Cygni 1975 and Nova Herculis 1934. Both of these stars reached 1st magnitude.

SUPERNOVAE

More than 100 times more luminous than a nova, and capable of outshining all the stars in a typical galaxy, a supernova is the violent, convulsive explosion of a massive star. Two types of explosion can create supernovae. The first occurs in a binary system when a white dwarf siphons off more matter from a companion star than it can support. The dwarf implodes and then rebounds violently from a rigid core in a massive explosion.

The second type of explosion arises when a massive star runs out of fuel. By the end c its life, a star with at least eigl times the mass of the Sun has produced iron in its core via nuclear-fusion processes. Since nuclear reactions using iron consume energy instead c producing it, the star's intern furnace shuts down. Unable t keep the core hot enough, th gas in the core can no longer support the weighty outer layers, and the star caves in o itself. The material crushes in on the core and then rebound producing the explosion.

Both types of explosion destroy the original star. Mos of the star's outer material is ejected, creating an expandin shell of gas called a supernova remnant. If it survives the blas the collapsed core of the star may also remain in the form of a spinning neutron star (which might be visible as a pulsar) or a black hole.

Naked-eye supernovae are very rare events—SN 1987A, in the Large Magellanic Cloud occurred almost 400 years afte the previous one!

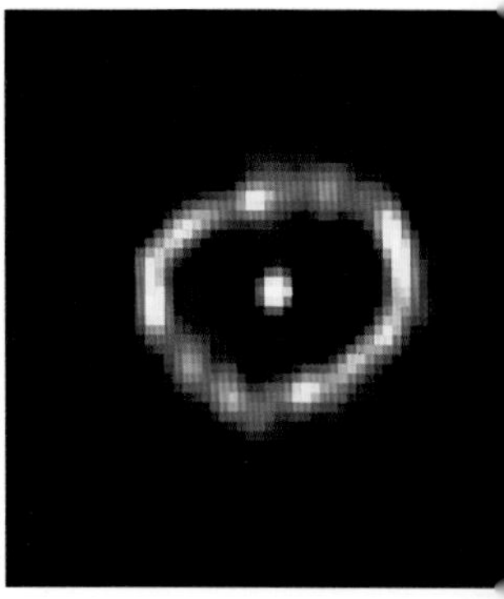

NOVA CYGNI 1992 *in Cygnus erupted on 19 February 1992. A Hubble image taken 467 days later (left) shows a ring of gas around the star. Seven months later (right), the ring had expanded.*

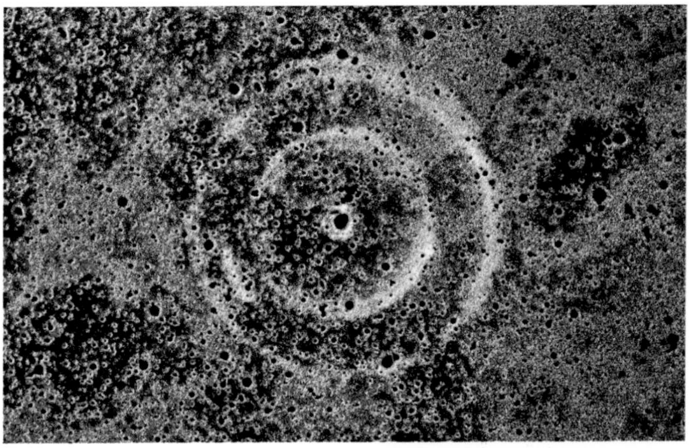

WO RINGS *(right) appeared around upernova 1987A a year after the ex- losion. These "light echoes" are sheets f dust lit up by the supernova. Nearby ars look black in this processed image.*

EARCHING FOR XPLOSIONS

Before a nova or supernova an be studied, of course, it nust be found. Searches should e methodical and thorough. f you are sweeping an area of ky for novae, say, the sweeps nust overlap so that no area is nissed. Dark, clear skies are specially important.

Nova hunters can use an nstrument as basic as a pair of x 50 binoculars. Some 20 to 5 novae are discovered each ear in our galaxy, most with inoculars or a wide-field elescope eyepiece. Nearly all ovae occur in a 20 degree rip centered on the galactic quator, which runs along the and of the Milky Way, so this the best place to search. The Magellanic Clouds have also ielded a few nova discoveries, but novae in other galaxies are generally too faint to see.

Star charts showing all stars down to at least magnitude 8 are a must. Over time, you will come to know the star patterns within your chosen viewing field. If you see a star that "should not be there," it may be a nova, although you might be looking at a known asteroid or variable star at its peak. Check against a good star catalog or sky-charting program. At their maxima, novae look distinctly reddish or yellow.

Binoculars are of little use if your goal is to find supernovae. A number of supernovae can be seen each year in other galaxies, but since they usually peak below magnitude 14, you will need at least an 8 inch (200 mm) telescope.

To conduct a simple search, you survey the sky, galaxy by galaxy, looking for "new" stars with the help of photographs and charts. A faint star in or near a galaxy that does not appear on the chart or photograph could be a supernova. Accurate charts showing the magnitudes of the field stars in and around the galaxy can be obtained from the AAVSO or *The Supernova Search Charts and Handbook* (see p. 276).

SUPERNOVA 1987A *began as one of many faint stars in the Large Magellanic Cloud (above left), but briefly became 10,000 times brighter in 1987 (left).*

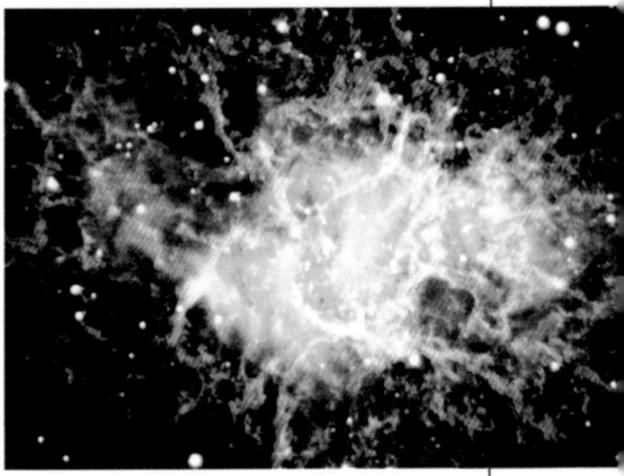

THE CRAB NEBULA *in Taurus is a remnant of a supernova seen in 1054.*

If you think you have seen a nova or supernova, note its position and estimate its magnitude (see p. 166). Always double-check your observation, and have it confirmed by another competent observer, an observing organization, or an observatory. Give them the date and time of your observation, the instrument used, the position and brightness of the object, and the observing conditions. If your observation is confirmed, you can report it to the International Astronomical Union's Central Bureau for Astronomical Telegrams in Massachusetts (see p. 276).

We are all in the gutter,

but some of us are looking

at the stars.

OSCAR WILDE (1854–1900),
British playwright, poet, and novelist

DEEP-SKY OBSERVING

Locating and observing deep-sky objects exposes you to the most remarkable and exotic treasures the universe has to offer.

When we look past the brightest stars in the sky, some of the specks of light we see are star clusters, nebulas, and galaxies. Astronomers have cataloged thousands of these deep-sky objects, many of which can be seen using binoculars and small to medium telescopes.

OBSERVING TIPS

No matter what instrument you use, dark, clear skies are essential for deep-sky astronomy. Some galaxies are only slightly brighter than the normal background skyglow, so you need all the contrast you can muster from your instrument and your eyes. Before you begin observing, give your eyes 15 or 20 minutes to become dark-adapted. Then use a red-filtered flashlight to study star charts without ruining your night vision.

If you live in or around the city, you can effectively boost the contrast by using light pollution reduction filters on your telescope (see p. 56). These are especially useful for nebulas, many of which are normally invisible from cities.

You can also gain a little contrast with the technique of averted vision. By looking slightly to one side of a deep-sky object, you avoid a "low sensitivity" patch on your retina.

For especially low-contrast objects, use averted vision and the "jiggle" method. Simply give the telescope tube a light tap when you think a nebula is in the field. The eye can detect contrast differences more easily if the image moves slightly.

Another hint is to keep both eyes open when looking through the eyepiece. Squinting shut the unused eye causes strain that can affect the active eye's vision. If you are distracted by the passive eye's close-up view of the telescope tube, simply cover the eye with your hand or an eyepatch. To completely eliminate visual distractions, you can drape your head in a dark hood. Your breath may fog the eyepiece in cold weather, so keep the back end of the hood open.

Finally, concrete, asphalt, and other hard surfaces absorb heat during the day and slowly reradiate it at night, creating local air turbulence. Always try to observe from a grassy area.

HOW BRIGHT?

Point sources such as stars have apparent magnitudes that indicate their visual brightness. The light of an extended object such as a cluster, nebula,

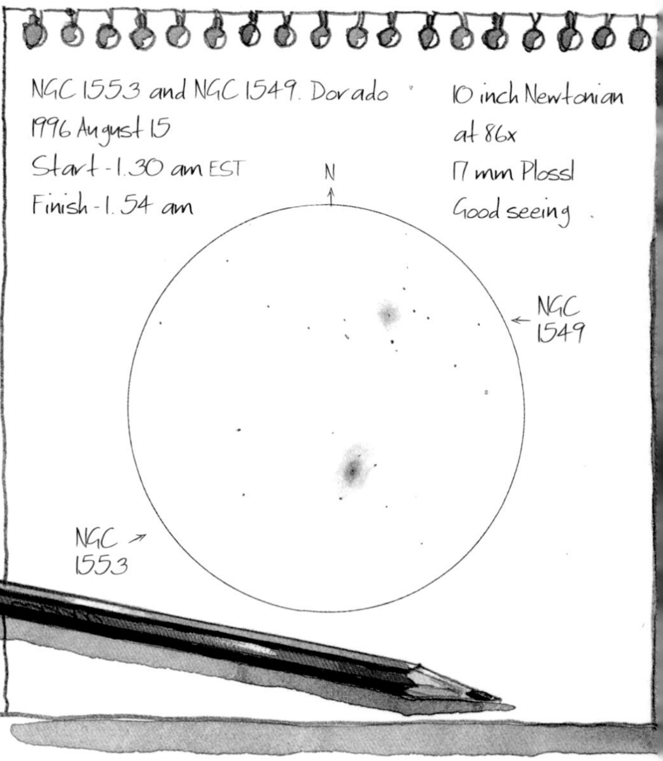
NGC 1553 and NGC 1549. Dorado
1996 August 15
Start - 1.30 am EST
Finish - 1. 54 am
10 inch Newtonian
at 86x
17 mm Plossl
Good seeing
N
NGC 1549
NGC 1553

A LARGE TELESCOPE *shows the spiral arms of NGC 2997 (top). Even with a small or medium scope, you can make satisfying galaxy sketches (right).*

THE VEIL NEBULA *(left) can hardly be seen through an unfiltered scope, but a light pollution reduction filter reveals its structure. A globular cluster (below).*

r galaxy, however, is spread ut over its entire area, so ou also need to consider the ize of the object, indicated y its apparent diameter. The inwheel Galaxy (M33) in riangulum has an apparent nagnitude of 5.5, which vould suggest a fairly bright bject, but it has a large pparent diameter of 62 arc-ninutes, twice that of the Full Aoon. This means the galaxy as low "surface brightness" nd is surprisingly hard to see.

Objects that have large di-meters and faint magnitudes— s many galaxies do—require ow magnification and atten-ion to contrast distinctions, vhile objects with small diam-ters and bright magnitudes— ıch as planetary nebulas— emand greater magnification nd accurate positioning of ne telescope's field of view.

Another factor that will ffect the brightness of a deep-ky object is its altitude. If a alaxy, nebula, or star cluster low over the horizon, its ght must traverse more layers f our atmosphere than it vould if it were overhead. 'our best views will be when ne object is high in the sky.

)EEP-SKY CATALOGS

ll of the brighter star clusters, ebulas, and galaxies are ssigned numbers in deep-sky atalogs (see p. 71). Be aware ıat many objects have two esignations: the Great Nebula in Orion, for example, is known as M42 and also as NGC 1976. Catalog numbers can help avoid the confusion that arises with common names. There are three objects sometimes known as the Pinwheel Galaxy. If you wish to specify the one in Ursa Major, you can call it M101 (or NGC 5457), and astronomers around the world will know which one you mean.

RECORDING METHODS

Deep-sky observers are often so inspired by what they see that they wish to record it in sketches or photographs.

To sketch a deep-sky object, draw a circle representing the field of view. With a soft pencil, mark bright field stars and roughly sketch the object while at the eyepiece. You can refine the sketch later. Note details such as the date, the equipment used, and the seeing conditions.

To photograph most deep-sky objects, you need to attach either a conventional camera (see p. 66) or a CCD camera (see p. 68) to your telescope.

THE WEBB SOCIETY

One of the most respected deep-sky observing organizations is the Webb Society. Founded in 1967, the society is named after the Reverend Thomas William Webb (1807–85), an eminent British amateur astronomer and author of *Celestial Objects for Common Telescopes*, first published in 1859. The book was a landmark work for amateurs wanting to explore the universe through their own telescopes. Webb (below) encouraged them to venture into deep space, writing that "in leaving our Sun and his attendants in the background, we are only approaching more amazing regions, and fresh scenes will open upon us of inexpressible and awful grandeur."

In the same spirit, the Webb Society today encourages the serious study of deep-sky objects. It's worldwide membership includes both professional and amateur astronomers. Through its various sections—Double Stars, Nebulas and Clusters, Galaxies, and the Southern Sky—the society provides a forum where amateurs can present their observations. Results of the society's work are published in their periodicals.

Clusters *and* Nebulas

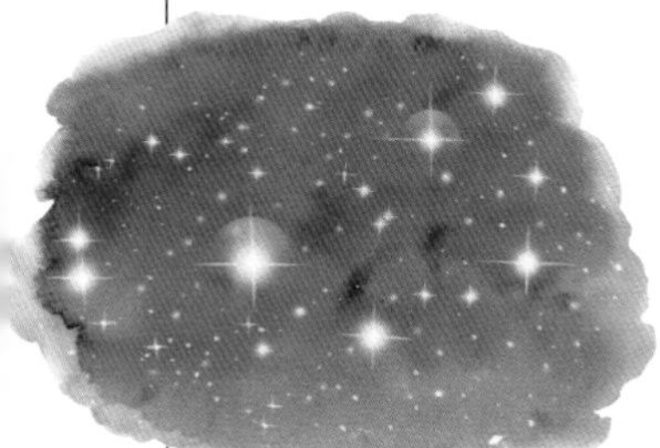

The plane of our galaxy is a stellar crucible in which coalescing clouds of dust and gas give rise to stars and star clusters by the millions.

The raw material of stars can be found in nebulas—tenuous clouds of dust and gas. Stars that condense from the same nebula may become bound by each other's gravity into an open cluster. For a few hundred million years, they move together through the Milky Way like a school of fish. Eventually, tidal forces and interactions with other interstellar clouds disperse the stars.

Open Observing

Also known as galactic clusters, open star clusters come in many shapes and sizes. A rich cluster may comprise several thousand stars, while a sparse one may contain only a dozen or so. Some clusters, such as the Pleiades (M45) in Taurus, are large and bright enough to see with the naked eye. Most, however, require a telescope.

Large open clusters, such as the Beehive (M44) in Cancer, are best viewed using the low magnification of a wide-angle eyepiece. Small, compact clusters, such as the Jewel Box (NGC 4755) in Crux, can stand much higher magnifications. By increasing the magnification, you gain contrast, but you should avoid increasing it so much that the cluster becomes large and sparse. Aim for a balance between image scale, contrast, and detail.

A star cluster's overall visual presence is determined by its concentration—how compact or loose it is—as well as the distribution and individual magnitudes of its stars. Compact clusters made up of faint stars near the resolution limit of your telescope will appear nebulous, while large rich clusters can be hard to distinguish from background stars.

Cities of Stars

In contrast to open clusters, globular clusters are tight, spherical aggregations of tens of thousands to hundreds of thousands of stars. They contain very old stars and formed before the disk of our galaxy took shape. Globular clusters are actually distributed in a spherical halo around the Milky Way, but from our perspective in the disk of the galaxy, they seem concentrate around the galactic center, in the constellations of Sagittariu and Ophiuchus (see p. 254).

Of the 150 globular cluster that have been cataloged, onl a few can be glimpsed with the naked eye. In binoculars, most appear as mere smudges of light. Through a 4 to 6 inc (100 to 150 mm) telescope, they take on more definition, looking like cottony globes with a sprinkling of outlying stars; with high magnification you may even be able to resolve individual stars within the core. A 10 inch (250 mm

THE JEWEL BOX *in Crux (above) is one of the Southern Hemisphere's most beautiful open clusters, while the stunning Pleiades Cluster (top) is a favorite in both hemispheres. The globular M13 (right) is also known as the Hercules Cluster.*

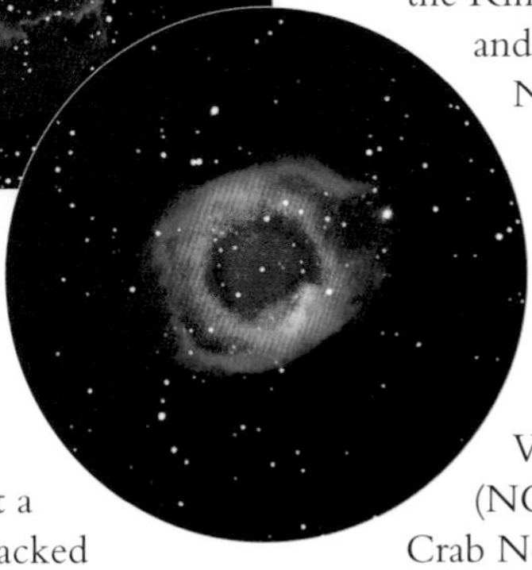

THE LAGOON NEBULA *(left) is visible to the naked eye on a dark night, but you need a telescope to see its dust lane. The Helix Nebula (below) is the closest planetary nebula to Earth.*

cope aimed at a dark sky will eveal a spectacular granular phere with strands of stars adiating out from the center. \ broadband light pollution eduction filter can help you ocate faint globulars, as will verted vision (see p. 170).

When you find a globular, notice its shape. Some will ook irregular, like a freeze-rame of bees swarming cha-otically around a hive. Rich, apidly rotating clusters can exhibit a distinct flattening. Examine the globular's disk or subtleties such as star hains, voids (apparent holes), or threads of dark dust.

One of the brightest globu-ars is M13 in Hercules, which s high overhead in Northern Hemisphere skies in late July. The grandest globular cluster of hem all, however, is Omega Centauri (NGC 5139), best seen from the Southern Hemisphere around mid-year. It contains about a million stars packed into a region that measures about 150 light-years across.

The Stuff of Stars

Nebulas can be placed into two broad categories: bright and dark. Bright nebulas are glowing clouds of dust and gas that would be invisible if they were not associated with stars.

A bright nebula can be simply a reflection nebula, where starlight reflects off the nebula's dust, like a lighthouse beam illuminating a fog bank. When a nebula is near a very hot star, however, the intense ultraviolet radiation from that star excites, or ionizes, atoms within the gas, causing the cloud itself to emit light and become visible as an emission nebula. Emission nebulas are apparent in star-forming regions—good examples are the Great Nebula in Orion (M42) and the Lagoon Nebula (M8). Emission/reflection nebulas are a mixture of the two nebula types.

In photographs, reflection nebulas are a cool blue color and emission nebulas glow red, but our eyes are not sensitive enough to detect these colors through a telescope.

Planetary nebulas, such as the Ring Nebula (M57) and the Dumbbell Nebula (M27), are the remains of stars that puffed away their outer envelopes in the final phases of their lives. The Veil Nebula (NGC 6960) and the Crab Nebula (M1) are both supernova remnants—the gaseous shards of stars that ended more dramatically by exploding (see p. 168).

Dark nebulas are clouds of dust and gas dense enough to obscure the light from background stars. They are most visible when silhouetted against bright nebulas—two examples are the famous Horsehead Nebula (IC 434) and the dark lanes in the Lagoon Nebula (M8). Infrared telescopes often reveal the presence of new stars within these dark clouds.

Observing Nebulas

Any telescope in the 4 to 6 inch (100 to 150 mm) range can show prominent nebulas in dark, clear skies, but the light grasp of larger scopes is a definite advantage when looking for less substantial nebulas. Once you have located a nebula, boost your magnification for greater contrast until the image begins to degrade. Narrowband light pollution reduction filters can be very helpful, even in dark skies.

A thousand little wisps,

aint nebulae,

Luminous fans and milky

treaks of fire ...

he Torch-Bearers,

LFRED NOYES (1880–1958),

nglish poet

GALAXIES

Through a telescope, galaxies appear as delicate whorls of soft light, some barely visible, others distinct and surprisingly symmetrical.

COMPANIONS *In 1850, Lord Rosse sketched M51 and NGC 5195 (top). The Andromeda Galaxy has two elliptical companions: M32 and M110 (above). The face-on spiral M100 (right) and the Sombrero Galaxy (below right), an edge-on spiral, are in the Virgo Cluster.*

Galaxies are immense, remote systems of gas, dust, and billions of stars held together by gravity. They come in a wide variety of shapes and sizes—from pinwheels, spheres, and footballs to shapeless clouds.

Astronomers once referred to some galaxies as "spiral nebulae" because they were spiral shaped and could not be resolved into stars. By the mid-1920s, however, powerful telescopes had revealed that the brightest of these fuzzy objects contained stars of their own. In 1926, the American astronomer Edwin Hubble sorted galaxies into three broad categories: spirals, barred spirals, and ellipticals. With a few modifications, his system is still used today.

SPIRAL STRUCTURE

The most common galaxies in the universe are dwarf ellipticals and irregulars, but by far the largest proportion of bright galaxies—about 75 percent—are spirals. These systems range from 15,000 to 150,000 light-years in diameter and may contain several hundred billion stars in a flattened disk. Within the disk, spiral arms appear to emerge from a bright central nucleus, traced out by young, hot stars and bright emission nebulas, like lights around a Christmas tree. Open star clusters and interstellar dust and gas are distributed throughout the disk. The Andromeda (M31) and the Whirlpool (M51) galaxies are typical spirals.

In a barred spiral, the bright stars and ionized gas of the nucleus extend for thousands of light-years from each side of the center in a straight "bar." From the end of each bar, the arms wrap back around the nucleus, as normal spiral arms do. In obvious cases, each bar looks something like a scythe blade. Approximately one-third of the spirals exhibit a bar-like structure, and some astronomers suspect that all spirals contain at least a weak bar running through the disk. NGC 1365, in Fornax, i sometimes known as the Grea Barred Spiral (see p. 195).

ELLIPTICALS

Elliptical galaxies tend to be shaped like footballs or spheres The largest, and rarest, ellipticals have a diameter of at least 100,000 light-years, and may contain more than 10,000 billion stars. Of the sky's brightes

alaxies, large ellipticals—such s M84 and M86 in Virgo— nake up about 20 percent. Auch more common are the aint dwarf ellipticals, which ontain just a few million stars nd may be no more than ,000 light-years across. The ompanions orbiting the Andromeda Galaxy, M32 and M110, are dwarf ellipticals.

)THER SHAPES ND TYPES

Galaxies with shapes that fall etween the elliptical and piral categories—such as NGC 1201 in Fornax—are alled lenticular galaxies. .enticulars have a central ucleus as spiral galaxies do, ut, like the ellipticals, show ittle or no evidence of spiral tructure. They contain mostly ld stars and very little gas.

A fifth category contains hose galaxies with no regular tructure—irregular galaxies. Only 5 percent of bright galaxies can be called irregular. These sprawling conglom- erations of stars may have onspicuous bars, or even ints of spiral structure. Intense ursts of star formation domi- nate the appearance of some rregulars. Look for "hotspots" of bright young clusters and onized gas. The Magellanic Clouds, easily seen with the

THE BARRED SPIRAL *nature of M95 in Leo is revealed only in telescopes of 10 inches (250 mm) or more.*

naked eye from the Southern Hemisphere, are irregular galaxies that contain many star-forming regions.

A galaxy that appears to have suffered a severe disturbance is known as a peculiar galaxy. One member of this class is the elliptical NGC 5128.

Active galaxies are those that have unusually energetic cores. Quasars are extremely distant objects believed to be the cores of active galaxies. They stand out only because they are exceptionally bright.

GALAXY CLUSTERS

If we step back and look at the universe of galaxies, we see that they, like stars, congregate into clusters.

Our Milky Way galaxy belongs to the Local Group. This cluster also contains the Andromeda Galaxy, the Pinwheel Galaxy (M33), and the Magellanic Clouds, but most of its 30 or so members are dwarf ellipticals and irregulars.

The nearest large cluster, the Virgo Cluster, contains 2,500 galaxies. Though it lies some 65 million light-years away, it covers more than 45 degrees of sky (see p. 236).

OBSERVING GALAXIES

Whether a galaxy is challenging or easy to observe depends on several factors, not the least of which is the galaxy's apparent magnitude. But magnitude means little or nothing unless you also consider its apparent diameter (see p. 171). Face-on galaxies have low surface brightness, which can make them difficult to locate against a bright sky background. Conversely, the light of edge-on spiral galaxies is concentrated in a narrow bar-shaped feature, so they have greater contrast.

With a 4 to 8 inch (100 to 200 mm) telescope, you can find many galaxies and see the prominent features of bright ones. But to see a faint spiral's dusty arms or nebulas, you will need a larger telescope, not to mention very dark skies. A 12 to 16 inch (300 to 400 mm) telescope provides spectacular views of galaxies down to about magnitude 14.

Without a doubt, galaxies can be challenging to observe. But what other endeavor extends your vision across such great gulfs of space or allows you to look back millions of years into the past? The quest to behold the diverse and enigmatic galaxies, each shining with the combined light of billions of stars, is deep-sky observing at its most profound.

COLLISION *with a spiral galaxy may ave created the dark lane that crosses he elliptical NGC 5128 (Centaurus A).*

I stood upon that silent hill
And stared into the sky until
My eyes were blind with stars and still
I stared into the sky.

The Song of Honour,
Ralph Hodgson (1871–1962), British poet

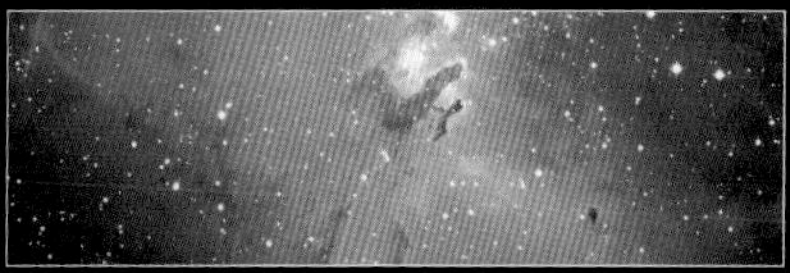

CHAPTER SIX
STARHOPPING GUIDE

Using *the* Starhopping Guide

For newcomers and experts alike, a good map and clear instructions are a great start to navigating your way through the night sky.

The 20 starhops on the following pages cover a selection of some of the best known areas across the night sky. They are intended as an introduction to the technique of starhopping, with the hope that you will move on from here to discover for yourself some of the multitude of other amazing objects appearing elsewhere in the sky.

To help plan future tours, look in the Resources Directory on page 274. It lists further reading that will provide information about where you can find many more starhopping destinations.

*The **starting point** for each starhop is an easy-to-find bright star.*

TOWARD THE HEART OF OUR GALAXY
M25 (IC 4725) Move 5 degrees north and 1 degree west of M22 to take a look at M25. This scattered open cluster, about 1/2 degree across, is an easy object for binoculars and small telescopes. Near the center of M25 is the yellowish Cepheid variable U Sagittarii, which varies from magnitude 6.3 to 7.1 over 6.75 days.
THE STELLAR NURSERY IN M16
In 1995, the Hubble Space Telescope returned an eerie picture of the nebulosity of M16, showing huge columns of gas that are several light-years in length (below). A close look at the surfaces of the giant pillars shows what scientists have called evaporating gaseous globules, or EGGs. Ultraviolet light from nearby stars is stripping some of the gas from the columns and exposing the EGGS, in which stars are forming. We see the globules because they are more dense than the rest of the nebula, and so are not "blown away" as easily. However, the gradual loss of material feeding the developing stars is thought to play an important role in limiting their size. We can only wonder how many more stars are forming deep within these huge columns, gathering material in private and becoming larger and larger.

__Color photographs__ supplement the text by showing some of the objects described on the starhop.

Each __object__ on the starhop is numbered to correspond with the 1 degree telescope fields of view (circles) on the main sky chart.

The __starhopping text__ takes you on a guided tour of the night sky and is frequently cross-referenced to other relevant chapters in the book. It provides important observer-based information about each of the objects on the starhop, along with tips for finding fainter objects and some interesting historical facts.

The illustrated __feature box__ takes a closer look at one of the more interesting objects on the starhop, elaborating on aspects of the science behind it.

KEY TO ICONS

naked eye · binoculars · telescope

*The **main sky chart** covers a portion of the night sky and includes all the objects described on the starhop, along with many others to look for.*

*The **hemispheres** each starhop is visible from.*

*A **sky compass** runs parallel with the RA lines on the grid of each map to help you orient your view of the sky to the chart. The declination lines form the horizontal part of the grid.*

*The **Milky Way** appears shaded in lighter blue.*

***One degree telescope fields of view** highlight the objects on the starhop, and are numbered in order of the hop.*

*The **transit time** is the best time for night viewing.*

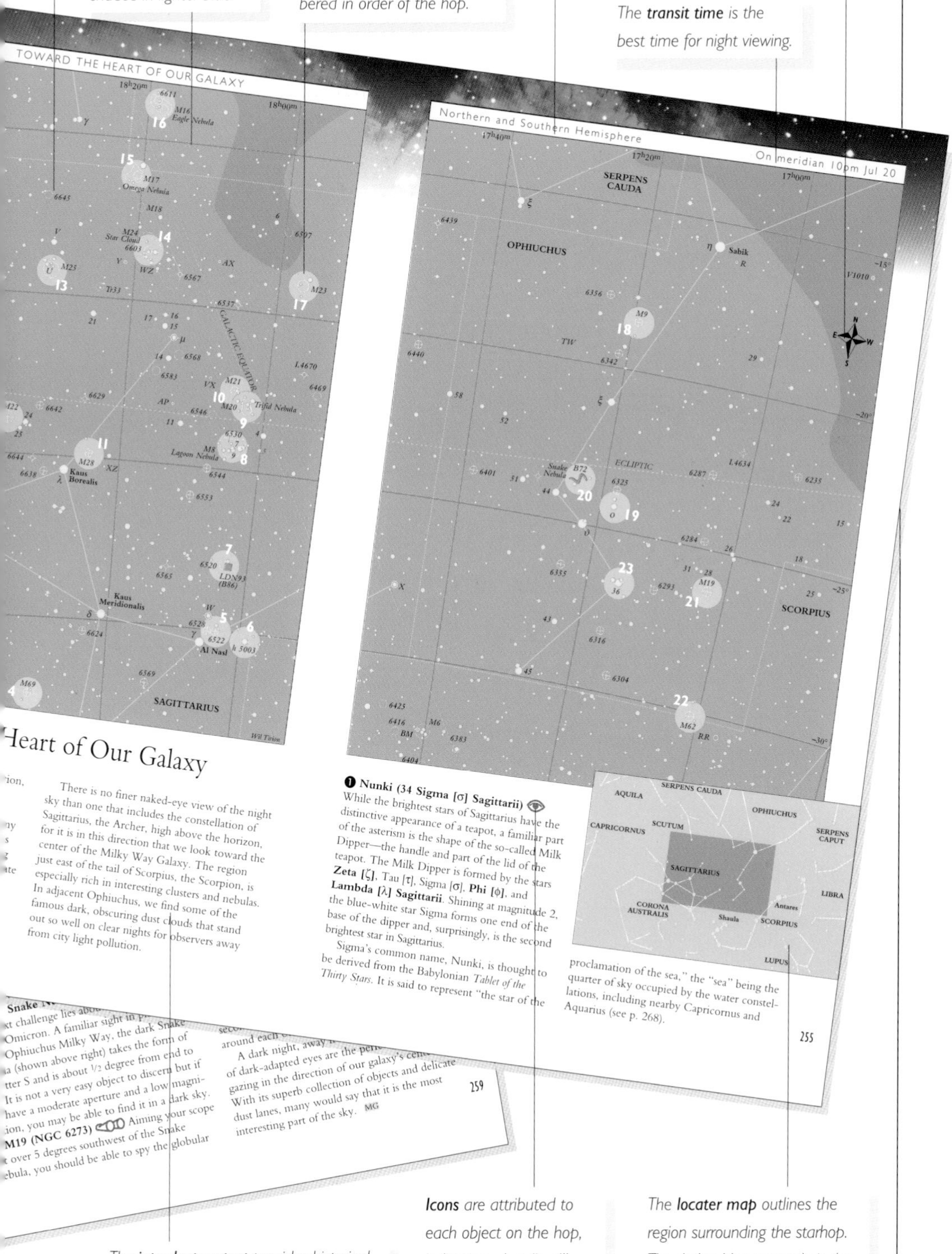
TOWARD THE HEART OF OUR GALAXY
Northern and Southern Hemisphere
On meridian 10pm Jul 20
SERPENS CAUDA
OPHIUCHUS
SCORPIUS
SAGITTARIUS
GALACTIC EQUATOR
ECLIPTIC
Kaus Borealis
Kaus Meridionalis
Al Nasl
Sabik
Heart of Our Galaxy
There is no finer naked-eye view of the night sky than one that includes the constellation of Sagittarius, the Archer, high above the horizon, for it is in this direction that we look toward the center of the Milky Way Galaxy. The region just east of the tail of Scorpius, the Scorpion, is especially rich in interesting clusters and nebulas. In adjacent Ophiuchus, we find some of the famous dark, obscuring dust clouds that stand out so well on clear nights for observers away from city light pollution.
Nunki (34 Sigma [σ] Sagittarii)
AQUILA
SCUTUM
CAPRICORNUS
CORONA AUSTRALIS
SERPENS CAPUT
LIBRA
LUPUS
Antares
Shaula
255
259

*The **introductory text** provides historical, mythological, or general information about the constellations on the starhop.*

***Icons** are attributed to each object on the hop, indicating what "tool" you need to observe the most important features.*

*The **locater map** outlines the region surrounding the starhop. The darker blue rectangle is the same 30 x 20 degree area shown on the main sky chart.*

How *to* Starhop

For novice stargazers, navigating your way among the stars and searching for faint deep-sky objects can be daunting, until you learn how to starhop.

Basically, starhopping is a technique that uses bright stars as a guide to finding fainter objects. All you need to starhop is an idea of how much sky you can see (the field of view) in your finderscope and telescope eyepieces, and how that view will compare to the same amount of sky on your star chart.

Plan your observing session before it gets dark so that you spend less time under the stars trying to figure out what to look for and more time actually observing. The 20 starhops on the following pages are a great way to get started, by taking you on guided tours of some well-known areas.

How Much Sky Can You See?

To start out, you will need a variety of templates of some sort to represent the fields of view of your binoculars or finderscope, and of your telescope eyepieces. Illustrators' plastic circle templates work well, as they come in a wide variety of sizes to choose from and are available from most artists' and office supply stores.

To find out what diameter circle represents your finderscope field of view on the sky charts, center your finder on an easy-to-locate star, such as Mizar (Zeta [ζ] Ursae Majoris) (see p. 232). Southern Hemisphere skywatchers can use brilliant Alpha [α] Centauri (see p. 246). Look at the fainter stars around Mizar or Alpha Centauri, and study the view. Notice which stars you can see along the edges of the field, then find those same stars on your chart. Center your template circle on the same bright star on the map and try out different diameter circles until you find the one that closely matches your finderscope view.

To locate fainter objects, it helps to have an idea of how large the object will appear to be in your telescope eyepiece. The apparent size of an object is its angular diameter, usually measured in minutes or seconds of arc. Measure the field of view of your eyepiece by using a bright star located close to the celestial equator, such as Mintaka (Delta [δ] Orionis), and follow these simple steps: place the star on the eastern edge of your view, time, in seconds, how long it takes for the star to drift across the center of the field and exit on the opposite (western) side, divide this time by four. The result represents the angular diameter of your field of view in arcminutes.

You should multiply the answer by 60 to obtain your field diameter in arcseconds—many faint objects and the separations of most double stars are smaller than an arcminute across. Measure the field of view for all of your

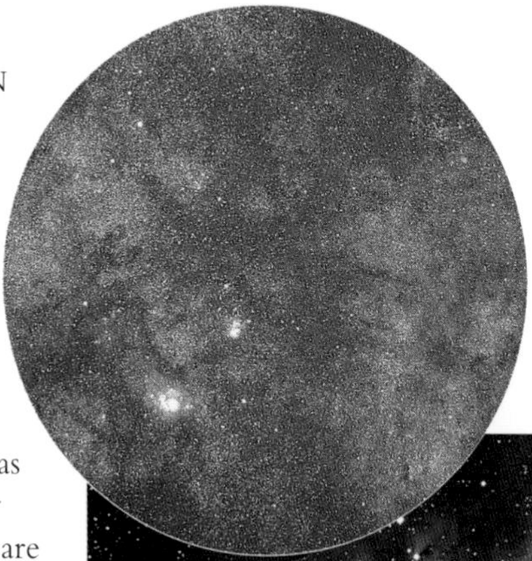

THE TRIFID NEBULA *appears as a small, faint patch of light in a binocular field view (left), but is revealed as a colorful nebula in the photograph below.*

RED FLASHLIGHT *(left) is essential*
· reading star charts under a dark sky.
·e key to symbols (below) applies to
the main sky charts in this chapter.

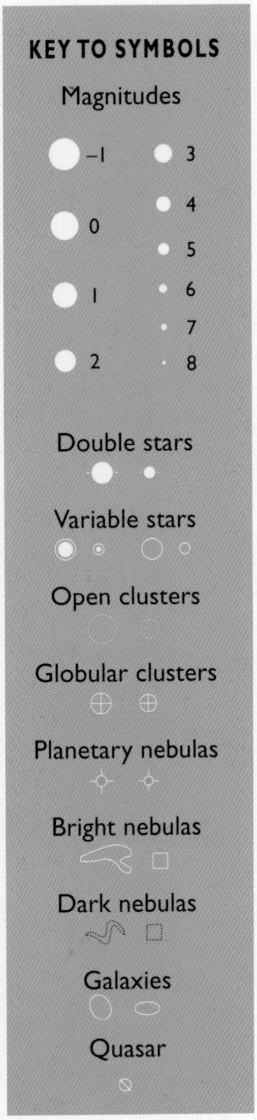
KEY TO SYMBOLS
Magnitudes
–1
0
1
2
3
4
5
6
7
8
Double stars
Variable stars
Open clusters
Globular clusters
Planetary nebulas
Bright nebulas
Dark nebulas
Galaxies
Quasar

yepieces, along with the ınderscope, and record this ıformation in your obser- ation logbook.

JSING THE SKY CHARTS

t is important to have a feel ɔr the size of your chart in elation to what you are ɔoking at. All the main sky harts in the Starhopping Guide are to the same scale ıbout 30 x 20 degrees). The

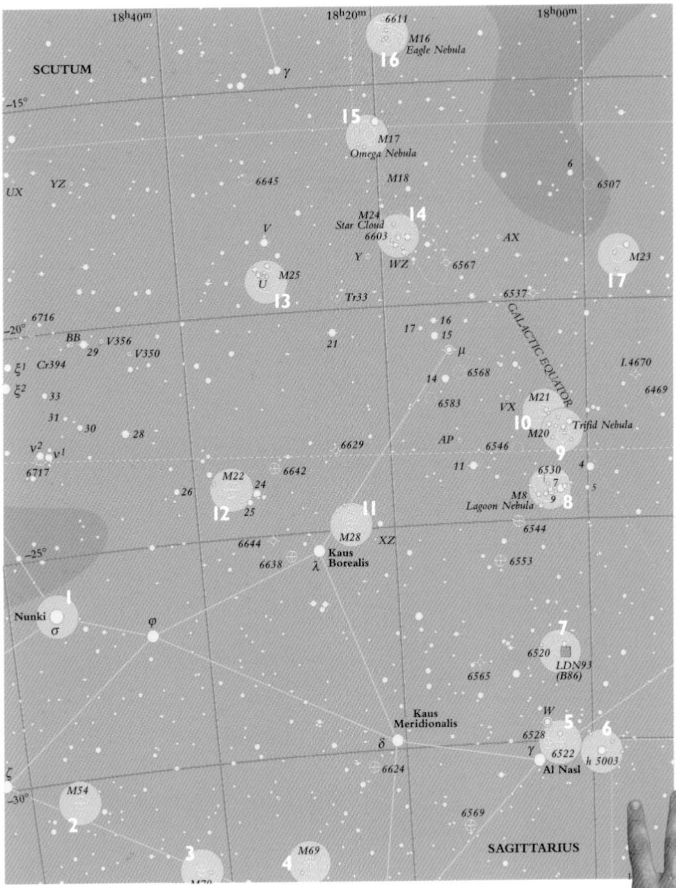
18h40m
18h20m
18h00m
SCUTUM
M16
Eagle Nebula
M17
Omega Nebula
M18
M24
Star Cloud
M25
M23
GALACTIC EQUATOR
M21
Trifid Nebula
M20
M8
Lagoon Nebula
M22
M28
Kaus Borealis
Nunki
LDN93 (B86)
Kaus Meridionalis
Al Nasl
M54
M69
SAGITTARIUS

span of your outstretched hand at arm's length measures close to 20 degrees—equivalent to the area represented by the main sky charts' vertical edge. Look for the brightest stars on your chart and match them in orientation to the same arrangement of stars appearing in the sky.

Starhopping is like moving down the link of a chain, with each link represented by your field-of-view circle. On your chart, each selected object is centered in a 1 degree field—about the same size as a low-power eyepiece—but remember that each eyepiece's field size will be different. Estimate how many fields to move your telescope in making each hop, using your template circle. As you move the telescope, look out for the patterns of fainter stars along the hop.

One of the most difficult aspects of starhopping is being sure to move your telescope

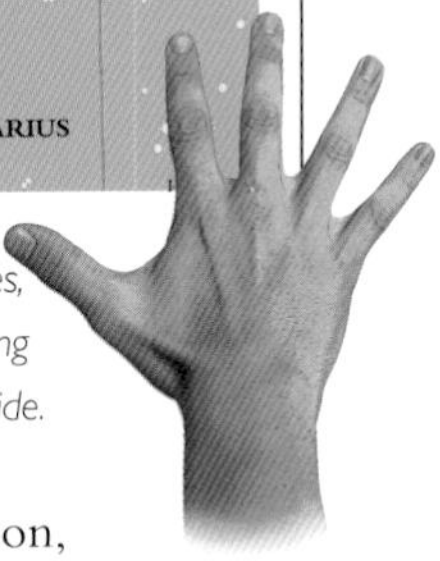

COMPARE *sky distances, relative to your chart, using your hand as a rough guide.*

in the right direction, as the view of the sky through a telescope is often upside down or back to front (see p. 50). The easiest way to orient yourself is to nudge your scope toward the north (or south) celestial pole to see where stars enter the northern (or southern) edge of the field. Nudge your scope at right angles to this to find east and west. A compass symbol is also marked on the grid of each main sky chart to help keep you on track.

The stars on the star charts are marked from magnitude –1 down to magnitude 8. The deep-sky objects are marked to as faint as magnitude 12.5.

Now that you know how to starhop, you are ready to embark on any of the guided sky tours featured. RG

KEY MAPS

Finding some constellations can be difficult, depending on your location, but a map of the whole sky is a great way to orient yourself to the overall picture.

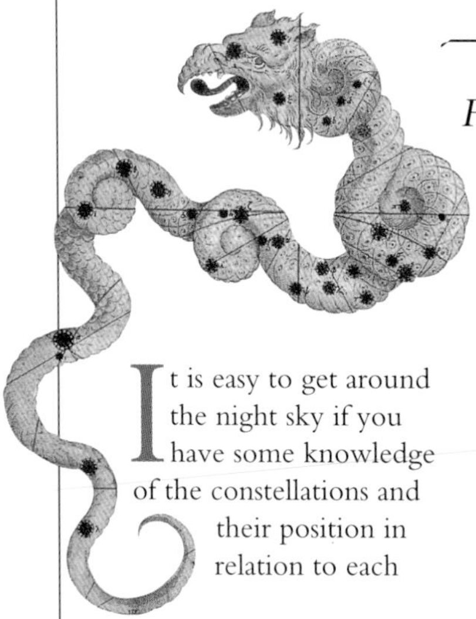

DRACO, *the Dragon, is one of the Northern Hemisphere's circumpolar constellations.*

It is easy to get around the night sky if you have some knowledge of the constellations and their position in relation to each other. A naked-eye view of the sky is the best way to learn the positions of the constellations, regardless of whether you own a telescope or not. When you are familiar with the constellations, you can then move on to particular regions, such as those in the starhops.

The key maps below represent the entire night sky. The projection used here halves the celestial sphere from top to bottom (celestial north to south). The north celestial and south celestial poles are at +90° and –90°, respectively. To avoid the

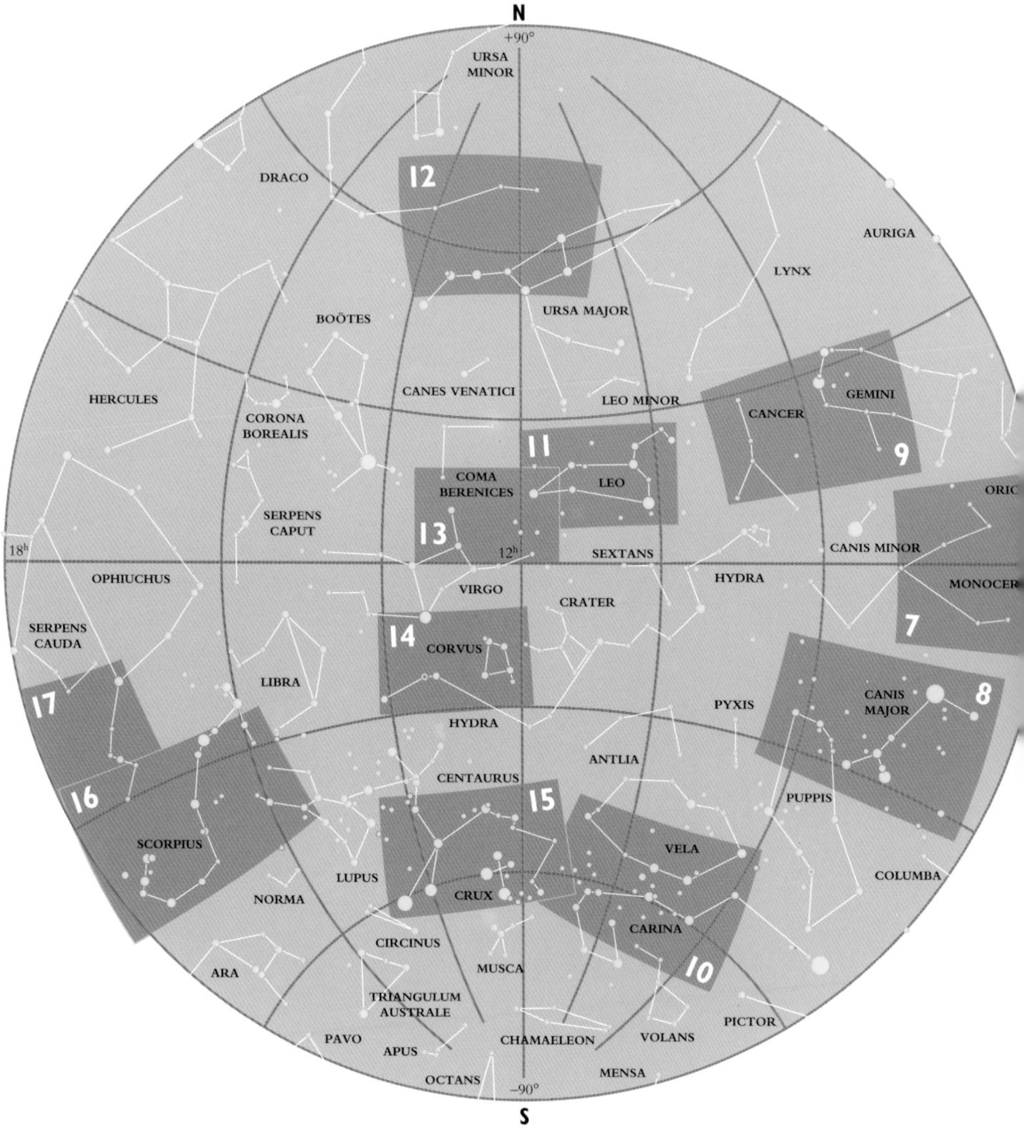
N
+90°
URSA MINOR
DRACO
12
AURIGA
LYNX
URSA MAJOR
BOÖTES
HERCULES
CANES VENATICI
LEO MINOR
GEMINI
CANCER
CORONA BOREALIS
11
9
COMA BERENICES
LEO
SERPENS CAPUT
13
CANIS MINOR
E
18h
12h
SEXTANS
OPHIUCHUS
HYDRA
MONOCER
VIRGO
CRATER
SERPENS CAUDA
14
CORVUS
7
LIBRA
17
CANIS MAJOR
8
PYXIS
HYDRA
ANTLIA
16
CENTAURUS
15
PUPPIS
SCORPIUS
VELA
COLUMBA
LUPUS
CRUX
NORMA
CARINA
CIRCINUS
ARA
MUSCA
10
TRIANGULUM AUSTRALE
PICTOR
PAVO
VOLANS
APUS
CHAMAELEON
MENSA
OCTANS
–90°
S

THE STARHOPS

istortion of the constellation atterns that occurs as the phere narrows toward the two oles, the key maps have been resented in stereographic projection. The only "distortion" rom this projection comes rom an increase in scale oward the edge of each map, giving them a more rounded appearance, making the star patterns easier to recognize.

The starhops are represented by the darker blue patches on the key maps, and are numbered from 1 to 20. They are ordered west to east across the sky according to the times that they come into view in the evening throughout the year. These regions have been chosen for their variety of objects as well as their location. One or two starhops might be too far north or south for you to observe but most of them can be enjoyed by everyone.

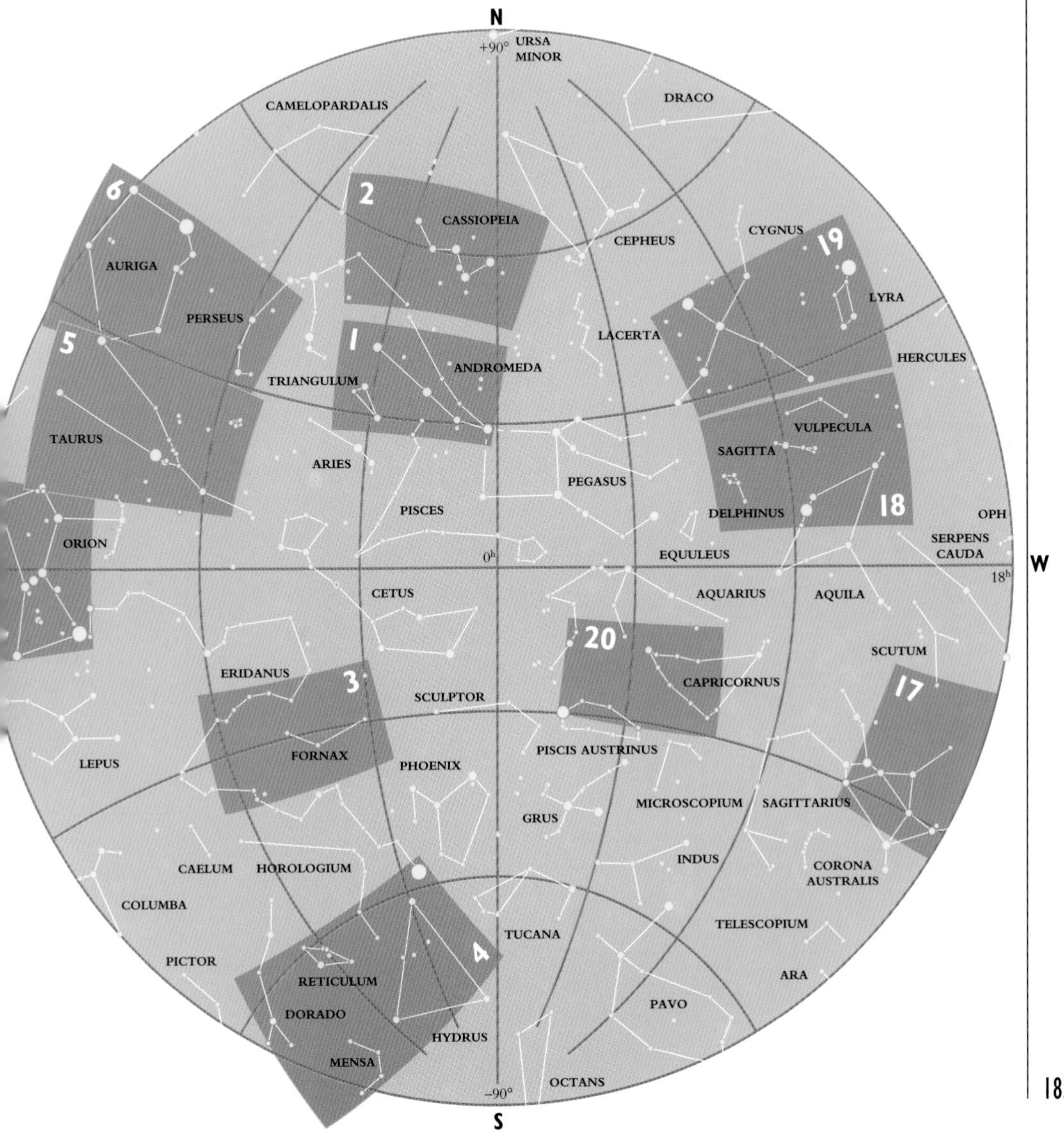

N
+90°
URSA MINOR
DRACO
CAMELOPARDALIS
CASSIOPEIA
CEPHEUS
CYGNUS
AURIGA
PERSEUS
LYRA
LACERTA
HERCULES
ANDROMEDA
TRIANGULUM
VULPECULA
SAGITTA
TAURUS
ARIES
PEGASUS
PISCES
DELPHINUS
OPH
SERPENS CAUDA
ORION
0h
EQUULEUS
W
18h
CETUS
AQUARIUS
AQUILA
SCUTUM
ERIDANUS
SCULPTOR
CAPRICORNUS
LEPUS
FORNAX
PHOENIX
PISCIS AUSTRINUS
MICROSCOPIUM
SAGITTARIUS
GRUS
INDUS
CAELUM
HOROLOGIUM
CORONA AUSTRALIS
COLUMBA
TELESCOPIUM
TUCANA
PICTOR
ARA
RETICULUM
PAVO
DORADO
HYDRUS
MENSA
OCTANS
−90°
S
1
2
3
4
5
6
17
18
19
20

A GALAXY HUNT AROUND ANDROMEDA

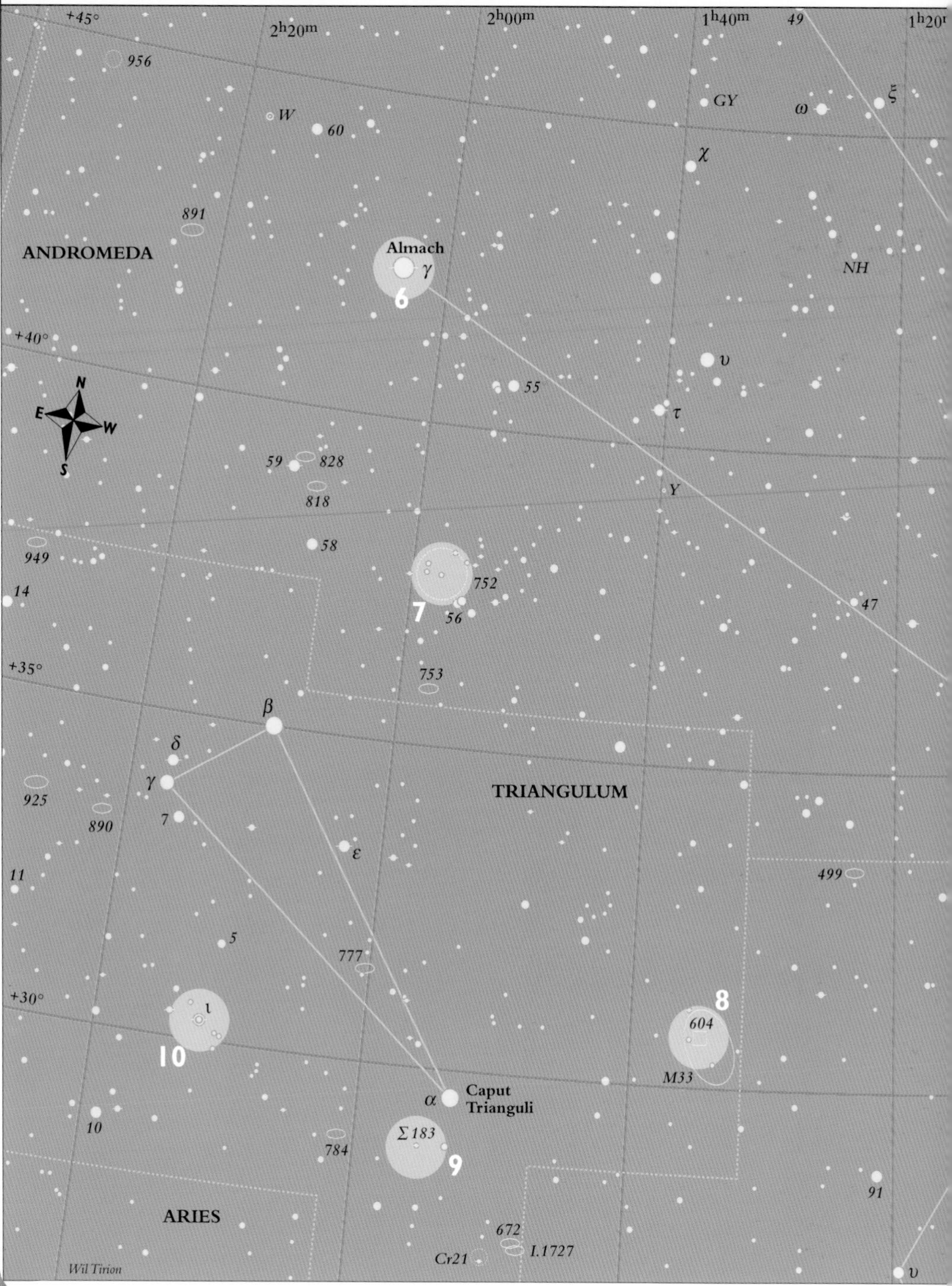
2h20m
2h00m
1h40m
+45°
+40°
+35°
+30°
ANDROMEDA
TRIANGULUM
ARIES
Almach
Caput
Trianguli
M33
Σ183
Cr21
I.1727
6
7
8
9
10
Wil Tirion

A Galaxy Hunt around Andromeda

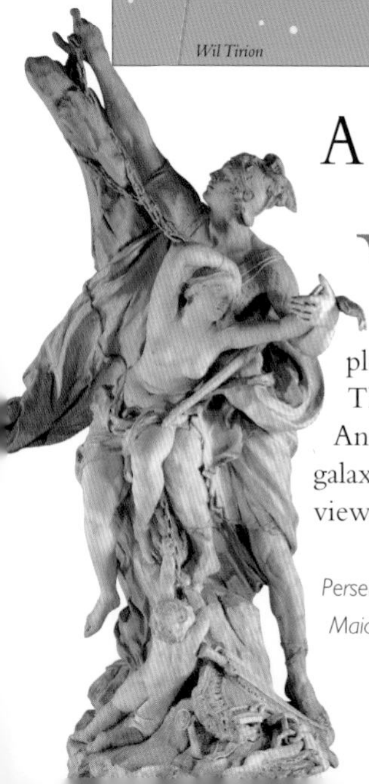

Visible from both hemispheres, this star-hop offers skywatchers plenty of celestial treasures. The tour highlights are the Andromeda and Pinwheel galaxies, providing unforgettable views in a clear, dark sky.

Perseus rescuing Andromeda, the Chained Maiden, by Pierre Puget (1620–94).

The Andromeda Galaxy, in the constellation of Andromeda, is one of the brightest and most accessible galaxies for observers, and has been wel known since ancient times. The earliest extant drawing of it is on a sky chart published in AD 96 by the Persian astronomer Abderrahaman al-Suf. The Pinwheel Galaxy lies in the neighboring constellation of Triangulum, the Triangle.

Keep in mind, though, that these galaxies are not the sum total of the area. It is also rich in many other interesting deep-sky objects.

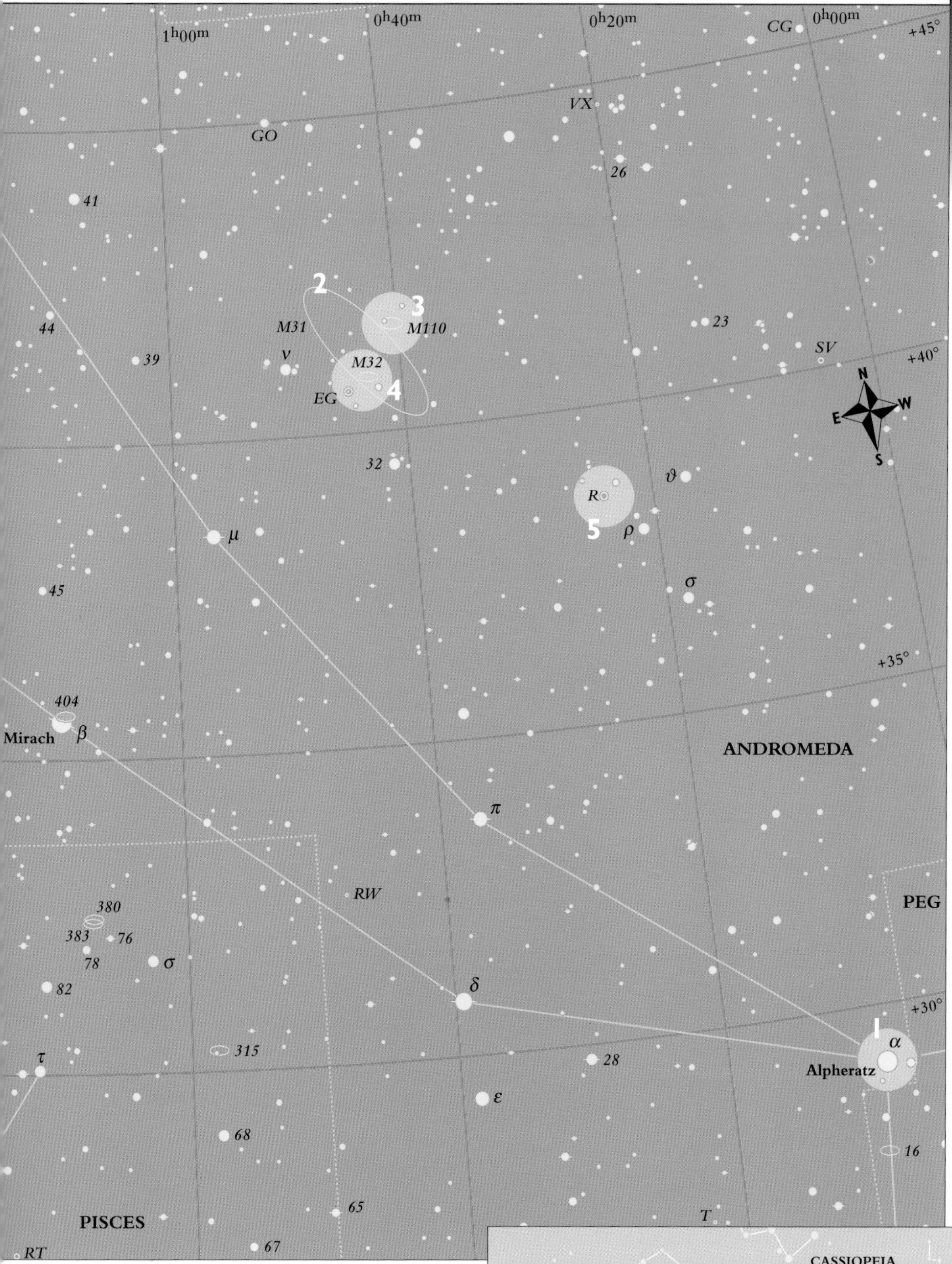
1h00m
0h40m
0h20m
0h00m
CG
+45°
VX
GO
26
41
2
3
M31
M110
44
39
V
M32
EG
4
23
SV
+40°
N
E
W
S
32
R
5
ρ
μ
σ
45
+35°
404
Mirach
β
ANDROMEDA
π
RW
PEG
380
383
76
78
σ
82
δ
+30°
1
α
Alpheratz
τ
315
28
ε
68
16
65
T
PISCES
RT
67

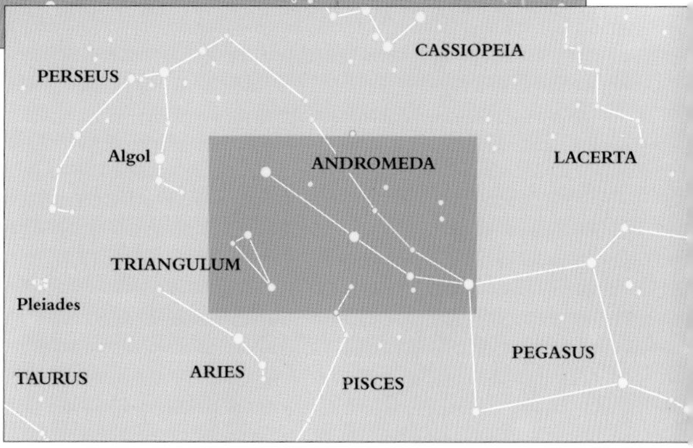
PERSEUS
CASSIOPEIA
Algol
ANDROMEDA
LACERTA
TRIANGULUM
Pleiades
PEGASUS
TAURUS
ARIES
PISCES

❶ Alpheratz (21 Alpha [α] Andromedae)

Alpheratz marks the northeastern corner of the Great Square of Pegasus, shown in the ocater map (right). Our starhop begins with this nagnitude 2.1 blue-white double star, the rightest in the Andromeda constellation. Using scope of 8 inches (200 mm) or more, you might e able to spot the very faint magnitude 11.3 econdary star, 82 arcseconds from the primary.

Alpheratz was once known as Delta [δ] Pegasi ecause it was considered to be a part of the onstellation of Pegasus before being renamed nd allocated to Andromeda. You might see it abeled as Delta [δ] Peg on some sky charts.

To find the Andromeda Galaxy, use the stars **Mirach (43 Beta [β] Andromedae)** and **37 Mu [μ] Andromedae**, located in the twin chains of stars northeast of Alpheratz, as pointers.

❷ The Andromeda Galaxy (M31, NGC 224) 👁 The spectacular Andromeda Galaxy is the most distant object we can see with the naked eye. As you gaze at the soft glow of the closest spiral galaxy to the Milky Way, you are looking across about 2,200,000 light-years. M31 is a large, bright object measuring 160 x 50 arcminutes. In a dark sky it is clearly visible to the unaided eye, appearing as a broad, hazy-white cloud. However, city observers need binoculars to catch a satisfying view. In a 4 inch (100 mm) telescope, the core of M31 seems flattened, flowing evenly into the disk of the galaxy. A telescope fitted with a low-power, wide-field eyepiece is an advantage when observing M31 because of its sheer size. A higher power eyepiece will reveal only its nucleus.

M31 (above) is vastly more distant than the open cluster NGC 752 (left), its neighbor in the Andromeda constellation.

In 1924, Edwin Hubble proved that M31 was not a part of the Milky Way, helping to determine the size of the universe. He did this by comparing the brightness of Cepheid variables in M31 and the Milky Way, finding that they were dimmer in M31 because they were so much farther away.

❸ M110 (NGC 205) ⊂◎ Within ½ degree of M31 is one of its two companion galaxies, M110, glowing at magnitude 8. In a 4 inch (100 mm) scope, M110 appears as an elongated patch of light, brightening toward its core.

M110 is the last object appearing on Charles Messier's list, having been added in 1967.

❹ M32 (NGC 221) ⊂◎ M32 is a compact elliptical galaxy about 1 degree southeast of M110. Also a companion galaxy of M31, M32 appears as an oval patch of light in a 4 inch (100 mm) scope, but is best viewed through larger telescopes.

EG Andromedae ⊂◎ Half a degree farther southeast is the cataclysmic variable red giant EG Andromedae. Its magnitude ranges from 7.1 down to 7.8 in a complicated cycle.

❺ R Andromedae ⊂◎ Southwest of EG is R Andromedae, a long-period Mira variable well known for its broad-ranging cycle, moving from magnitude 5.3 down to 15 over a period of about 409 days. It is sometimes visible in binoculars, but at minimum, is a challenge for an 8 inch (200 mm) telescope. The stars **24 Theta [θ] Andromedae 25 Sigma [σ] Andromedae**, and **27 Rho [ρ] Andromedae** form a triangle west of R, that helps locate the star when it is faint.

❻ Almach (57 Gamma [γ] Andromedae) ⊂◎ Sweep your scope about 12½ degrees east to Almach, a fine, color-contrast close double star—one of the best doubles for small-telescope observers. The magnitude 2.3 golden primary contrasts with its magnitude 5.1 greenish blue companion, lying 10 arcseconds away.

M33, the Pinwheel Galaxy, is the biggest and brightest galaxy in the Local Group, after the Milky Way and the Andromeda Galaxy.

7 NGC 752 Shift about 5 degrees south of Almach to NGC 752. The 60 stars in this open cluster can be seen with the unaided eye from a dark site. With a finderscope or binoculars, you may be able to see chains and knots of stars forming a twisted X within this rich cluster.

8 The Pinwheel Galaxy (M33, NGC 598) About 7 degrees southwest of NGC 752 is the beautiful Pinwheel Galaxy, across the border in Triangulum. Although bright, at magnitude 5.5, this spiral galaxy is hard to see because it appears face-on and is spread over a large patch of sky. It appears as a fuzzy glow through binoculars in a dark sky, but you will need a scope of 8 inches (200 mm) and a wide-field eyepiece to see more of the galaxy, including its diffuse, twisting arms.

NGC 604 The splotch of nebulosity sitting prominently on the northeastern tip of the Pinwheel Galaxy is NGC 604, shining more brightly than other parts of M33's spiral arms. NGC 604 is an oval-shaped emission nebula of glowing hydrogen gas where new stars are born.

9 Struve [Σ] 183 About 4 degrees southeast of M33 is the bright yellow multiple star **Caput Trianguli (2 Alpha [α] Tri)**. Use it as a pointer to locate the much dimmer double star Struve 183, about 1 degree southeast. A fine close double, Struve 183 has a magnitude 7.7 light yellow primary and a magnitude 8.4 blue secondary, 6 arcseconds away.

10 6 Iota [ι] Tri Four degrees northeast of Struve 183 is the stunning color-contrast double 6 Iota [ι] Tri. The primary is a magnitude 5 yellow star and the secondary is a magnitude 6.5 pale blue star, shining 3.8 arcseconds away.

Take the time to re-explore the highlights of our tour, along with the many other deep-sky objects surrounding them in this area. RG

RADIO MAPPING OF M31

In 1912, American astronomer V. M. Slipher made the first spectrogram of M31. The blueshifts and redshifts of the spectrum lines of different parts of the galaxy showed that it was rotating.

This image (left), based on the 21 centimeter wavelength by the Westerbork Synthesis Radio Telescope, also shows that M31 is rotating and that there is a high concentration of cold hydrogen gas in the galaxy's outer arms. The color enhancement of the radio image uses yellow and red to indicate the areas that are rotating away from us, while green and blue indicate areas that are in systematic motion toward us. From our vantage point, the Andromeda Galaxy is rotating in a clockwise direction.

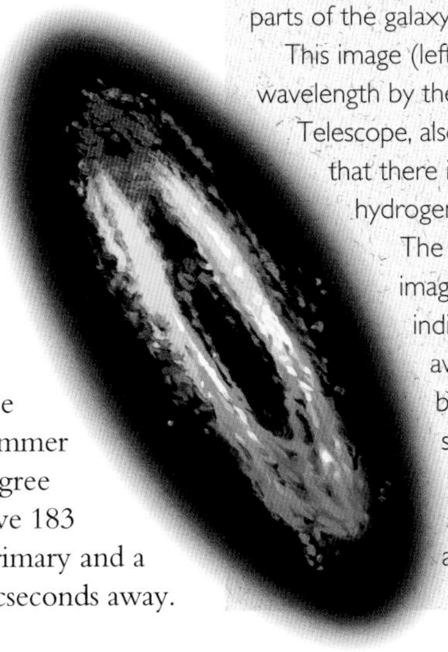

CLUSTERS AROUND CASSIOPEIA AND PERSEUS

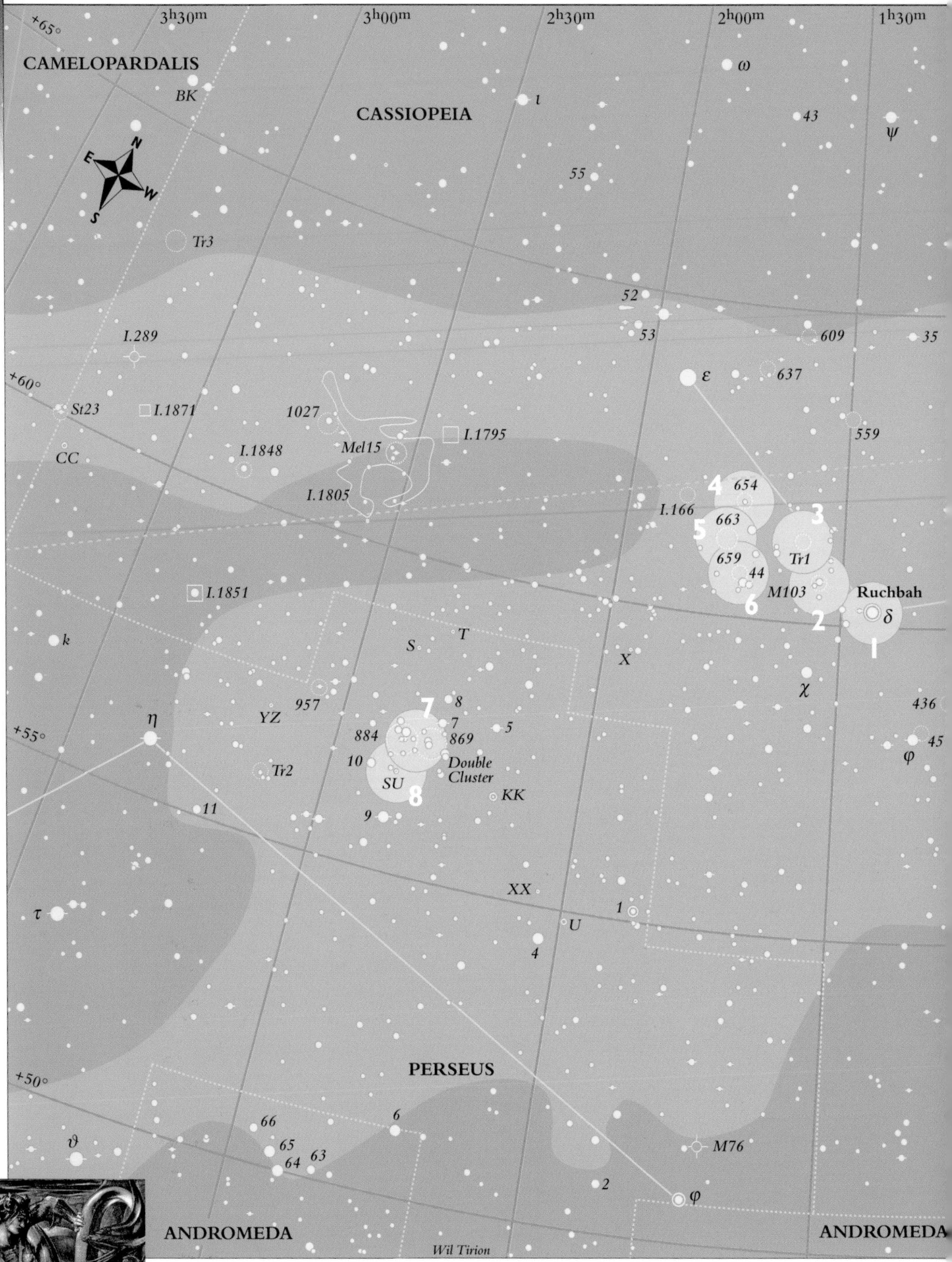
3h30m
3h00m
2h30m
2h00m
1h30m
+65°
+60°
+55°
+50°
CAMELOPARDALIS
CASSIOPEIA
PERSEUS
ANDROMEDA
ANDROMEDA
BK
Tr3
I.289
St23
I.1871
1027
Mel15
I.1795
I.1848
I.1805
CC
I.1851
Ruchbah
M103
Tr1
654
663
659
44
I.166
609
637
559
35
52
53
55
43
957
YZ
884
869
Double Cluster
SU
KK
Tr2
XX
U
M76
Wil Tirion

Clusters around Cassiopeia and Perseus

Unfortunately, the spectacular open clusters in this part of the Milky Way are not visible to stargazers in mid-southern latitudes. For northern observers, however, Cassiopeia is a circumpolar constellation that never sets. The best months to explore this region are September through February. This area of the night sky is especially rewarding for binocular observers.

Cassiopeia, the Queen of Ethiopia, and Perseus, the Hero, are central figures in the lore connected to several other mythological characters for whom nearby constellations are named. Cassiopeia sits on her throne in the form of an easy-to-locate W-shaped asterism of bright northern stars. To her southeast is the Y-shaped figure of her son-in-law Perseus.

The figure of Perseus in a detail from The Doom Fulfilled, *a painting by English artist Edward Burne-Jones (1833-98).*

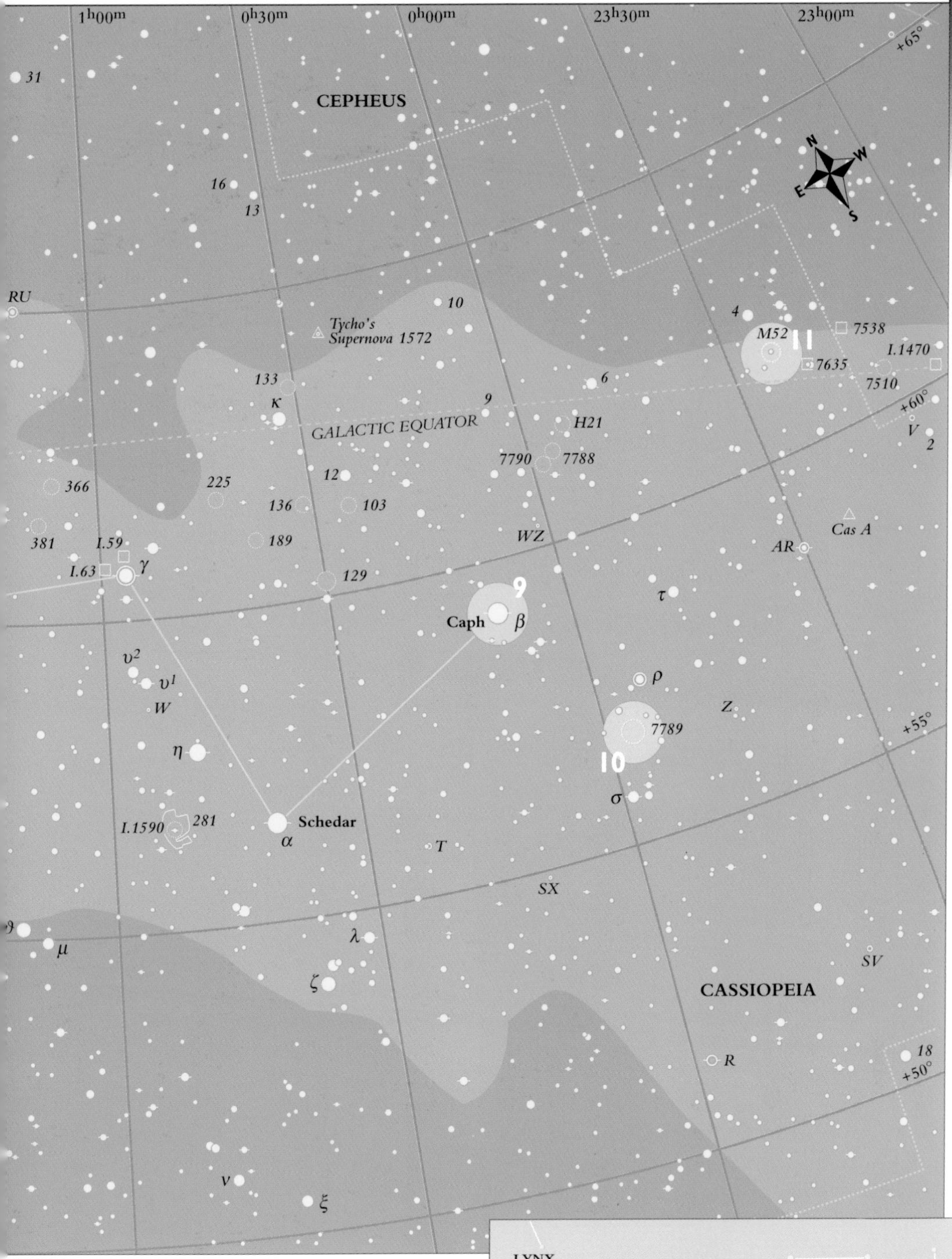
1h00m
0h30m
0h00m
23h30m
23h00m
+65°
CEPHEUS
31
16
13
RU
Tycho's Supernova 1572
10
4
M52
11
7538
I.1470
7635
7510
133
κ
6
9
+60°
V
2
GALACTIC EQUATOR
H21
7790
7788
12
366
225
136
103
381
I.59
189
WZ
Cas A
AR
I.63
γ
129
9
τ
Caph
β
υ2
υ1
W
ρ
Z
7789
+55°
η
10
σ
I.1590
281
Schedar
α
T
SX
μ
λ
SV
ζ
CASSIOPEIA
R
18
+50°
ν
ξ

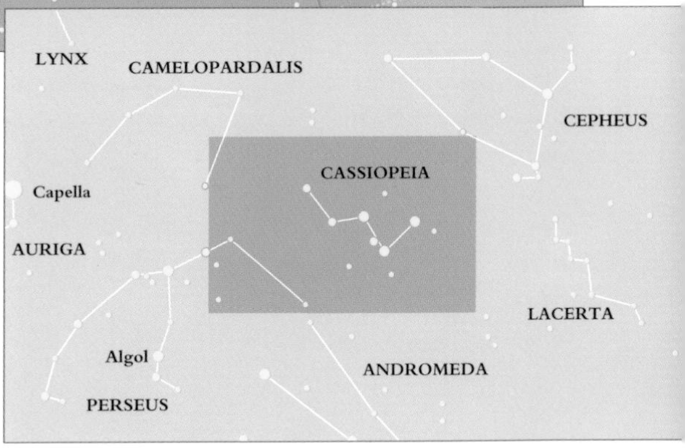
LYNX
CAMELOPARDALIS
CEPHEUS
CASSIOPEIA
Capella
AURIGA
LACERTA
Algol
ANDROMEDA
PERSEUS

1 Ruchbah (37 Delta [δ] Cassiopeiae)

Our tour of rich Milky Way open clusters begins at Ruchbah, the magnitude 2.7 blue-white star forming a part of the eastern downstroke of Cassiopeia's highly recognizable W shape. Ruchbah is thought to lie 45 light-years from us. This beautiful star is the primary in a multiple system as well as an eclipsing-binary variable. A faint secondary star revolves around Ruchbah, partially eclipsing it, causing Ruchbah's magnitude to dim very slightly over a 760-day period.

Ruchbah's name originates from the Arabic *Al Rukah*, in relation to its position as "the knee" of the woman in the chair.

About 1/2 degree northeast of Ruchbah, within the same binocular or finder field, is the U-shaped gathering of open clusters: M103, Trumpler 1, NGC 654, NGC 663, and NGC 659.

The Double Star Cluster (above)—NGC 869, at right, and NGC 884, at left—is a favorite target for Northern Hemisphere observers. Appearing close in the sky, they are not really associated.

❷ **M103 (NGC 581)** Shift your scope about 1 degree northeast of Ruchbah to locate M103. Pierre Méchain discovered this loose, magnitude 6.2 open star cluster in 1781. The brighter stars within the cluster seem to form the shape of an arrowhead. A 4 inch (100 mm) scope can resolve many of the fainter stars, some of which are colored. M103 may not actually be a star group that is bound by gravity, but simply a grouping of about 60 stars that appear as a scattered cluster from our vantage point.

❸ **Trumpler 1** Looking 1/2 degree farther northeast from M103, you should be able to spot Trumpler 1. This poor open cluster is a little difficult to pick out from the rich Milky Way star field it appears to be a part of. The small clump of stars seems to have a bright streak running through it that cannot be properly resolved, even with a 6 inch (150 mm) telescope.

❹ **NGC 654** About 1 degree northeast from Trumpler 1 is NGC 654. This cluster consists of about 60 stars in a loose association, but is still an easy object to see with binoculars appearing as a hazy glow of tiny, faintly resolved stars.

❺ **NGC 663** Nudge 1/2 degree south of NGC 654 to the open cluster NGC 663. The 80 stars in this cluster give it a roundish look, somewhat like a cross between an open and a globular cluster, through a 6 inch (150 mm) telescope.

SUPERNOVA REMNANT CASSIOPEIA A

When Cassiopeia A (3C 461) exploded as a supernova some 9,700 years ago, a gigantic circular shell of gas raced out into space. Today, this still-expanding supernova remnant is too faint to see with amateur equipment, but its powerful energy is detectable at radio wavelengths. The radio output emission is created by high-speed electrons spiraling around magnetic field lines as the expanding cloud collides with thin gas between the stars.

Radio images of the cloud (right), the brightest radio source outside of the Solar System, show gas racing away from the spot where the star exploded. By calculating this speed and the distance traveled by the gas since the explosion, astronomers estimate that the light from the explosion reached Earth around 1680, creating a 5th magnitude star. No record exists of anyone noticing this short-lived supernova in Cassiopeia.

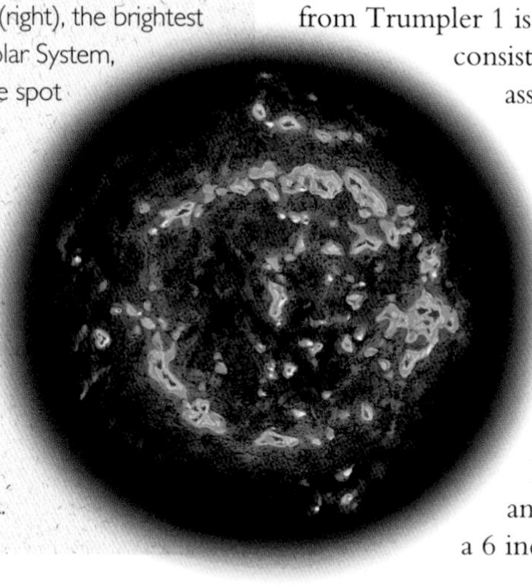

the rich star field (above), the Bubble Nebula (NGC 7635) ppears at lower right to the open cluster M52, seen at upper left. IGC 7789 (right) is a rich open cluster, slightly larger than M52.

NGC 659 Some 1/2 degree southwest of NGC 663 lies NGC 659, an X-shaped cluster of about 40 stars. The magnitude 5.8 golden-ellow double star **44 Cassiopeiae**, which is robably not a member of the cluster, lies close by.

The Double Star Cluster (NGC 884 Chi [χ] Persei} and NGC 869 {h Persei}) Return to Ruchbah, then hop about 8 de-rees southeast to the Double Star Cluster in erseus. This magnificent conglomeration of right stars has been well known since ancient imes. Because of their size, these two 100-plus nember clusters are best viewed with binoculars or a wide-field telescope eyepiece. Careful study of each cluster will reveal the many double-, nultiple-, and variable-star systems they contain.

SU Persei Some 1/2 degree southeast of the center of NGC 884 is the semi-regular ariable SU Persei, which is actually a member of NGC 884. This pulsating red supergiant aries from magnitude 9.4 down to 10.8 during cycle that lasts about 533 days.

Caph (Beta [β] Cassiopeiae) Sweep ome 17 degrees west and slightly north of SU ersei to Caph, a wide double star and a short-eriod pulsating variable of a rare type known as Delta Scuti. Its variability is hard to detect visually ecause its magnitude of 2.2 fluctuates very lightly during a cycle that lasts a couple of hours.

⑩ NGC 7789 Hop 4 degrees southwest to the broad open cluster NGC 7789. This large cluster is located between the yellow semi-regular variable **7 Rho [ρ] Cassiopeiae** and the wide, white multiple-star system **8 Sigma [σ] Cassiopeiae**. Use binoculars or a wide-field eyepiece to view the cluster, because it covers an area roughly the apparent size of the Moon.

⑪ M52 (NGC 754) Located about 6 de-grees northwest from NGC 7789 is M52, an open star cluster. The 200 members of this cluster are located on the western border of Cassiopeia. A fine string of stars extending like an arm in an east-west direction across the cluster is a good test for 4 inch (100 mm) scopes or larger.

In areas filled with relatively easy binocular targets, such as this, it is a pleasure to amble around the many "families" of open clusters. RG

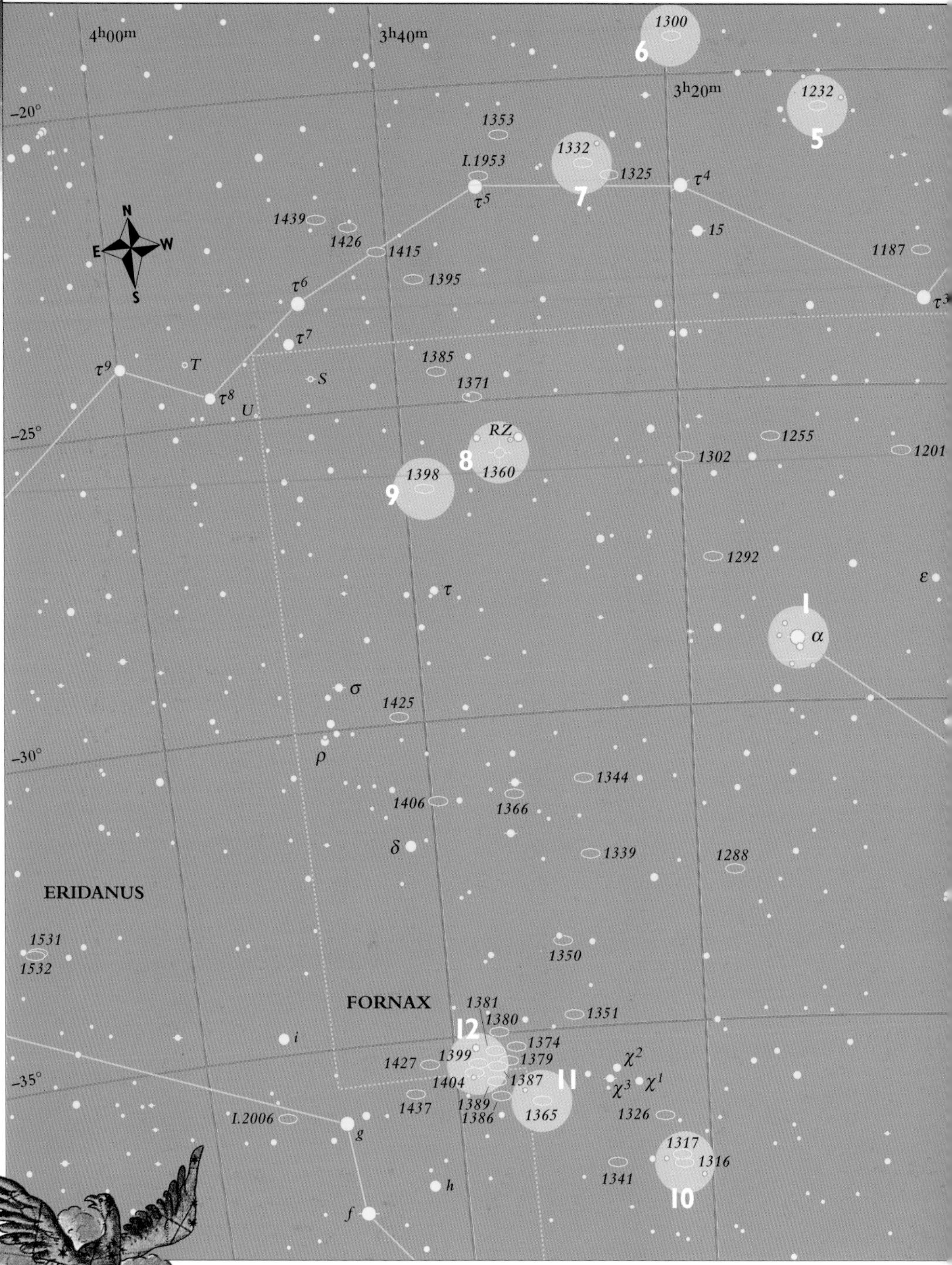
4h00m
3h40m
3h20m
−20°
−25°
−30°
−35°
N
E
W
S
1300
6
1232
5
1353
1332
I.1953
1325
7
τ4
τ5
1439
1426
1415
15
1187
1395
τ6
τ3
τ7
1385
τ9
T
S
1371
τ8
U
RZ
1255
1201
8
1302
1398
1360
9
1292
ε
τ
1
α
σ
1425
ρ
1344
1406
1366
δ
1339
1288
ERIDANUS
1531
1532
1350
FORNAX
1381
1380
1351
12
i
1374
χ2
1427
1399
1379
1404
1387
11
χ3
χ1
1437
1389
1386
1365
1326
I.2006
g
1317
1316
h
1341
10
f

A Galaxy Feast in the Furnace

Although it can be seen from both hemispheres, this part of the sky is much more accessible to Southern Hemisphere skywatchers. Northerners are best advised to view this region during winter evenings, when it is at its highest point in the sky.

The constellation of Fornax was introduced by the eighteenth-century French astronomer Nicolas Louis de Lacaille. He originally named it Fornax Chemica, the Chemical Furnace, but today it is simply known as the Furnace.

Although this patch of sky may appear fairly barren at first, throughout Fornax and in parts of neighboring Eridanus, the River, you are looking toward a region rich in galaxies. The Fornax Galaxy Cluster, a relatively nearby cluster of galaxies, is one of the highlights of this starhop. You need at least a 4 inch (100 mm) telescope to see many of the star systems in this area.

The flaming Furnace burns beside the Phoenix and the River in Carel Allard's eighteenth-century Celestial Planisphere.

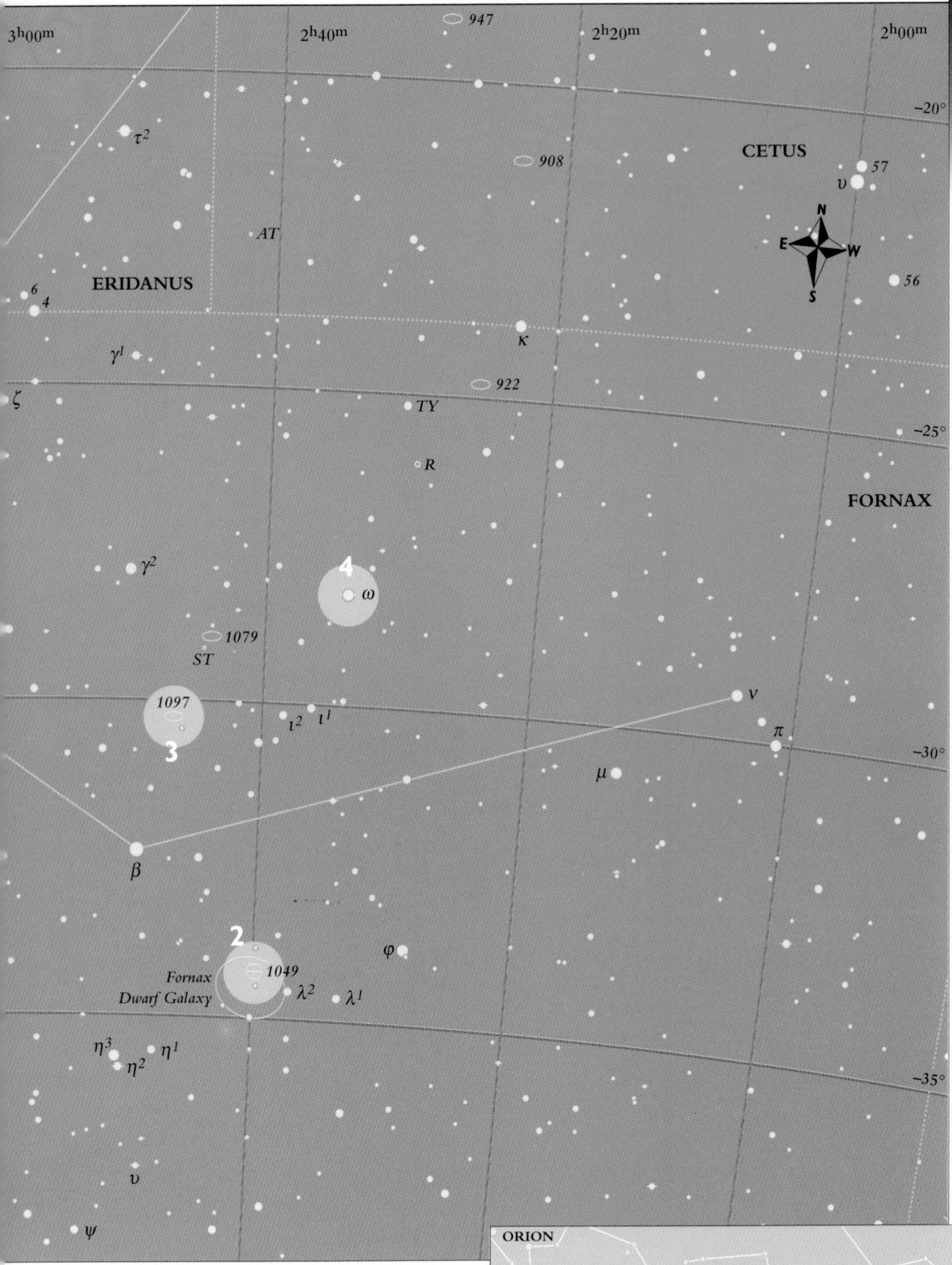
3h00m
2h40m
2h20m
2h00m
947
-20°
CETUS
908
57
56
N
E
W
S
ERIDANUS
AT
922
TY
R
-25°
FORNAX
4
ω
1079
ST
1097
3
ν
π
-30°
μ
β
2
φ
1049
Fornax
Dwarf Galaxy
-35°

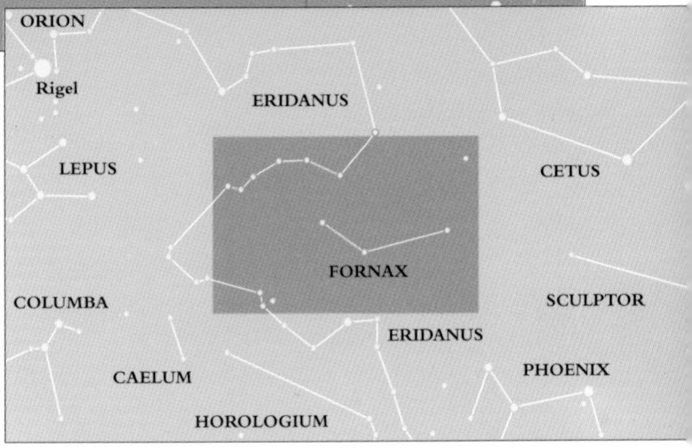
ORION
Rigel
ERIDANUS
LEPUS
CETUS
FORNAX
COLUMBA
SCULPTOR
ERIDANUS
CAELUM
PHOENIX
HOROLOGIUM

1 Alpha [α] Fornacis Our starting ›oint on this journey is the star Alpha Fornacis, he brightest star in Fornax, although at 4th nagnitude it is not what you would call ›rominent. This is an interesting binary star, vith its magnitude 7 companion, also a deep ,ellow color, shining about 5 arcseconds away. The exact period during which the stars orbit :ach other is not altogether certain, but the :urrently accepted value is about 314 years.

The difference in brightness between the two :omponents makes this a difficult object for mall scopes. On a steady night with good eeing, a 3 inch (75 mm) scope should show the secondary fairly easily. The secondary star is also suspected of being variable, possibly fading to as low as magnitude 8 at minimum, when you may have some difficulty seeing it.

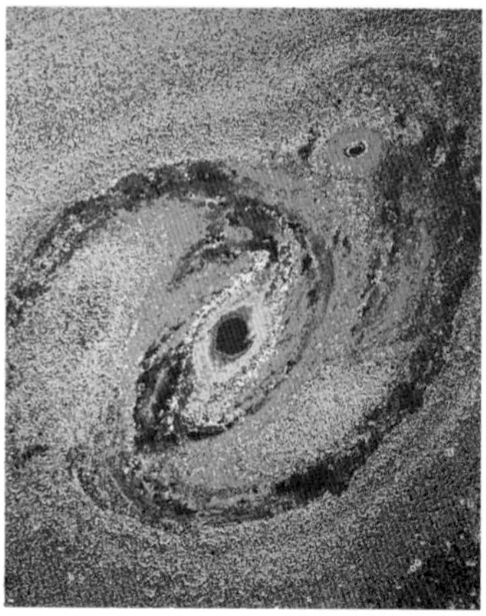

A portion of the Fornax Cluster of galaxies (above). The computer-enhanced image of NGC 1097 (left) shows its spiral structure.

❷ **NGC 1049** About 9 degrees southwest from Alpha, you will find the star **Beta [β] Fornacis**. Use it as a pointer to help locate NGC 1049, 3 degrees farther southwest. This globular cluster calls for a larger aperture because it shines at only 13th magnitude. An 8 inch (200 mm) scope may just show it as a fuzzy blob about 20 arcseconds across. It appears faint because it is not a part of our galaxy, belonging instead to the dim Fornax System, a 1 degree wide dwarf galaxy that looms so large it cannot be recognized visually, even though it is within the Local Group of galaxies.

❸ **NGC 1097** Shift 4 degrees north and a little east to NGC 1097, a barred spiral galaxy that is easy enough to see with a small scope. It shines at magnitude 9.3 and has a very bright nucleus. Its elongated form should be revealed through a 4 inch (100 mm) scope or larger.

❹ **Omega [ω] Fornacis** Just over 3 degrees northwest of NGC 1097 is Omega Fornacis, a double star and an easy object for any small telescope. Its 5th and 7th magnitude components glow 11 arcseconds apart.

❺ **NGC 1232** To find NGC 1232, hop back to Alpha Fornacis, then move just over 8 degrees north and a little west, across the border into Eridanus. This large but faint spiral galaxy is not an easy object for small scopes and its total magnitude of 9.9 is somewhat deceptive. As with many objects on this tour, an 8 inch (200 mm) telescope will reveal more of the galaxy.

❻ **NGC 1300** To locate NGC 1300, nudge your scope 2½ degrees east and a little south of NGC 1232, where you will find the 4th magnitude star **16 Tau^4 [τ^4] Eridani**. Just over 2 degrees north of Tau^4 is the 10th magnitude barred spiral galaxy NGC 1300. This beautiful galaxy has a bright core and is easy to see through a 4 inch (100 mm) scope. In a dark sky, a hint of the arms can be detected in 12 inch (300 mm) telescopes.

❼ **NGC 1332** Return to Tau^4 before moving 1½ degrees east and a little north to the elongated elliptical galaxy NGC 1332. Shining at magnitude 10.3, this fuzzy patch has a fairly bright nucleus and is not hard to see through a 4 inch (100 mm) telescope.

❽ **NGC 1360** About 4½ degrees south and a little east, back in Fornax, is NGC 1360. This planetary nebula, about 6 arcminutes across, offers an interesting change from observing the many galaxies in this region. Its magnitude 11 central star can be distracting, but the nebula is not a difficult object in a 6 inch (150 mm) scope.

'he elliptical shape of NGC 1360 (above) becomes obvious to he observer only when it is seen through a large telescope.

9 NGC 1398 Gently guide your scope degree southeast of NGC 1360 to NGC 1398, glowing at magnitude 9.7. This barred spiral galaxy is not difficult to spot, but is best suited to inch (150 mm) scopes and larger.

10 NGC 1316 The next hop is quite some listance away. First, shift 10 degrees south to the ittle triangle of stars **Chi^{1,2,3} [χ^1, χ^2, and χ^3] Fornacis**. About 2 degrees southwest of this riangle is the 9th magnitude spiral NGC 1316, he brightest galaxy in the Fornax Cluster. A 3 inch (75 mm) telescope shows it clearly as a uzzy patch of light. NGC 1316 is also known as the radio source Fornax A. See if you can spot he 11th magnitude galaxy **NGC 1317**, just 5 arcminutes north of NGC 1316.

11 The Great Barred Spiral (NGC 1365) Retrace your steps to the Chi^{1,2,3} triangle, then hop degree east to find the nagnificent NGC 1365, favorite among deep-ky observers. This 9th nagnitude galaxy is the best example of a barred piral in the southern sky. Easily found in 4 inch (100 mm) elescopes because of its prominent central region, NGC 1365 has a bright bar that s visible with an 8 inch (200 mm) telescope.

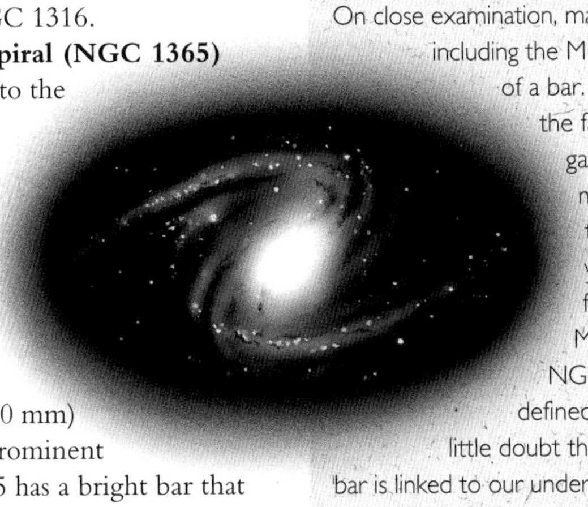

12 NGC 1399 Using a wide-field eyepiece, nudge 1 degree northeast of NGC 1365 to the heart of the Fornax Cluster. With a 4 inch (100 mm) scope you should easily be able to see quite a few faint galaxies, the brightest two being NGC 1399 and **NGC 1404**, both ellipticals, lying only a few arcminutes apart.

The Fornax region has the effect of making you feel very small. Several of the galaxies we have seen on this tour belong to the Fornax Cluster, our close neighbor in the universe. MG

BARRED SPIRAL GALAXIES

In a barred spiral galaxy, the spiral arms seem to originate from the ends of a "bar," composed of stars, gas, and dust, that crosses the nucleus. The bar relates to dynamic conditions within a galaxy. On close examination, many spiral galaxies—including the Milky Way—show the trace of a bar. The ratio of the mass of the faint halo surrounding a galaxy to that of its disk may play a role, as calculations suggest that galaxies with less massive haloes form bars more quickly. Many barred spirals, such as NGC 1365 (left), have well-defined arms, and there seems little doubt that an explanation of the bar is linked to our understanding of spiral structure.

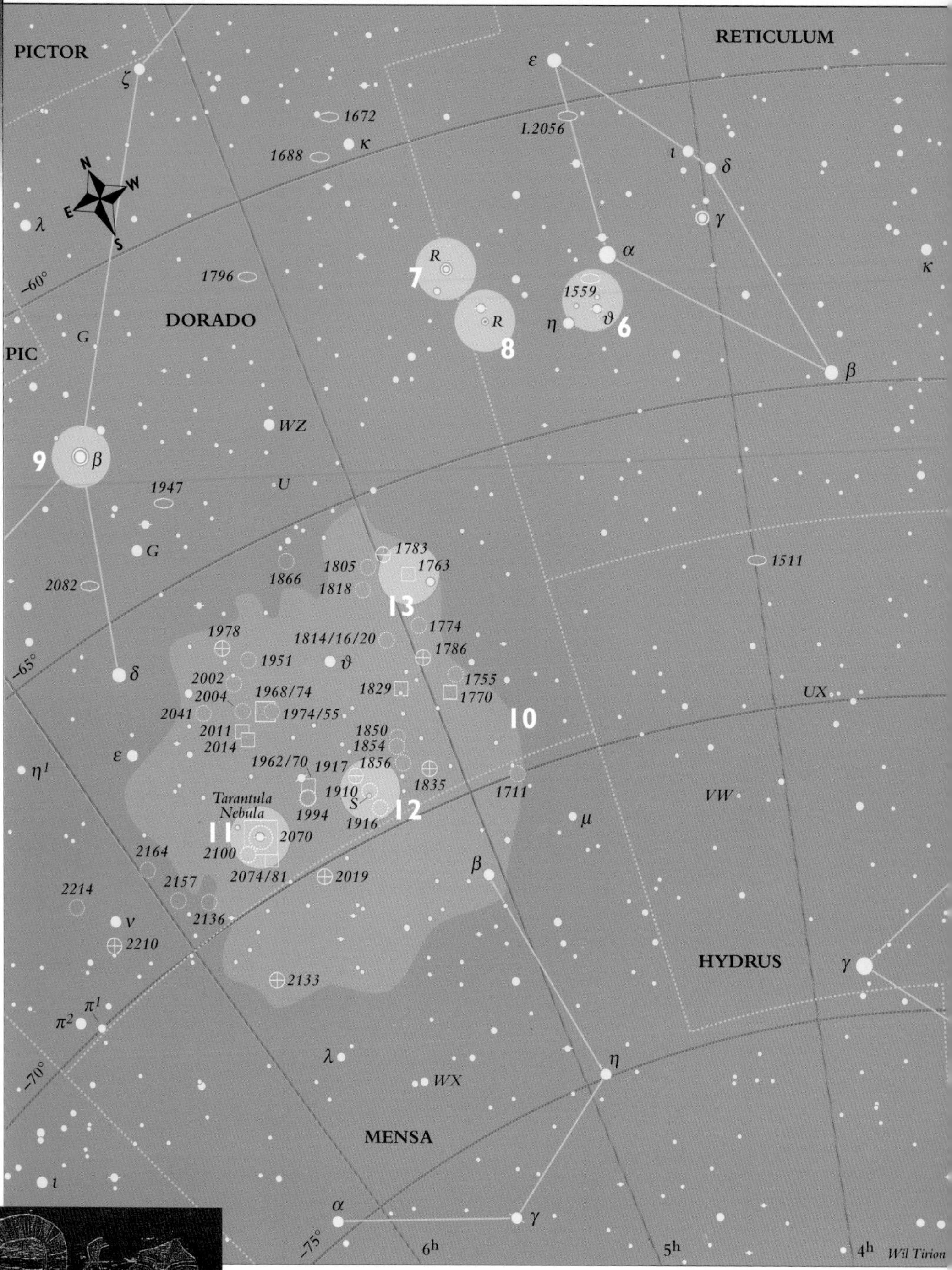
PICTOR
RETICULUM
DORADO
PIC
MENSA
HYDRUS
1672
1688
I.2056
1796
1559
1947
2082
1511
1783
1763
1805
1866
1818
1774
1978
1814/16/20
1951
1786
2002
2004
1968/74
1829
1755
1770
2041
1974/55
2011
2014
1850
1854
1856
1962/70
1917
1835
1711
Tarantula Nebula
1910
1994
1916
2070
2100
2164
2074/81
2019
2214
2157
2136
2210
2133
WZ
WX
UX
VW
-60°
-65°
-70°
-75°
6h
5h
4h
Wil Tirion

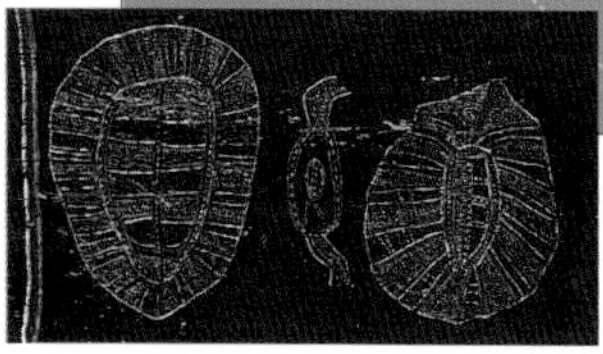

Venturing into the Far South

This is our southernmost starhop, so unfortunately this region of the sky is probably not visible for skywatchers in the Northern Hemisphere. Even if you were observing within 15 degrees of the equator, you could still only expect to see a few of the objects.

The highlights of this area are the Magellanic Clouds, originally known as the Cape Clouds because they were associated with being as far south as the Cape of Good Hope. Portuguese seamen saw these two large patches of sky during voyages made in the fifteenth century. The explorer Ferdinand Magellan later described them and they were duly named in his honor, although no one knew what it was they were looking at.

Four centuries later, astronomers finally recognized the existence of galaxies other than our own, realizing that the Magellanic Clouds were two of the nearest galaxies to the Milky Way.

An Australian Aboriginal bark painting depicting the Large and Small Magellanic Clouds, from Groote Eylandt, circa 1954.

Southern Hemisphere On meridian 10pm Dec 10

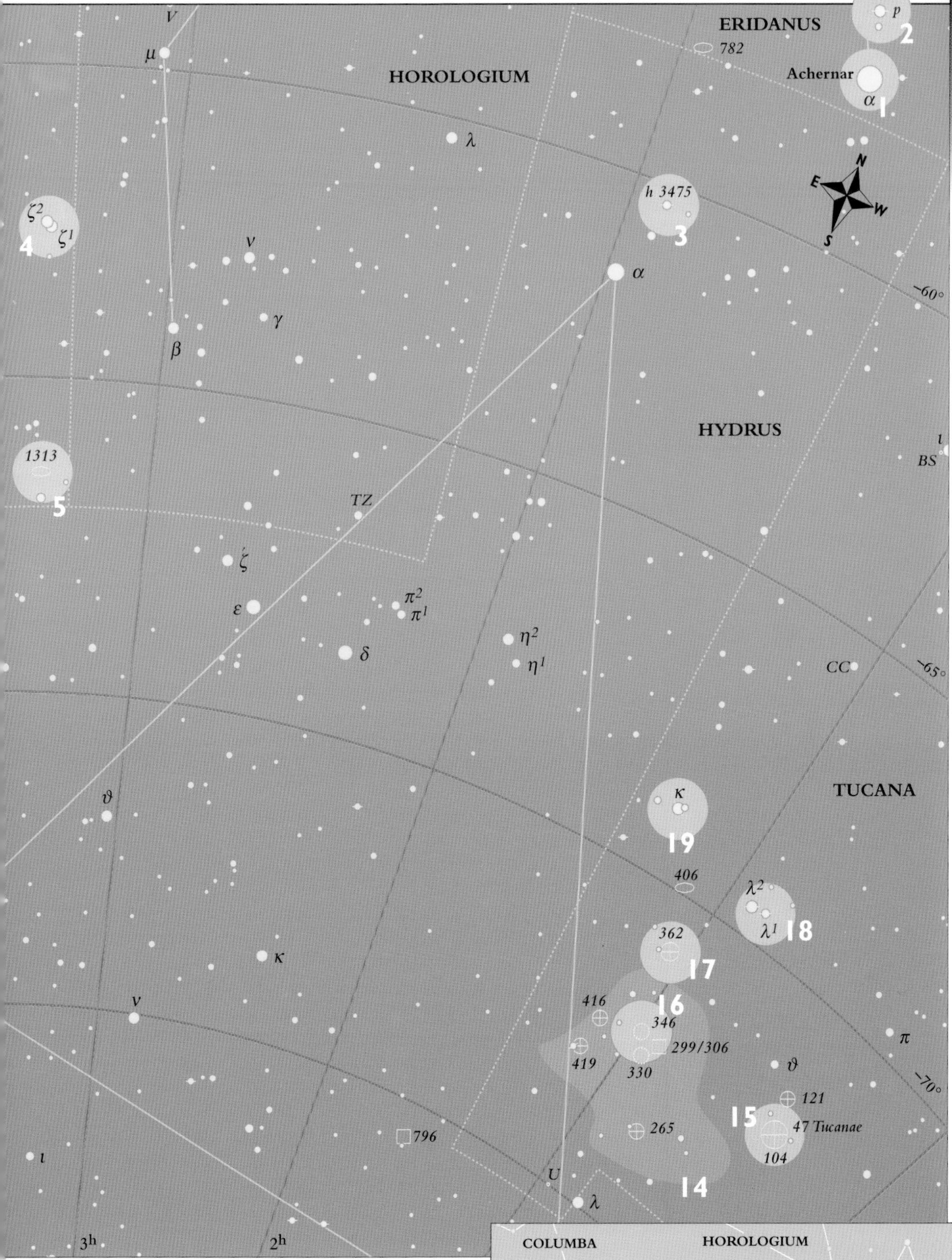
ERIDANUS
782
Achernar
HOROLOGIUM
h 3475
HYDRUS
TUCANA
1313
TZ
BS
CC
406
362
416
346
299/306
419
330
121
47 Tucanae
104
265
796
U
N
E
W
S
–60°
–65°
–70°
3h
2h
1
2
3
4
5
14
15
16
17
18
19

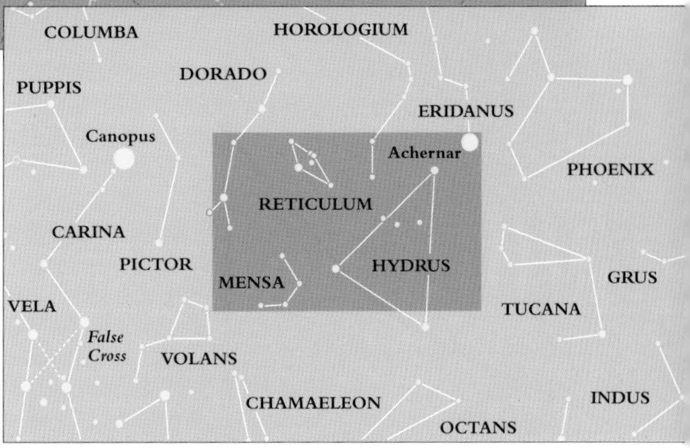
COLUMBA
HOROLOGIUM
DORADO
PUPPIS
ERIDANUS
Canopus
Achernar
PHOENIX
RETICULUM
CARINA
PICTOR
HYDRUS
MENSA
GRUS
VELA
TUCANA
False Cross
VOLANS
CHAMAELEON
INDUS
OCTANS

1 Achernar (Alpha [α] Eridani) Our starting point for this starhop is the brilliant magnitude 0.5 bluish white Achernar, the ninth brightest star in the night sky and the brightest in the constellation of Eridanus, the River.

Achernar is the most southerly of the very bright stellar beacons, standing out in an otherwise barren patch of sky. Finding it is easy—it is about as far from the south celestial pole as the Southern Cross is on the opposite side.

The name Achernar comes from the Arabic meaning "the end of the river," and was originally attributed to Theta [θ] Eridani. Theta was so named because it appeared to mark the end of the River for ancient astronomers, whose view of the sky was limited by their Northern Hemisphere locations. We now know Theta as Acamar, which is a corruption of its former name.

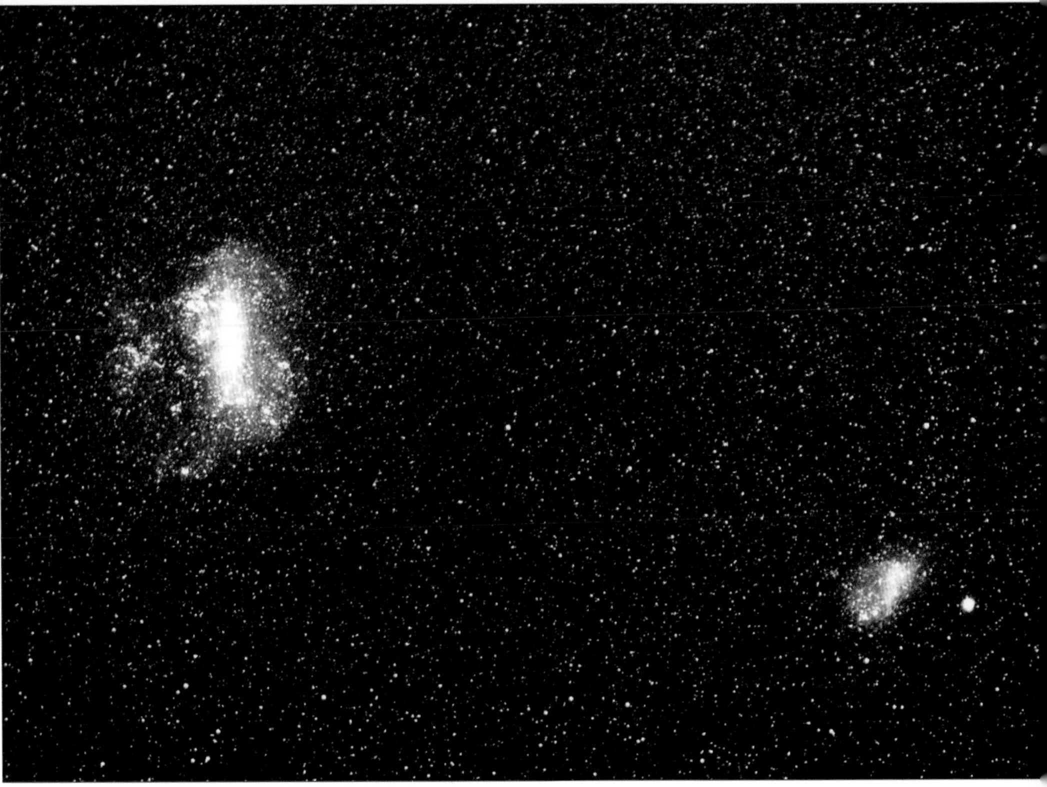

❷ p Eridani (Dunlop 5) Aim your scope toward Achernar, before nudging it just over 1 degree north to locate the double star p Eridani, upstream from Achernar. This is a wonderful double for any size telescope, and a favorite among many observers. Shining at magnitude 5.8, its yellow-white components are equal in brightness, lying 10.4 arcseconds apart.

❸ h 3475 Return to Achernar, then move to a point about two-thirds of the way to **Alpha [α] Hydri** to find h 3475. The stars of this double are much closer together than those of p Eridani, with its two 7th magnitude components being 2.4 arcseconds apart. It is a good test for a 3 inch (75 mm) scope on a still night. This is one of the many double stars discovered by John Herschel during his stay in South Africa in the 1830s.

The LMC and the SMC (top) are classified as irregular galaxies. Despite its chaotic-looking appearance, the fainter NGC 1313 (above) has sufficient order to be classified as a barred spiral.

❹ Zeta [ζ] Reticuli To find Zeta Reticuli, hop to Alpha Hydri, then shift west about 9 degrees along the way to **Alpha [α] Reticuli**. Zeta is a wide double star that is easy to separate with binoculars, its two 5th magnitude component stars, Zeta[1] and Zeta[2], lying 5 arcminutes apart. Each is a magnitude 4.7 yellowish star. These stars are remarkably similar to the Sun, a fact that has prompted speculation about the possibility of life existing on orbiting planets, although this is not known for sure.

❺ NGC 1313 Moving poleward 4 degrees will bring NGC 1313 into view. This is a well-known barred spiral galaxy near the corner of the constellation of Reticulum, the Reticule. It has a total magnitude of 9.4, but appears quite large and faint. An 8 inch (200 mm) scope or larger is recommended for viewing this galaxy, but even with a 12 inch (300 mm) telescope, the bar is only vaguely visible.

❻ Theta [θ] Reticuli To the northeast of NGC 1313 and just southeast of Alpha Reticuli, Theta Reticuli shines faintly to the unaided eye. Theta is a pleasing double star comprising 6th and 8th magnitude components, separated by about 4 arcseconds. With a 3 inch (75 mm) scope, you should have little difficulty separating them, but the higher the magnification, the better the view.

NGC 1559 In the same field, about 1/2 degree north of Theta, is the barred spiral galaxy

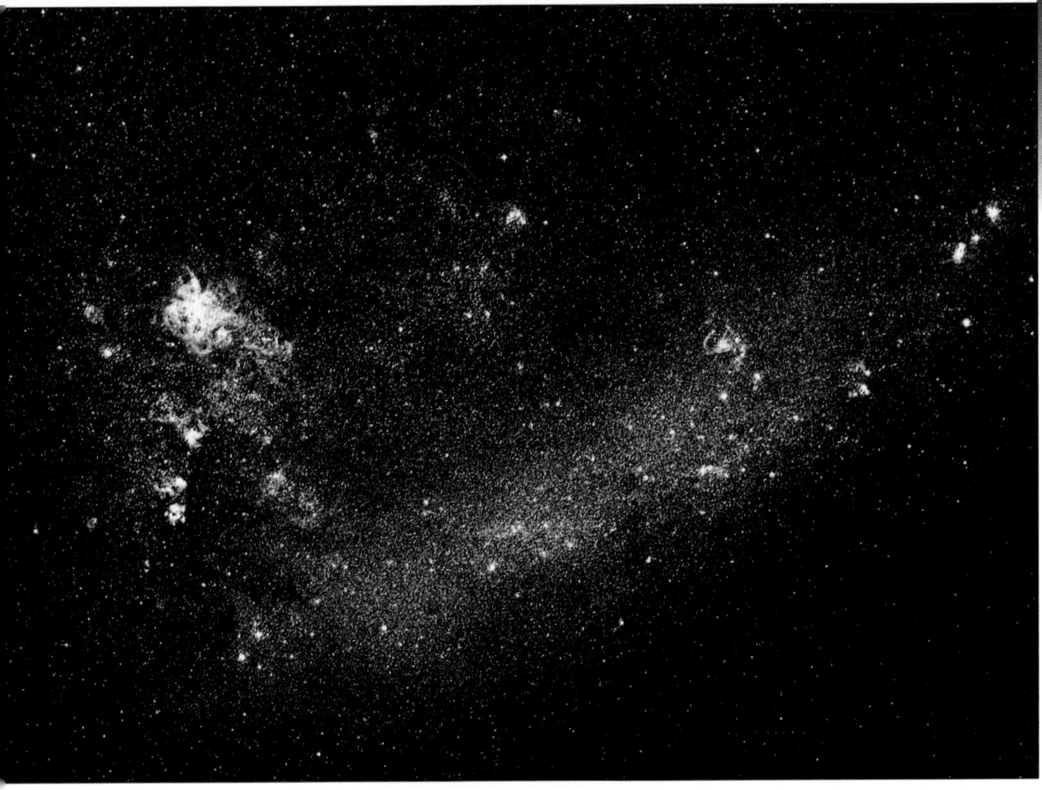

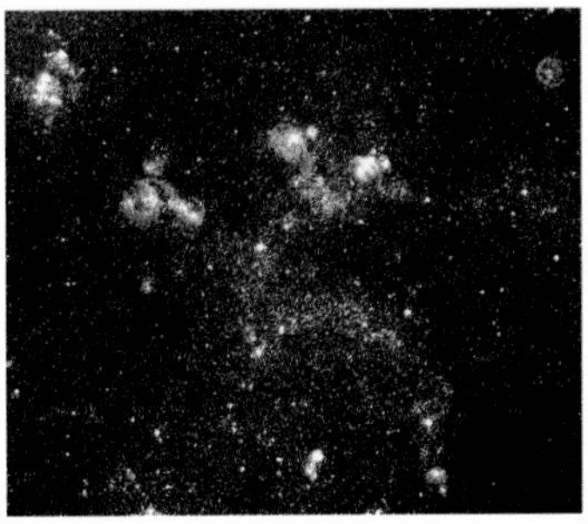

The pink, glowing clouds (left) are just some of the many star-forming regions scattered throughout the LMC (above).

NGC 1559. Although it is about a magnitude fainter in total brightness than NGC 1313, this galaxy is more compact, appearing as a broad, fuzzy line. A 4 inch (100 mm) scope, or possibly even a smaller one, should clearly show the galaxy. The brightening toward its core is easier to see with larger scopes.

7 R Doradus Moving toward the border with Dorado, 2 degrees to the northeast of NGC 1559, is R Doradus, just inside the constellation of Dorado, the Swordfish. R Doradus is a vivid orange-red color, not unlike a fainter view of the planet Mars. It is a semi-regular variable star with a period that fluctuates from cycle to cycle, averaging out at about 338 days. R Doradus is an easy binocular target because it only varies between magnitudes 4.8 and 6.6.

8 R Reticuli Moving back about 1 degree southwest is R Reticuli. Unlike R Doradus, this Mira-type variable can drop to magnitude 14 at minimum, so you will not see it with binoculars then. However, at maximum, around every 278 days, it can reach magnitude 6.5, which makes checking to see just how bright it is a worthwhile effort. It has a ruddy look that is obvious through binoculars, and it lies 12 arcminutes south of a 6th magnitude star.

9 Beta [β] Doradus Swing 7 degrees east to the Cepheid variable Beta Doradus, which fluctuates between magnitude 3.5 and 4.1 over a period of 9.84 days. It is thought to lie 1,700 light-years away.

10 The Large Magellanic Cloud (LMC) Far more distant, but seen just a few degrees poleward of Beta, lies the Large Magellanic Cloud. Easily visible to the unaided eye as a "cloud" of light several degrees across, the LMC is a fine sight in binoculars, which help show the elongated, patchy appearance of this irregular galaxy. Lying about 160,000 light-years from the Milky Way, the LMC is the second closest galaxy to our own—the closest known being a dwarf galaxy in Sagittarius that was detected in 1994.

11 The Tarantula Nebula (NGC 2070) Near the eastern end of the the LMC is the wonderful Tarantula Nebula. Being the brightest object in the LMC, the Tarantula Nebula is visible to the unaided eye—a remarkable fact, given that it is in another galaxy. Even through a 3 inch (75 mm) scope, some of its intricate spider-like structure can be seen, while the view in an 8 inch (200 mm) is stunning. Within the nebula's dazzling heart is a dense and mysterious cluster of brightly glowing supergiants known as R136. NGC 2070 is more than 1,000 light-years across, making it the largest known diffuse nebula, some 30 times the size of the Great Nebula in Orion.

The wispy tendrils of the Tarantula Nebula are clearly seen in this image (above), taken just after nearby SN1987A blazed into view

⓬ **NGC 1910** Sweep your scope through the LMC, heading slowly west for 3 degrees, until you come across NGC 1910. One of the many clusters in the LMC, this compact group contains the bluish white variable star **S Doradus**, which is the prototype of its class. S Doradus varies irregularly between magnitude 9 and 11, the variations being the result of the periodic shedding of the star's outer layers.

⓭ **NGC 1763** To spot the bright nebula NGC 1763, head northwest, to the edge of the Large Magellanic Cloud. NGC 1763 is easily seen with binoculars, and a 3 inch (75 mm) scope reveals two bright, seemingly separate areas.

⓮ **The Small Magellanic Cloud (SMC)** Leaving the LMC and heading some 20 degrees west, you will come across its smaller counterpart—the SMC. Slightly more distant than the LMC and somewhat smaller, the SMC, also an irregular galaxy, is nevertheless an easy naked-eye target. Binoculars will not show nearly as much of the SMC as the LMC. This lesser galaxy has a more irregular form, appearing as a large blob of light, about 3½ degrees across, that fits neatly into a typical binocular field.

⓯ **47 Tucanae (NGC 104)** A chief item of telescopic interest in this region is the naked-eye spot of light about 3 degrees west of the SMC, 47 Tucanae. A 3 inch (75 mm) telescope shows a "granular" appearance, while a 6 inch (150 mm) reveals a multitude of stars. It is widely

SUPERNOVA 1987A

Astronomers around the world were excited when, on 24 February 1987, a new naked-eye star suddenly appeared within the Large Magellanic Cloud. It was the first supernova seen with the unaided eye since 1604, reaching magnitude 2.9 at its peak and remaining visible for several months. The object was originally a massive blue star called Sanduleak –69 202.

The star actually exploded some 165,000 years ago, the light having taken that long to reach us. The light was accompanied by a burst of neutrinos—tiny, elusive particles. Scientists had predicted that these would be produced in very large numbers during a supernova explosion.

The study of SN1987A continues today, with rings of light, shown in the illustration of the Hubble Space Telescope image (right), being the most recent surprises in the ongoing tale of the supernova's discovery.

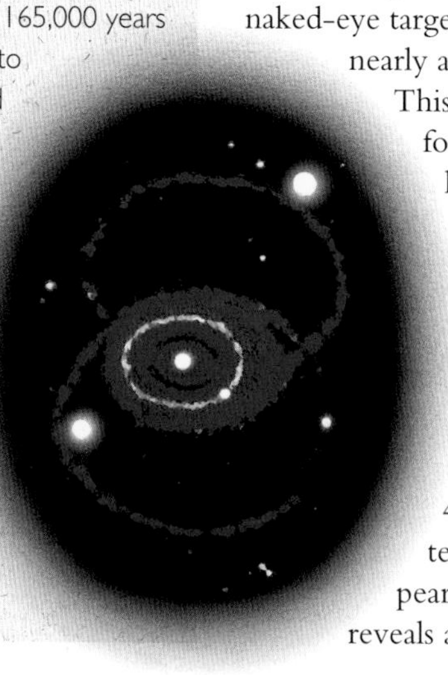

n the sky, the fabulous globular cluster 47 Tucanae (right) ıppears to lie close to the much more distant SMC (above).

ıccepted that 47 Tucanae is the most brilliant globular after Omega Centauri, although it has ı condensed core, giving it quite a different ıppearance. This is the astronomer Johann Bode's celebrated "ball of suns," also having been referred to as "a stupendous object" by John Herschel. Seemingly leading the SMC ıround the sky, 47 Tucanae is a superb globular belonging to the Milky Way Galaxy.

16 NGC 346 Move 3 degrees back to the outheast, into the SMC, to find NGC 346, an open cluster that lies within the SMC. About 5 arcminutes across, it is embedded in a region of nebulosity and appears as an easy-to-see fuzzy pot in a 3 inch (75 mm) scope. Its total brightness s close to that of a 10th magnitude star. See if you can pick out the nearby cluster **NGC 330**, which is visible in the same low-power field, ıbout 1/2 degree southwest of NGC 346.

17 NGC 362 Edge your scope 1 degree north to NGC 362. Seen as a round patch of ight through binoculars, this is a conspicuous globular cluster. It is hard to resolve through a 4 inch (100 mm) telescope, which shows only ı hint of its starry nature, but a 6 inch (150 mm) will show it reasonably well.

18 Lambda[1] [λ[1]] Tucanae (Dunlop-2) One and a half degrees farther north lies a little pair of naked-eye stars. The fainter of the two is Lambda[1] Tucanae. This star is a wide double that is easy to resolve in any size telescope. The 6th and 8th magnitude yellow-white components of Lambda[1] are separated by 20 arcseconds.

19 Kappa [κ] Tucanae The final object on our hop is another double star, Kappa Tucanae, 2 degrees east and a little north of Lambda[1]. Kappa is an easy pair in a 3 inch (75 mm) scope, especially on high power. Its 5th and 7th magnitude stars lie 5.4 arcseconds apart. Look for another 7th magnitude star, shining just a few arcminutes away in the same field.

Southern Hemisphere observers really are the envy of northerners for having the Magellanic Clouds all to themselves. It is a rewarding pastime simply to spend an evening browsing around this region with anything from the unaided eye to a large telescope. MG

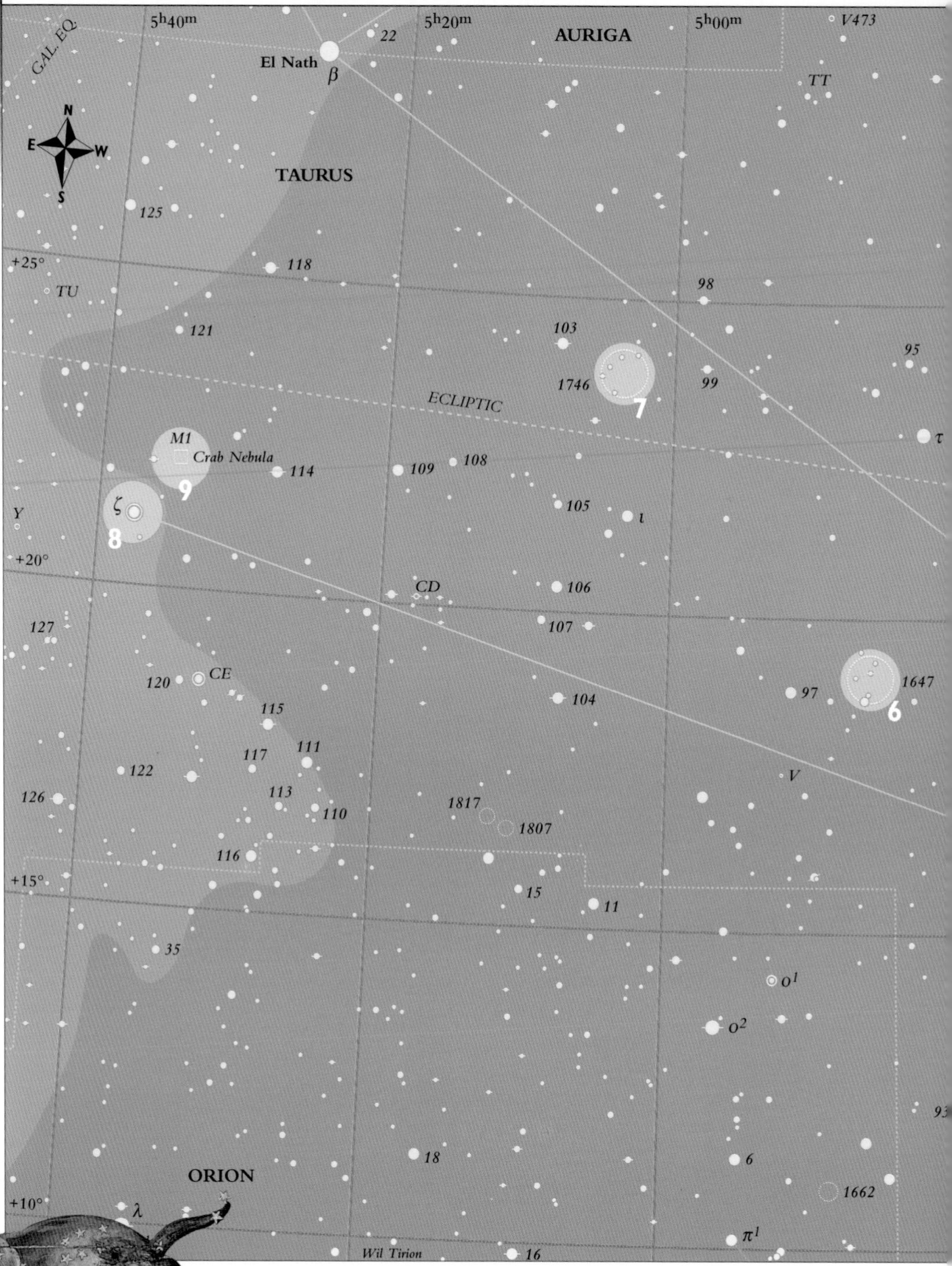

5h40m
5h20m
5h00m
GAL. EQ.
AURIGA
El Nath
β
22
V473
TT
N
E
W
S
TAURUS
125
+25°
TU
118
121
98
103
95
1746
99
ECLIPTIC
7
τ
M1
Crab Nebula
114
109
108
9
ζ
8
Y
105
ι
+20°
CD
106
127
107
120
CE
115
104
97
1647
6
117
111
122
126
113
110
1817
1807
V
116
+15°
15
11
35
o1
o2
93
18
6
ORION
1662
+10°
λ
π1
Wil Tirion
16

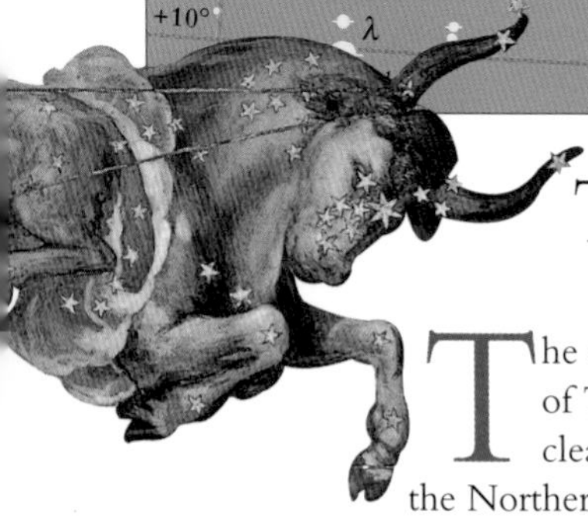

Taurus and the Seven Sisters

The zodiac constellation of Taurus, the Bull, is clearly visible from both the Northern and Southern hemispheres, from February through April.

According to legend, the red eye of mighty Taurus glares at Orion, the Hunter, as the bull guards the Pleiades, the Seven Sisters, from Orion's advances. Orion's heart is set on making Merope, one of the Pleiades, his wife. Her parents, the Titan Atlas and the Oceanid Pleione, watch closely from the edge of the Pleiades Cluster. Of the seven young sapphire-blue Pleiade sisters, only Merope married a mortal, the King of Corinth, so she hides her shame behind a wispy reflection nebula.

The Pleiades' half-sisters form the nearby Hyades Cluster, another fantastic naked-eye and binocular object on this starhop.

The well-known zodiacal figure of Taurus, the Bull, from a sixteenth-century fresco in the Villa Farnese, Caprarola, Italy.

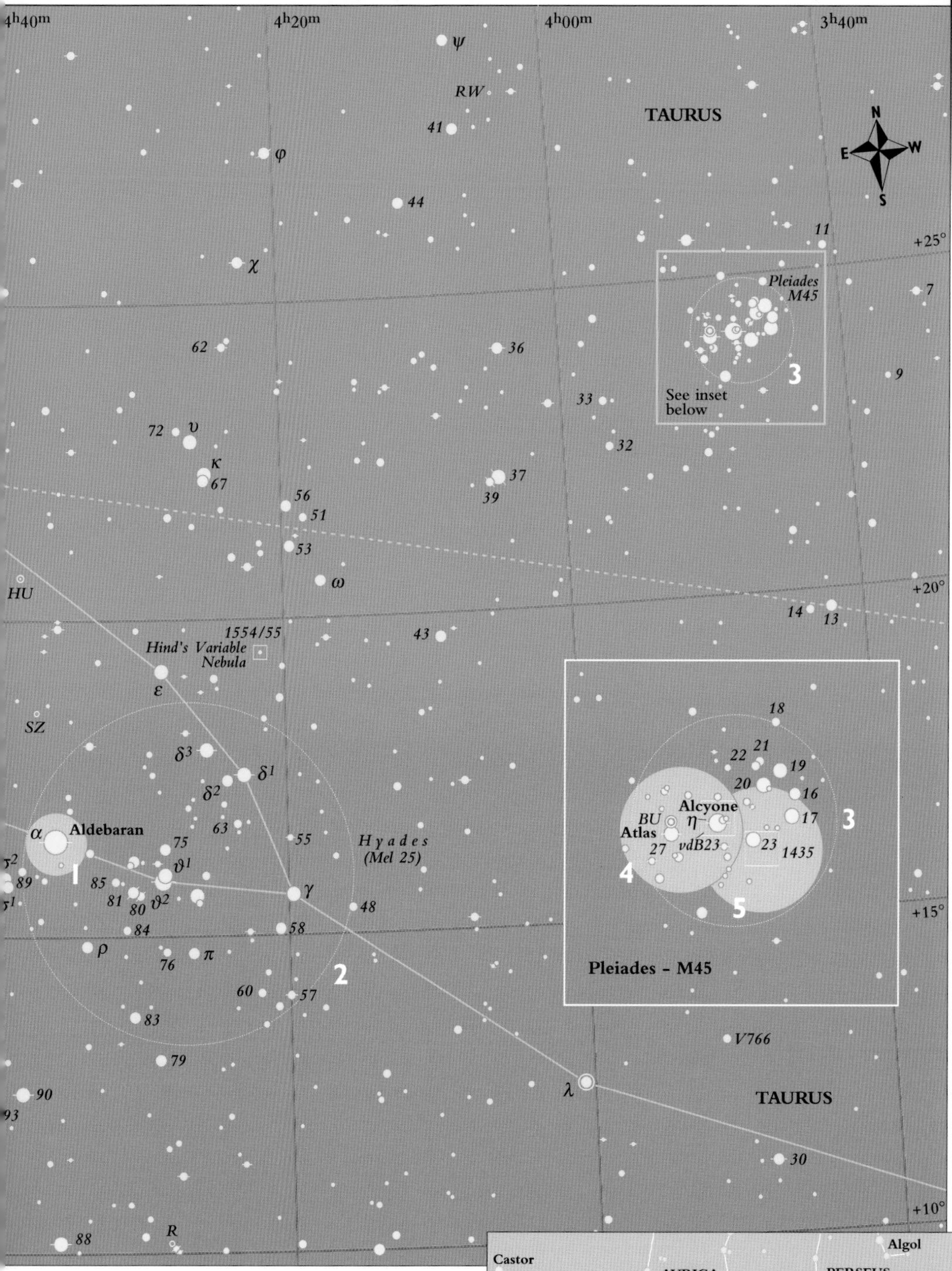
4h40m
4h20m
4h00m
3h40m
TAURUS
N
E
W
S
+25°
+20°
+15°
+10°
Pleiades
M45
See inset
below
3
Hind's Variable
Nebula
1554/55
HU
SZ
Aldebaran
Hyades
(Mel 25)
1
2
Alcyone
BU
Atlas
vdB23
1435
4
5
Pleiades - M45
V766
TAURUS

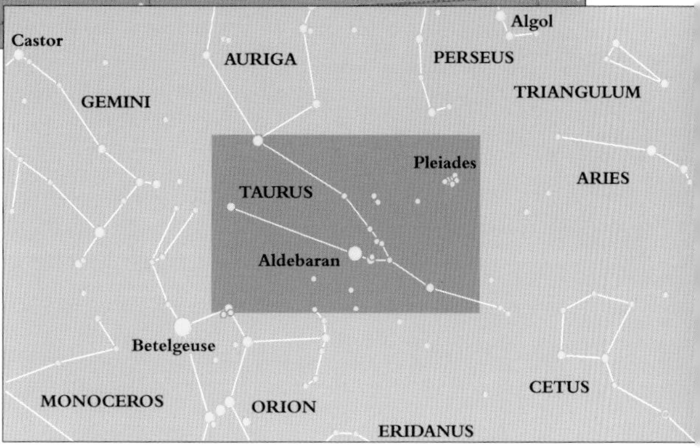
Castor
AURIGA
PERSEUS
Algol
TRIANGULUM
GEMINI
Pleiades
TAURUS
ARIES
Aldebaran
Betelgeuse
CETUS
MONOCEROS
ORION
ERIDANUS

Aldebaran (87 Alpha [α] Tauri) This
›ur of Taurus and the Seven Sisters begins at
ıe 1st magnitude orange giant Aldebaran, the
3th brightest star in the sky. Although it seems
› form the eastern tip of the Hyades Cluster,
ldebaran, just 61 light-years distant, is actually
ıuch closer to us. Like many other orange giants,
is possible that this star is also slightly variable.
ldebaran has five close companions, but they
:e faint and extremely difficult to observe
ırough amateur equipment.

At about 40 times larger than the Sun and
ıtrinsically some 125 times brighter, Aldebaran
'ould fill most of the area inside Earth's orbit.

The name Aldebaran is derived from the Arabic *Al Dabaran*, meaning "the follower," probably because this star follows the Pleiades through the night sky.

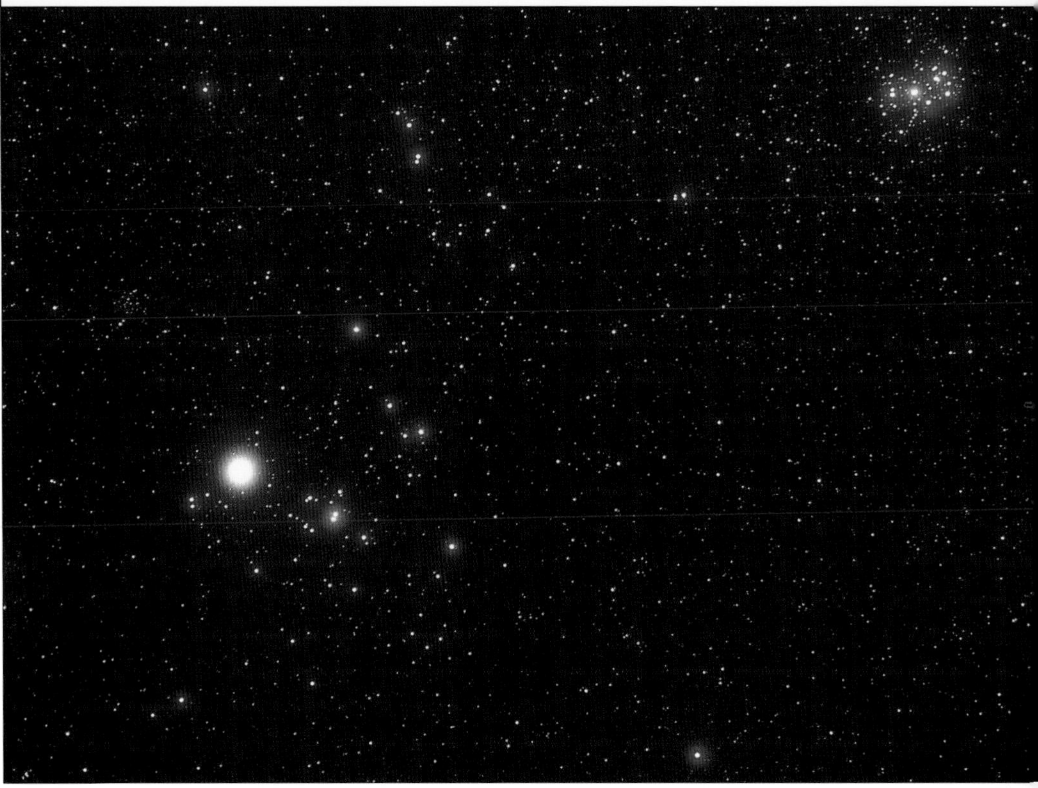

The Pleiades, at upper right, and the Hyades, beside Aldebaran a left (above), are nearby clusters appearing large in the sky.

2 The Hyades (Mel 25) This large (more than 4 degrees in diameter), rich, distinctly V-shaped open star cluster consists of a mixture of about 100 bright stars and many faint ones, including dozens of double and variable stars. The cluster, which is 150 light-years away, marks the head of the bull. It is named for the half-sisters of the Pleiades, being the daughters of Atlas and Aethra. The Hyades' distance to us provides a crucial step in measuring distances in the universe, as it is one of the closest clusters to us.

M45, THE SEVEN SISTERS

The ancient Greeks were not the only people to develop myths and legends about the origins of the cluster they named the Pleiades. The Australian Aborigines also integrated the skies they observed into a heritage of stories, known as the Dreamtime. The bark painting (below), by Wongu, a member of the Yolgnu tribe in Arnhem Land, depicts the Seven Sisters inside a canoe. The three stars of Orion's belt are shown in a line at the head of the canoe, to the right of the image, with the seven prominent stars in the middle representing the Pleiades, their wives. The lone fish inside the canoe, at center top, represents the nearby Hyades Cluster, while the fish swimming in the water are bright stars in the Milky Way.

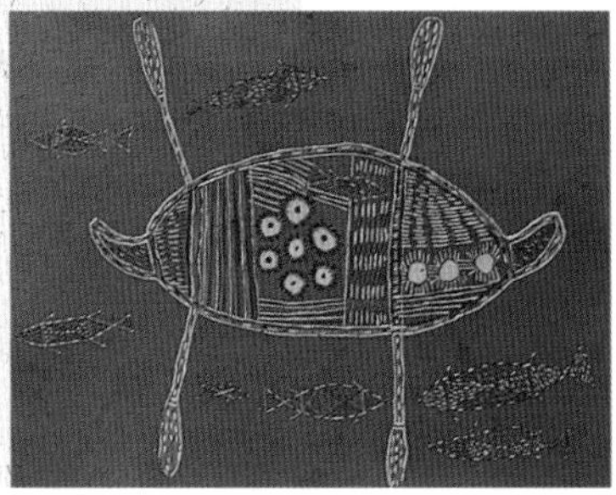

3 The Pleiades (M45, Mel 22) About 14 degrees northwest of Aldebaran is M45—the Seven Sisters—the brightest and most famous star cluster in the sky. M45 is a young open cluster dominated by youthful hot blue stars and enveloping nebulosity. The brightest part of the nebula surrounding the Pleiades is around the magnitude 4.1 star **Merope (23 Tauri)**. The Pleiades star nursery has been known since ancient times. Seven bright stars are visible to the naked eye from a dark site, although the cluster shows best through binoculars or a low-power telescope eyepiece, since it covers an area about four times the size of the Full Moon. It is essential to use high power when studying the fainter stars and patches of nebulosity.

4 Alcyone (25 Eta [ε] Tauri) The brightest member of the Pleiades is magnitude 2.9 Alcyone, a wide quadruple-star system embedded within the reflection nebula **van den Bergh 23**. The individual components of Alcyone's system are easy to separate with a small telescope.

Atlas (27 Tauri) This young blue giant star is a close double that usually requires a telescope of at least 10 inches (250 mm) and good seeing conditions to separate its magnitude 3.6 and 6.8 components,

which lie only 0.4 arcseconds away from each other.

5 Tempel's Nebula (NGC 1435) This is the reflection nebulosity in which blue-white Merope is embedded. The nebula appears to be nearly transparent so is quite hard to detect, but once you "see" it you might think that it looks as if someone has smeared white shoe polish on a pane of glass. This nebula is part of a series of small nebulas extending over much of the western side of the Pleiades.

6 NGC 1647 (Mel 26) Return to Aldebaran, then hop about 3½ degrees northeast to NGC 1647. This open star cluster contains about 25 uniformly bright 8th magnitude stars gathered in a loose grouping along the line of the western horn of the bull. More than 200 additional fainter stars are part of this rich cluster. As you study the individual stars, try to pick out some of the many close double stars.

7 NGC 1746 (Mel 28) Look for the open star cluster NGC 1746 about 5½ degrees northeast of NGC 1647 and just 1 degree southwest of the magnitude 5.5 blue-white double star **103 Tauri**. NGC 1746 has a dense central region of about 20 stars, along with about 30 more in knots scattered about the cluster's core.

8 123 Zeta [ζ] Tauri Shift your scope about 8 degrees southeast to Zeta Tauri, the last bright star in the southern horn of the bull. Zeta is an extremely close binary star whose components cannot be separated with amateur scopes. It is also a variable, belonging to the rare Gamma Cassiopeiae-type class. Zeta lies about 520 light-years away from us.

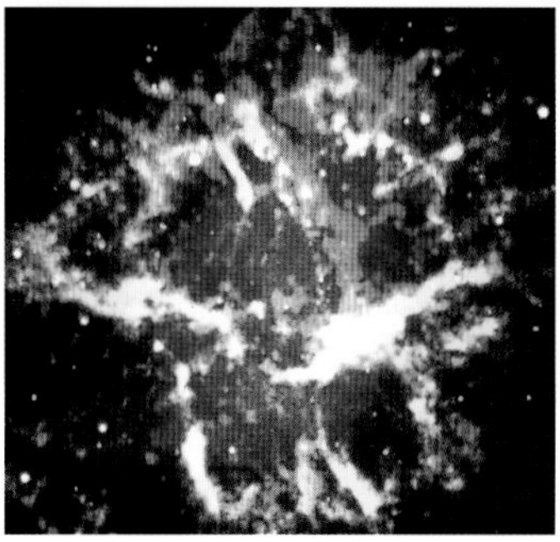

The Pleiades (above) and the Crab Nebula (left) are highlights in Taurus.

9 The Crab Nebula (M1, NGC 1952) Nudge about 1 degree northwest of Zeta to the famous Crab Nebula, lying about 6,500 light-years from us. The nebula is visible with small telescopes but can be somewhat disappointing. Details in the cloud can be detected in 10 inch (250 mm) scopes or larger.

First discovered by British amateur astronomer John Bevis in 1731, M1 is the gaseous supernova remnant of the explosion of a star witnessed and recorded by Chinese astronomers in 1054 (see also p. 169). The explosion was so bright that the star was visible in daylight for a period of 23 days.

In the core of the Crab Nebula, a tiny spinning neutron star flashes a beam of energy on and off 30 times a second. This "star," called a pulsar, is all that remains of the original star that exploded as a supernova so long ago. This was the first pulsar to be detected visually.

This patch of sky has intrigued and entranced skywatchers since ancient times because its highlights, the Pleiades and Hyades, are so big and bright, as well as being rich in mythology. RG

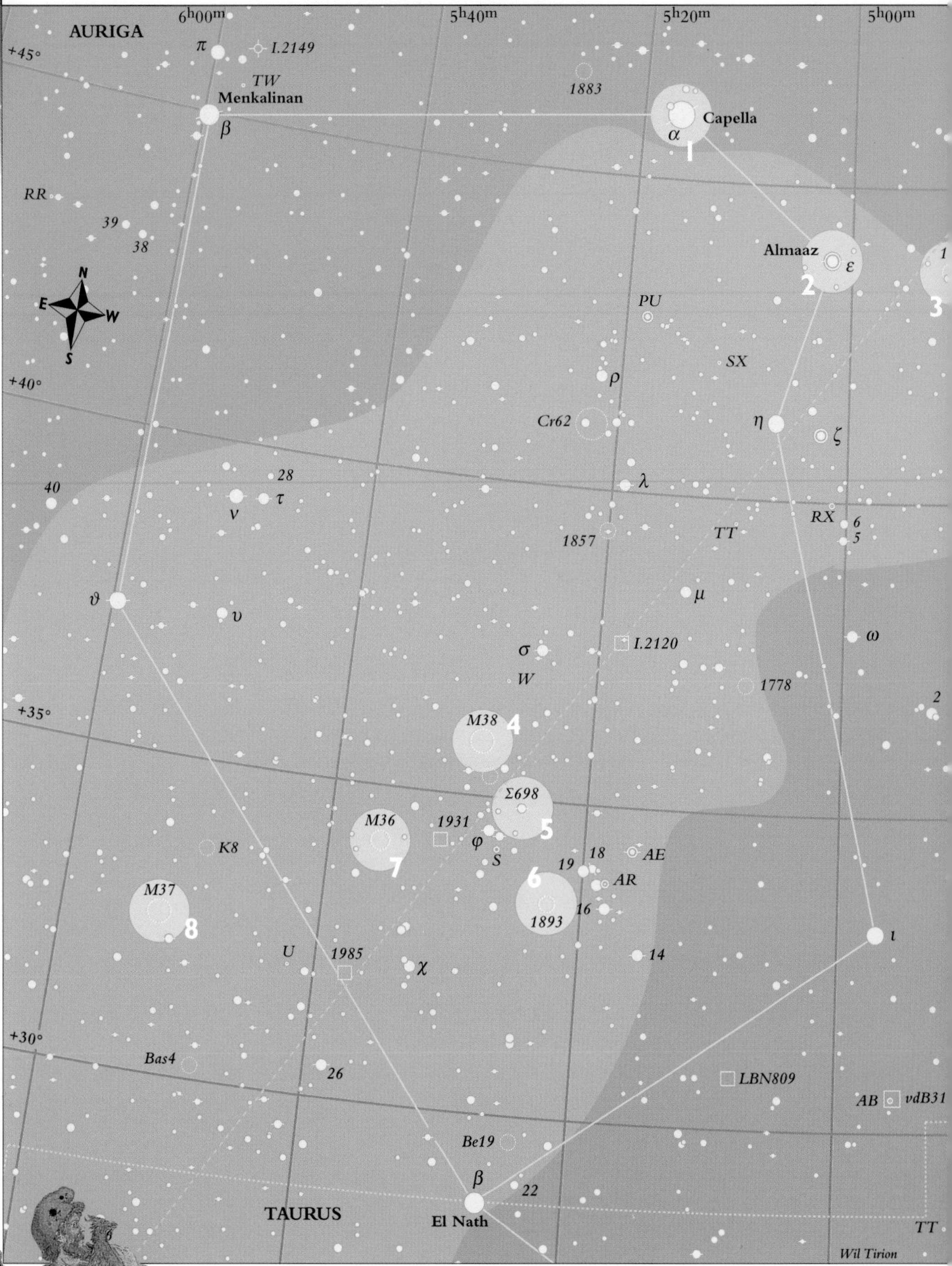
AURIGA
6h00m
5h40m
5h20m
5h00m
+45°
+40°
+35°
+30°
I.2149
TW
Menkalinan
1883
Capella
RR
39
38
Almaaz
PU
SX
Cr62
28
40
RX
TT
1857
I.2120
1778
W
M38
Σ698
M36
1931
K8
AE
AR
M37
1893
U
1985
Bas4
26
LBN809
AB
vdB31
Be19
22
TAURUS
El Nath
TT
Wil Tirion

Riding with the Charioteer

The rich star fields of the Milky Way in the pentagon-shaped constellation of Auriga, the Charioteer, hug the galactic equator, forming the centerpiece of this area of sky. It is visible to skywatchers in both the Northern and Southern hemispheres.

Auriga, the Charioteer, carrying radiant Capella on his back, from Johann Bayer's seventeenth-century star atlas, Uranometria.

The route of this hop includes some of the finest open star clusters in the northern half of the sky. Cutting diagonally across Auriga is a string of open star clusters, including M36, M37 and M38. In addition to these highlights, many other star clusters lie among a variety of objects.

To the ancient Greeks and Romans, the constellation of Auriga represented either a charioteer or a herd of goats. In the herd, bright white Capella is the mother of the flock and her three starry "kids" graze close by.

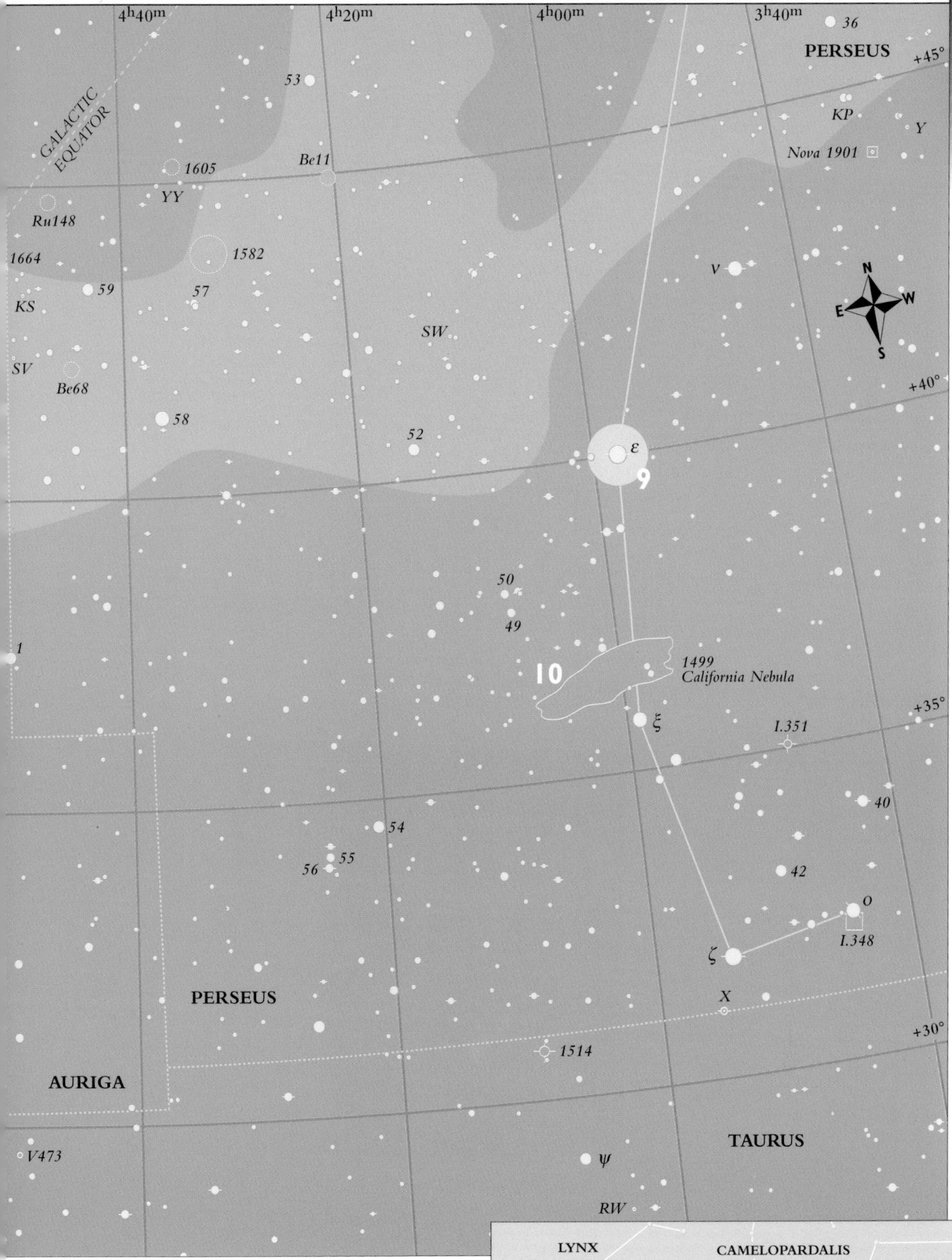
4h40m
4h20m
4h00m
3h40m
PERSEUS
+45°
GALACTIC EQUATOR
53
36
KP
Y
Nova 1901
1605
Be11
YY
Ru148
1664
1582
59
57
V
KS
SW
SV
Be68
+40°
58
52
ε
9
50
49
1
10
1499
California Nebula
ξ
+35°
I.351
40
54
55
56
42
o
I.348
ζ
PERSEUS
X
+30°
1514
AURIGA
TAURUS
V473
ψ
RW

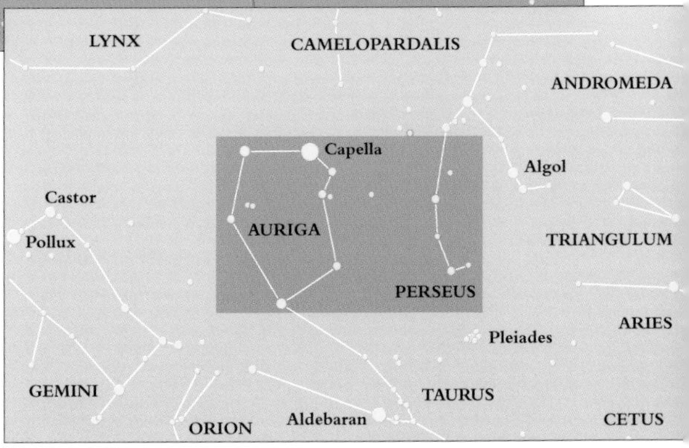
LYNX
CAMELOPARDALIS
ANDROMEDA
Capella
Algol
Castor
AURIGA
Pollux
TRIANGULUM
PERSEUS
Pleiades
ARIES
GEMINI
TAURUS
ORION
Aldebaran
CETUS

1 Capella (13 Alpha [α] Aurigae) This tarhop begins with the sixth brightest star in the ky, the magnitude 0.08 golden-yellow giant Capella. This brilliant spectroscopic multiple star ies only 45 light-years from Earth. Capella's four omponents are too close together to be separated hrough conventional telescopes.

Capella A and Capella B are binary yellow ompanions that form one of the star systems naking up Capella. The A and B stars are stimated to be 90 and 70 times more luminous han the Sun, respectively. The other binary star ompleting the multiple system is Capella H, onsisting of a pair of red dwarf stars.

In many ancient myths, Capella, the Goat Star, is represented as a she-goat, carried over the shoulder of the charioteer. It is also sometimes referred to as the Little She-goat.

M37, M36, and M38 (left to right, above) lie within the Milky Way in Auriga (left).

❷ Almaaz (Epsilon [ε] Aurigae) About 3½ degrees southwest of Capella is Almaaz, an eclipsing binary. Intrinsically, this magnitude 3.8 white star is one of the most luminous individual stars we know of. It is partially eclipsed by a mysterious companion object every 27 years, the eclipse itself lasting about 2 years. The next eclipse is due to start in 2009. During the eclipse, Almaaz's brightness drops by about ½ magnitude. Many theories have been posed as to the nature of the eclipsing object. One theory holds that it is an almost transparent infrared supergiant star that is so large it would fill most of the area inside the orbit of Saturn. It is also possible that it may be simply a cloud of dust and gas.

❸ NGC 1664 Nudge 2 degrees west of Almaaz to the faint open star cluster NGC 1664. Use low power for this rather sparse 30-member cluster, which appears like a string of tiny jewels on a chain, or an elongated hollow diamond. The brightest star, of about 7th magnitude, is located near the southern end of the chain.

❹ M38 (NGC 1912) Scanning some 10 degrees to the southeast along the galactic equator you will find M38, the dimmest of the three Messier objects in the center of Auriga. This cross-like open cluster, consisting of about 160 stars of magnitude 10 or fainter, was discovered by the Italian astronomer Giovanni Hodierna in the seventeenth century. He also discovered the open clusters M36 and M37.

❺ Struve [Σ] 698 The double star Struve 698 lies about 1 degree southwest of M38. This beautiful little double star consists of two red stars of magnitudes 6.6 and 8.7, separated by 31 arcseconds. They were first measured by astronomer F. G. W. Struve in 1831. The components should be easy enough to resolve through a 4 inch (100 mm) telescope.

❻ NGC 1893 Another 1½ degrees farther southwest is NGC 1893, an elongated open star cluster comprised of about 40 to 60 members. Most of the component stars of the cluster are faint, but they still stand out in the foreground of the surrounding star fields.

❼ M36 (NGC 1960) Sweeping some 3 degrees east of NGC 1893 is M36, an open star cluster of about 60 young blue and white stars, with a slight condensation of stars near its center. Search through the cluster to find the numerous double stars within it.

❽ M37 (NGC 2099) Moving on about 4 degrees southeast from M36 lies M37. This

The California Nebula (above) is a large, faint cloud of gas amid the stars of Perseus, including bright Minkib at center bottom.

very rich, magnitude 6.2 open star cluster, the best in Auriga, contains about 170 bright stars, as well as hundreds of fainter stars scattered throughout. The majority of the stars are young blue-white giants and supergiants. The brighter ones form a rough trapezoid shape with a belt of fainter stars cutting across the geometric figure.

9 **45 Epsilon [ε] Persei** Return to M38, then sweep about 17 degrees west and a little north to Epsilon Persei. The primary for this double-star system is a magnitude 2.9 blue-white giant. The greenish magnitude 8.1 secondary is 8.8 arcseconds away and can be seen through scopes of 6 inches (150 mm) or larger. The difference in brightness between the two components makes them quite difficult to separate.

10 **The California Nebula (NGC 1499)** Hop 4 degrees due south from Epsilon Persei to **Minkib (46 Xi [ξ] Persei)**, the 4th magnitude blue-white star located on the southwestern fringe of the California Nebula. Minkib is the illuminating power for the nebula, the last object on our hop. This elusive and faint reflection nebula appears as a huge, elongated patch of grayish wispy reflection nebulosity. It is named for its vague resemblance to the state of California, USA, but you need to use your imagination to see that. The nebula has a very low surface brightness, and is best seen with a nebula filter (see p. 56). Using low power and a wide-field eyepiece at a dark site is also a great help to observers.

Ride with the Charioteer, scanning the depths of the Milky Way Galaxy for the many other star clusters and double stars to be found. RG

STELLAR EVOLUTION: CAPELLA AND THE SUN

The life of a star can last from a few million to more than a hundred billion Earth years. Its life span is directly related to the amount of hydrogen fuel it contains and the rate at which it converts this fuel to helium. Part of our Sun's life cycle is shown in the illustration below. Currently about five billion years old (below, top), the Sun will grow to become a yellow giant (below, right), Capella's present size. The Sun is expected to use up the hydrogen fuel in its core in about another five billion years, before swelling up as a red giant (below, left) then fading away to become a white dwarf star.

Capella is 13 times larger than the Sun. It is already well on the way to becoming a red giant star.

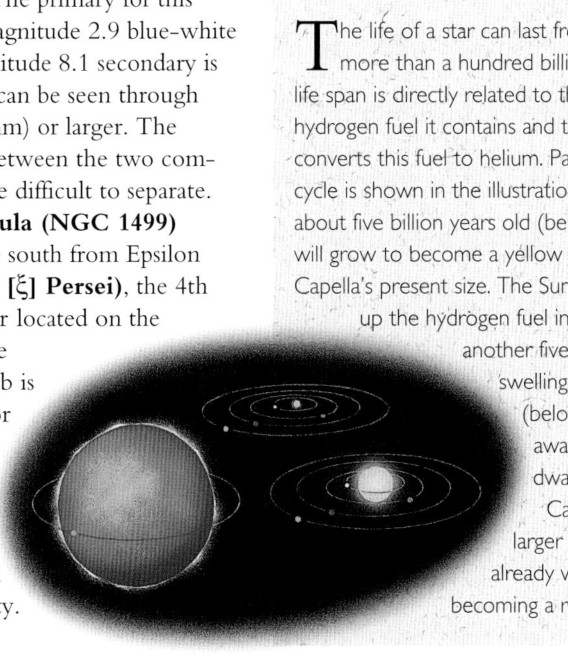

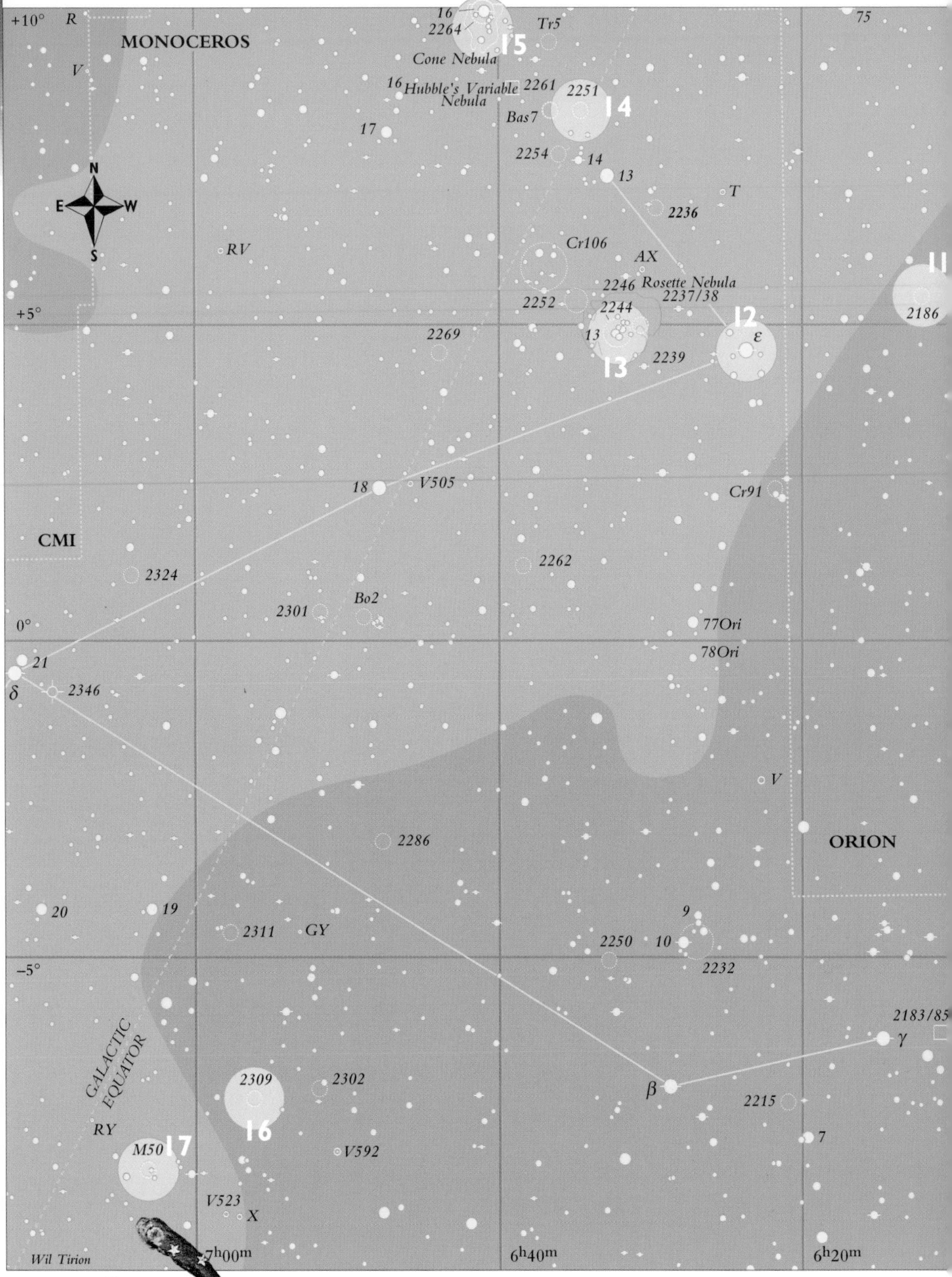

MONOCEROS
CMI
ORION
Cone Nebula
Hubble's Variable Nebula
Rosette Nebula
2237/38
GALACTIC EQUATOR
+10°
+5°
0°
–5°
7h00m
6h40m
6h20m
Wil Tirion
15
14
13
12
11
16
17
M50
2264
2261
2251
2254
2236
2246
2244
2252
2239
2269
2186
Cr106
Cr91
Tr5
Bas7
2262
2324
2301
Bo2
2346
2286
2311
2250
2232
2183/85
2215
2309
2302
V505
V592
V523
77Ori
78Ori

The Stellar Nursery of Orion

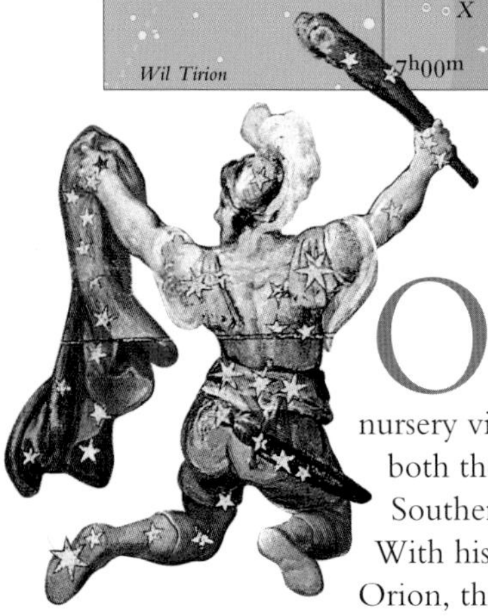

Orion straddles the celestial equator, making its star nursery visible to observers in both the Northern and Southern hemispheres. With his mighty club raised, Orion, the Hunter, dominates the night sky early in the year as he prepares to battle Taurus, the Bull. Behind Orion's back is the mystical Monoceros, the Unicorn, partly obscured by a maze of Milky Way star fields.

The Hunter's starry sword dangles from his three-star belt. The sword harbors the Great Nebula in Orion, where gas covering the sky throughout the constellation is unmasked by the blazing light of young, hot blue-white stars. Even younger stars lie hidden within the gas clouds around it, revealed by their infrared glow.

Orion, the Hunter, as depicted in a sixteenth-century fresco painted on the walls of the Villa Farnese, Caprarola, Italy.

Northern and Southern Hemisphere | On meridian 10pm Jan 20

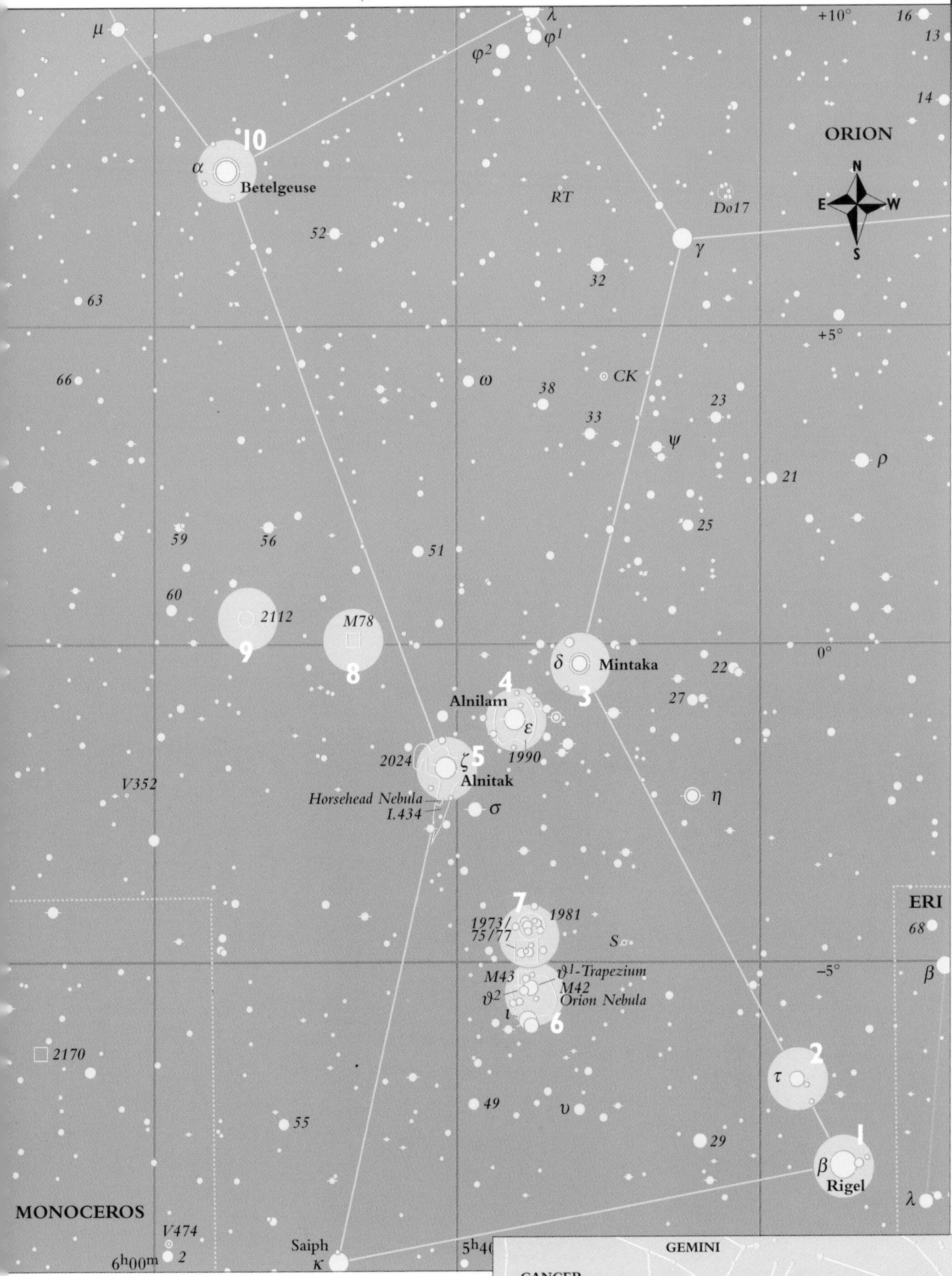
ORION
N
E
W
S
+10°
+5°
0°
–5°
Betelgeuse
Mintaka
Alnilam
Alnitak
Rigel
Saiph
Horsehead Nebula
I.434
M78
2112
2024
1990
1973/
75/77
1981
M43
ϑ1-Trapezium
M42
Orion Nebula
Do17
RT
CK
V352
V474
2170
ERI
MONOCEROS
6h00m
1
2
3
4
5
6
7
8
9
10

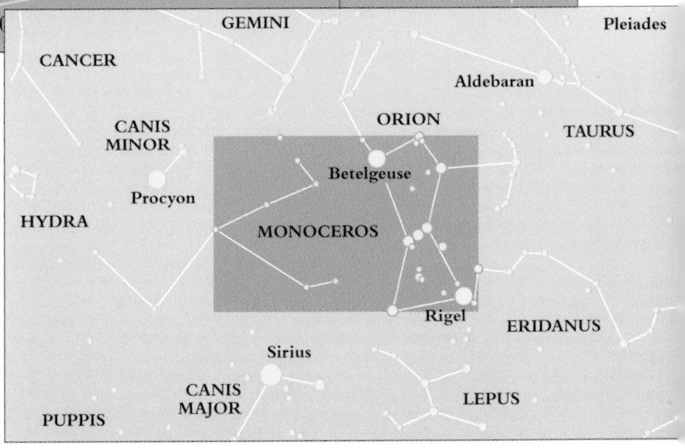
GEMINI
Pleiades
CANCER
Aldebaran
ORION
TAURUS
CANIS
MINOR
Betelgeuse
Procyon
HYDRA
MONOCEROS
Rigel
ERIDANUS
Sirius
CANIS
MAJOR
LEPUS
PUPPIS

❶ Rigel (19 Beta [β] Orionis) Our journey around the stellar nursery in Orion begins at Rigel, the seventh brightest star in the sky, marking Orion's left (western) foot. Rigel is a brilliant magnitude 0.1 bluish white supergiant. Estimated to lie some 900 light-years away, Rigel is about 40 times the size of the Sun and its actual luminosity is about 57,000 times greater. It would fill most of the area inside the orbit of Mercury!

Although Rigel is clearly visible to the naked eye, a 6 inch (150 mm) telescope is needed to separate the magnitude 6.7 bluish companion from the primary star.

Rigel's name is derived from the Arabic *Rijl Jauzah al Yusra,* meaning "left leg of the giant." It has been said that honors and splendor would befall those who were born under this star.

2 20 Tau [τ] Orionis About 2 degrees northeast of Rigel is the magnitude 3.6 double star Tau Orionis. This is an easy-to-observe, wide double, with its magnitude 3.6 blue primary shining 35 arcseconds from the much fainter secondary.

3 Mintaka (34 Delta [δ] Orionis) Follow Orion's left leg north to the westernmost of the three bright belt stars—Mintaka. This is a magnitude 2.2 white double star shining just south of the celestial equator. In telescopes of 4 inches (100 mm) or more, Mintaka's bluish magnitude 6.7 companion is visible 52 arcseconds directly north of it.

4 Alnilam (46 Epsilon [ε] Orionis) Follow the line of the belt east from Mintaka to Alnilam, enveloped in the faint emission/reflection nebula **NGC 1990**. Alnilam is a young magnitude 1.7 blue-white supergiant with an intrinsic luminosity of about 40,000 times that of the Sun. Alnilam's name comes from the Arabic *Al Nitham*, meaning "the string of pearls."

5 Alnitak (50 Zeta [ζ] Orionis) The third of the belt stars, Alnitak is a double star, shining at magnitude 1.8 and illuminating much of this region. The elusive Horsehead Nebula (see Box) is within the long stretch of nebulosity IC 434, located about 1/2 degree south of Alnitak.

M42 and M43 (above) form the pink glow in the sword of the Orion constellation (left).

6 The Great Nebula (M42, NGC 1976) Shift about 4 degrees southward, following the stars that mark Orion's sword, to discover the wonderful Great Nebula in Orion. This glowing, irregular cloud is one of the most impressive sights for skywatchers, as more astonishing details unfold with every increase in aperture. The Great Nebula was first seen telescopically by Italian astronomer Nicholas Peiresc in 1611. M42 is an emission nebula visible to the naked eye in dark skies, appearing as a soft fuzzy spot in the middle of Orion's sword. It is about 30 light-years in diameter, and estimated to be some 1,600 to 1,900 light-years away from us. There are dozens of variable stars within it.

The vibrant red colors of the nebula show up beautifully in photographs, but are too faint to be seen with telescopes. However, some observers have reported seeing a slight pale-green tint to it through an 8 inch (200 mm) scope.

The turbulent appearance of M42 is borne out by spectroscopic studies of the cloud, which show that gas is racing in different directions within it, indicating the localized areas where gas is condensing into new stars.

The glimmering nebula NGC 1977 (above) is another illuminated portion of the gas that envelops Orion.

M42 is considered to be one of the most beautiful objects in the heavens, full of wreaths of swirling gas. It is a sight you will undoubtedly want to return to many times.

The Trapezium (41 Theta^1 [θ] Orionis) The Trapezium is a group of four young, bright, hot white stars at magnitudes 5.1, 6.7, 6.7, and 8.0. Theta^1 is the main source of the strong ultraviolet light that causes the abundant hydrogen gas in M42 to glow. It is only possible to separate Theta's four additional faint, magnitude 11 to 16 components in telescopes of 15 inches (380 mm) or larger.

M43 (NGC 1982) Slightly northeast of the Trapezium is the diffuse nebula M43. This circular emission nebula appears to be attached to the northern side of M42, although it is not actually a part of it. An 8th magnitude star is the source of the light that makes M43 visible to us. The dark patch lying between M42 and M43 is known as the Fishmouth.

7 NGC 1973, 1975, and 1977 About 1/2 degree north of M43 is the triple nebula of NGC 1973, 1975, and 1977. This glowing triangle of nebulosity consists of three separate emission/reflection nebulas that appear to be connected. Most of the energy exciting these wispy nebulas to glow is sourced from the multiple star 42 Orionis.

NGC 1981 Move on about 25 arcminutes north from NGC 1973, 1975, and 1977 to find the star cluster NGC 1981. This loose open star cluster consists of about twenty magnitude 8 to 10 stars. Most of the members of the cluster are young white stars.

THE HORSEHEAD NEBULA (B33)

An interesting, but exceedingly difficult, object to observe is the famous Horsehead Nebula (below). The Horsehead is a thick black cloud of dust and gas that can be seen only because it blots out some of the light coming from the faintly glowing streamers of the diffuse emission nebula IC 434. The Horsehead is estimated to be 1 light-year across and composed of a thin haze of dust in non-luminous gas.

You need a clear dark sky to have any chance of observing this elusive object—probably the most challenging you will encounter. To locate the nebula, scan the area about halfway between Alnitak and the 11th and 12th magnitude stars directly south of it. An H-beta filter that screws into your eyepiece may help you to observe it.

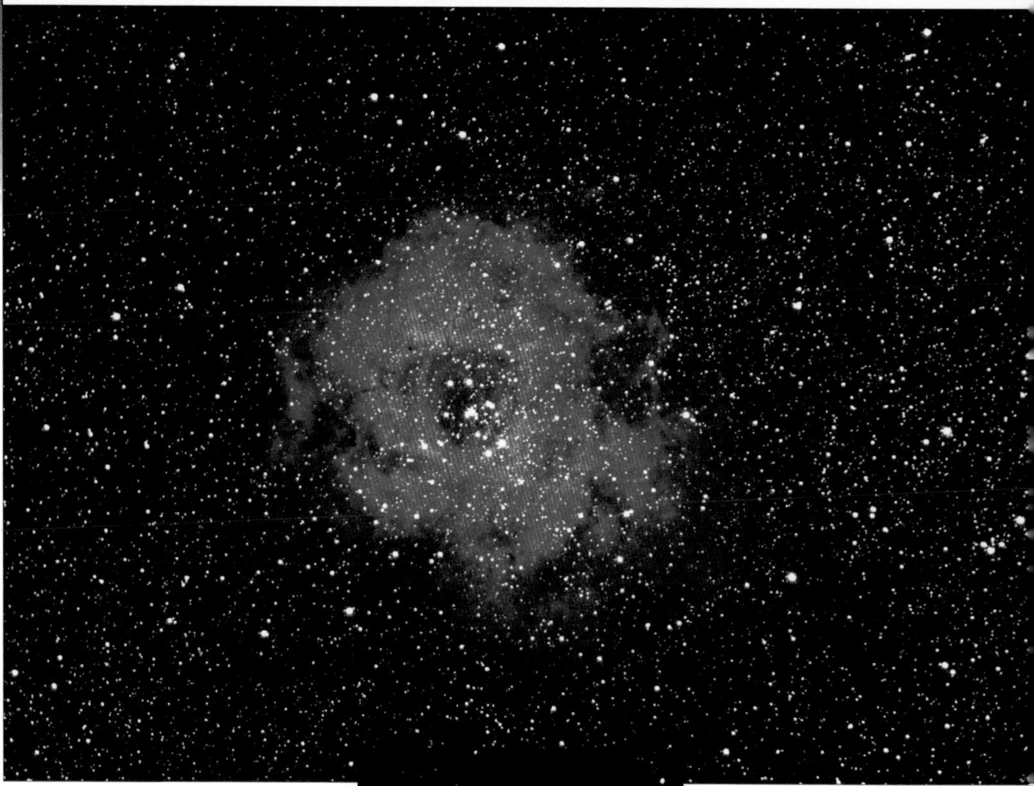

The Rosette Nebula (above) surrounds the star cluster NGC 2244 (below). They lie near the red supergiant Betelgeuse (left).

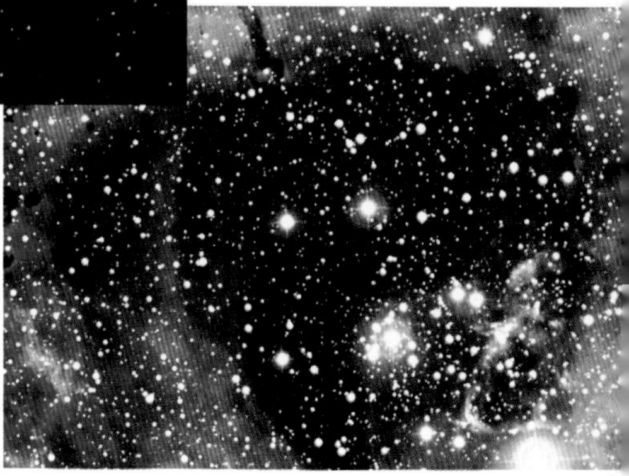

8 M78 (NGC 2068) Return to Alnitak then hop about 2½ degrees northeast to M78. This small, circular nebula softly glows at magnitude 8.3 in reflected light from two young, hot blue-white stars embedded within it. M78 is a small, illuminated portion of a large, dark nebula that partly encircles much of Orion's belt and sword area. Other illuminated patches form an arc known as Barnard's Loop, after the American astronomer Edward Emerson Barnard (see p. 61).

9 NGC 2112 Two degrees east of M78 is the small open star cluster NGC 2112. The 50 members of this compressed open cluster shine at a combined magnitude of 9.1. It is a challenge to pick out from the surrounding star fields.

10 Betelgeuse (58 Alpha [α] Orionis) This cool red supergiant lies 7 degrees north of NGC 2112, and is quite possibly the largest star in our part of the Milky Way. Betelgeuse is the only 1st magnitude star known to vary significantly in brightness, with an irregular period lasting about 6 years. Estimates indicate that at its normal size, it would fill the area within the orbit of Mars, and that when it swells up, its diameter would equal Jupiter's orbit. Betelgeuse is also the only star, besides the Sun, with surface features that have been seen by astronomers.

11 NGC 2186 The open star cluster NGC 2186 lies about 4 degrees southeast of Betelgeuse. This is a large, loose open star cluster enveloped in a rich Milky Way star field. Most of the 30 members of the cluster are 9th to 11th magnitude and can be seen with small scopes.

12 8 Epsilon [ε] Monocerotis The 4th magnitude star about 3 degrees east of NGC 2186 is Epsilon Monocerotis, our next target. Epsilon is a triple-star system in which the three pale yellow-white to bluish white components are easy to separate through a small telescope. The A star shines at magnitude 4.3, the B star at magnitude 6.7, 13 arcseconds away, and the C star at magnitude 12.7.

13 The Rosette Nebula (NGC 2237) The faint circular mass of gas 2 degrees east of Epsilon is our next stop. This doughnut-like emission nebula includes the open star cluster NGC 2244. The nebula is very large at about

30 arcminutes in diameter. Because it is so diffuse, t can be difficult to locate in telescopes smaller han 8 inches (200 mm). The Rosette is estinated to be at a distance of about 2,600 lightyears from us. This would make it some 55 lightears in diameter—almost twice the size of M42.

NGC 2244 This open star cluster, which s enveloped by the Rosette Nebula, appears in he hole of the doughnut. Its brightest apparent nember is the 6th magnitude yellow giant 12 Monocerotis. Some astronomers think that his star is a foreground object and not actually art of the cluster, since the other members ppear to be mostly young white stars.

14 NGC 2251 Hop about 4 degrees north o NGC 2251. This elongated open star cluster contains about 30 stars located mostly in a wisted string. The cluster shines with a combined magnitude of about 7.3, but most of the tars are fainter than magnitude 12.

15 The Christmas Tree Cluster (NGC 2264) Slide about 2 degrees northeast to the open tar cluster NGC 2264. The riangular shape of this open tar cluster strongly suggests ts common name, given y Lowell Observatory stronomer Carl Otto Lampand. There is a faint nebula urrounding the cluster, and the black **Cone Nebula** intrudes visually into this at the southern end. This effect is too faint to be seen in telescopes smaller than 12 inches (300 mm). The brightest star in the cluster, S Monocerotis, is an intensely luminous, magnitude 4.6 blue-white double.

16 NGC 2309 Take a long hop, about 17 degrees south, to NGC 2309. About 40 stars make up this magnitude 10.5 compact open star cluster. It appears to be centered on a magnitude 11.5 star, in front of a very attractive field.

17 M50 (NGC 2323) About 2 degrees southeast of NGC 2309 is the last object on our starhop, the easy-to-find, rich open star cluster M50. This beautiful magnitude 6.3 cluster contains about 50 stars, appearing as a mottled patch of light barely visible to the unaided eye in dark skies. The cluster is roughly diamond shaped and is best viewed with binoculars or a low-power telescope eyepiece. A bright yellow-orange star holds center stage slightly southwest of the core. Two parallel arms of stars extend out from the core.

The Orion area is another patch of the night sky that you could spend years exploring and yet still not see all of the objects that are to be found. Our tour has only touched on some of the highlights. RG

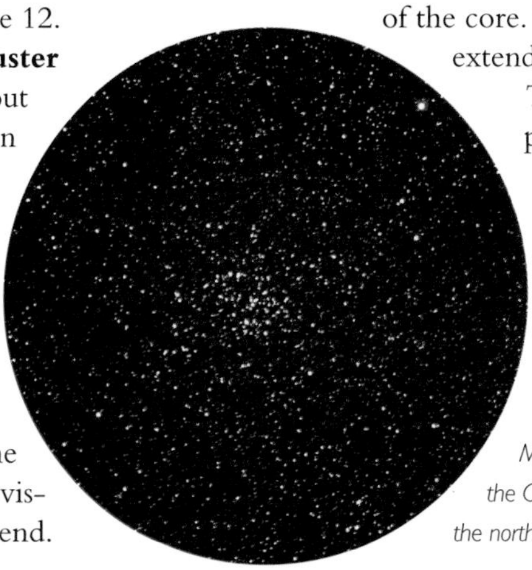

M50 (left) lies in Monoceros, along with the Christmas Tree Cluster, which extends to the north of the Cone Nebula (above).

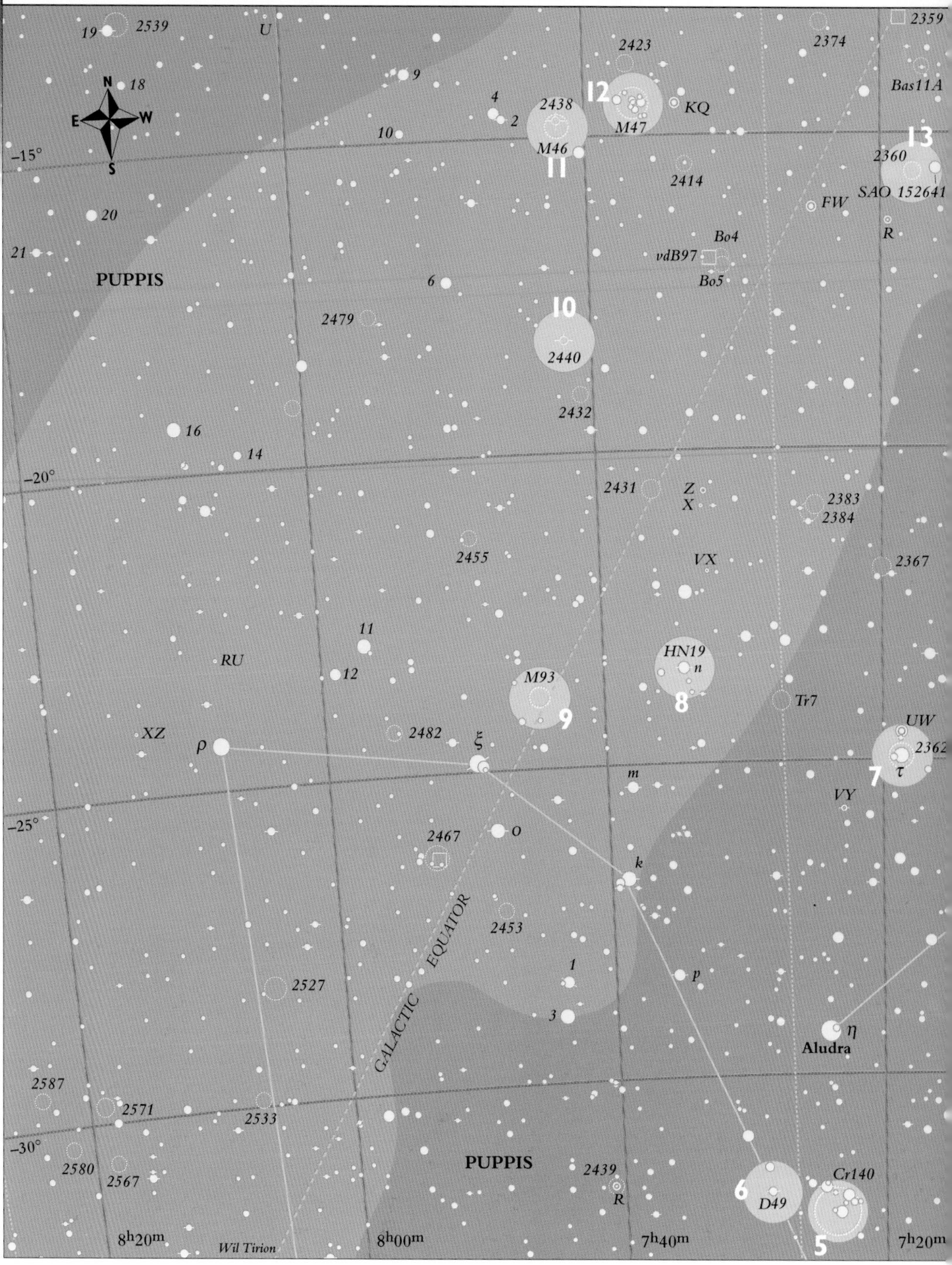
N
E
W
S
PUPPIS
PUPPIS
–15°
–20°
–25°
–30°
8h20m
8h00m
7h40m
7h20m
Wil Tirion
GALACTIC EQUATOR
M46
M47
M93
2438
2423
2374
2359
Bas11A
2360
SAO 152641
2414
FW
R
KQ
Bo4
vdB97
Bo5
2440
2432
2479
2539
2431
2383
2384
2455
2367
VX
Z
X
HN19
Tr7
UW
2362
VY
2482
2467
2453
2527
2587
2571
2533
2580
2567
2439
D49
Cr140
Aludra
RU
XZ
U
1
2
3
4
5
6
7
8
9
10
11
12
13

The Dog Star and Surrounds

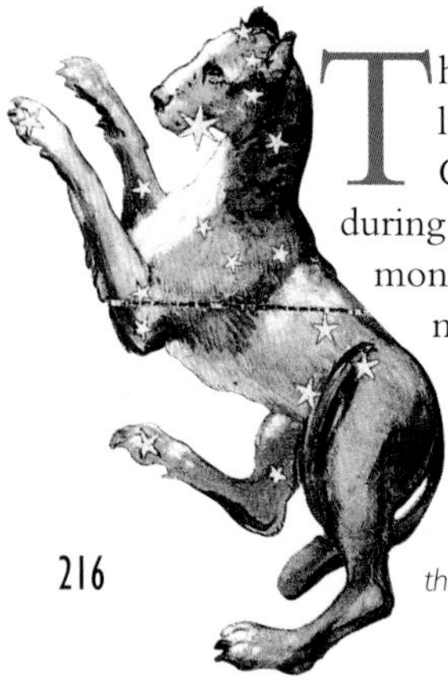

The best views of the constellation of Canis Major, the Great Dog, are to be had during the evenings of the early months of the year. This prominent, easy-to-find constellation lies on the southwestern side of the Milky Way.

Canis Major, the Great Dog, is one of the most striking of all the constellations.

Although visible from both hemispheres, southern observers will have a better view of it.

Farther north of this sky chart is the constellation of Canis Minor—which, as you might suspect, is the Little Dog. In classical mythology, the two canines are said to have attended Orion. Moving into the band of the Milky Way near Canis Major, we cross the border into neighboring Puppis, the stern of the ship Argo. This region contains many interesting objects for binoculars and small telescopes.

Northern and Southern Hemisphere | On meridian 10pm Feb 10

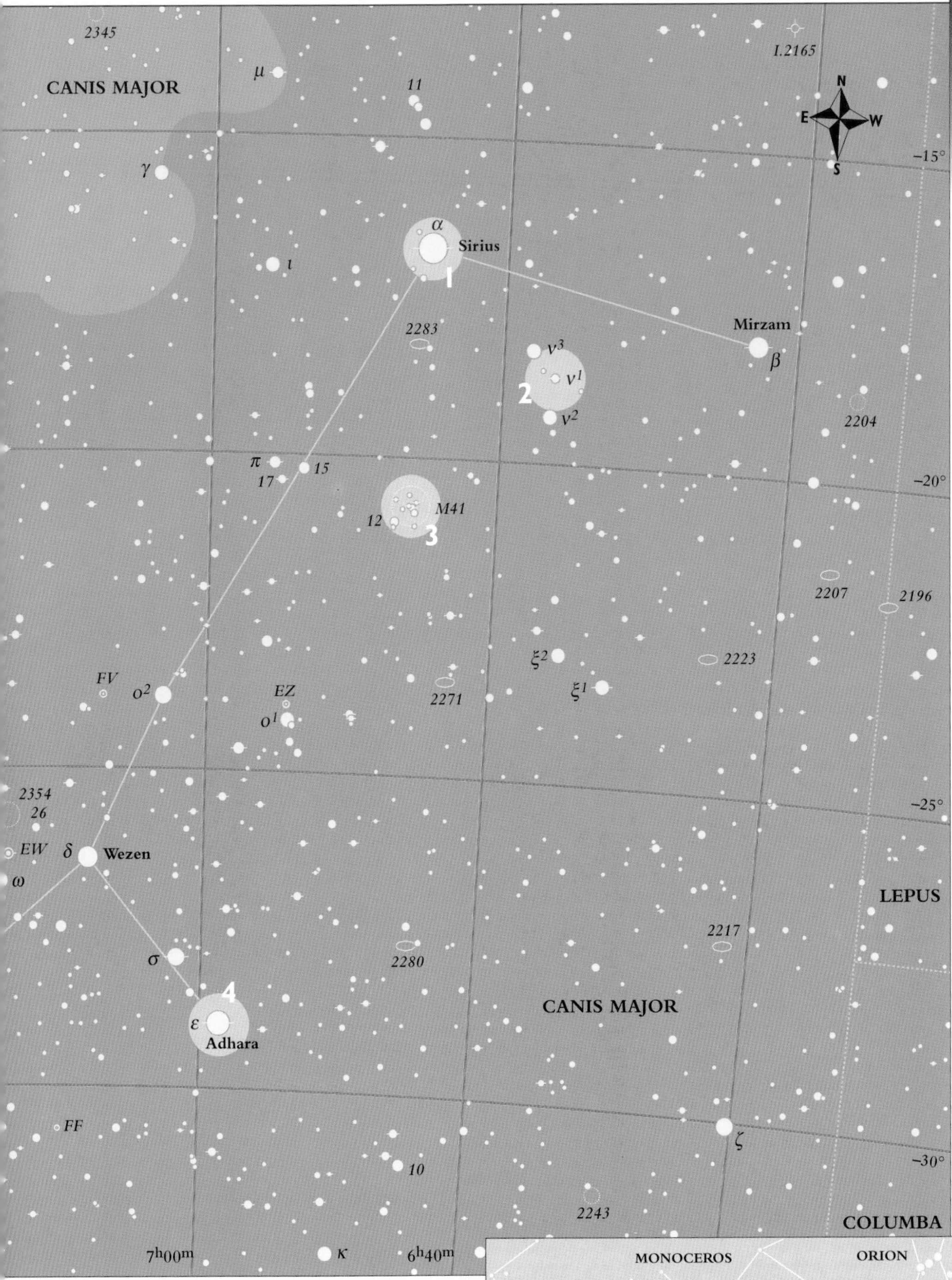
2345
CANIS MAJOR
μ
11
I.2165
N
E
W
S
−15°
γ
α
Sirius
ι
1
2283
Mirzam
ν3
ν1
β
2
ν2
2204
π
17
15
−20°
M41
12
3
2207
2196
ξ2
2223
FV
o2
EZ
ξ1
2271
o1
2354
26
−25°
EW
δ
Wezen
ω
LEPUS
2217
σ
2280
4
CANIS MAJOR
ε
Adhara
FF
ζ
−30°
10
2243
COLUMBA
7h00m
κ
6h40m

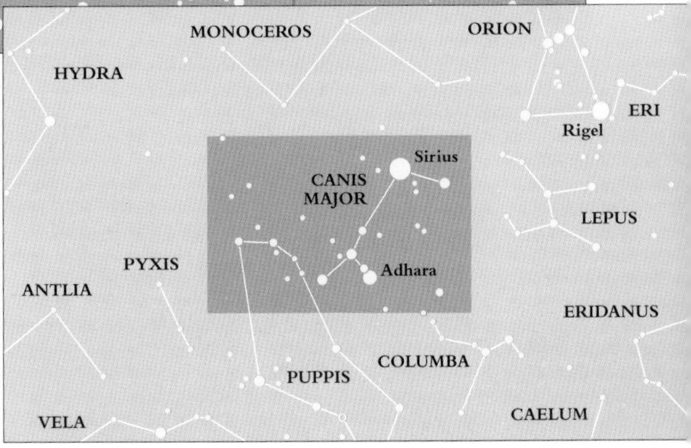
MONOCEROS
ORION
HYDRA
ERI
Rigel
Sirius
CANIS MAJOR
LEPUS
PYXIS
Adhara
ANTLIA
ERIDANUS
COLUMBA
PUPPIS
VELA
CAELUM

❶ Sirius (9 Alpha [α] Canis Majoris)

The three stars of Orion's belt point toward Sirius, the Dog Star—the brightest star in the sky, at magnitude −1.4. Sirius has a magnitude 8.7 companion, Sirius B, that orbits it every 50 years. Sirius B, or "the Pup," as it is often called, currently lies only about 4 arcseconds away, and is very difficult to see even with large amateur scopes. Sirius B is a famous example of a white dwarf star, which is only about the size of the Earth and extremely dense. A cubic inch of its material would weigh about 2 tons!

The brilliance of Sirius has given rise to countless stories through the ages. In ancient Egypt, Sirius was known as the Nile Star because at the time of year when it rose just before dawn, it heralded the annual flooding of the Nile River, an important event in Egyptian life.

The constellations of Canis Major and Puppis span much of this slice of the Milky Way (above). Sirius glows brightly on the left.

❷ 6 Nu^1 [ν^1] Canis Majoris To find Nu^1 Canis Majoris, nudge your scope 3 degrees southwest from Sirius. Nu^1 is the middle of a group of three stars—the others, predictably, are Nu^2 and Nu^3. Nu^1 is a fine double that is easy to separate in any size scope. Its components, at magnitudes 5.8 and 8.5, lie 17.5 arcseconds apart.

❸ M41 (NGC 2287) A 3 degree hop southeast brings us to the star cluster M41. This magnificent star cluster is visible to the unaided eye as a patch of light 4 degrees almost due south from Sirius. Many of its stars can be seen with binoculars, but a wide-field telescope eyepiece is best, as the cluster covers more than 1/2 degree.

❹ Adhara (21 Epsilon [ε] Canis Majoris) Swing your scope 8 degrees south and a little east to find the bright double star Epsilon Canis Majoris. Epsilon is, curiously, the second brightest star in Canis Major. At magnitude 1.5, it is a half magnitude brighter than Beta. Epsilon can be resolved using high magnification in 3 inch (75 mm) scopes and larger. Its magnitude 7.4 companion lies 7.5 arcseconds away.

❺ Collinder 140 This cluster, about 6 degrees southeast of Epsilon, can be seen with the naked eye and is a fine binocular object. It contains about 30 stars, the brightest at magnitude 5.4.

❻ Dunlop 49 Puppis Adjust your scope about 1 degree east and slightly north of our last stop to find Dunlop 49, across the border in Puppis. This is a superb double star for any small scope, because its magnitude 6.5 and 7.2 stars are almost 9 arcseconds apart.

WHITE DWARF STARS

Sirius B, Sirius A's companion (below), was the first white dwarf to be identified. Incredibly, Sirius B has a mass almost equal to that of our Sun, but a diameter of only about 19,000 miles (30,500 km)—less than three times that of Earth.

We now know that a white dwarf is the collapsed core of a star—the last stage in the evolution of a star that was originally up to about eight times as massive as the Sun. In 1931, it was discovered that a white dwarf star cannot contain more than 1.4 solar masses. However, stars lose a great deal of mass during their lives by shedding material, most visibly in the form of planetary nebulas late in their life cycle.

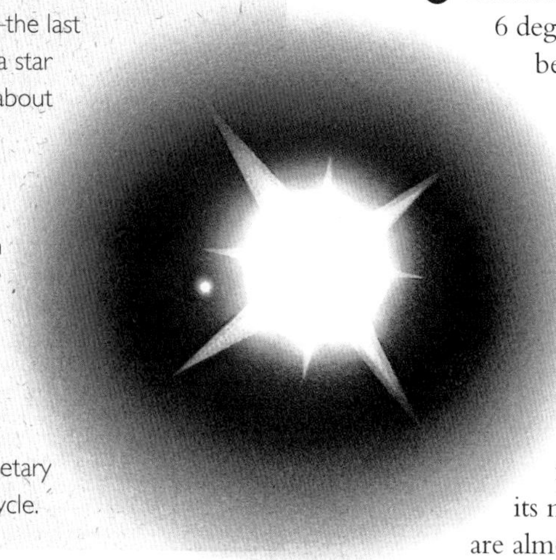

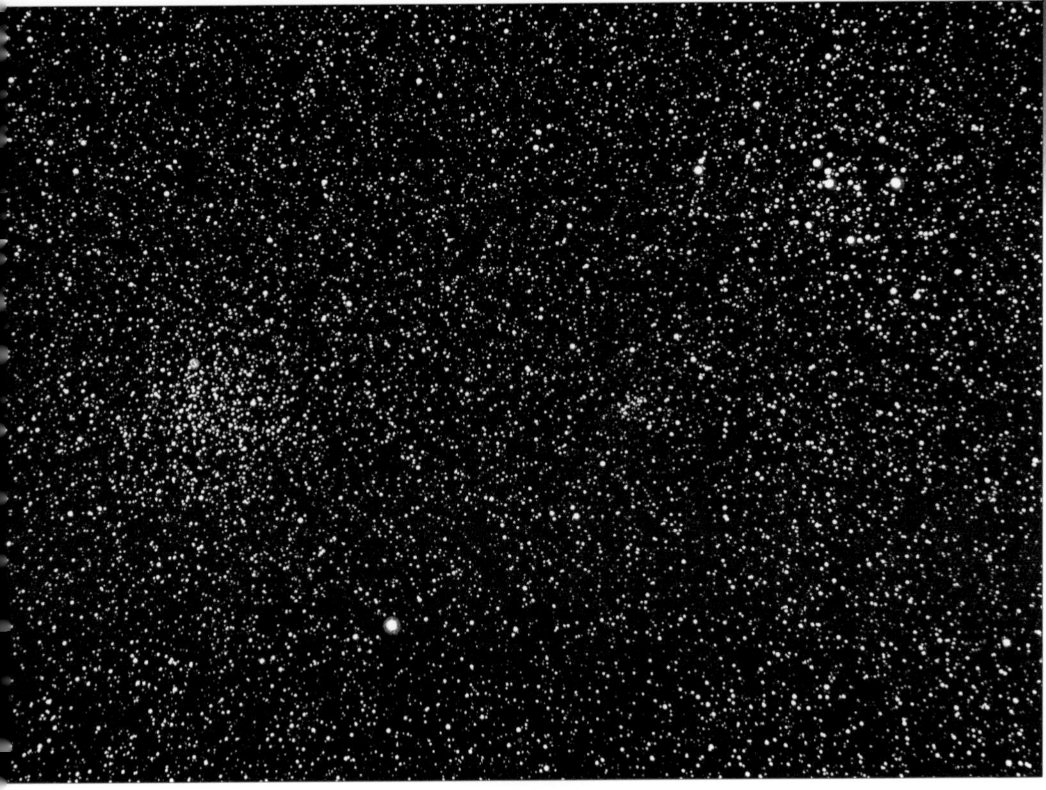

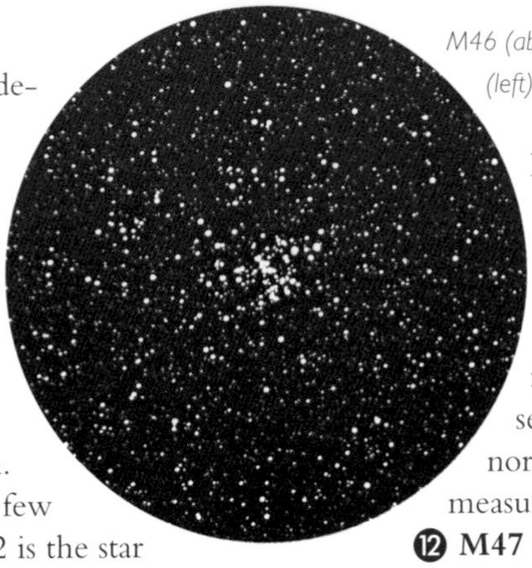

M46 (above, left), M47 (above, right), and M93 (left) are the highlights of northern Puppis.

7 NGC 2362 The star **30 Tau [τ] Canis Majoris**, 7 degrees northwest of Dunlop 49, lies at the heart of tightly packed NGC 2362. At magnitude 4, the cluster is one of the real gems of the sky for all size scopes. At 8 arcminutes across, this cluster contains some 60 stars, many of which pop into view when using averted vision.

8 n Puppis (HN19) A few degrees northeast of NGC 2362 is the star n Puppis, a fine double for any size telescope to separate. Its magnitude 6 components are almost equal in brightness, separated by 9.6 arcseconds.

9 M93 (NGC 2447) Shift your gaze 2½ degrees east to the star cluster M93. Covering about ⅓ degree, this cluster is easy to see with binoculars, its brightest members forming a compact group near the middle. With a 3 inch (75 mm) telescope, you should be able to see a number of stars of magnitude 8 and fainter.

10 NGC 2440 About 5½ degrees north and a little west of M93 lies NGC 2440, a planetary nebula shining at magnitude 10.8. A 3 inch (75 mm) scope reveals its somewhat irregular shape. NGC 2440 covers an area of about 14 x 32 arcseconds. Its bluish color is quite obvious, especially in larger telescopes.

11 M46 (NGC 2437) Hop 3½ degrees farther north to the open star cluster M46. Discovered by Messier in 1771, M46 contains a large number of fairly faint stars, which Messier thought were a nebulous patch. Switch to high magnification to view the magnitude 10 planetary nebula **NGC 2438**, which is seen against the backdrop of the northern part of M46. NGC 2438 measures about 1 arcminute across.

12 M47 (NGC 2422) Switch back to low magnification to observe M47, lying about 1½ degrees west of M46. Unlike M46, this open cluster is visible to the naked eye, although it is best viewed using binoculars. It contains several stars of around magnitudes 6 and 7, as well as many others that are fainter.

13 NGC 2360 Lying 4½ degrees west of M47 is NGC 2360, the last object on our starhop. It is located just east of the magnitude 5 star **SAO 152641**. Visible as a small patch of light through binoculars, NGC 2360 is a cluster comprising about 80 stars of around magnitude 10 and fainter. A beautiful object through a 4 inch (100 mm) telescope, the cluster looks somewhat like crystals of spilt table salt.

Tonight's tour of the Dog Star and surrounds has brought us back almost to Sirius, shining like a welcoming beacon. This part of the Milky Way has revealed an interesting variety of objects, most notably the many star clusters. MG

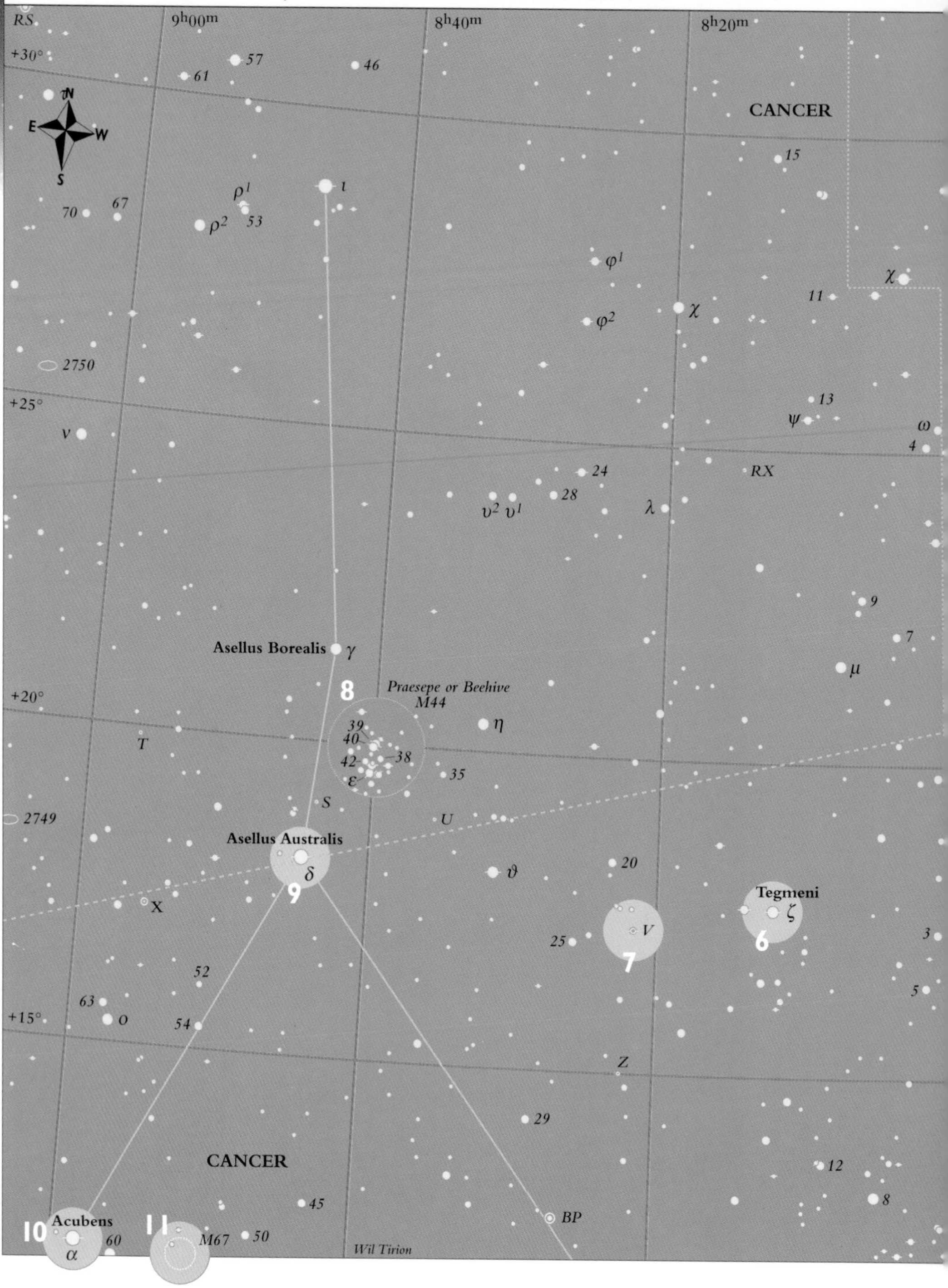
9h00m
8h40m
8h20m
+30°
+25°
+20°
+15°
N
E
W
S
CANCER
Asellus Borealis
Praesepe or Beehive
M44
Asellus Australis
Tegmeni
Acubens
M67
2750
2749
RS
RX
BP
8
9
7
6
10
11
Wil Tirion

Jewels in Cancer and Gemini

The constellations on this hop along the zodiac are clearly visible from both the Northern and Southern hemispheres. The legends of Cancer, the Crab, and Gemini, the Twins, come from Greek mythology. The goddess Hera sent Cancer to kill her enemy Hercules, who crushed the crab with his mighty club. Saddened by the crab's demise, Hera placed it among the stars to honor its valiant but ill-fated efforts. Gemini represents the twin sons of Leda, the Queen of Sparta. One was fathered by her husband, King Tyndareus, and the other by Zeus while in the form of a swan (represented by Cygnus). The twins, Castor and Pollux, served aboard Jason's ship, the Argo, in the legendary voyage of the Argonauts.

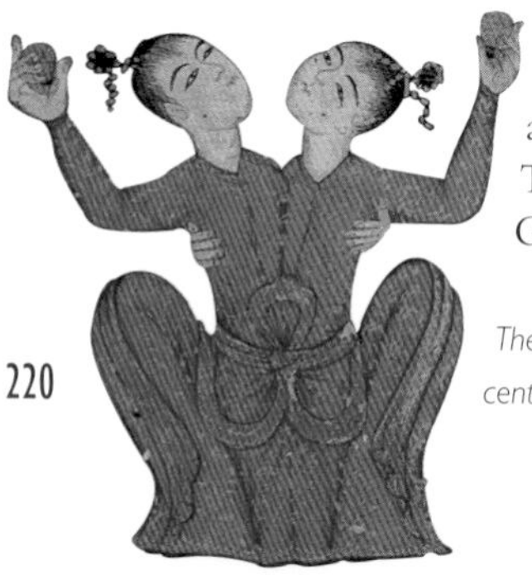

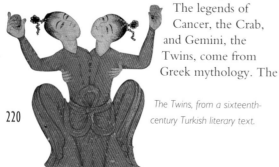

The Twins, from a sixteenth-century Turkish literary text.

Northern and Southern Hemisphere | On meridian 10pm Feb 20

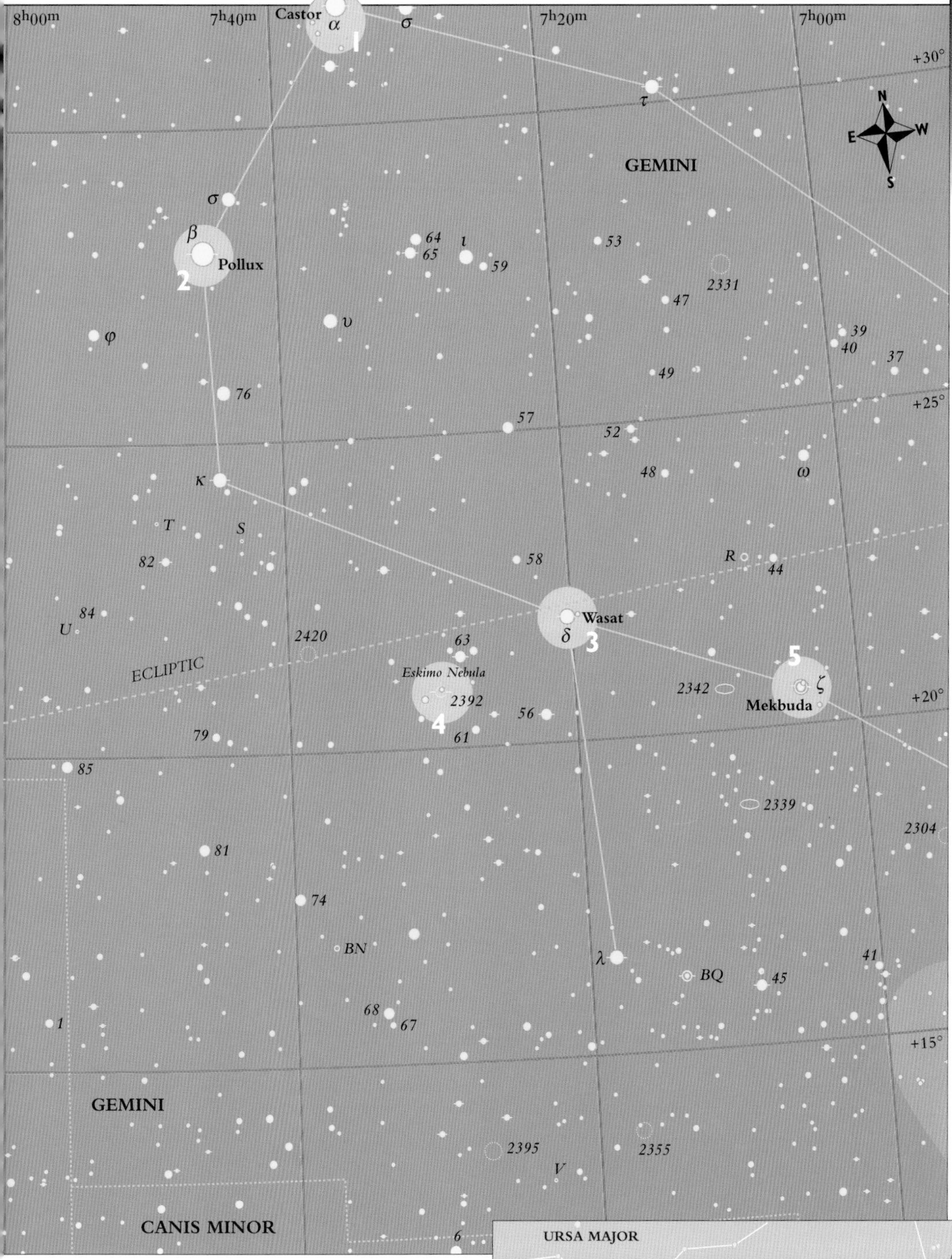

8^h00^m
7^h40^m
7^h20^m
7^h00^m
+30°
+25°
+20°
+15°
N
E
W
S
Castor
Pollux
GEMINI
Wasat
Mekbuda
Eskimo Nebula
ECLIPTIC
GEMINI
CANIS MINOR
2331
2420
2392
2342
2339
2304
2395
2355
BN
BQ

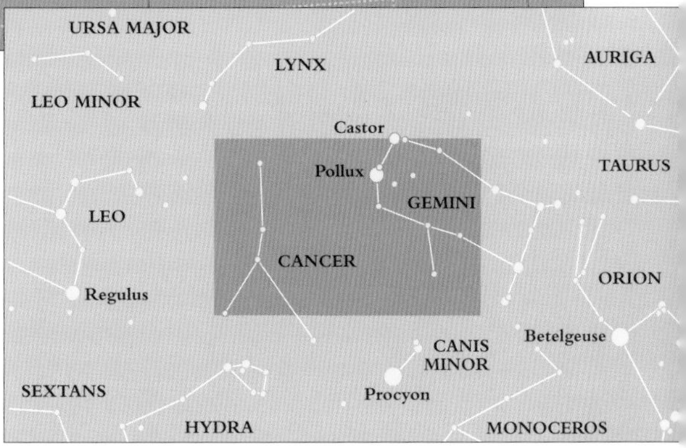

URSA MAJOR
LYNX
AURIGA
LEO MINOR
Castor
Pollux
GEMINI
TAURUS
LEO
CANCER
ORION
Regulus
CANIS MINOR
Betelgeuse
SEXTANS
Procyon
HYDRA
MONOCEROS

❶ Castor (66 Alpha [α] Geminorum)

Our tour begins at one of the best known and brightest doubles, the 1st magnitude yellowish star, Castor, the Horseman, at the eastern end of Gemini. In fact, Castor is a multiple star shining with a combined visual magnitude of 1.6.

The magnitude 1.9 A and magnitude 2.9 B stars, themselves both spectroscopic binaries, currently lie 3 arcseconds apart and can be separated in scopes of 10 inches (250 mm) or larger. The B star orbits the A star over a period of about 400 years. The faint red dwarf C star (designated YY Geminorum), also a spectroscopic binary, completes the entire Castor system.

A spectacular bright shower of meteors, called the Geminid meteors, emanates from a point near Castor each year, reaching its maximum around 14 December (see also p. 158).

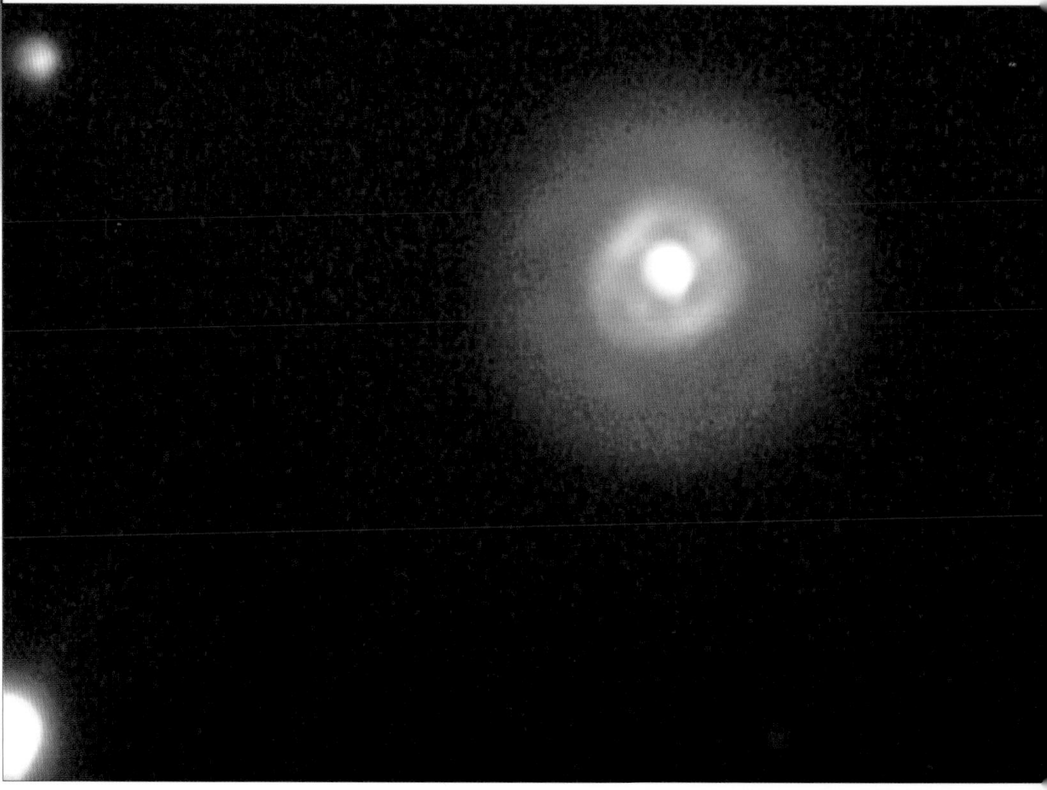

❷ Pollux (78 Beta [β] Geminorum) Four and a half degrees southeast of Castor is the other twin, Pollux. This is the sky's 17th brightest star, shining as a magnitude 1.2 golden-yellow giant. Pollux is about 45 light-years from Earth and intrinsically 35 times brighter than the Sun.

❸ Wasat (55 Delta [δ] Geminorum) Move your scope about 8 degrees southwest to the double star Wasat, a magnitude 3.5 yellow star with a faint red dwarf companion at magnitude 8.2. This star system lies some 55 light-years away with an orbital period believed to be more than 1,000 years long. In 1930, Clyde Tombaugh discovered Pluto near this binary system.

❹ The Eskimo Nebula (NGC 2392) About 2 degrees southeast of Wasat is the bright planetary nebula NGC 2392, discovered by William Herschel in 1787. It is intense and compact, looking somewhat like a large, fuzzy bluish green star, making an attractive contrast to the magnitude 8.3 orange star nearby, which has a similar brightness. As with most planetary nebulas, high magnification is essential for good viewing. You can only expect to see the eskimo's "face" in photographs taken through large scopes.

❺ Mekbuda (43 Zeta [ζ] Geminorum) Sweep some 6 degrees west of NGC 2392 to Mekbuda, one of the brightest variable stars. This yellow pulsating Cepheid variable has a period of 10.16 days during which its magnitude fluctuates between 3.6 and 4.2. With binoculars, you might also note a magnitude 7.6 companion star nearby, which is actually unrelated to Zeta.

The Eskimo Nebula (top) and the Praesepe (above) are opposites in terms of the level of magnification that reveals them best.

❻ Tegmeni (16 Zeta [ζ] Cancri) Hop about 16 degrees east and slightly south of Mekbuda to the binary star Tegmeni, across the border in Cancer. The B component of this system orbits the A star over a period of 59.6 years. Both components are yellowish main-sequence stars with nearly equal magnitudes of 5.6 and 5.9.

❼ V Cancri This red giant Mira-type variable lies 2 degrees east of Tegmeni. This is one of many variables whose full cycle can be followed in scopes of less than 8 inches (200 mm). The normal magnitude range of V Cancri is 7.9 to 12.8, during a period of 125 days. At maximum, it would fill the volume of Mars's orbit.

❽ The Praesepe/Beehive Cluster (M44, NGC 2632) Shift your scope 5 degrees northeast to the Praesepe, a large, naked-eye open

The open star cluster M67 (above) is often overlooked in favor of its larger cousin, the Praesepe.

star cluster that has been known since ancient times. It covers about 1½ degrees of sky and is some 520 light-years distant. The Praesepe's several hundred scattered stars show best through binoculars or a finderscope because the cluster is quite broad. At higher magnification, it is quite disappointing because the view is too close.

9 Asellus Australis (47 Delta [δ] Cancri) Nudge 2 degrees southeast of the center of the Praesepe to Asellus Australis, a magnitude 4.3 yellow optical double star. The Romans gave this star its name, the Southern Donkey. Asellus Australis and **Asellus Borealis (43 Gamma [γ] Cancri)** are known collectively as the Donkeys.

10 Acubens (65 Alpha [α] Cancri) About 6 degrees southeast of Asellus Australis is the magnitude 4.3 white wide double star Acubens, about 130 light-years away. Its name is derived from the Arabic *Al Zubanah*, meaning "the claw," because it marks one of the crab's claws. The secondary star in this system shines at magnitude 11.8 and is visible through 3 inch (75 mm) telescopes or larger.

11 M67 (NGC 2682) Use the bright stars Asellus Australis and Acubens to help locate M67, about 2 degrees west of Acubens. This cluster, the last object on our tour, was discovered by the German astronomer Johann Gottfried Koehler between 1772 and 1779. In dark skies, this densely packed magnitude 6.1 open cluster can be seen with the naked eye, but it needs a small scope or binoculars to really appreciate it. M67 contains some 500 stars of magnitudes 10 to 16 and many more fainter ones. M67 is estimated to be four to five billion years old—one of the oldest open clusters known.

Cancer and Gemini include some interesting star fields, beckoning you to mine their collection of clusters and double and variable stars. RG

EXTRASOLAR PLANETS

Only recently have astronomers detected clear evidence of planets orbiting distant Sun-like stars. One of the first of a handful of extrasolar planets to be discovered was the planet-size companion of 55 Rho^1 [ρ] Cancri, found by a team from the Universities of California and San Francisco. The illustration below compares the orbit of Mercury around the Sun (top) on a similar scale to Rho^1 Cancri and its planet (bottom). Rho^1 is a magnitude 5.9 yellow star with a planetary companion, about 0.8 times the mass of Jupiter, that orbits Rho^1 over just 14.7 days.

Extrasolar planets are far too faint to be seen visually, even by the HST, but they reveal their presence by the tiny wobbles their gravity induces in the motions of each star.

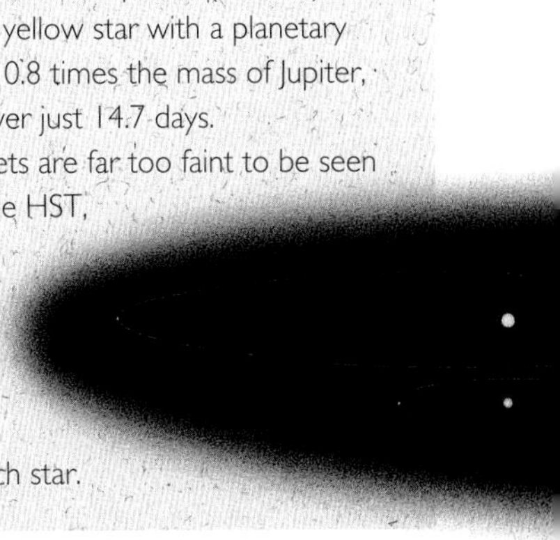

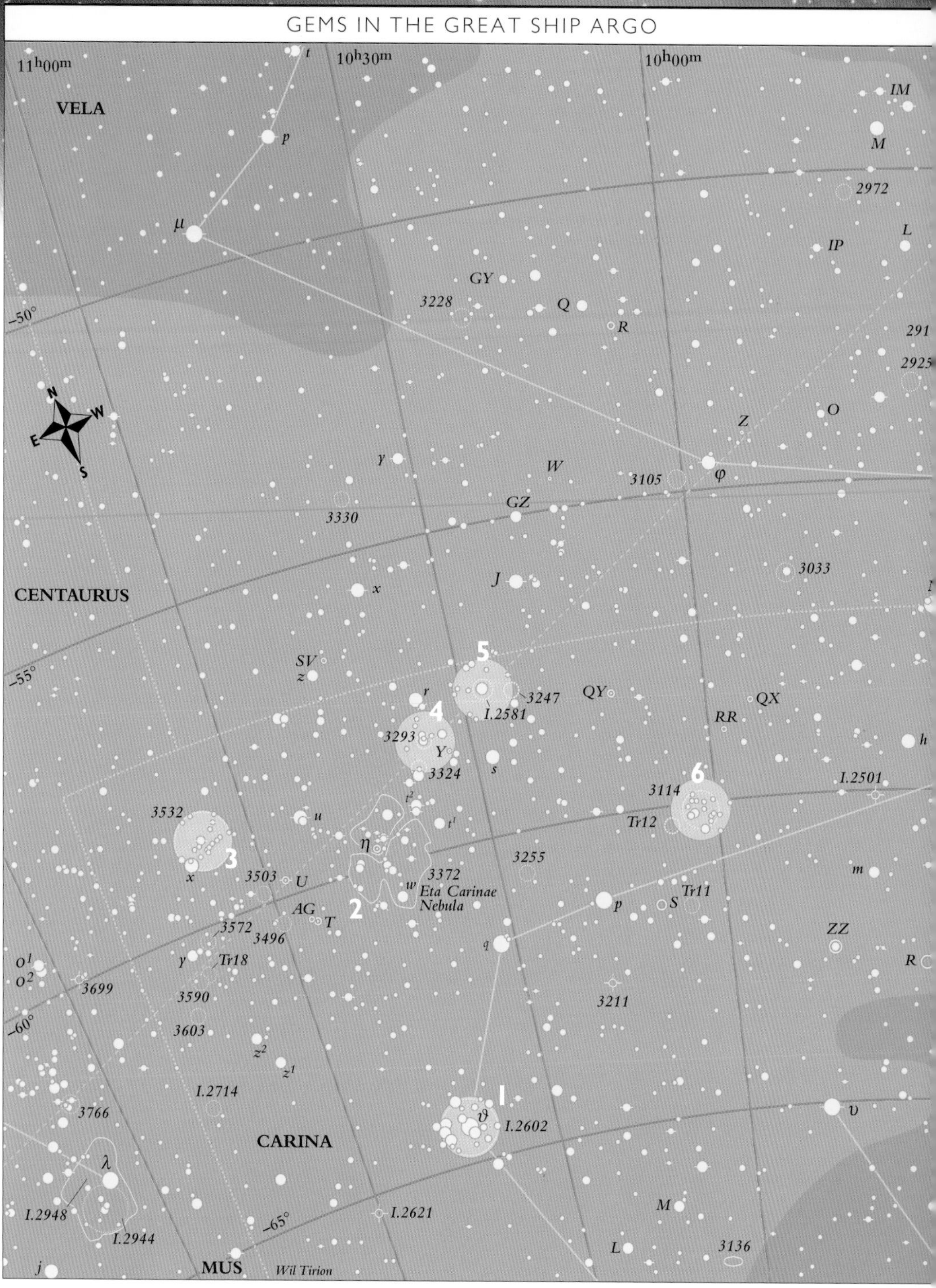
GEMS IN THE GREAT SHIP ARGO
11^h00^m
10^h30^m
10^h00^m
VELA
CENTAURUS
CARINA
MUS
-50°
-55°
-60°
-65°
3228
3330
3105
3033
2972
3247
I.2581
3293
3324
3532
3503
3372
Eta Carinae
Nebula
3255
3114
Tr12
Tr11
I.2501
3572
3496
Tr18
3699
3590
3603
3211
I.2714
3766
I.2602
I.2948
I.2944
I.2621
3136
Wil Tirion

Gems in the Great Ship Argo

Northern Hemisphere skywatchers are disadvantaged here because many objects in this patch of sky are not seen at all from latitudes above about 30 degrees north. Carina, the Keel, and Vela, the Sails, were two of the four constellations formed when the huge constellation of Argo Navis was subdivided by the French astronomer Nicolas Louis de Lacaille in the eighteenth century. The other two are Puppis, the Stern; and Pyxis, the Compass. Straddling the constellations of Carina and Vela is the so-called False Cross, which newcomers to the southern skies can easily confuse with the Southern Cross.

The Great Ship Argo from a fresco in the Villa Farnese.

With binoculars or a telescope, you can expect to see many fine objects within this part of the Milky Way Galaxy, including several of the best star clusters in the entire night sky.

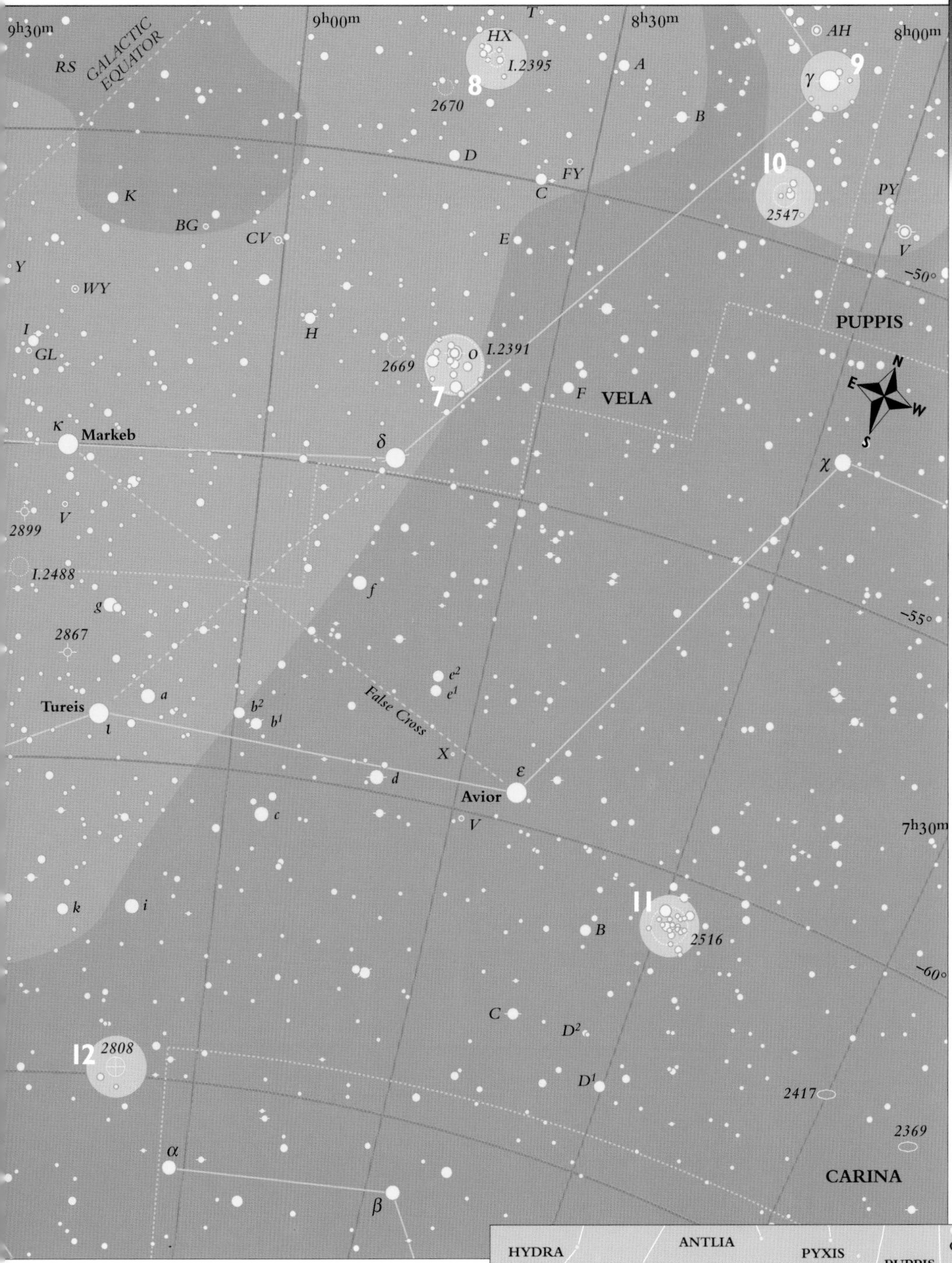

9h30m
9h00m
8h30m
8h00m
7h30m
GALACTIC EQUATOR
RS
HX
I.2395
2670
T
A
AH
B
D
FY
C
PY
K
BG
CV
E
2547
V
-50°
Y
WY
I
GL
H
2669
O
I.2391
PUPPIS
F
VELA
N
E
W
S
κ
Markeb
δ
χ
2899
V
I.2488
f
g
2867
-55°
e2
e1
Tureis
a
b2
b1
ι
False Cross
X
ε
d
Avior
c
V
k
i
B
2516
-60°
C
D2
2808
D1
2417
2369
α
CARINA
β
7
8
9
10
11
12

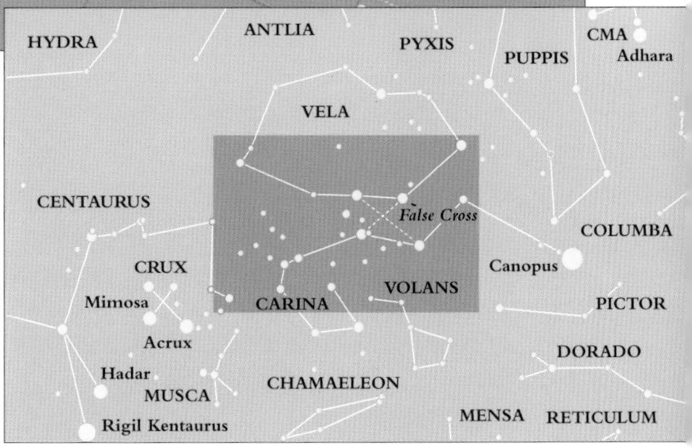

HYDRA
ANTLIA
PYXIS
PUPPIS
CMA
Adhara
VELA
CENTAURUS
False Cross
COLUMBA
CRUX
Canopus
Mimosa
CARINA
VOLANS
PICTOR
Acrux
DORADO
Hadar
MUSCA
CHAMAELEON
Rigil Kentaurus
MENSA
RETICULUM

❶ Theta [θ] Carinae and IC 2602 Our starting point on this starhop through the Great Ship Argo is the 3rd magnitude star Theta Carinae. Even with the unaided eye, you can easily see that Theta is more than just a single star. With binoculars, you are afforded a superb view, and this closer inspection allows you to put the character of the object into perspective. It is a central star surrounded by a number of scattered 5th magnitude and fainter stars forming the open star cluster IC 2602.

Theta Carinae, which covers a full degree of sky, has also been called the Southern Pleiades by some Southern Hemisphere observers.

To find the next stop on this starhop, the star Eta Carinae, shift your scope about 4½ degrees due north of IC 2602. Eta is enveloped in the dramatic and beautiful Eta Carinae Nebula.

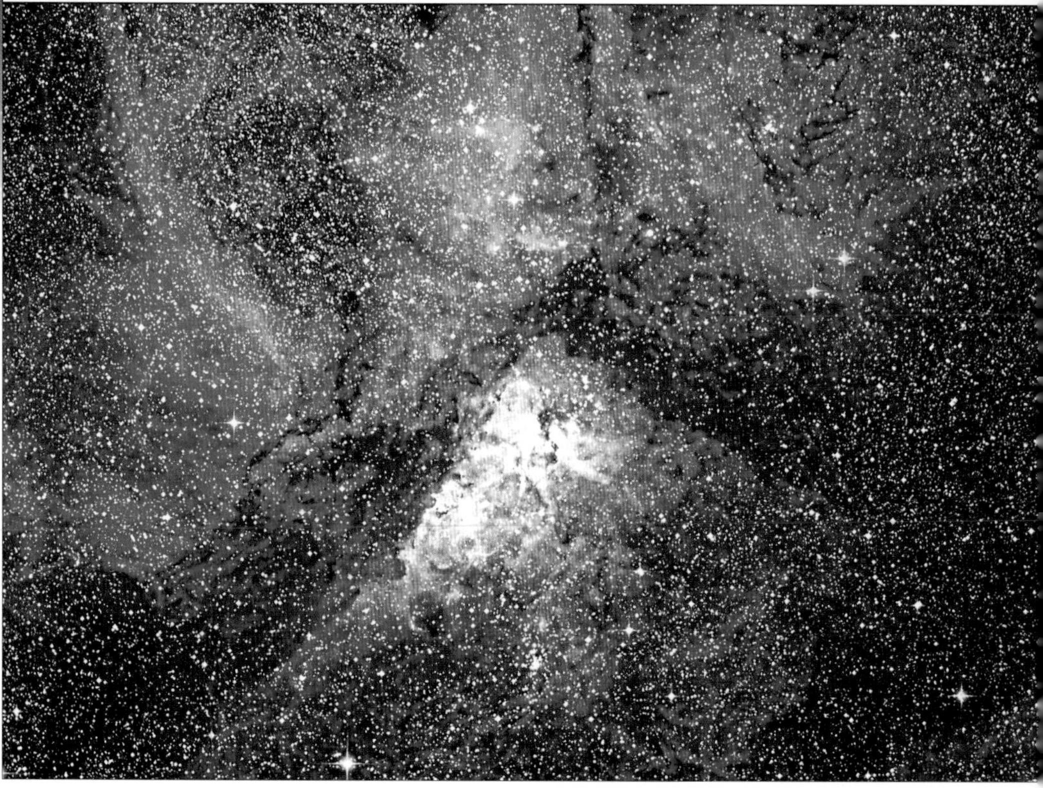

❷ **Eta [η] Carinae** This is the brightest star in a field that is truly beautiful, even in scopes of less than 3 inches (75 mm). The star appears orange-red and is embedded in the superb **Eta Carinae Nebula**. Eta is a variable star—one of the most luminous in the Milky Way. In 1843, it outshone every star in the night sky except Sirius. A 4 inch (100 mm) scope will show Eta surrounded by a small "blob" of red light—the nebulous patch called the Homunculus (see Box).

❸ **NGC 3532** Turning about 3 degrees east and slightly north of Eta, the open cluster NGC 3532 is revealed. John Herschel, who cataloged thousands of celestial objects when observing from South Africa in the 1830s, considered NGC 3532 to be the finest cluster he had ever seen. It appears as a fuzzy patch to the naked eye and is a superb object with binoculars. Any size scope will resolve a large number of stars, which are best viewed using low power.

❹ **NGC 3293** The compact cluster NGC 3293 is 4 degrees west and a little north of NGC 3532. Through binoculars, this cluster appears as a tiny bright spot, but a telescope view reveals a stunning group of several dozen stars of different colors. A 3 inch (75 mm) scope with moderate magnification gives a fantastic view. An orange star near one edge of the cluster is especially attractive.

❺ **IC 2581** Nudge your scope about 1 degree west and slightly north to bring IC 2581 into view. Gathered around quite a distracting magnitude 5 star, IC 2581 is a cluster of some

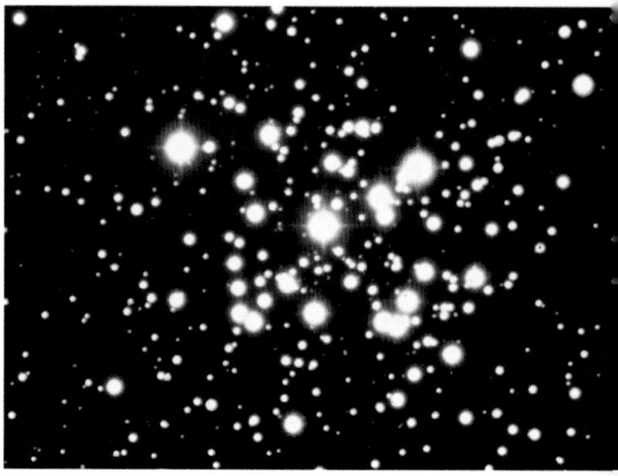

Eta Carinae Nebula (top) and NGC 3293 (above) are two of the highlights of the Milky Way in Carina.

25 stars that are much more obvious with high magnification. You should be able to resolve this cluster well with a 4 inch (100 mm) telescope.

❻ **NGC 3114** Shift your focus 4 degrees southeast to locate the next cluster, NGC 3114. Binoculars do a fine job of revealing this magnificent cluster of 7th magnitude stars and fainter covering an area of more than 1/2 degree. To the naked eye, it appears as a 4th magnitude patch of light. Low magnification is essential when viewing this cluster through a telescope.

❼ **IC 2391** Almost 2 degrees north and a little west of **Delta [δ] Velorum** is IC 2391, also known as the Omicron Velorum cluster. Its members are loosely packed around the magnitude 3 bluish white **Omicron [o] Velorum**. It is a better subject for binoculars than for a scope.

NGC 3532 (above) is one of the finest open clusters in the sky, lying about 1,300 light-years away from us.

8 **IC 2395** Head north about 5 degrees rom IC 2391 to IC 2395. You can spot this luster in binoculars, but it is better suited to a elescope. It is not exactly a rich group, but still one worthy of attention, and a 3 inch (75 mm) cope, or even smaller, provides a good view.

9 **Gamma [γ] Velorum** A little more han 6 degrees west of IC 2395 is Gamma Velorum. This is a spectacular double star whose nagnitude 2 and 4 components are 41 arcseconds part and can be separated with binoculars. Any mall scope will show two other stars, of magnitudes 8 and 9, about 90 arcseconds away.

10 **NGC 2547** This attractive open luster, about 2 degrees south of Gamma, contains tars with a moderate range in brightness. A 4 inch (100 mm) scope provides a good view of the cluster, which covers about 1/3 degree.

11 **NGC 2516** Eleven degrees farther south of NGC 2547 is NGC 2516. This cluster is a ine sight in binoculars. The long axis of the False Cross points toward it, as if to draw attenion to it, and it is easily located a few degrees outhwest of **Epsilon [ε] Carinae.** The 1/2 degree wide cluster contains a number of stars of nagnitude 7 and fainter.

12 **NGC 2808** A little south of the halfway point between NGC 2516 and Theta Carinae is NGC 2808. This globular cluster shows up well as a glowing spot of light in binoculars. A 6 inch (150 mm) scope or larger is needed to start to resolve it well, but a hint of its stellar nature can be obtained with smaller telescopes.

Whenever you fix your gaze toward the starry sky, you will always remember the fine objects in this area—especially the great clusters. MG

THE ETA CARINAE NEBULA (NGC 3372)

The Eta Carinae Nebula is one of the finest areas of nebulosity in the southern sky—a visually stunning object whose complex structure is an impressive sight. Photographs show that the nebula is very detailed and extensive, covering some 4 square degrees of sky. This nebula is carved into two glowing halves by a dark dust lane, one of many that seem to divide the nebulosity into a number of areas of glowing light. The brightest of these "islands" of light contains the dark, irregular, and elongated mass that is the Keyhole Nebula (below). It covers quite a small area and appears close to the star Eta.

The star Eta (seen at lower left) is surrounded by a bright patch of nebulosity known as the Homunculus. It gets its name from the fact that, close up, its peanut shape appears to vaguely resemble the body of a man. This gas was ejected during the 1843 outburst.

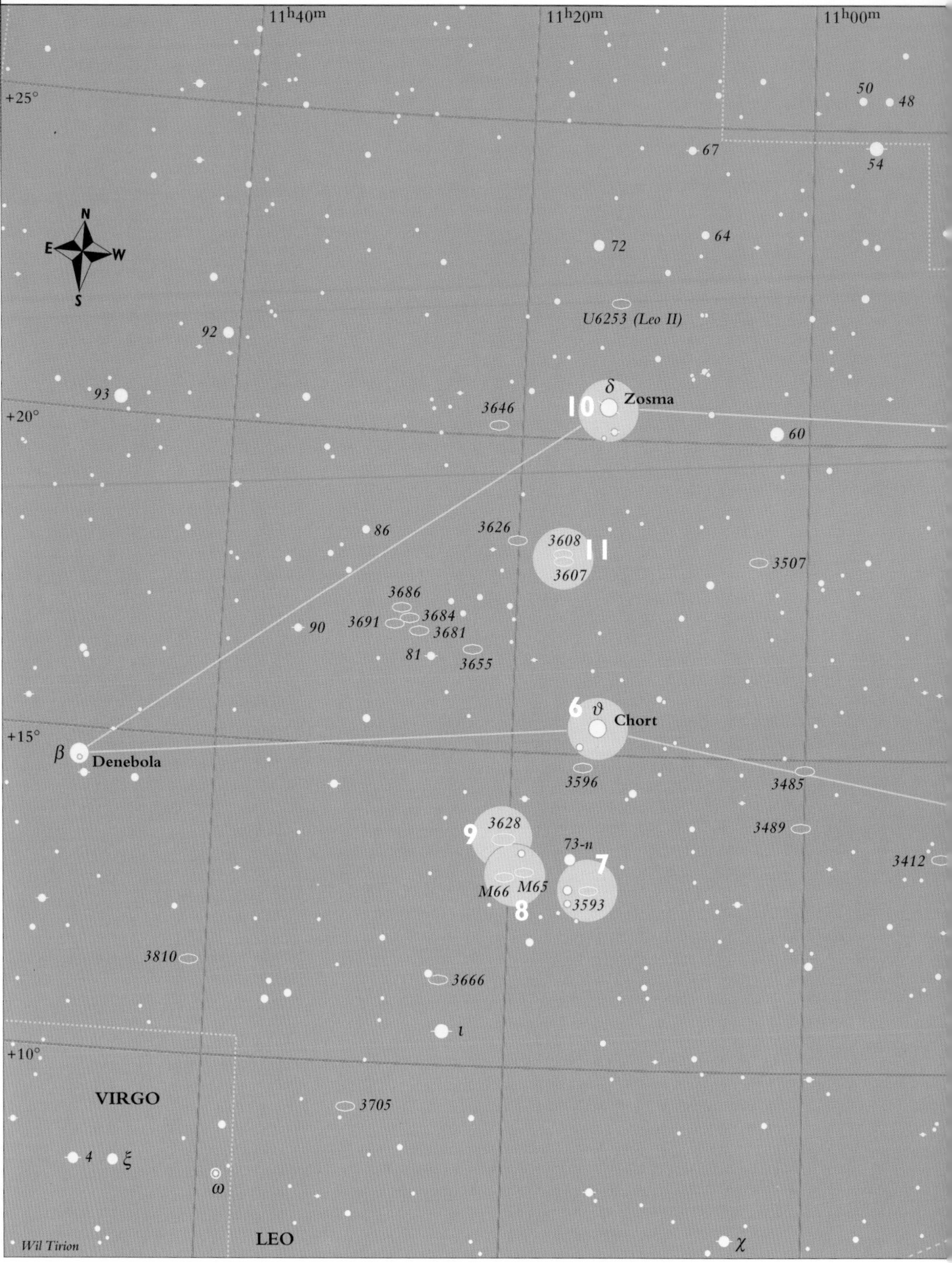
11h40m
11h20m
11h00m
+25°
+20°
+15°
+10°
N
E
W
S
50
48
67
54
64
72
U6253 (Leo II)
92
93
δ
Zosma
10
3646
60
86
3626
3608
11
3607
3507
3686
3691
3684
3681
90
81
3655
6
ϑ
Chort
β
Denebola
3596
3485
3628
9
3489
73-n
7
3412
M66
M65
8
3593
3810
3666
ι
VIRGO
3705
4
ξ
ω
Wil Tirion
LEO
χ

Roaming around the Lion

Like all ancient zodiac constellations, Leo, the Lion, lies on the Sun's path through the sky, the ecliptic. It is easily visible from both the Northern and Southern hemispheres. There are plenty of bright galaxies, double stars, and variables in this region. Leo is easy to locate because this part of the sky is dominated by the backward-question-mark, or sickle, asterism that marks the lion's head and chest, and the triangle of stars that marks its hindquarters.

Leo, from a fresco in the Villa Farnese, painted in 1575.

According to ancient Greek legend, Leo was the cave-dwelling Nemean Lion choked to death by the mighty Hercules. Because no weapon could penetrate Leo's skin, Hercules used the lion's own razor-sharp claw to remove its pelt. He made the pelt into a cloak to protect himself and carried it with him into the heavens.

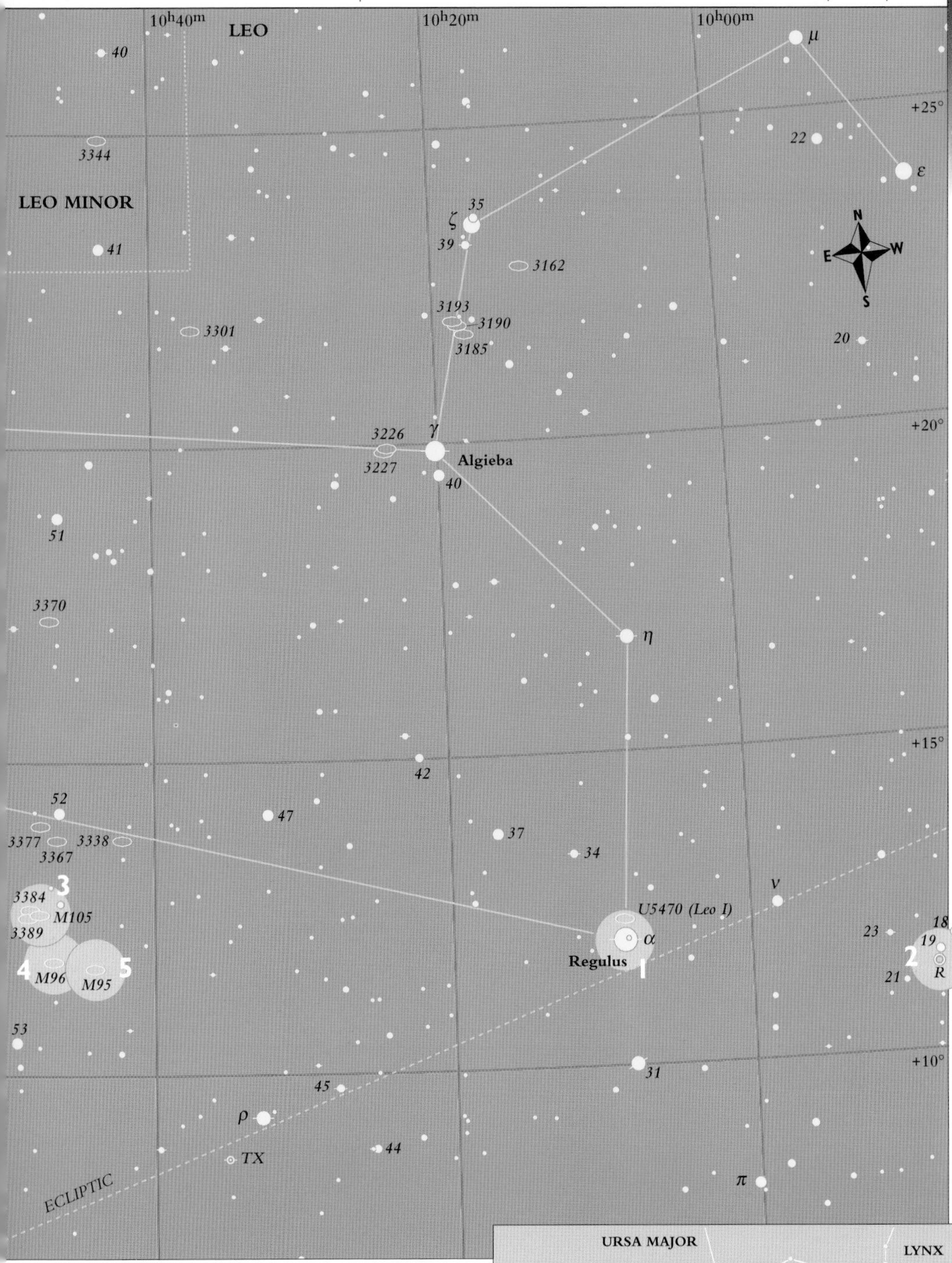
10h40m
10h20m
10h00m
LEO
LEO MINOR
+25°
+20°
+15°
+10°
N
E
W
S
40
3344
41
3301
35
ζ
39
3162
3193
3190
3185
μ
22
ε
20
3226
3227
γ
Algieba
40
51
3370
η
42
52
47
37
34
3377
3338
3367
3384
M105
3389
M96
M95
3
4
5
U5470 (Leo I)
α
Regulus
1
v
23
18
19
21
R
2
53
31
45
ρ
TX
44
π
ECLIPTIC

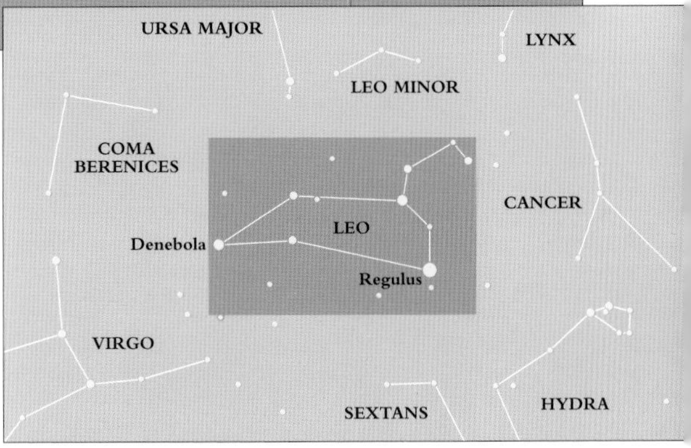
URSA MAJOR
LYNX
LEO MINOR
COMA BERENICES
CANCER
LEO
Denebola
Regulus
VIRGO
SEXTANS
HYDRA

❶ Regulus (32 Alpha [α] Leonis) This starhop begins at Regulus, the blue-white heart of Leo, and the 21st brightest star in the sky. It is located at the base of the sickle-shaped asterism that marks the lion's chest and mane and, at magnitude 1.3, is the primary of a very close multiple-star system. Regulus is 5 times the diameter of our Sun and 160 times as luminous.

The Polish astronomer Nicolaus Copernicus gave Regulus its Latin name, which means "little king," referring to the belief that this star ruled the heavens. This celestial imagery shares an apt symmetry with the commonly held notion of the lion as king of the beasts on Earth.

Hop about 5 degrees west to the magnitude 5.6 reddish star **18 Leonis** and magnitude 6.4 yellow-white **19 Leonis**. Use these two stars to locate R Leonis in the same low-power eyepiece view.

Several galaxies in Messier's list, including M65 (left), are found in Leo (above).

❷ R Leonis In 1782, the Polish astronomer J. A. Koch first recorded the variability of this pulsating red giant Mira-type variable, one of the earliest to be discovered. The magnitude of R Leonis cycles from 4.4 down to fainter than 11th magnitude. Although it is normally a sharp red star, R Leonis can appear a deep shade of purple during its 312-day cycle.

❸ M105 (NGC 3379) Return to Regulus, then hop about 10 degrees east to find **52 Leonis**, a magnitude 5.4 yellowish star. About 1½ degrees south of 52 Leonis is a triangle of hazy light formed by three galaxies—M105 and NGC 3384 and 3389. At magnitude 9.2, M105 is the brightest and largest of the three. Typical of elliptical galaxies, M105 appears as a fuzzy ball that you cannot focus into a sharp point of light.

NGC 3384 The second galaxy in the group is NGC 3384, also an elliptical, which appears as an elongated oval with a bright nucleus.

NGC 3389 The third galaxy, NGC 3389, is fainter and harder to find. Appearing as a pale, thin streak of light, it is a magnitude 11.8 spiral that seems to glow evenly from edge to edge.

❹ M96 (NGC 3368) About 1 degree south of M105 is the spiral galaxy M96. This silver-gray beauty appears in a relatively starless field, so it is easy to spot. The intense core of the galaxy is much sharper and brighter than the wispy wreath of its arms.

❺ M95 (NGC 3351) This magnitude 9.7 barred spiral galaxy, adjacent to M96, has a fairly bright core surrounded by wispy arms. Its nucleus appears almost stellar, with a fuzzy fringe around it. M95 is not as big or as bright as M96. Even in scopes larger than 10 inches (250 mm), it is difficult to see the faint arms of the galaxy, but a 4 inch (100 mm) shows the gradual brightening toward the core.

❻ Chort (70 Theta [θ] Leonis) Sweep about 8 degrees northeast of M95 to blue-white Chort. This stunning jewel marks the hips of the Lion and is the western apex of the triangle of stars marking the rear haunches of Leo.

❼ NGC 3593 Nudge about 3 degrees south of Chort to find NGC 3593, a magnitude 11.3 spiral galaxy with a bright and elongated core. The outline of the galaxy's arms is quite diffuse, blending into the background at the edges.

❽ M65 (NGC 3623) This beautiful magnitude 9.9 spiral galaxy is at least 200,000 light-years distant from M66, although the famous pair appear in the same low-power eyepiece view.

The edge-on spiral galaxy NGC 3628 (above) is located quite close to M65 (left) and clearly shows a warp effect in its disk.

M66 (NGC 3727) At magnitude 9.9, M66 is brighter and a bit shorter than M65. M66 is another spiral galaxy, with a finger of light extending from its southern tip. Many of its dark areas can be seen with a 4 inch (100 mm) scope.

9 NGC 3628 This elongated, pencil-slim spiral galaxy is at the northern apex of the richly studded triangle that includes M65 and M66. The edges are quite grainy on this magnitude 9.9 object. It is halfway between two magnitude 10 stars to the north and south, but because of its low surface brightness and being seen edge-on, it is not easy to locate.

10 Zosma (Delta [δ] Leonis) Sweep about 7 degrees northwest of NGC 3628 to blue-white Zosma, a magnitude 2.6 main-sequence star at the heart of a very open multiple system. These stars are easy to separate in any size scope because the magnitude 8.6 companion is located 191 arc-seconds from the primary.

11 NGC 3607 Move back some 2½ degrees south of our last stop to find the elliptical galaxy NGC 3607, a magnitude 10.9 object that looks like an out-of-focus star through a telescope.

NGC 3608 At about a magnitude fainter than NGC 3607, the elliptical galaxy NGC 3608 appears smaller, although astronomers believe that these galaxies are actually about the same size. NGC 3608 is roundish with a faint stellar core.

This is just a sampling of the night-sky wonders to be found in Leo. Additional faint galaxies, shown on the main sky chart, are within the body of the celestial king of beasts, waiting for you to explore them. RG

THE MESSIER MARATHON

In the Northern Hemisphere, a window of opportunity opens around the time of the vernal equinox, every March, when all 110 Messier objects can be observed on one night.

French astronomer Charles Messier and his collaborator, Pierre Méchain, were engaged in a quest to discover comets and, in the process, listed many other wonderful celestial objects in order to avoid confusing them with comets.

To complete a Messier Marathon, begin with the spiral galaxy M74 in Pisces, in the western sky at dusk. During the night, sweep from one Messier object to the next, ending with M30 in Capricornus, shortly before the onset of morning twilight.

The beautiful spiral galaxy M66 (left) is just one of the marathon objects to aim for while you are in Leo.

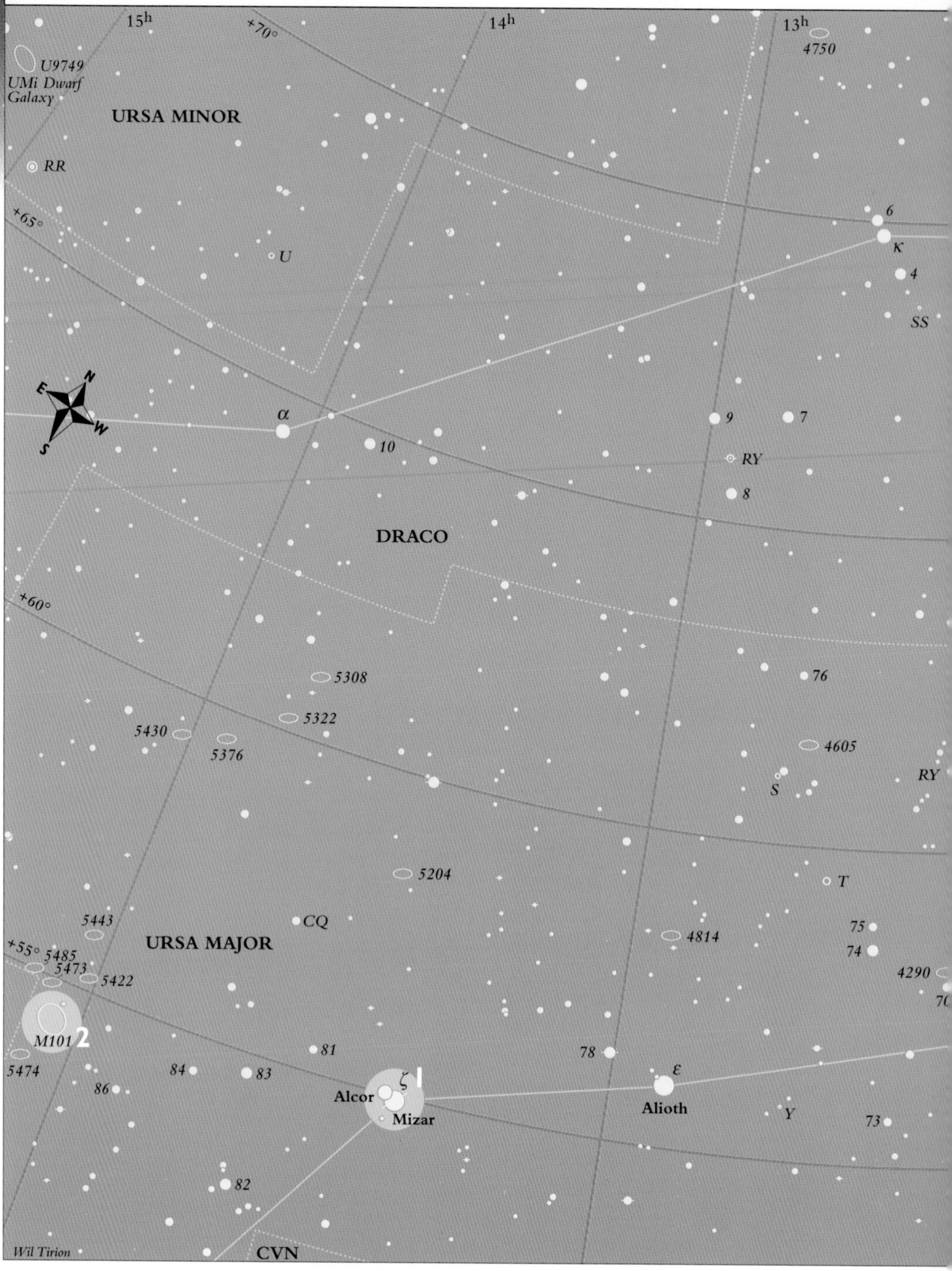
15h
14h
13h
+70°
+65°
+60°
+55°
U9749
UMi Dwarf Galaxy
URSA MINOR
RR
U
4750
6
κ
4
SS
N
E
S
W
α
10
9
7
RY
8
DRACO
5308
5322
5430
5376
76
4605
S
RY
5204
T
CQ
5443
URSA MAJOR
5485
5473
5422
4814
75
74
4290
M101
2
5474
81
78
ε
84
83
86
Alcor
ζ
1
Mizar
Alioth
Y
73
82
Wil Tirion
CVN

A Hop around the Big Dipper

One of the first star patterns that Northern Hemisphere skywatchers usually learn to recognize is the Big Dipper, in the constellation of Ursa Major, the Great Bear. Unfortunately for observers in mid-southern latitudes, this famous asterism is not visible. The seven bright stars of the circumpolar bowl and handle are part of what is known as the Ursa Major Moving Group, the closest open star cluster to us.

The Dipper stars are rich in mythology and have been known by various names throughout history. The lore that many cultures created for this constellation is based on that of a bear. However, the ancient Egyptians considered the asterism to be either a hippopotamus or a Nile River boat for the god Osiris. Ancient Romans saw the Dipper as seven oxen pulling a plough.

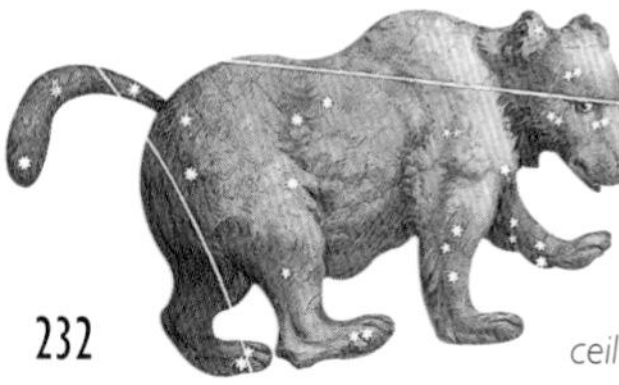

The Great Bear, depicted in a ceiling fresco from the Vatican.

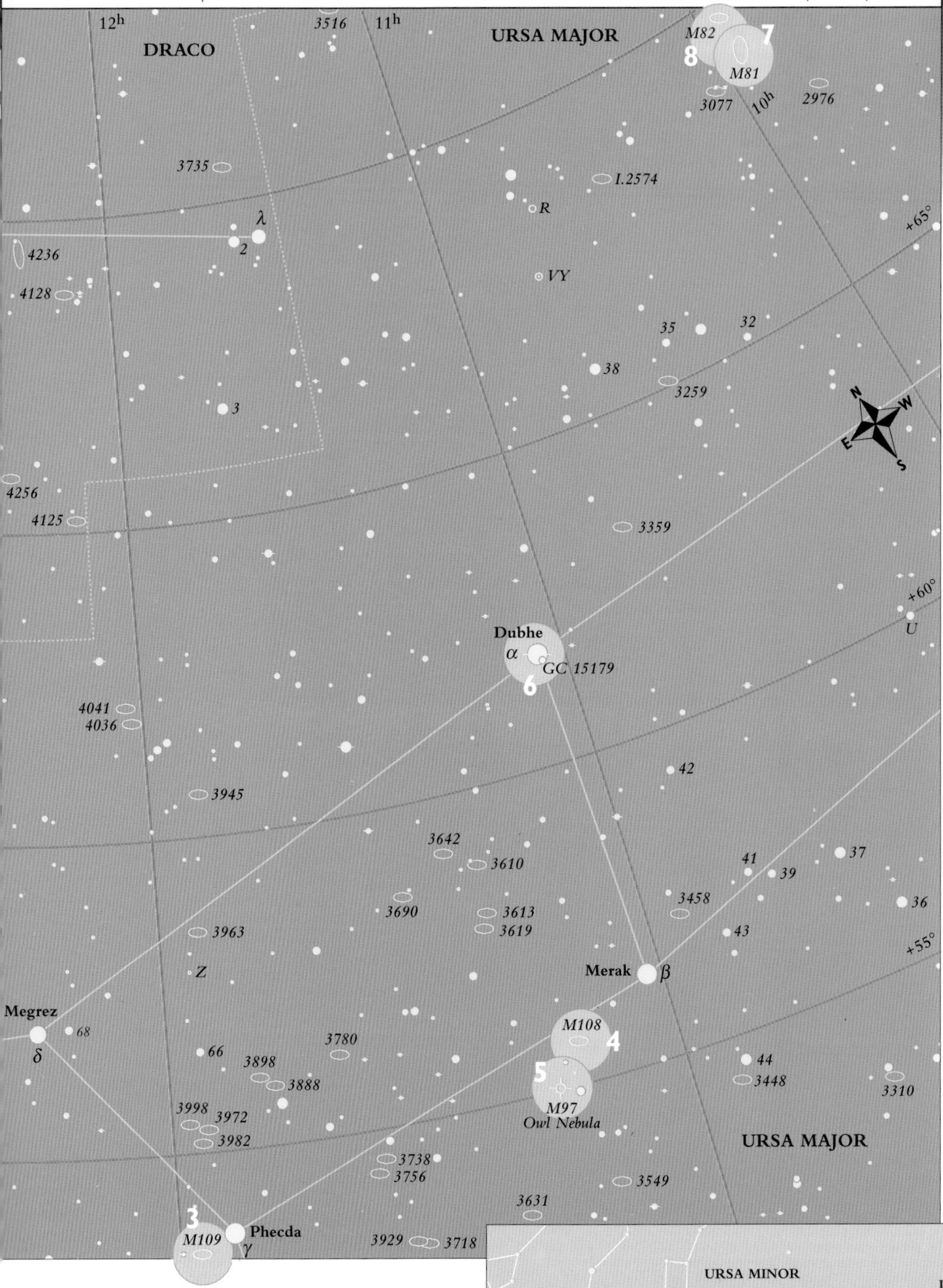
12h
11h
10h
DRACO
URSA MAJOR
M82
M81
7
8
3516
3077
2976
3735
I.2574
R
VY
λ
2
4236
4128
35
32
38
3259
3
+65°
N
W
E
S
4256
4125
3359
+60°
U
Dubhe
α
GC 15179
6
4041
4036
42
3945
3642
3610
41
39
37
3458
36
3690
3613
3619
43
3963
Z
Merak
β
+55°
Megrez
δ
68
66
3780
M108
4
3898
3888
5
44
3448
3310
3998
3972
3982
M97
Owl Nebula
URSA MAJOR
3738
3756
3549
3631
3
M109
Phecda
γ
3929
3718

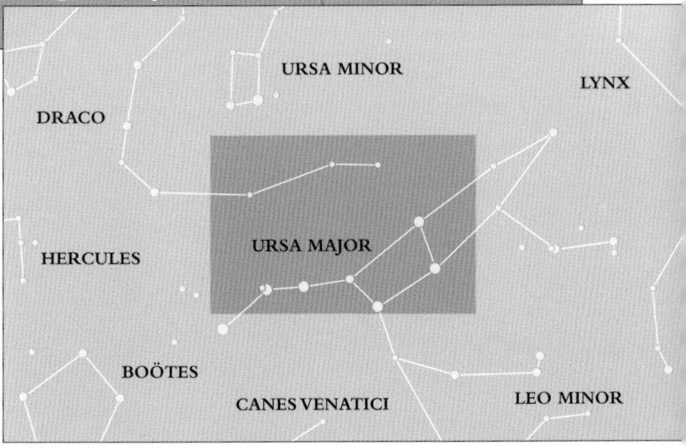
URSA MINOR
LYNX
DRACO
HERCULES
URSA MAJOR
BOÖTES
CANES VENATICI
LEO MINOR

❶ Mizar (79 Zeta [ζ] Ursae Majoris) and Alcor (80 Ursae Majoris) This tour begins at Mizar and Alcor, the famous pair that makes up the apparent wide double star in the middle of the Dipper handle. If your eyesight is sharp, you can separate the components with the naked eye, since they are 12 arcminutes apart. A pair of binoculars provides a great view of Mizar and Alcor, and with a small telescope, you can see that Mizar is actually a double star itself. Its 4th magnitude companion, known as Mizar B, lies about 14 arcseconds away. In 1650, Italian astronomer Giovanni Riccioli identified Mizar's two stars as the first true binary-star system.

M101, our next target on this starhop, is one of three Messier objects commonly known as the Pinwheel Galaxy. The other two "Pinwheels" are M99 (see p. 238) and M33 (see p. 187).

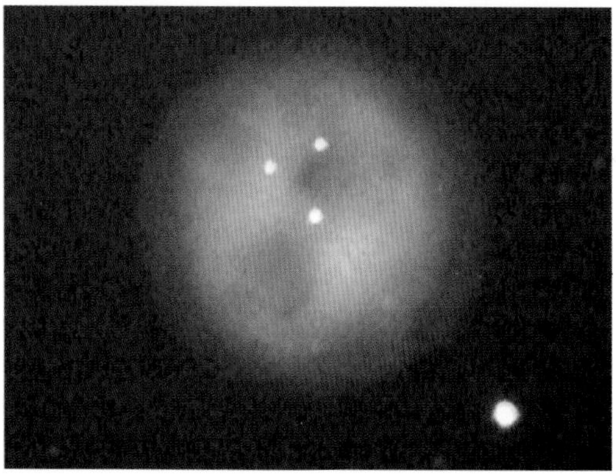

The Big Dipper (above) is home to a variety of unique objects including the Owl Nebula (left), a very large planetary nebula.

❷ The Pinwheel Galaxy (M101, NGC 5457) This gem of a spiral galaxy lies about 5 degrees east of Mizar and Alcor. M101 is one of the largest and finest face-on spirals visible, and although it is fairly bright, at around magnitude 9, it is also diffuse, making it hard to spot, even with a 10 inch (250 mm) scope. It is best viewed in a dark sky using low magnification and a wide field in order to detect the knotty arms loosely extending outward from the core.

Pierre Méchain discovered M101 in 1781. He also found an object sometimes listed as M102, considered to be a missing Messier object. However, in a letter dated 1783, Méchain wrote that he thought M102 was just a re-sighting of M101.

❸ M109 (NGC 3992) Hop over to the southeastern corner of the Dipper's bowl to **Phecda (64 Gamma [γ] Ursae Majoris)** and use this magnitude 2.4 white star, "the thigh of the Bear," to locate the barred spiral M109 about 1/4 degree east of it. The arms, visible through the extended grayish nebulous haze of the galaxy, are best seen on a very dark night using a 10 inch (250 mm) telescope or larger.

❹ M108 (NGC 3556) To find M108, hop first to the bright, white **Merak (48 Beta [β] Ursae Majoris)**. If you draw an imaginary line between Merak and Phecda, M108 is about 1 1/2 degrees southeast from Merak along this line. M108 is a milky-white, edge-on spiral galaxy that appears flat, with no sign of the central bulge typical of most other spiral galaxies. M108 glows at magnitude 10.1. It has a mottled texture with about four bright spots visible within its arms. Visible through a 10 inch (250 mm) telescope, an elongated dust lane runs through the long axis of the galaxy.

❺ The Owl Nebula (M97, NGC 3587) The Owl Nebula is found about 1/2 degree farther southeast from M108. This large, but diffuse, magnitude 12 planetary nebula was discovered by Pierre Méchain in March 1781. Although the nebula appears to us about the same size as Jupiter, it is estimated to be 3 light-years in diameter, making it one of the largest planetary nebulas. You should be able to see it through a 4 inch (100 mm) scope in a dark sky, but because of its low surface brightness you need a 10 inch (250 mm) telescope or larger to gain even a hint of the owl's dark eyes.

This CCD image of M81 clearly reveals its classic spiral arms surrounding a bright central nucleus.

6 Dubhe (50 Alpha [α] Ursae Majoris) To the Arabs, this magnitude 1.8 golden-yellow close double star was *Thar al Dubb al Akbar*, meaning "the back of the Great Bear." Dubhe and Merak form the western end of the Dipper's bowl and point the way to Polaris, the North Star. Dubhe also has a magnitude 7 bluish companion, **GC 15179**, lying 6.3 arcminutes from the primary, which is easily separated in small scopes. A line between Phecda and Dubhe will point you toward our next targets, the galaxies M81 and M82, which are located about 10 degrees northwest of Dubhe.

7 M81 (NGC 3031) The bright core of M81 appears distinctly elliptical, at the center of a strikingly symmetrical overall structure that is easily visible with binoculars. Lurking within the bright core are most of the galaxy's 250 billion stars. The two spiral arms of the galaxy appear to be quite diffuse because they have fewer stars tracing them out.

M81 and M82 lie about 38 arcminutes apart, easily appearing in the same low-power telescope or binocular field. They are considered by many to be one of the finest pairs for observers.

8 M82 (NGC 3034) Smaller and dimmer than M81, M82 is classified as a peculiar galaxy (see Box). In scopes smaller than 15 inches (380 mm), M82 looks like an unusually amorphous, edge-on spiral, although a spiral structure has not been confirmed, even in photographs taken at the largest Earth-bound telescopes or with the Hubble Space Telescope.

Sweep your scope in and around the bowl of the Big Dipper to observe and enjoy the many other faint galaxies waiting to be found. RG

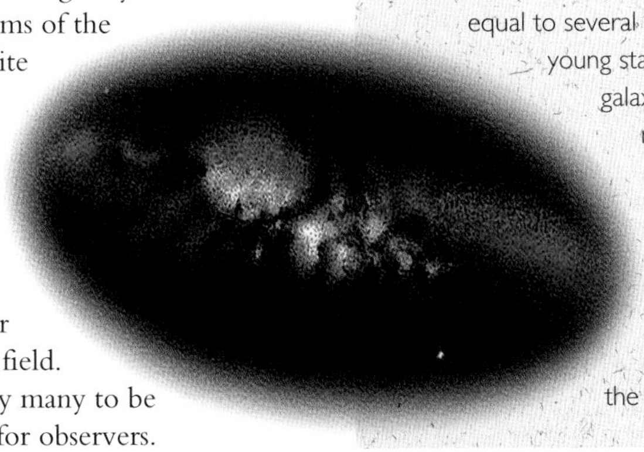

STAR OUTBURST IN M82

M82 is now thought to be a nearby example of a starburst galaxy. Astronomers attribute the burst of star formation in M82 to a possible encounter with M81, its companion spiral galaxy, approximately 100 million years ago. This may have severely disrupted M82's gas clouds, igniting the starburst that involved a mass of material equal to several million Suns. The massive young stars that formed within the galaxy's nucleus gave rise to raging winds of hot gas, which, combined, made a powerful galactic wind. This wind created the filamentary structures (left) that were once thought to be the products of an explosion.

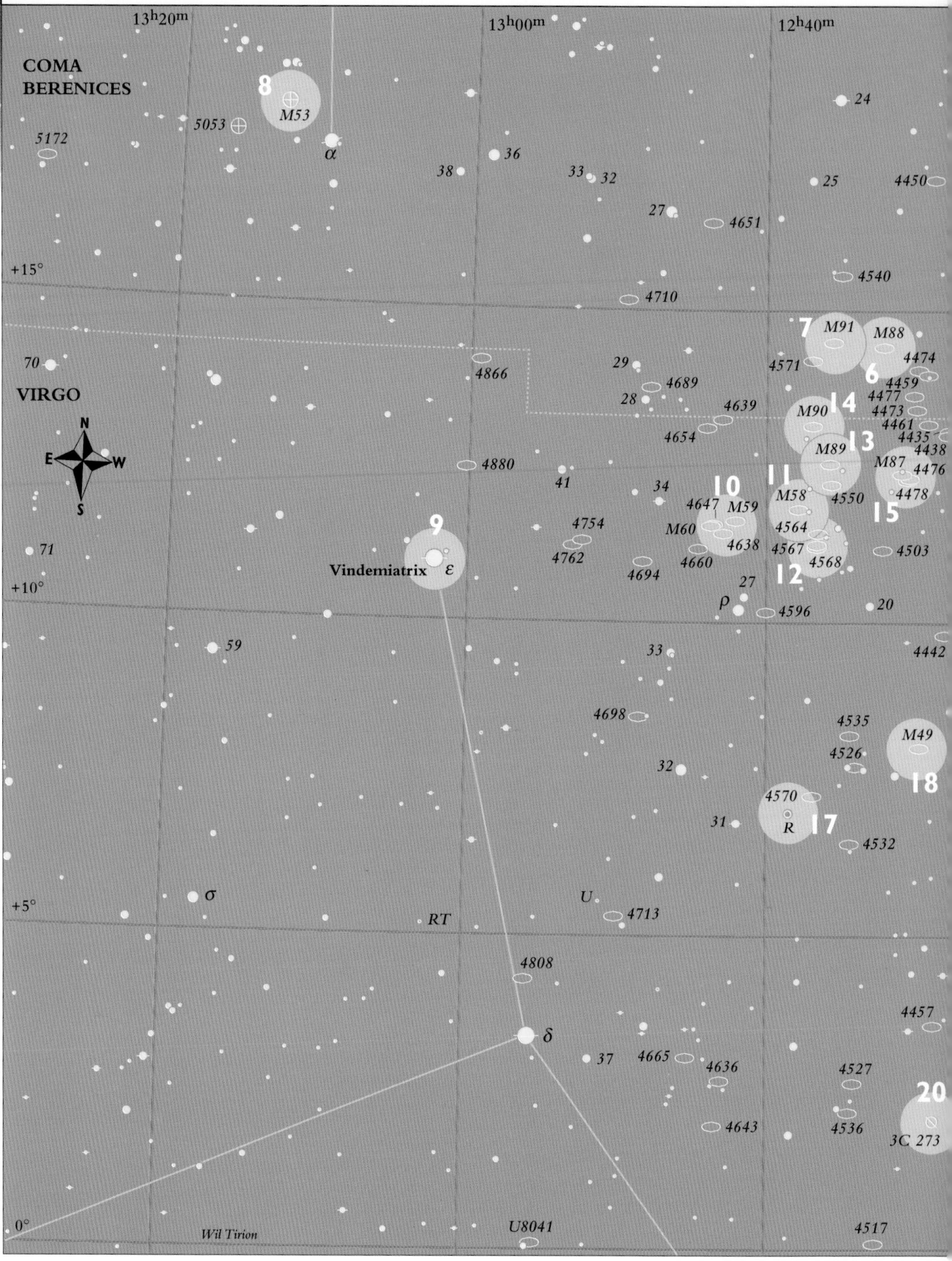
13h20m
13h00m
12h40m
COMA BERENICES
8
M53
5053
5172
α
38
36
33
32
24
25
4450
27
4651
+15°
4540
4710
7
M91
M88
70
VIRGO
4866
29
4689
4571
4474
6
4459
4477
28
4639
M90
14
4473
4461
4654
4435
M89
13
4438
N
E
W
S
4880
M87
4476
41
34
10
11
M58
4550
4478
4647
M59
15
9
4754
M60
4564
71
4638
4567
4503
4762
4660
4568
Vindemiatrix
ε
4694
27
12
+10°
ρ
4596
20
59
33
4442
4698
4535
M49
4526
32
18
4570
31
R
17
4532
σ
+5°
U
RT
4713
4808
δ
4457
37
4665
4636
4527
20
4643
4536
3C 273
0°
Wil Tirion
U8041
4517

The Virgo-Coma Cluster of Galaxies

From April through June, skywatchers are afforded the best opportunity to observe about 250 of the 3,000 "island universes" located within the Virgo Cluster of Galaxies. This cluster extends north and south through much of the constellations of Virgo, the Virgin, and Coma Berenices, the Hair of Berenice. This region is clearly visible from both the Northern and Southern hemispheres.

The majority of galaxies in the Virgo Cluster are some 65 million light-years from us. This means that the light from these distant multitudes of stars has been traveling, uninterrupted, through space for approximately 65 million years. Throughout this sky tour, you will be looking back into time and seeing what these galaxies were like about the time the dinosaurs died off on Earth. Amazing.

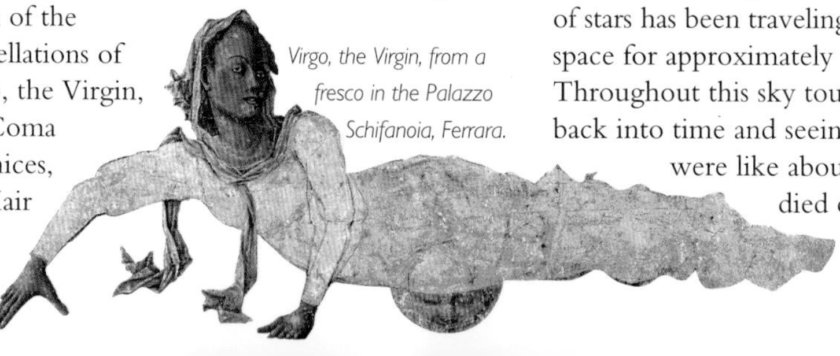

Virgo, the Virgin, from a fresco in the Palazzo Schifanoia, Ferrara.

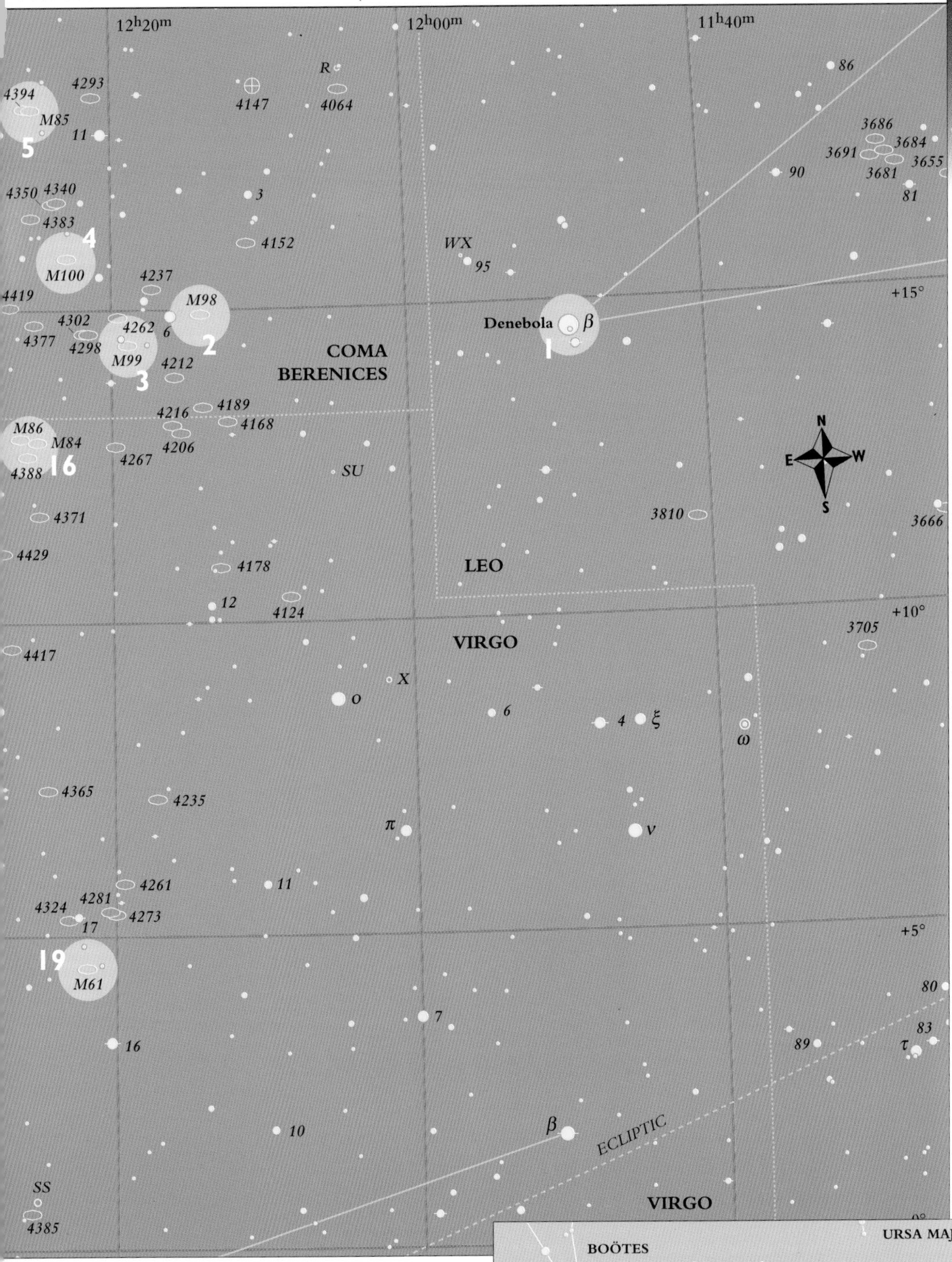

12h20m
12h00m
11h40m
COMA
BERENICES
LEO
VIRGO
Denebola
ECLIPTIC
+15°
+10°
+5°
M85
M100
M98
M99
M86
M84
M61
4394
4293
4147
4064
4350
4340
4383
4152
4237
4419
4302
4377
4298
4262
4212
4216
4189
4168
4206
4267
4388
4371
4429
4178
4124
4417
4365
4235
4261
4281
4324
4273
3686
3684
3691
3655
3681
3810
3666
3705
SU
WX
SS
4385
N
E
W
S

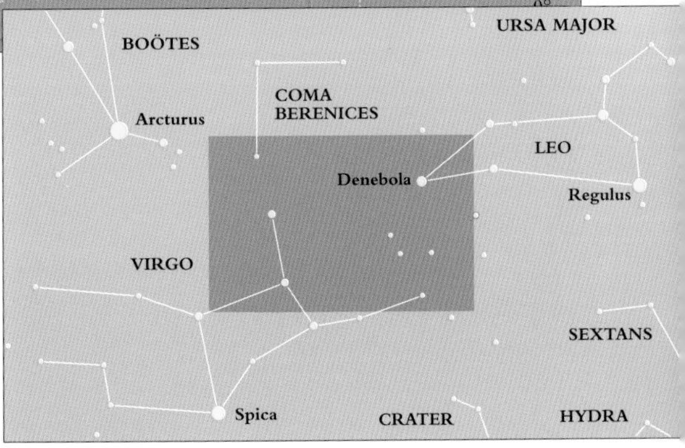

BOÖTES
URSA MAJOR
COMA
BERENICES
Arcturus
LEO
Denebola
Regulus
VIRGO
SEXTANS
Spica
CRATER
HYDRA

❶ Denebola (94 Beta [β] Leonis) To locate the galaxies on the northwestern side of Virgo and Coma Berenices, aim your scope toward our starting point, the easy-to-locate Denebola, a magnitude 2.1 blue-white giant on the eastern edge of Leo. On this starhop, we will be looking directly out of the disk of the Milky Way Galaxy, where the lack of stars, dust, and gas allows observation of objects located millions of light-years beyond our galaxy's bounds.

To early Arab astronomers, Denebola was called *Al Dhanab al Asad*, which denoted its position as the tuft of hair on the lion's tail and from which its modern name is derived.

From Denebola, hop about 6½ degrees east to the magnitude 5 whitish star **6 Comae**, to help you find the spiral galaxies M98 and M99, which all appear within the same finder view.

The center of the Virgo Cluster (above) includes M86 and M84. M100 (left) lies just a few degrees north of the center.

2 M98 (NGC 4192) On the evening of 15 March 1781, French astronomer and surveyor Pierre Méchain discovered M98, M99, and M100. The edge-on spiral galaxy M98 appears as a thin patch of light with a slight bulge visible near its nucleus. Although M98 may appear to observers as a member of the Virgo Cluster of Galaxies, it is actually located about halfway between the Sun and the cluster itself.

3 The Pinwheel Galaxy (M99, NGC 4254) Nudge your scope 1 degree southeast of 6 Comae to bring M99 into view. In 4 inch (100 mm) scopes or larger, the sweeping arms in this roundish face-on galaxy will become visible. At 50,000 light-years in diameter, M99 is about half the size of the Milky Way Galaxy.

4 M100 (NGC 4321) Return to 6 Comae, then hop about 2 degrees northeast. Two 5th magnitude stars heading in the same direction will point you toward M100—a large, face-on spiral galaxy glowing at magnitude 9.6. At 7 arcminutes across, it presents the largest apparent size of any galaxy in the Virgo Cluster. In 8 inch (200 mm) telescopes or smaller, it can resemble a dim globular cluster, appearing circular with very faint arms. It is relatively easy to see because of its brilliant star-like core.

5 M85 (NGC 4382) Almost 2 degrees north of M100 is the magnitude 4.7 yellow double star, **11 Comae**. Use it to direct you to M85, about 1 degree northeast. A circular patch of light glowing at magnitude 9.2, M85 is estimated to be 100 to 400 billion times more massive than the Sun.

NGC 4394 In the same low-power telescope eyepiece as M85 is the magnitude 10.9 barred spiral galaxy NGC 4394. This small, round spiral galaxy is about 3.9 arcminutes across with a brightly shining nucleus that appears almost stellar.

6 M88 (NGC 4501) The easiest way to find M88 is to return to 6 Comae then hop back to M99. M88 and our next target, M91, are at the same declination as M99, so with an equatorially mounted telescope, center M99 in your eyepiece, then turn off the motor drive, if your scope is equipped with one, and wait 14 minutes or so for M88 to drift into your eyepiece field of view. M88 appears elongated through a telescope because we are looking at a spiral galaxy tilted to our line of sight. Its core seems almost stellar, shining brightly for such a distant object.

7 M91 (NGC 4548) Leave your scope fixed for three more minutes after sighting M88, and the barred spiral galaxy M91 will also make an appearance, slipping quietly into your 1 degree

field. The slightly fainter M91 has low surface brightness, so it is more difficult to see than M88. The thin bar of its barred spiral structure is only visible in 12 inch (300 mm) telescopes or larger.

8 M53 (NGC 5024) Shift almost 9 degrees northeast of M91 to **42 Alpha [α] Comae.** One degree farther northeast is the magnitude 7.7 globular cluster M53, lying 65,000 light-years from us. In February 1775, German astronomer Johann Bode discovered this small, roundish globular cluster, describing it as "a new nebula, appearing through the telescope as round and pretty lively." At 3 arcminutes in diameter, M53 is a relatively easy target for any size telescope. The glowing halo of stars encircling the tightly compacted core of the cluster is a delight through a 4 inch (100 mm) telescope.

9 Vindemiatrix (47 Epsilon [ε] Virginis) About 8 degrees south of M53 lies the easy-to-locate magnitude 3 star Vindemiatrix. This yellow-white giant glows as a bright light in the night sky. From ancient Roman times, the early morning rising of Vindemiatrix in late August marked the time to commence harvesting the grapes. Its name is derived from the Latin word *Vindemitor*, which means "grape gatherer."

10 M60 (NGC 4649) Shift about 4½ degrees west and slightly north of Vindemiatrix to locate M60, one of the largest elliptical galaxies known. A halo seems to surround the bright core of this magnitude 8.8 galaxy. The mass of M60 is estimated to be equal to an amazing 1 trillion Suns, making it comparable in mass to M49.

NGC 4647 This spiral galaxy and M60 appear to be interacting even though NGC 4647 is moving away from the Sun, while M60 is in approach. Although faint, at magnitude 12, NGC 4647 has a bright core and appears roundish and slightly smaller than M60.

M59 (NGC 4621) Also in the same low-power eyepiece view is M59, a magnitude 9.8 elliptical galaxy that appears oval with a bright core.

The Coma Berenices Cluster (top) is much more distant than M99 (right) in the Virgo Cluster or M53 (center) in the Milky Way.

M87 (above), a giant elliptical galaxy surrounded by thousands of globular clusters, lies near the heart of the Virgo Cluster of Galaxies.

⑪ **M58 (NGC 4579)** Swing about 1 degree west of M59 to M58, a magnitude 9.8 barred spiral galaxy. In an 8 inch (200 mm) scope, it appears elongated and faint, due to its low surface brightness. M58 is almost a twin, in terms of its size and mass, to the Milky Way Galaxy.

⑫ **The Siamese Twins (NGC 4567/68)** Nudge your scope about 1/2 degree southwest of M58 to the Siamese Twins. This unusual pair consists of two seemingly attached galaxies, though at magnitude 10.8, NGC 4568 is slightly brighter than its "twin," and appears larger.

⑬ **M89 (NGC 4552)** M89 is an elliptical galaxy almost 1 1/2 degrees north of the Siamese Twins. It appears as a round, glowing patch with a bright center through an 8 inch (200 mm) scope.

⑭ **M90 (NGC 4569)** Continuing about 1/2 degree northeast of M89 is the graceful M90, appearing as an elongated oval with a bright core. This beautiful magnitude 9.5 spiral galaxy is about 80,000 light-years in diameter. You should also be able to see M90's dust lanes through a 10 inch (250 mm) telescope or larger.

⑮ **M87 (NGC 4486, 3C 274)** Hop 1 1/2 degrees southwest of M90 to bring M87 into view. This is a giant elliptical galaxy surrounded by more than 4,000 globular clusters. It is estimated to contain more than an incredible 1 trillion solar masses. We see it glow at magnitude 9.6, as a bright, fuzzy patch of light.

What you see is not all you get with this monster in the sky. Images taken by the Hubble Space Telescope reveal evidence for a black hole in the core of M87, surrounded by a mass of stars

VIRGO'S BIG HEART, M87

Among more than 1,000 systems within the vast Virgo Cluster of Galaxies is the giant elliptical galaxy M87, one of the largest known. It covers an area of about 4 x 3 arcminutes and is surrounded by an extraordinary collection of more than 4,000 globular clusters. In a 6 inch (150 mm) scope, you will see it clearly as a magnitude 9 glow with a definite nucleus, but in photographs taken with professional equipment, it is possible to see a jet of very hot gas emanating from the nucleus. This was discovered by Heber Curtis at the Lick Observatory in 1918. The jet is about 4,000 light-years long and is now known to emit strong radio waves (right) and X-rays. Today, the jet is believed to be a high-speed beam of particles, probably associated with a central black hole.

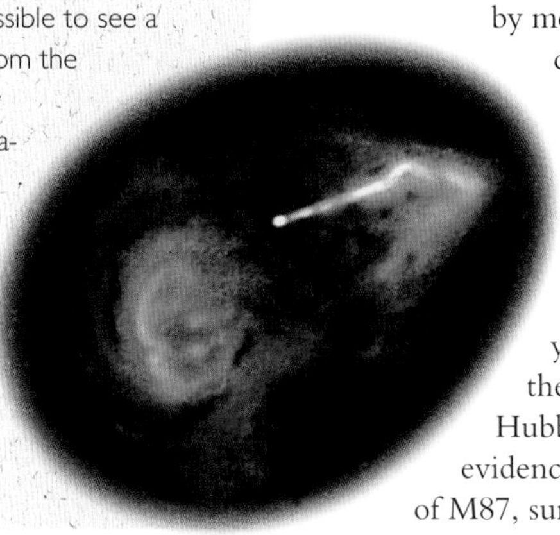

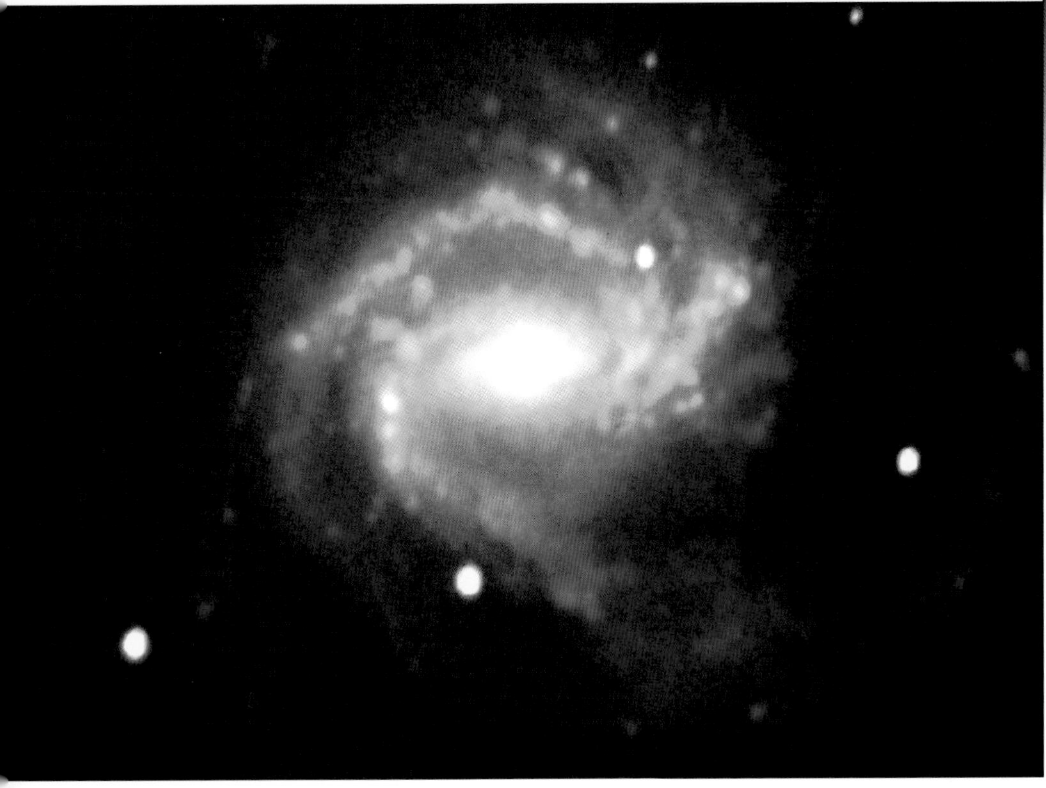

packed hundreds of times more tightly than in a normal elliptical galaxy. This galaxy is also known as Virgo A, because it is one of the earliest sources of radio waves discovered in the sky.

⑯ **M86 (NGC 4406)** Located a little more than 1 degree northwest of M87 is M86, at magnitude 9.8. This slightly oval galaxy brightens a little toward its core. M86 may be as close to us as 20 million light-years, or as far away as 42 million. Despite the general expansion of the universe and unlike the majority of galaxies in the Virgo Cluster, which are moving away from us, M86 is moving toward us.

M84 (NGC 4374) Within the same telescope-eyepiece view and lying directly west of M86 is another elliptical galaxy—M84. It may be located on the western edge of the core of the Virgo Cluster. At magnitude 9.9, M84 appears near-circular and, like M86, has a noticeable brightening toward its nucleus.

⑰ **R Virginis** Swing your scope some 4½ degrees southeast of M86 and M84 toward the red giant Mira-type variable R Virginis. The magnitude of R Virginis ranges from 6.2 down to 12 over a period of about 145 days. It was discovered to be a variable star in 1809.

⑱ **M49 (NGC 4472)** Our next stop is the giant elliptical galaxy M49, lying about 2 degrees northwest of R Virginis. It appears to us as a 9th magnitude oval with an almost even luminosity across most of its surface.

⑲ **M61 (NGC 4303)** Moving on about 3 degrees southwest of M49, we come to the wide, yellow double star **17 Virginis**. About ½ degree farther south from 17 Virginis is the large, magnitude 9.7 face-on spiral galaxy M61. The arms and dust lanes of this galaxy are visible through a 10 inch (250 mm) telescope.

⑳ **3C 273** Coming to the end of our journey through the Virgo Cluster is the really neat object, 3C 273, about 2½ degrees southeast of M61. This is the brightest known quasar and probably the most distant object visible in an 8 inch (200 mm) telescope. It appears as just one of many faint stars through a telescope eyepiece.

There are many more interesting galaxies to observe in the Virgo-Coma area, dozens more than we could possibly cover here, so, chart in hand, move on to enjoy discovering for yourself some of the other "island universes." RG

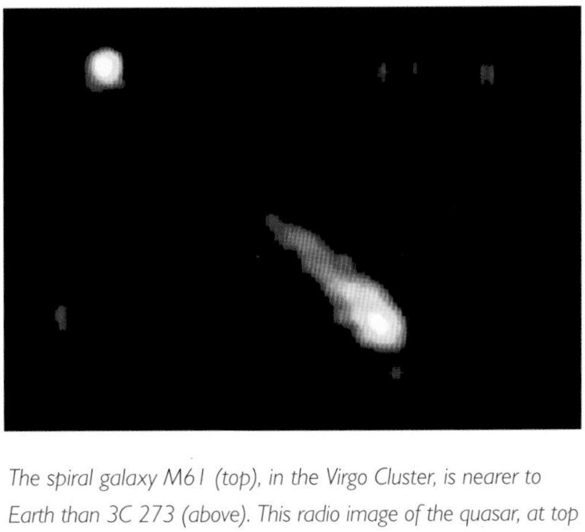

The spiral galaxy M61 (top), in the Virgo Cluster, is nearer to Earth than 3C 273 (above). This radio image of the quasar, at top left, shows a jet of material, at center right, emerging from its core.

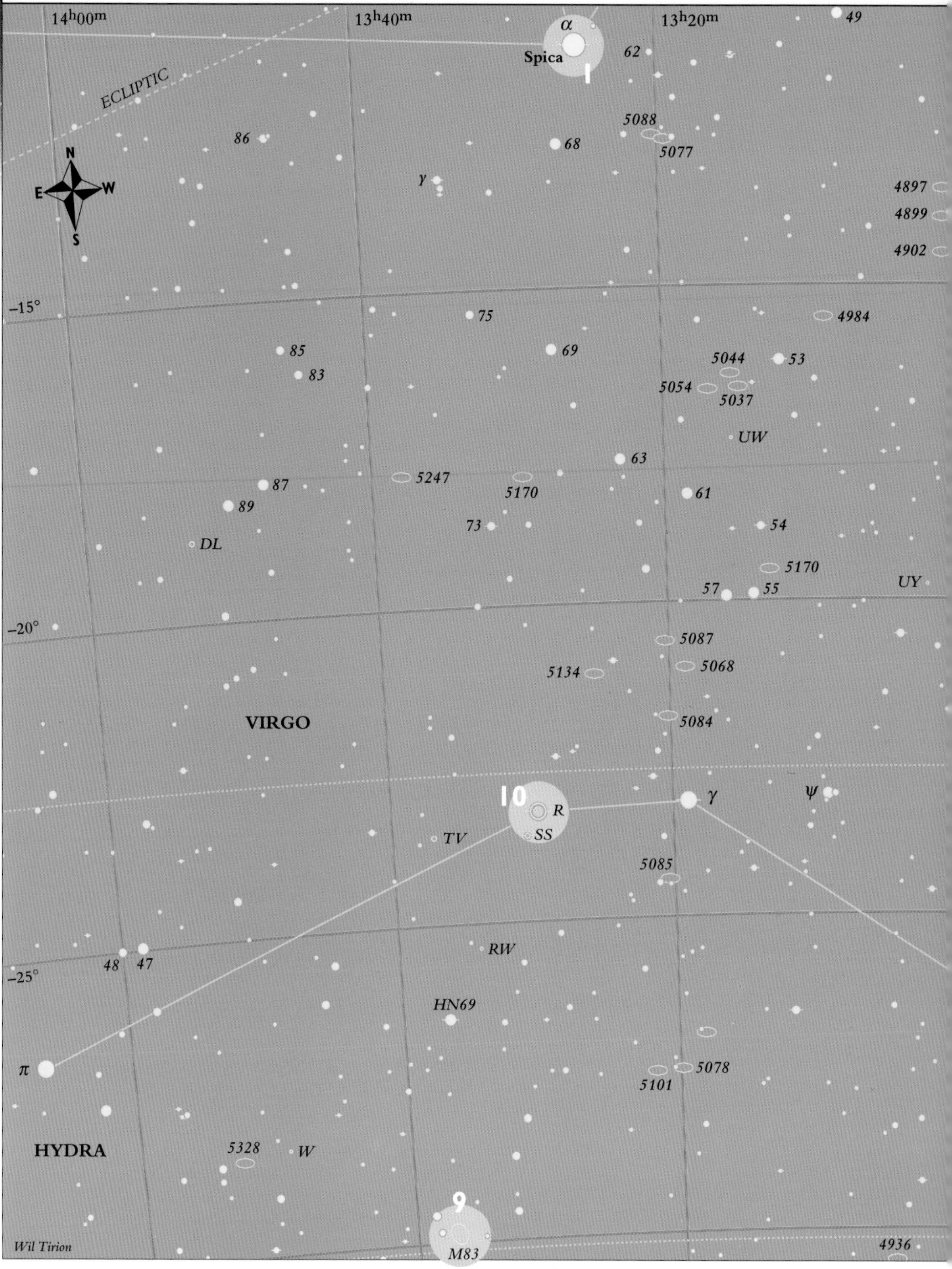
14h00m
13h40m
13h20m
α
Spica
62
49
1
ECLIPTIC
86
5088
68
5077
N
E
W
S
γ
4897
4899
4902
–15°
75
4984
85
83
69
5044
53
5054
5037
UW
63
5247
5170
87
89
61
73
54
DL
5170
UY
57
55
–20°
5087
5068
5134
VIRGO
5084
10
R
SS
γ
ψ
TV
5085
RW
–25°
48
47
HN69
π
5078
5101
HYDRA
5328
W
9
M83
4936
Wil Tirion

A Stroll around the Sombrero

Although visible from both hemispheres, this part of the sky is in the southern half of the celestial sphere, so is a little better suited to Southern Hemisphere observers. However, fine views can be had in the evenings from April to June from most parts of the world. Some 30 degrees south of the Virgo Cluster of Galaxies is the point where the Virgin meets the eastern end of the long constellation of Hydra, the Sea Serpent. Hydra and Virgo are the two largest constellations in the celestial sphere, but the third constellation on our chart for this starhop is one of the smallest—Corvus, the Crow. In mythology, Corvus was sent by Apollo to collect water with a cup. Instead, it returned with a watersnake. All three were banished to the sky, with the cup becoming the nearby constellation of Crater.

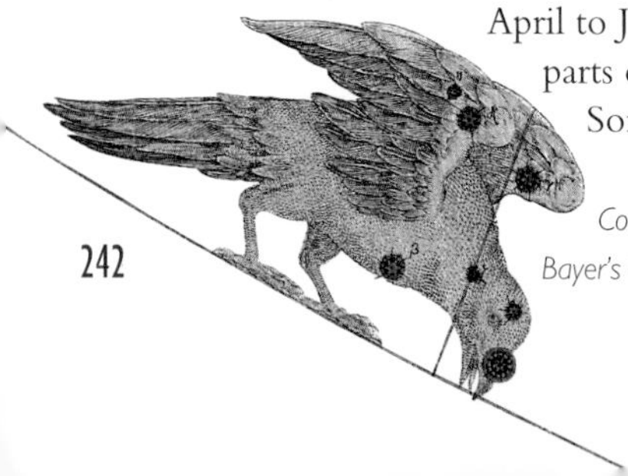

Corvus, the Crow, from Bayer's Uranometria (1639).

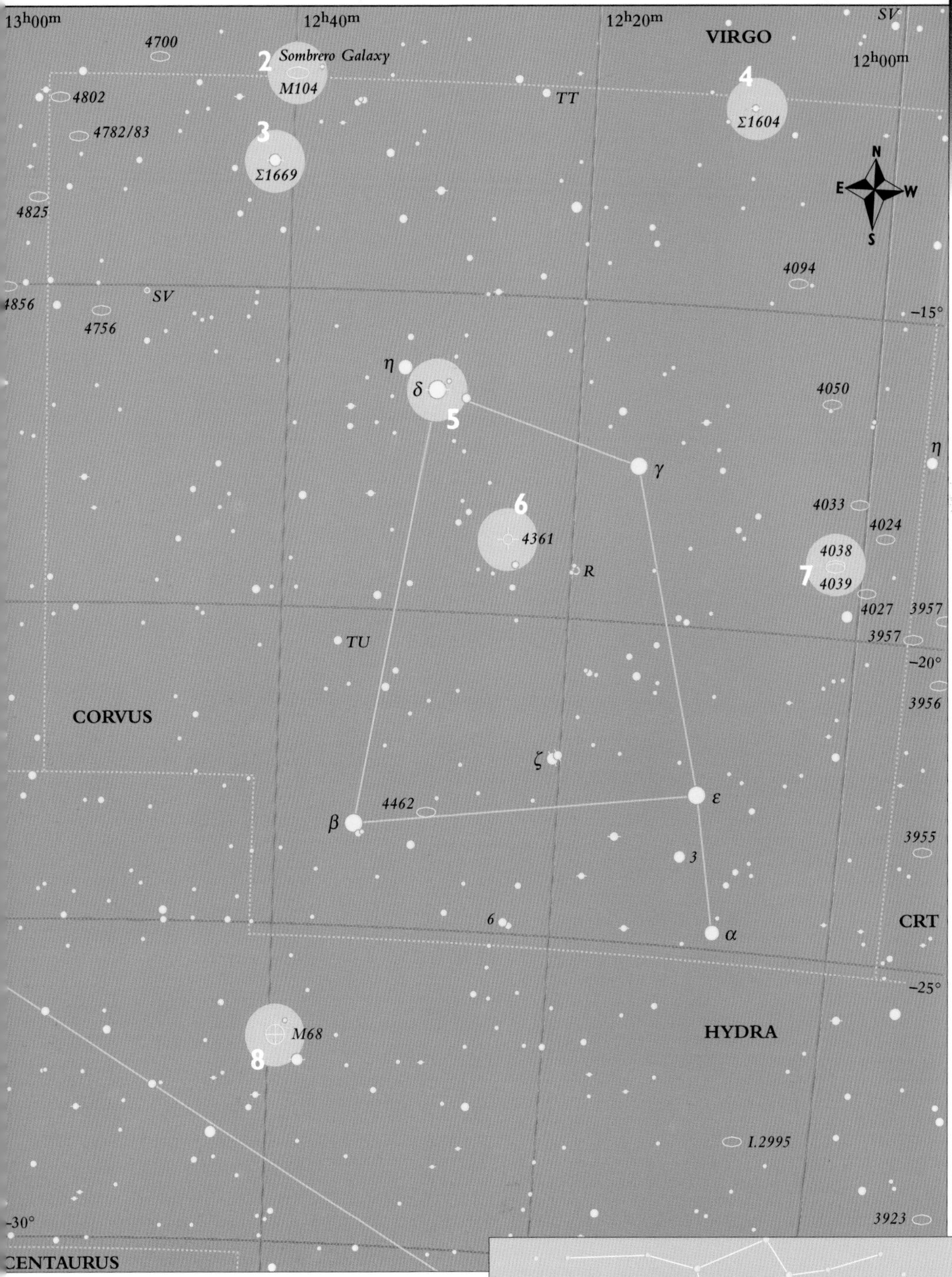
13h00m
12h40m
12h20m
12h00m
VIRGO
SV
4700
2 Sombrero Galaxy
M104
4802
4782/83
3
Σ1669
TT
4
Σ1604
4825
N
E
W
S
4094
4856
SV
4756
−15°
η
δ
5
4050
η
γ
6
4361
4033
4024
4038
7
4039
R
4027
3957
3957
TU
−20°
3956
CORVUS
ζ
ε
4462
β
3955
3
6
α
CRT
−25°
M68
8
HYDRA
I.2995
−30°
3923
CENTAURUS

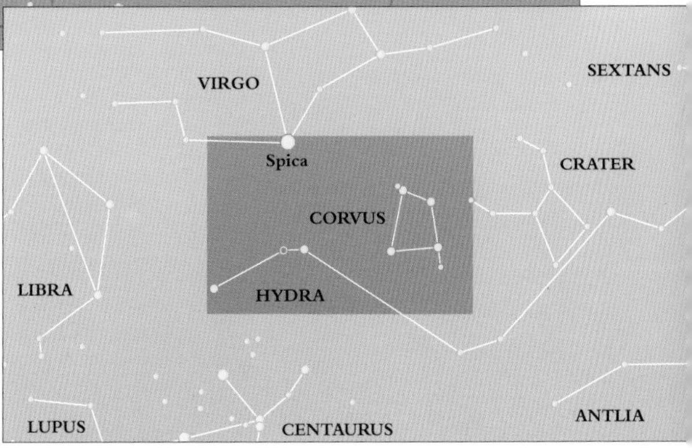
VIRGO
SEXTANS
Spica
CRATER
CORVUS
LIBRA
HYDRA
LUPUS
CENTAURUS
ANTLIA

❶ Spica (67 Alpha [α] Virginis)

Attention is drawn to this region of the sky by he brilliant, magnitude 1 bluish white Spica, which is where we begin this starhop. Spica is a louble star but, unfortunately, you will not have he chance of separating the components with a elescope or binoculars, since Spica is a spectro-copic binary (see p. 165). The two stars lie less han 20 million miles (32 million km) apart.

These stars also form an eclipsing-binary ystem, in which one star slightly eclipses the other, resulting in small variations in the bright-ness of Spica. The two components revolve around each other every 4 days.

Surprisingly, Spica does not form a part of the figure of Virgo. In some legends, Virgo becomes Demeter, the Corn Goddess, and Spica is depicted as an ear of wheat she is holding.

The Sombrero Galaxy (above) is the brightest of the Virgo galaxies with a distinctive dust band crossing its middle.

❷ The Sombrero Galaxy (M104, NGC 4549) Aim your scope at Spica, before heading 11 degrees due west to find M104. This famous 8th magnitude spiral galaxy is easily visible, appearing as a fuzzy patch of light through a 3 inch (75 mm) scope. With larger telescopes, it is not difficult to see how it came by its common name. The tightly coiled spiral arms result in a strong resemblance to the brim of that famous Mexican hat. Look for the prominent dark lane that runs along the length of the galaxy.

❸ Struve 1669 [Σ 1669] Heading $1\frac{1}{2}$ degrees to the south and just a little east of the Sombrero, we come to Struve 1669, in Corvus. This double star is listed in the catalog of Frederich von Struve, a prolific discoverer of such objects in the early 1800s. Its 6th magnitude yellowish components are almost equal in brightness at magnitudes 6 and 6.1, and, with a separation of 5 arcseconds, the object is a fine but fairly close double for small scopes. A 10th magnitude star can be seen about 1 arcminute from Struve 1669.

❹ Struve 1604 [Σ 1604] To find our next target, Struve 1604, another of Struve's doubles, first return to M104, then shift your scope just over 7 degrees in a due westerly direction. The brightest component is of 7th magnitude, with a 9th magnitude attendant 9 arcseconds away. There is another 9th magnitude star lying some 19 arcseconds farther, making the grouping an attractive sight for any small telescope.

INTERACTING GALAXIES

As galaxies drift through space, there are times when two or more of them can approach so closely that they exert a strong gravitational pull on each other, with the resulting tidal forces causing interesting effects. The pair NGC 4038 and NGC 4039 is a superb example of this (below). Named the Ring-Tail Galaxy because of its shape, it is also known as the Antennae since the material has been expelled in two directions as a result of the interaction.

Many other examples of interacting galaxies are known; for example, in Canes Venatici, the galaxy NGC 5195 appears to have brushed past the Whirlpool Galaxy (M51, NGC 5194). Even our own Milky Way Galaxy is suffering from the tidal effects of an interacting system—the nearby Magellanic Clouds have slightly distorted its shape.

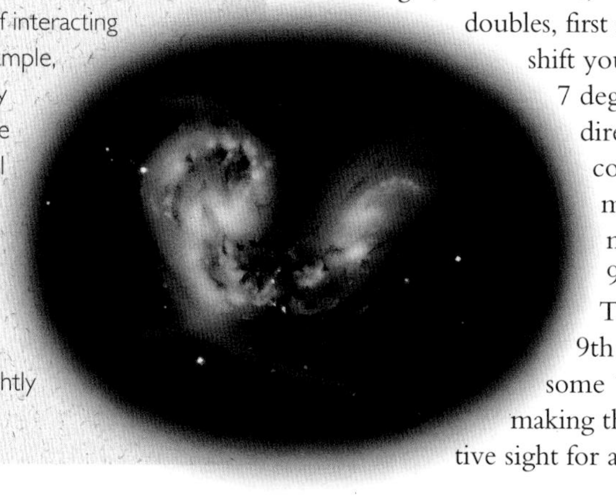

This patch of sky (above) shows Corvus at right, and Spica to the upper left. M83 (right) is a classic spiral galaxy lying in Hydra.

5 Algorab (7 Delta [δ] Corvi) Head almost 7 degrees southeast to the relatively bright star Delta Corvi. Shining at 3rd magnitude, Delta Corvi has a 9th magnitude companion that is fortunately 24 arcseconds away, making it easily visible in a 3 inch (75 mm) scope. Delta lies at the northeastern corner of the deformed rectangle that forms the main pattern in Corvus.

6 NGC 4361 Try heading into the rectangle's interior, moving your scope southwest 2½ degrees, to find the 10th magnitude planetary nebula NGC 4361. This needs a 4 inch (100 mm) scope or larger to be seen well. It has a distinctly irregular disk shape and covers a total of about 45 x 110 arcseconds. NGC 4361 was discovered by William Herschel in 1785.

7 The Ring-Tail Galaxy (NGC4038/9) Cross northwest to **Gamma [γ] Corvi**. Use this magnitude 2.5 star as a midway point to help find NGC 4038/9 (see Box). Nudge your scope about 3½ degrees farther southwest from Gamma to find a patch of light in the field. These interacting galaxies shine at magnitude 10.7. The pair takes the form of the letter C. The best size scope for revealing detail is a 6 inch (150 mm).

8 M68 (NGC 4590) Moving on to **Beta [β] Corvi**, about 9 degrees southeast, head 3½ degrees farther south and slightly east to find M68, across the border into Hydra. This magnitude 8 globular cluster was discovered by Messier in 1780. A 4 inch (100 mm) scope is necessary to begin to resolve the cluster into stars.

9 M83 (NGC 5236) Head 13 degrees southeast to locate M83, an 8th magnitude spiral galaxy—one of the easiest to see with a small scope and visible in 7 x 50 binoculars. A 6 inch (150 mm) scope reveals it as a fuzzy elliptical patch of light with a small, very bright core.

10 R Hydrae Moving 6½ degrees north and a little west we come to R Hydrae, just over 2 degrees east of **Gamma [γ] Hydrae**. This is one of the easiest long-period Mira-type variables to locate. It varies between magnitudes 4 and 10 over 390 days, making it fairly easy to see, even at minimum.

This starhop includes two of the best galaxies for large and small scopes—the Sombrero and M83—along with many other faint galaxies that seem to form a path between them. MG

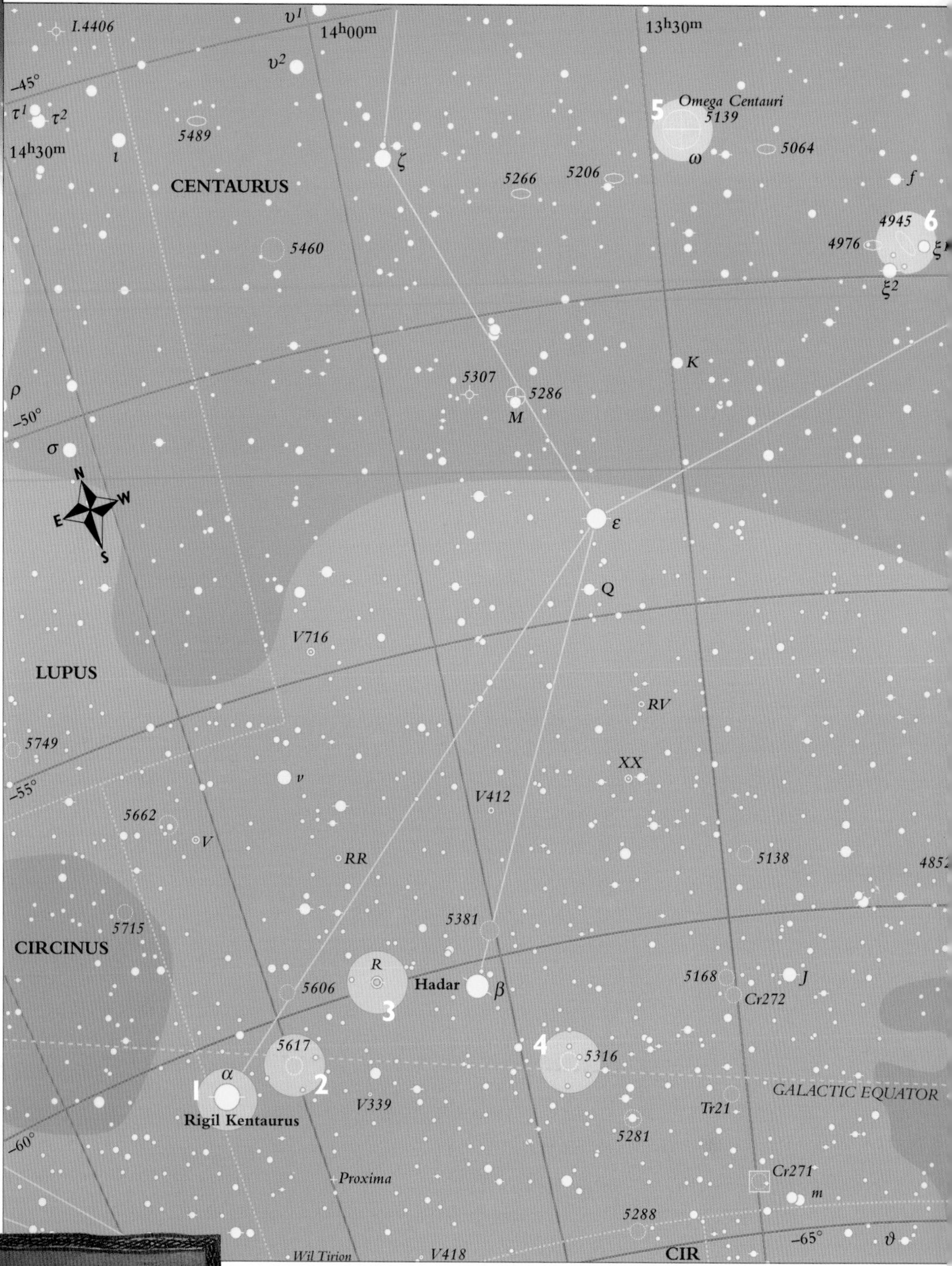
I.4406
υ1
14h00m
13h30m
υ2
–45°
τ1
τ2
5489
14h30m
ι
CENTAURUS
ζ
Omega Centauri
5139
5
ω
5064
5266
5206
f
4945
6
4976
ξ1
ξ2
5460
K
5307
5286
M
ρ
–50°
σ
N
W
E
S
ε
Q
V716
LUPUS
RV
5749
XX
υ
–55°
V412
5662
V
RR
5138
4852
5381
5715
CIRCINUS
R
5168
J
5606
Hadar
β
Cr272
3
5617
4
5316
α
2
1
GALACTIC EQUATOR
V339
Rigil Kentaurus
Tr21
5281
–60°
Proxima
Cr271
m
5288
–65°
ϑ
Wil Tirion
V418
CIR

Closing In on Crux

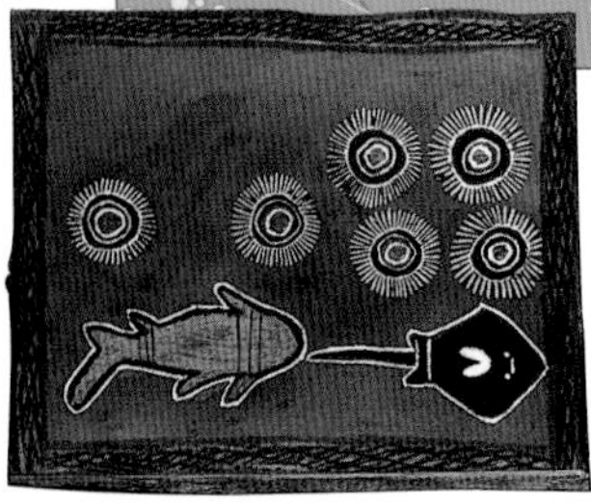

This part of the sky is far more accessible to observers in the Southern Hemisphere. However, those living south of about 20 degrees north can still catch a glimpse of these objects, low in the sky, during the evenings from May through July. Crux, the Southern Cross, and parts of nearby Centaurus, the Centaur, contain a number of interesting and visually stunning objects for one of our most southerly starhops.

An Australian Aboriginal bark painting depicts the stars of the Southern Cross (a stingray) being chased by the Pointers (a shark).

In 1517, Andrea Corsali described a group of stars he saw in the southern sky as "a marveylous crosse in the myddest of fyve notable starres." Today, we know it as the Southern Cross, and although it is the smallest of all the constellations, it is a prominent group of stars. The Southern Cross is famous, among other things, for its appearance on the flags of several countries.

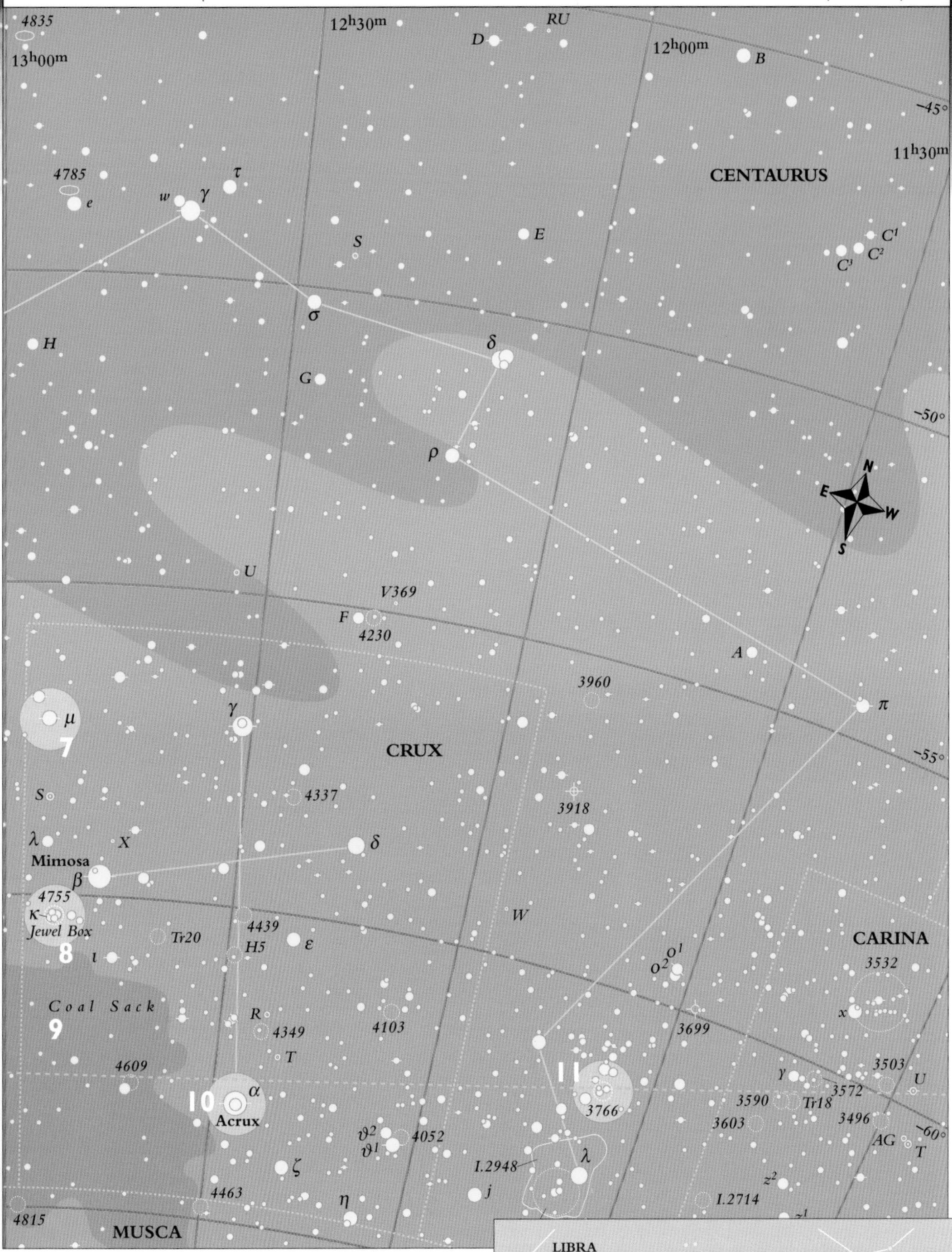
12h30m
12h00m
13h00m
11h30m
-45°
-50°
-55°
-60°
CENTAURUS
CRUX
CARINA
MUSCA
Mimosa
Jewel Box
Coal Sack
Acrux
N
E
W
S
7
8
9
10
11
4835
4785
4230
3960
4337
3918
4755
Tr20
4439
H5
4349
4103
4609
3699
3532
3503
3572
3590
Tr18
3603
3496
3766
4052
I.2948
I.2714
4463
4815
V369
AG

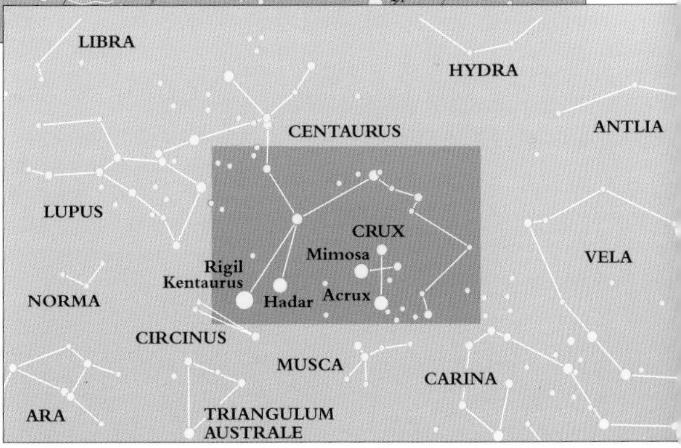
LIBRA
HYDRA
CENTAURUS
ANTLIA
LUPUS
CRUX
VELA
Mimosa
Rigil Kentaurus
NORMA
Hadar
Acrux
CIRCINUS
MUSCA
CARINA
ARA
TRIANGULUM AUSTRALE

❶ Rigel Kentaurus (Alpha [α] Centauri)

Our starting point for this starhop is the famous Alpha Centauri, the brightest star in Centaurus and, indeed, the third brightest in the entire sky. This is a spectacular double star for any size telescope. The two component stars, of magnitudes 0.0 and 1.2, are yellow and yellow-orange respectively, the brighter of the two being quite similar to our Sun. They orbit their common center of gravity every 80.1 years, varying in apparent separation from about 2 to 22 arcseconds. Currently, they are 16 arcseconds apart. Just 4.3 light-years distant, Alpha Centauri is the nearest star system to the Sun.

To make things really easy, the Southern Cross is clearly identified by the two Pointer stars, Alpha and **Hadar (Beta [β] Centauri)**, forming a line toward Crux in a westerly direction.

CLOSING IN ON CRUX

2 NGC 5617 Move your scope just over 1 degree west of Alpha, in the direction of the blue giant Beta (β) Centauri, the other pointer, to locate NGC 5617. Shining at magnitude 6, this open cluster is visible with binoculars, and is also a good subject for small telescopes. A 4 inch (100 mm) scope on low power provides a good view of the cluster, which is about 10 arcminutes across.

3 R Centauri Aim your scope to a point two-thirds of the way from Alpha to Beta Centauri, and then shift its position 1/2 degree north, to locate R Centauri. This is a Mira-type variable star whose magnitude ranges between 5.3 and 11.8 over 546 days. At maximum, it can be seen with the unaided eye. A small scope will show the star's beautiful red color at its brightest.

4 NGC 5316 Direct your scope toward Beta Centauri, before hopping 2 degrees southwest to NGC 5316. This open cluster looks similar to NGC 5617 in binoculars, appearing as a patch of light. A small scope provides a good view of its attractive field, with a 4 inch (100 mm) resolving quite a number of stars.

5 Omega [ω] Centauri (NGC 5139) To find our next target, first follow a line about 7 1/2 degrees from Beta to **Epsilon [ε] Centauri**. Extend it another 6 1/2 degrees to locate Omega Centauri. This is the finest of all the globular clusters in the sky (see Box). It can be seen with the unaided eye as a 4th magnitude spot of light, and stands out well in binoculars. A 3 inch (75 mm) scope begins to resolve it, and the view through an 8 inch (200 mm) is stunning.

Alpha Centauri (left) is one of the two stars pointing to Crux in the Milky Way (above).

6 NGC 4945 From Omega, adjust your telescope's position just over 4 degrees to the southwest to locate NGC 4945. This magnitude 9 edge-on barred spiral galaxy appears as a slender streak of light about 20 arcminutes long, less than 1/2 degree east of the star **Xi^1 [ξ^1] Centauri**. A wide-field eyepiece will capture the galaxy, along with Xi^1 and the brighter **Xi^2 [ξ^2] Centauri**, an easy double star. NGC 4945 is visible through a 4 inch (100 mm) scope, even though it has low surface brightness, but is better suited to larger telescopes.

7 Mu [μ] Crucis Swing your scope just over 7 degrees to the south and a little west, to close in on the star Mu Crucis, a showpiece wide double. Its bluish white 4th and 5th magnitude stars are separated by 35 arcseconds. Even a steadily held pair of 7 x 50 binoculars will separate them, but the view through a telescope is an opportunity not to be missed.

8 The Jewel Box (NGC 4755) Three degrees south of Mu is another showpiece object, the Jewel Box. A small scope will show many separate stars in the cluster. However, it is quite

The Jewel Box (above) is justly famous for its attractive contrasting colors formed into a compact, A-shaped cluster.

compact, so moderate magnification is a help. The cluster is shaped roughly like the letter A, with one of its brightest stars, a red supergiant, being very obvious in a field of contrasting colors.

9 The Coal Sack Continuing south a few more degrees, you will enter the south-eastern corner of Crux, where you should be able to see the Milky Way looking very dark. This is the area known as the Coal Sack. The most famous of the dark nebulas, the Coal Sack is a cloud of gas and dust, about 6 degrees across, which obscures the stars beyond it. Even in light-polluted skies, the Milky Way star field appears fairly barren in this area, but fine views of the Coal Sack can be obtained from dark, country sites where it stands out wonderfully.

10 Acrux (Alpha [α] Crucis) Heading 2 degrees west of the Coal Sack, Alpha Crucis pops into view—the star at the foot of the Southern Cross. This is a celebrated double star, whose components, at magnitudes 1.4 and 1.9, lie 4.4 arcseconds apart and form a very long-period binary system. A 3 inch (75 mm) telescope with high magnification shows them well.

11 NGC 3766 Just over 6 degrees west of Alpha lies NGC 3766. In binoculars, this open cluster appears as an obvious fuzzy spot of light in a field very rich in stars. The view through a 4 inch (100 mm) scope is quite beautiful. Two of its brightest stars are a distinctive reddish orange.

There is little to outshine the view of this part of the night sky, especially on a moonless night away from bright city lights. MG

A STAR ATTRACTION

Omega Centauri appeared as a star in Ptolemy's catalog nearly 2,000 years ago. We now know it as the most spectacular globular cluster in the entire sky, containing about a million stars of 11th magnitude and fainter. Omega Centauri's stars are packed so tightly that, near the center, they are typically only 1/10 of a light-year apart. At some 17,000 light-years away, it is also among the nearest globular clusters to Earth.

An interesting observation about Omega Centauri is that its stars vary significantly in their content of elements heavier than hydrogen and helium, suggesting that the stars differ in age and are not all about the same age, as was previously thought. The reasons for these variations remain unclear.

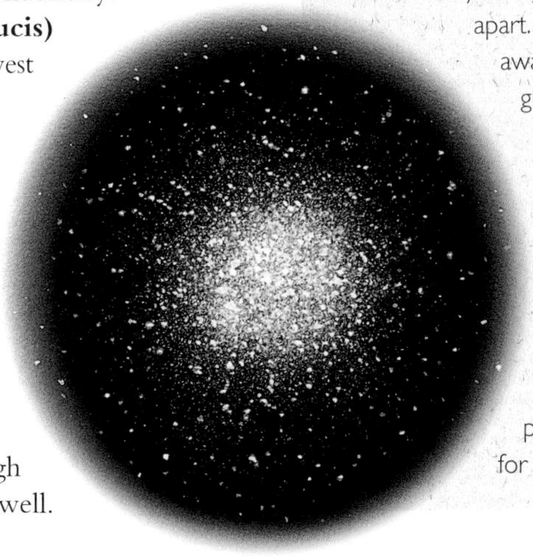

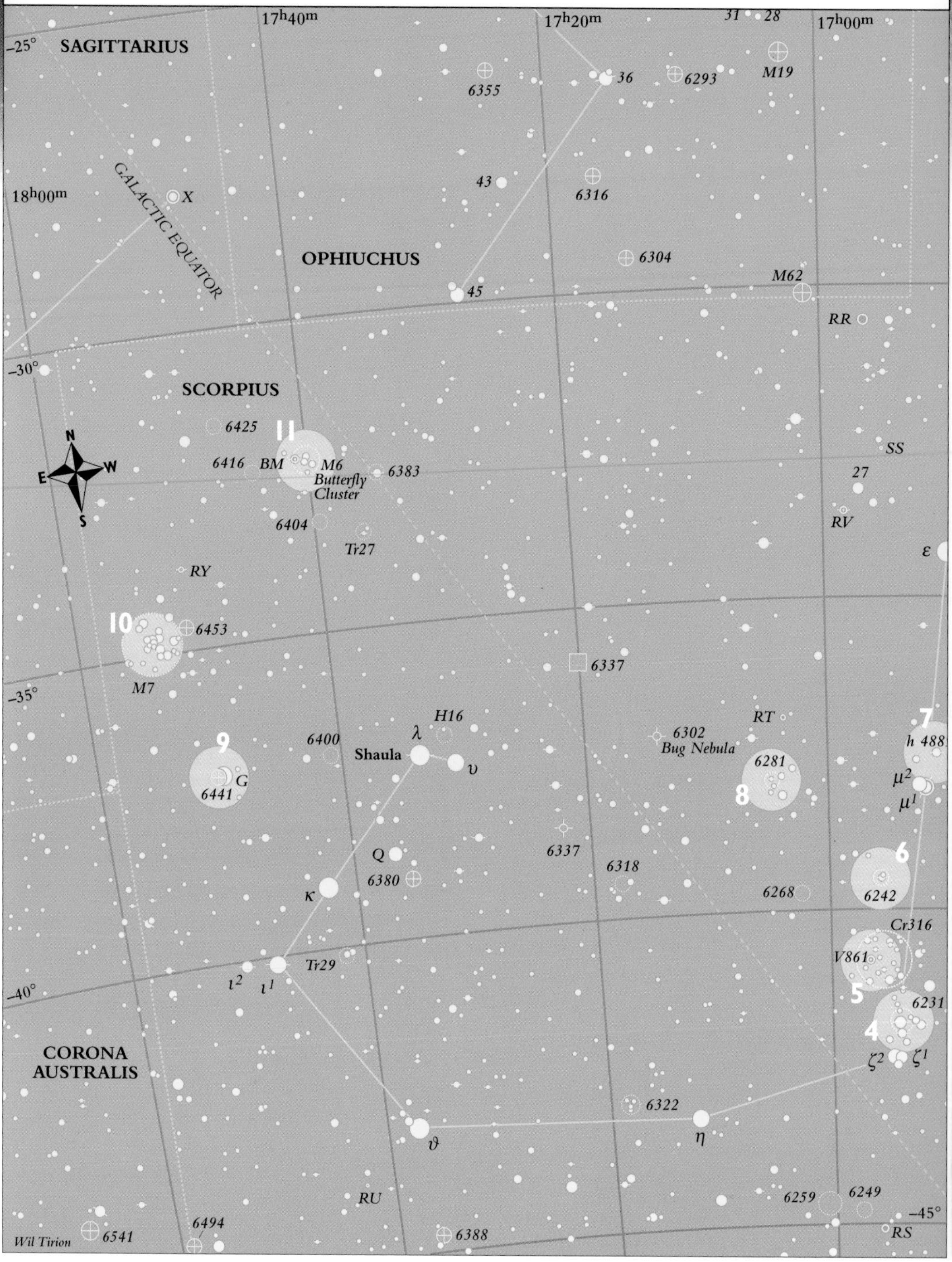

17h40m
17h20m
17h00m
18h00m
–25°
–30°
–35°
–40°
–45°
SAGITTARIUS
OPHIUCHUS
SCORPIUS
CORONA AUSTRALIS
GALACTIC EQUATOR
N
E
W
S
36
6293
M19
6355
43
6316
X
6304
45
M62
RR
6425
6416
BM
M6 Butterfly Cluster
6383
SS
27
RV
6404
Tr27
RY
ε
10
6453
6337
M7
H16
RT
7
6400
λ
Shaula
υ
6302 Bug Nebula
6281
h 4889
9
G
6441
8
μ2
μ1
Q
6337
6318
6
6380
κ
6268
6242
Cr316
Tr29
V861
ι2
ι1
5
6231
4
ζ2
ζ1
6322
ϑ
η
RU
6259
6249
6494
6541
6388
RS
Wil Tirion

Exploring the Scorpion

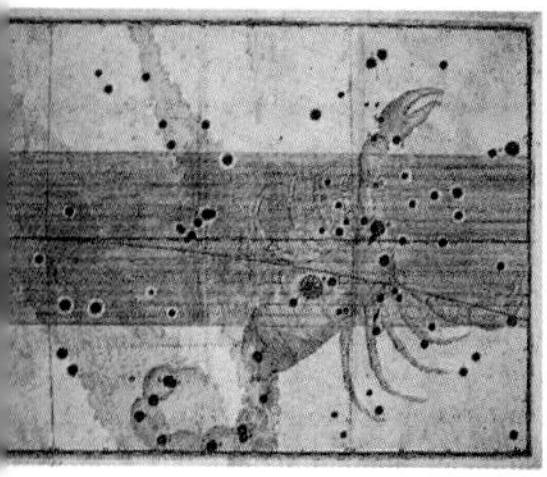

Scorpius, the Scorpion, as represented in the 1639 edition of Johann Bayer's illustrated star atlas, *Uranometria*.

Scorpius is at its best when high in the sky during the evenings of June, July, and August. At this time, binocular and telescope observers are treated to a view of many fine celestial objects spotted throughout a very rich part of the Milky Way.

One of the ancient legends tells us that Orion, the Hunter, died after being stung by a scorpion, and that the two were turned into constellations on opposite sides of the sky. Orion must flee the scorpion for eternity, so as Scorpius rises in the east, Orion sets in the west.

For those less familiar with the night sky, Scorpius is one of the easiest constellations to locate. Simply look for the bright, reddish orange star Antares that marks the Scorpion's heart, and the distinctive curving shape of Scorpius's body.

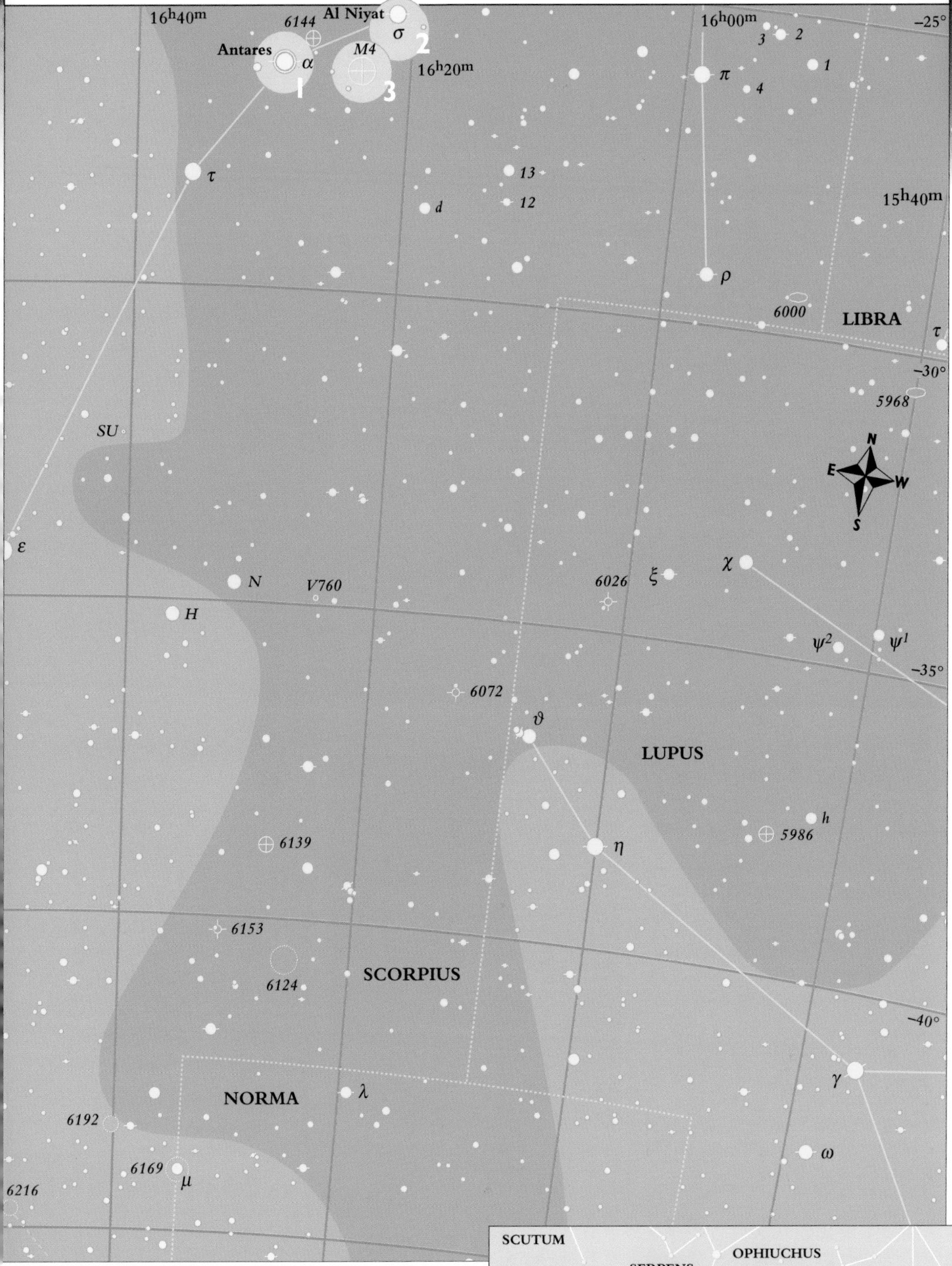
16h40m
6144
Al Niyat
Antares
M4
16h20m
16h00m
-25°
15h40m
LIBRA
-30°
5968
6000
SU
N
V760
H
6026
-35°
6072
LUPUS
6139
5986
6153
6124
SCORPIUS
-40°
NORMA
6192
6169
6216

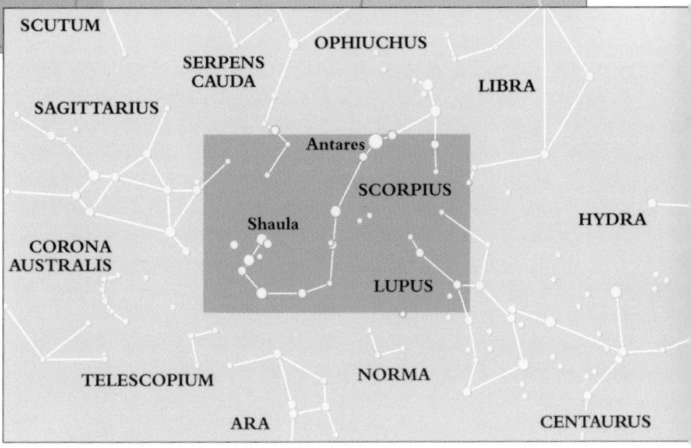
SCUTUM
SERPENS CAUDA
OPHIUCHUS
LIBRA
SAGITTARIUS
Antares
SCORPIUS
Shaula
HYDRA
CORONA AUSTRALIS
LUPUS
TELESCOPIUM
NORMA
ARA
CENTAURUS

1 Antares (Alpha [α] Scorpii) The name Antares means "rival to Mars" and it is easy to see how the name came about. Antares shines at 1st magnitude and looks remarkably like a slightly fainter version of the planet Mars.

Antares is a red supergiant star at a very late stage of its evolution. It is immense, with a diameter close to 300 times that of our Sun and is enveloped in a haze of nebulosity (see p. 252). Antares is also a famous, but difficult to separate, double star. The magnitudes of the two components are 1.2 and 5.4, with the secondary often reported as appearing greenish. This is due to a contrast effect; it is actually a star of spectral class B, so that by itself it would appear bluish white. Antares's components lie almost 3 arcseconds apart, so you need at least a 6 inch (150 mm) telescope to separate them.

A spectacular view of Scorpius (above). The nebulosity surrounding Antares (left).

2 Al Niyat (20 Sigma [σ] Scorpii) Move about 2 degrees northwest to look at 20 Sigma Scorpii, an easy but unequal double star, whose 3rd and 8th magnitude components are 20 arcseconds apart. They are easy to separate in a 3 inch (75 mm) telescope.

3 M4 (NGC 6121) Shift 1 degree south and a little back toward Antares to bring M4 into view. This is one of the great globular clusters in the sky. Easily seen in binoculars, it is nearly 1/2 degree in diameter. A 4 inch (100 mm) scope begins to resolve individual stars in the cluster, although larger telescopes will reveal many more stars.

4 NGC 6231 The best way to exit the Antares area is by running southeast along the scorpion's body until you reach the bend in its tail, where NGC 6231 is located, some 17 degrees from M4. It contains a number of bluish white supergiant stars. The cluster covers about 1/4 degree, but its brightest members form a more compact group. This is a fine open cluster for small scopes, being clearly visible to the naked eye as a bright spot of light just north of the star pair **Zeta^{1,2} [ζ^1, ζ^2] Scorpii**. Zeta^1, an orange star, and bluish white Zeta^2 are not a physical pair but are worth a quick look for their contrasting colors.

5 v861 Scorpii From NGC 6231 adjust your scope 1 degree north and about 1/2 degree east to encounter v861 Scorpii. Lying within the scattered 1 degree wide cluster **Collinder 316**, v861 Scorpii is a bluish star that varies in magnitude between 6.1 and 6.7 every 7.8 days. In mid-1978, the star was found to be a source of X-rays, and is now a well-known X-ray binary variable.

6 NGC 6242 Nudge your scope just under 1 1/2 degrees north and a little west through this very rich field to find NGC 6242. This open cluster is a pleasing binocular target in addition to being a rewarding sight through a 3 inch (75 mm) telescope. Its brightest star, near the southern end of the cluster, is an attractive orange-red color.

7 h 4889 One and a half degrees northwest of NGC 6242 is the brilliant, wide pair **Mu^{1,2} [μ^1, μ^2] Scorpii**. Another 1/2 degree farther is h 4889. This is a double whose 6th and 8th magnitude stars lie 7 arcseconds apart. Mu^1 and Mu^2 themselves are a fine naked-eye pair of stars that may be physically associated. In Polynesian legend, they were called "the inseparable ones."

8 NGC 6281 Sweep your telescope about 3 degrees east for a look at NGC 6281. Easily visible in binoculars, this is a 5th magnitude open cluster similar in size to NGC 6242,

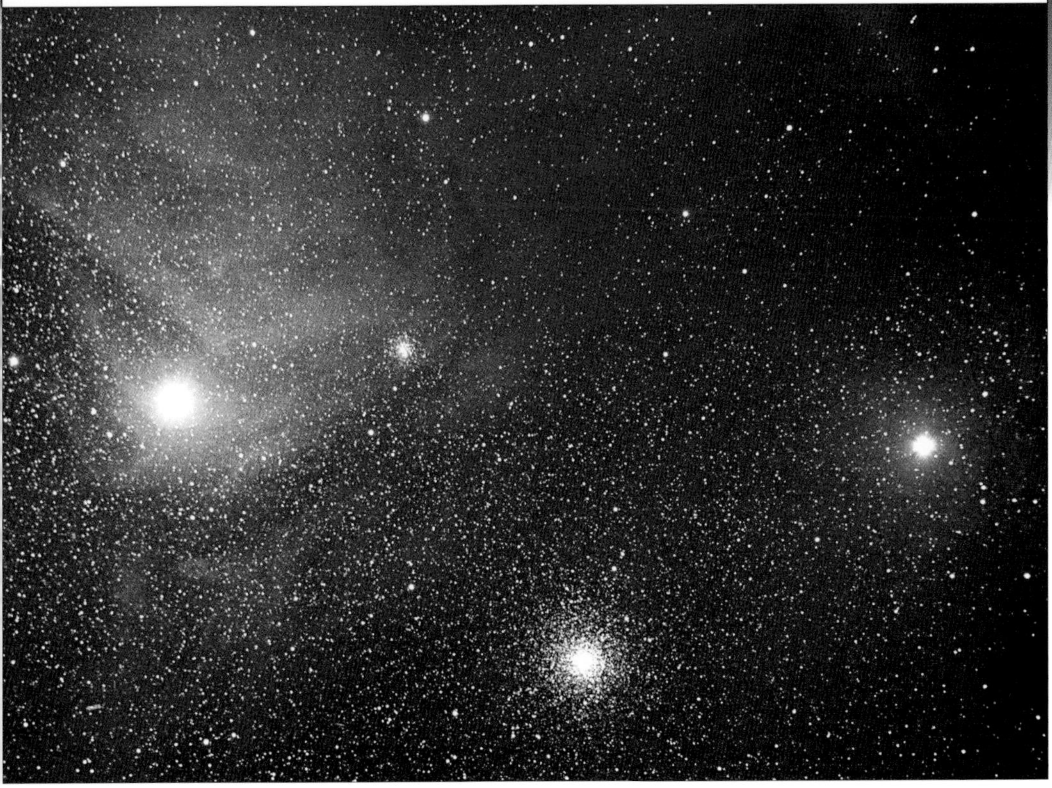

A number of impressive globular clusters (above), including M4, lie within the field around brilliant Antares, shown center left.

but containing fewer stars. It is also well worth a look in any size telescope.

❾ **NGC 6441** Hop just over 9 degrees east, past **Lambda [λ] Scorpii**, in search of NGC 6441. Although by no means bright, this globular cluster is quite easy to locate, being only a few arcminutes east of the yellow-orange 3rd magnitude star **G Scorpii**. The cluster requires quite a large scope to resolve its stars. A 10 inch (250 mm) scope reveals its granular appearance.

❿ **M7 (NGC 6475)** Shift 2 degrees to the north and a little east to find M7. Easily seen as a 1 degree wide, hazy patch of light with the naked eye, M7 is one of the best open clusters for binoculars, which reveal many of its stars. A wide-field eyepiece is essential for a good view through a telescope.

⓫ **The Butterfly Cluster (M6, NGC 6405)** Four degrees northwest of M7 we come to another gem, the open cluster M6. It is just possible to see it with the naked eye, but binoculars or a small telescope provide a far better view. The view through a telescope reveals its main stars arranged in a pattern resembling the body and wings of a butterfly, hence its common name.

BM Scorpii Near the eastern end of M6, and marking the trailing end of the butterfly's left wing, lies the orange semi-regular variable star BM Scorpii. It varies between magnitude 6.8 and 8.7 over a period of about 850 days.

The grouping of stars and star clusters around Scorpius in a dark sky is an unforgettable sight. It is easy to lose track of time while sweeping your binoculars through this fascinating area. MG

M7, AN ANCIENT OPEN CLUSTER

M7 is the southernmost object in Charles Messier's famous catalog of non-stellar objects. It is also one of the brightest. However, Messier observed it from Paris, where its maximum altitude of only about 6 degrees would have meant that he could not see it in its true splendor. The cluster contains about 80 stars scattered over more than 1 degree of sky. Studies of M7 show that it is about 220 million years old. This compares with only about 50 million years for its nearby companion M6 and 70 million years for the Pleiades in Taurus. There are few open clusters much older than M7, because as clusters age, they are dispersed by gravitational effects and encounters with interstellar clouds.

TOWARD THE HEART OF OUR GALAXY

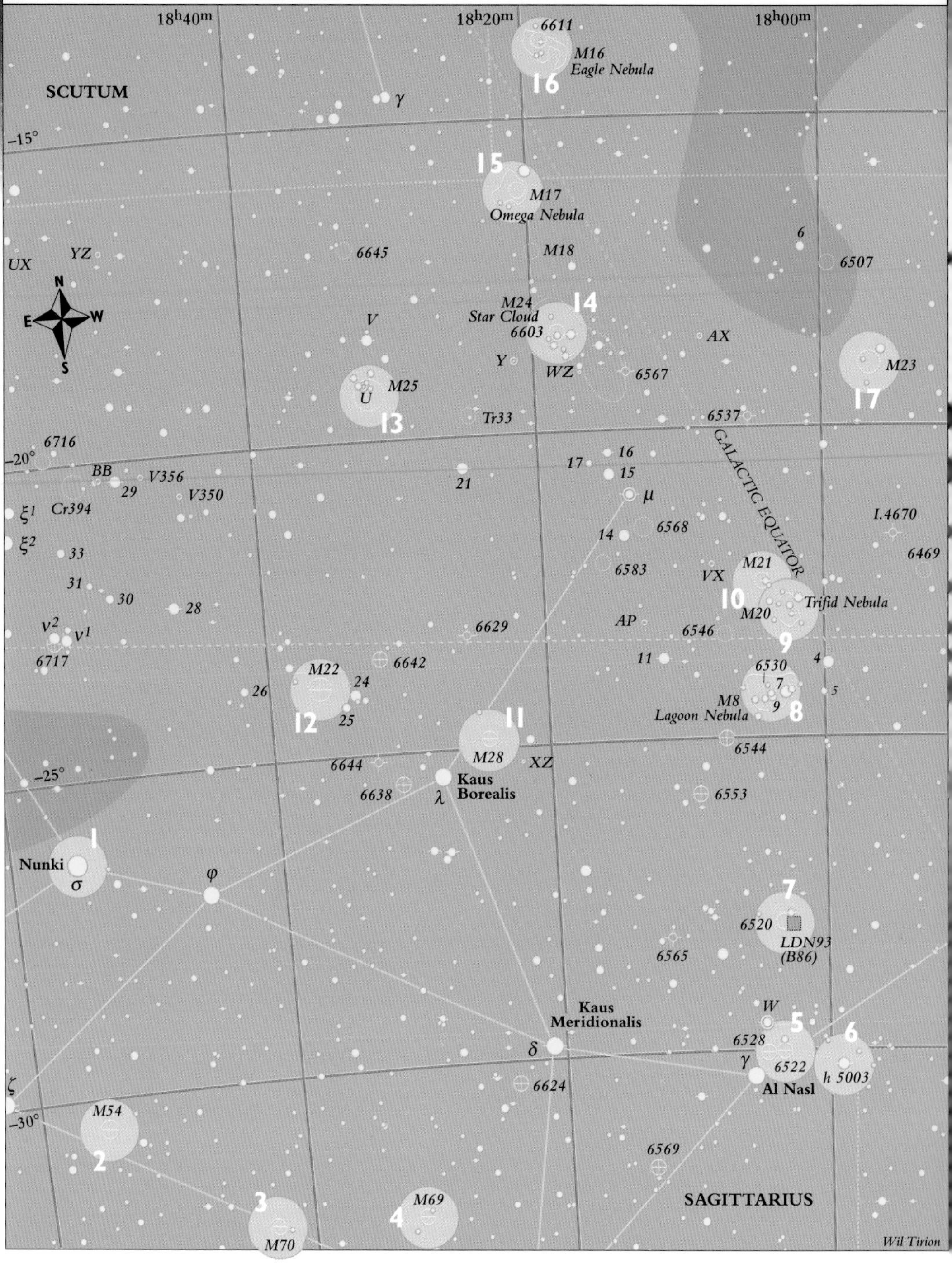
18h40m
18h20m
18h00m
SCUTUM
–15°
–20°
–25°
–30°
6611
M16
Eagle Nebula
16
γ
15
M17
Omega Nebula
M18
6
6507
UX
YZ
6645
N
E
W
S
V
M24
Star Cloud
6603
14
AX
Y
WZ
6567
M23
M25
U
13
Tr33
6537
17
6716
16
15
17
BB
V356
29
V350
21
μ
GALACTIC EQUATOR
ξ1
Cr394
14
6568
I.4670
ξ2
33
6583
VX
M21
6469
31
30
28
10
M20
Trifid Nebula
AP
6629
6546
9
ν2
ν1
6717
6642
11
6530
4
M22
24
7
5
26
M8
9
Lagoon Nebula
8
12
25
11
M28
6544
XZ
6644
Kaus
Borealis
6638
λ
6553
1
Nunki
σ
φ
7
6520
LDN93
(B86)
6565
Kaus
Meridionalis
W
5
6
6528
δ
γ
6522
h 5003
Al Nasl
ζ
6624
M54
2
6569
SAGITTARIUS
3
M69
4
M70
Wil Tirion

Toward the Heart of Our Galaxy

Being at a moderate southerly declination, this region passes overhead for those living in the Southern Hemisphere. However, northern skywatchers can spot many of the objects in this patch of sky during the evenings from late June to September.

Sagittarius, the Archer, from an ancient Arabic manuscript.

There is no finer naked-eye view of the night sky than one that includes the constellation of Sagittarius, the Archer, high above the horizon, for it is in this direction that we look toward the center of the Milky Way Galaxy. The region just east of the tail of Scorpius, the Scorpion, is especially rich in interesting clusters and nebulas. In adjacent Ophiuchus, we find some of the famous dark, obscuring dust clouds that stand out so well on clear nights for observers away from city light pollution.

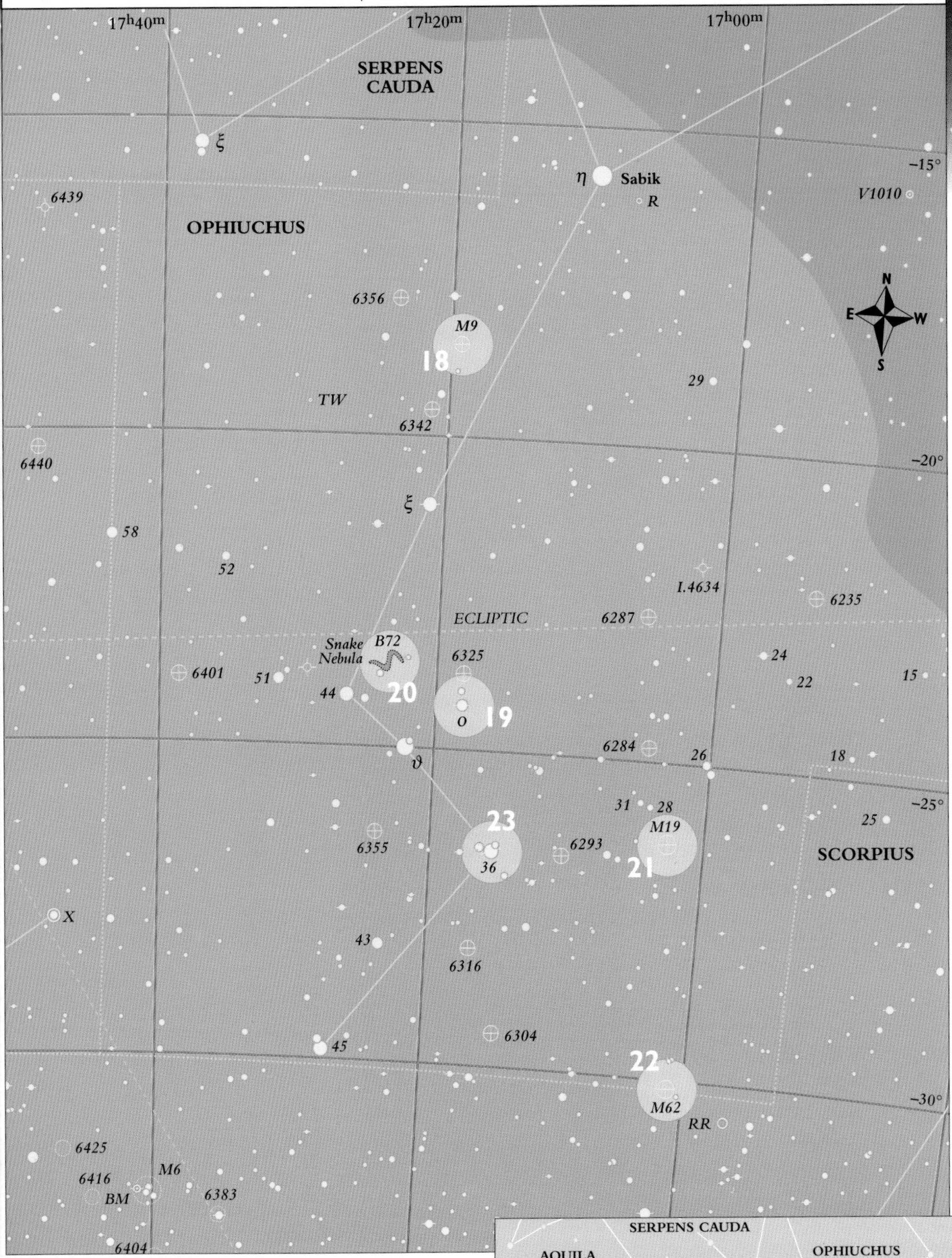
17h40m
17h20m
17h00m
SERPENS CAUDA
OPHIUCHUS
SCORPIUS
Sabik
R
V1010
−15°
−20°
−25°
−30°
N
E
W
S
6439
6356
M9
18
TW
6342
29
6440
58
52
I.4634
6235
ECLIPTIC
6287
Snake Nebula
B72
6325
24
22
15
6401
51
44
20
19
6284
26
18
31
28
M19
25
23
6355
36
6293
21
X
43
6316
6304
45
22
M62
RR
6425
6416
M6
BM
6383
6404

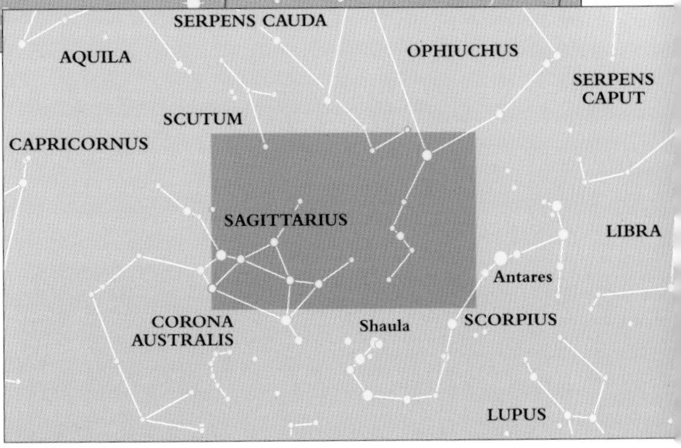
SERPENS CAUDA
AQUILA
OPHIUCHUS
SERPENS CAPUT
SCUTUM
CAPRICORNUS
SAGITTARIUS
LIBRA
Antares
CORONA AUSTRALIS
Shaula
SCORPIUS
LUPUS

❶ Nunki (34 Sigma [σ] Sagittarii)

While the brightest stars of Sagittarius have the distinctive appearance of a teapot, a familiar part of the asterism is the shape of the so-called Milk Dipper—the handle and part of the lid of the teapot. The Milk Dipper is formed by the stars **Zeta [ζ]**, Tau [τ], Sigma [σ], **Phi [ϕ]**, and **Lambda [λ] Sagittarii**. Shining at magnitude 2, the blue-white star Sigma forms one end of the base of the dipper and, surprisingly, is the second brightest star in Sagittarius.

Sigma's common name, Nunki, is thought to be derived from the Babylonian *Tablet of the Thirty Stars*. It is said to represent "the star of the proclamation of the sea," the "sea" being the quarter of sky occupied by the water constellations, including nearby Capricornus and Aquarius (see p. 268).

❷ **M54 (NGC 6715)** Center your telescope on Nunki and head about 4 degrees due south to M54, an 8th magnitude globular cluster that seems to have fallen out of the dipper's bowl. It can be seen in binoculars, but is very difficult to resolve. Even a 12 inch (300 mm) telescope shows it only as somewhat granular.

Sagittarius has many fine targets like the globular cluster NGC 6522 (above), and Barnard 86 and NGC 6520 (both left).

❸ **M70 (NGC 6681)** About 3 degrees southwest lies another globular cluster, M70. This is also an 8th magnitude object, but it can be resolved a little more easily than M54, with a 6 inch (150 mm) scope showing some of its stars. M70 became famous in 1995 when astronomers Alan Hale and Thomas Bopp, observing the cluster from different locations, discovered Comet Hale-Bopp (see p. 145), the Great Comet of 1997, in the same telescope field as the cluster.

❹ **M69 (NGC 6637)** The globular cluster M69 lies 2½ degrees due west of M70 and appears virtually as its twin. It requires a 6 inch (150 mm) telescope to begin to resolve it properly into individual stars. Unlike the previous two globulars, which were discovered by Charles Messier, M69 was first seen by astronomer Nicolas Louis de Lacaille in 1752.

❺ **NGC 6522** Almost 6½ degrees northwest of M69 lies **10 Gamma [γ] Sagittarii**, a bright star. Move across to it, and place brilliant Gamma in the southeastern corner of a wide-field view. You should also be able to see the 9th magnitude globular cluster NGC 6522. Some of the cluster's stars are visible through an 8 inch (200 mm) telescope.

NGC 6528 The magnitude 10 globular cluster, NGC 6528, is too difficult to resolve even with a 12 inch (300 mm) telescope.

❻ **h 5003** Hop 1 degree west of our last stop to h 5003. This delightful double star is a great object for any size scope. A 3 inch (75 mm) telescope, or even smaller, will show this orange and yellow pair of stars, at magnitudes 5 and 7, separated by 5.5 arcseconds.

❼ **NGC 6520** Return to NGC 6522 and NGC 6528, then shift 2 degrees due north to locate NGC 6520. This compact open cluster, visible in binoculars, is a pleasing sight in a 3 inch (75 mm) scope or larger. Between NGC 6520 and a 7th magnitude star a few arcminutes northwest lies the dark nebula **Barnard 86**, also known as LDN 93. A 4 inch (150 mm) telescope should reveal the nebula as an apparent gap in the star field around it.

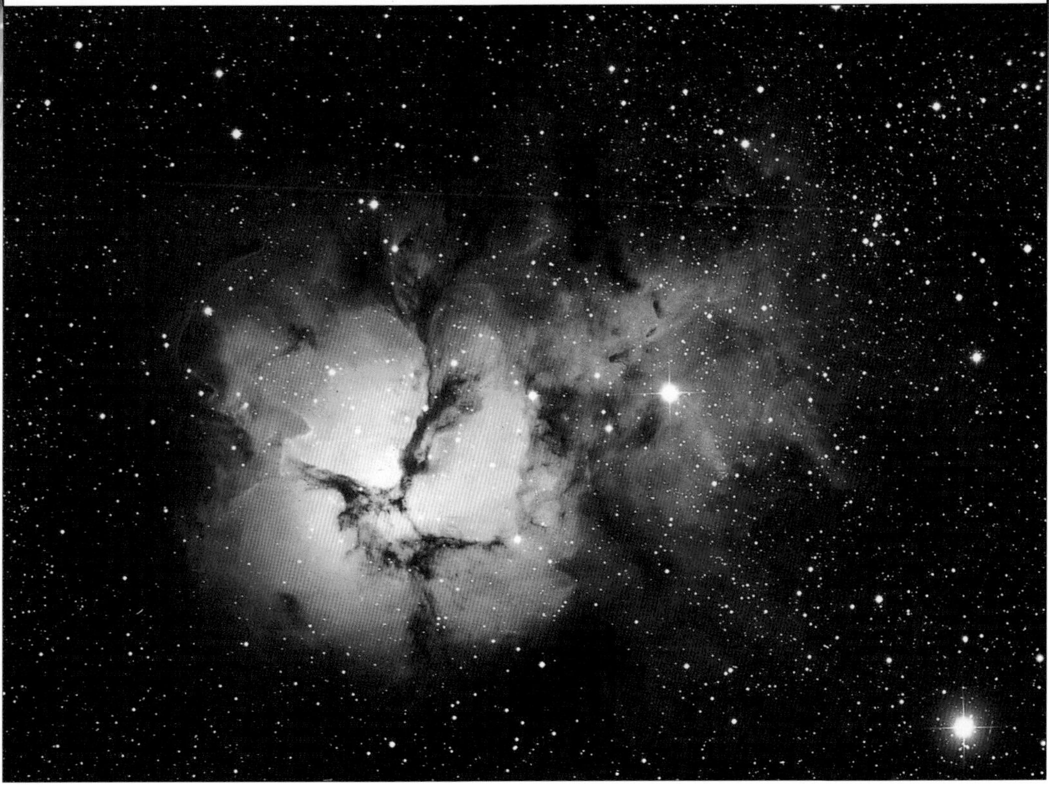

❽ The Lagoon Nebula (M8, NGC 6523) Ease your scope $3\frac{1}{2}$ degrees north to find the wonderful Lagoon Nebula. One of the few bright nebulas visible to the unaided eye, this can be seen as a milky-white spot of light against the backdrop of the Milky Way. Although it is one of the finest nebulas for a small telescope, binoculars also provide a superb view.

❾ The Trifid Nebula (M20, NGC 6514) Try your hand at the Trifid Nebula, $1\frac{1}{2}$ degrees north and just a little west of M8. This famous nebula is much fainter than the Lagoon Nebula and is best observed in a dark sky. There is always a feeling of delight to see the three dark lanes that give the Trifid Nebula its prominent shape. These dust lanes are visible in a 4 inch (100 mm) scope. It is helpful to use averted vision when viewing this nebula.

❿ M21 (NGC 6531) Nudge your scope just over $\frac{1}{2}$ degree northeast of the Trifid to spot M21. Covering almost $\frac{1}{4}$ degree, M21 is easily resolved in a 3 inch (75 mm) telescope, with its brightest stars being of 8th magnitude.

⓫ M28 (NGC 6626) Moving on about 6 degrees southeast, hop to the star Lambda [λ] Sagittarii, then backtrack 1 degree to pinpoint M28. Although visible in binoculars, it takes a 6 inch (150 mm) scope to begin to resolve the stars of this globular cluster. In smaller telescopes, M28 appears as a round, fuzzy glow.

M20, the Trifid Nebula (above), lies at center bottom of the view of the Milky Way (right) which includes M17 at the top.

⓬ M22 (NGC 6656) About 3 degrees northeast of M28 lies one of the few globular clusters clearly visible with the unaided eye—M22, a spectacular object at magnitude 5.1. A 3 inch (75 mm) telescope will show some of its brightest members, and the view through a 6 inch (150 mm) is unforgettable. You may be able to detect its slightly elliptical shape. The cluster has a diameter of more than 20 arcminutes and, at a distance of only about 10,000 light-years, is one of the closest of its type to us.

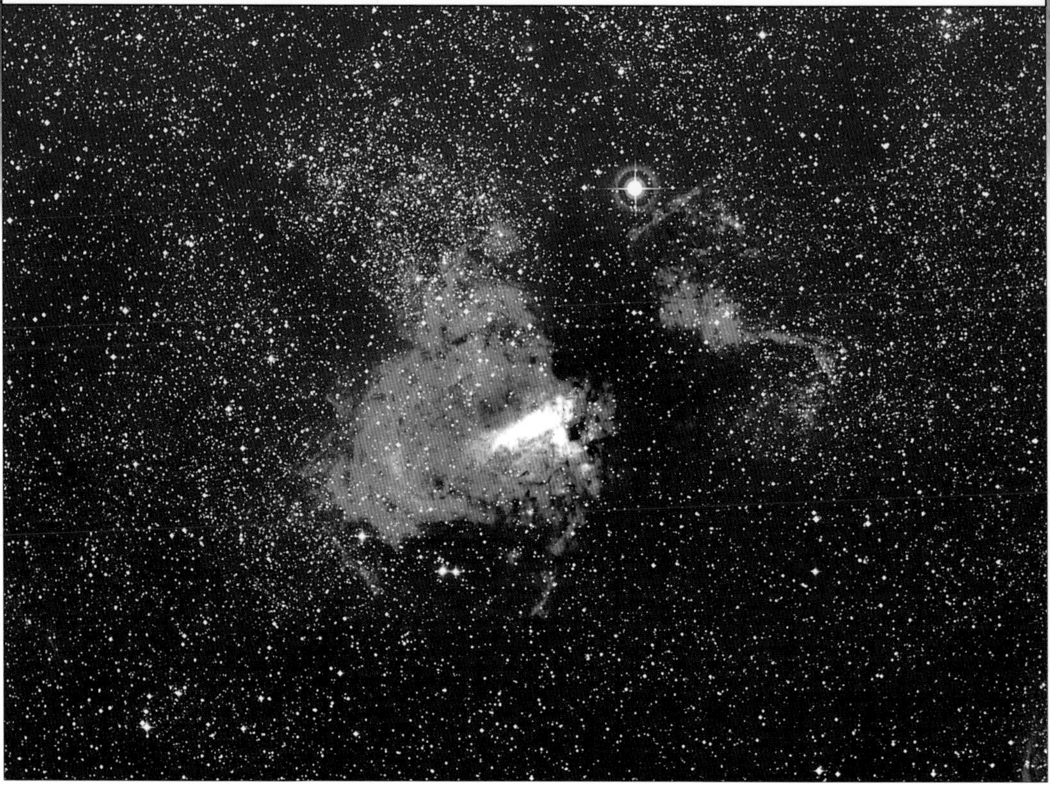

M17, the Omega Nebula (above) lies on the border of Sagittarius.

⑬ M25 (IC 4725) Move 5 degrees north and 1 degree west of M22 to take a look at M25. This scattered open cluster, about 1/2 degree across, is an easy object for binoculars and small telescopes. Near the center of M25 is the yellowish Cepheid variable **U Sagittarii**, which varies from magnitude 6.3 to 7.1 over 6.75 days.

THE STELLAR NURSERY IN M16

In 1995, the Hubble Space Telescope returned an eerie picture of the nebulosity of M16, showing huge columns of gas that are several light-years in length (below). A close look at the surfaces of the giant pillars shows what scientists have called evaporating gaseous globules, or EGGs. Ultraviolet light from nearby stars is stripping some of the gas from the columns and exposing the EGGS, in which stars are forming. We see the globules because they are more dense than the rest of the nebula, and so are not "blown away" as easily. However, the gradual loss of material feeding the developing stars is thought to play an important role in limiting their size. We can only wonder how many more stars are forming deep within these huge columns, gathering material in private and becoming larger and larger.

⑭ M24 To locate the next object on our hop, scan your eye 3 degrees west of M25. M24 is unique among Messier objects because it is a huge star cloud about 1 degree wide and 2 degrees long. Sweeping your scope through it will reveal a very rich region, including the open cluster **NGC 6603** near the northeastern end. The cluster is visible with a 4 inch (100 mm) scope, but can be resolved in larger telescopes.
⑮ The Omega Nebula (the Swan Nebula, M17, NGC 6618) If you look another 2 1/2 degrees farther north and a little east, M17 comes into view. This bright nebula is easy to see in binoculars, appearing as a small streak of light. In a 4 inch (100 mm) scope, its unusual shape looks a bit like the figure 2 with an extended baseline. Because of this pattern, M17 is called both the Swan Nebula and the Omega Nebula.
⑯ The Eagle Nebula (M16, NGC 6611) Across the border into Serpens, about 2 1/2 degrees north of M17, the combination of hazy star cluster and nebula, M16, appears. The cluster is obvious enough with binoculars, but you need a 6 inch (150 mm) scope and a dark sky to clearly see that it is actually immersed in a region of nebulosity called the Eagle Nebula (see Box). In larger scopes, the whole field is beautiful.
⑰ M23 (NGC 6494) Head 7 1/2 degrees southwest of M16, back into Sagittarius, for a view of M23. This open cluster is nearly 1/2 degree across. It is a fine sight in a 4 inch (100 mm)

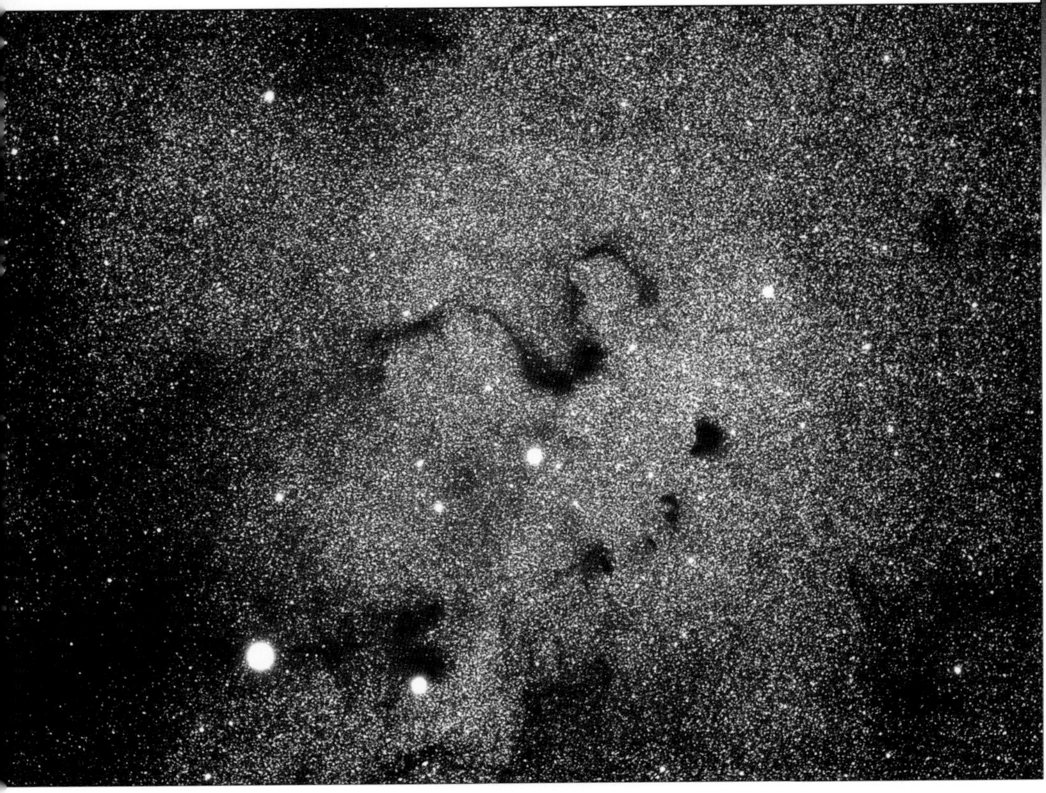

scope, which resolves quite a few stars. The straight and curved lines of stars within the cluster provide a delightful detail.

18 M9 (NGC 6333) Shift your scope 9 degrees farther west to find 8th magnitude M9, a globular cluster about 2½ degrees north of the star **Xi [ξ] Ophiuchi**. It appears fairly condensed and is visible in binoculars, although a 6 inch (150 mm) scope is needed to see some of its multitude of stars clearly.

19 39 Omicron [ο] Ophiuchi About 3 degrees south of Xi lies a small triangle of stars, the westernmost and faintest of the three being our next target. A beautiful double star for any small telescope, Omicron's colorful 5th and 7th magnitude orange and yellowish components are an easy 10 arcseconds apart, so should be no problem to visually separate with most telescopes.

20 The Snake Nebula (Barnard 72) The next challenge lies about 1½ degrees northeast of Omicron. A familiar sight in photographs of the Ophiuchus Milky Way, the dark Snake Nebula takes the form of the letter S and is about ½ degree from end to end. It is not exactly an easy object to discern, but if you use moderate aperture and a low magnification, you might be able to find it in a dark sky.

21 M19 (NGC 6273) Aiming your scope just over 5 degrees southwest of the Snake Nebula, you should be able to spy the globular

Dark dust lanes create different and dramatic effects in B72, the Snake Nebula (above), and M16, the Eagle Nebula (left).

cluster M19. At magnitude 7, M19 is an object that is easy to see with binoculars, appearing as a round, fuzzy spot. A 4 inch (100 mm) telescope will begin to show a granular appearance.

22 M62 (NGC 6266) Our next stop, the globular cluster M62, is about 4 degrees due south of M19. A little brighter but somewhat more difficult to resolve than M19, M62 appears as a more uniform haze of stars requiring an 8 inch (200 mm) telescope for a good view. However, a number of individual stars can just be resolved when using a 6 inch (150 mm) telescope or larger.

23 36 Ophiuchi Returning to the northeast, two-thirds of the way to **Theta [θ] Ophiuchi** is the double star 36 Ophiuchi. Shining like a pair of distant orange headlights, the 5th magnitude stars of 36 Ophiuchi are equal in brightness and color, separated by about 5 arcseconds. The two form a binary system, moving around each other about every 550 years.

A dark night, away from city lights, and a pair of dark-adapted eyes are the perfect formula for gazing in the direction of our galaxy's center. With its superb collection of objects and delicate dust lanes, many would say that it is the most interesting part of the night sky. MG

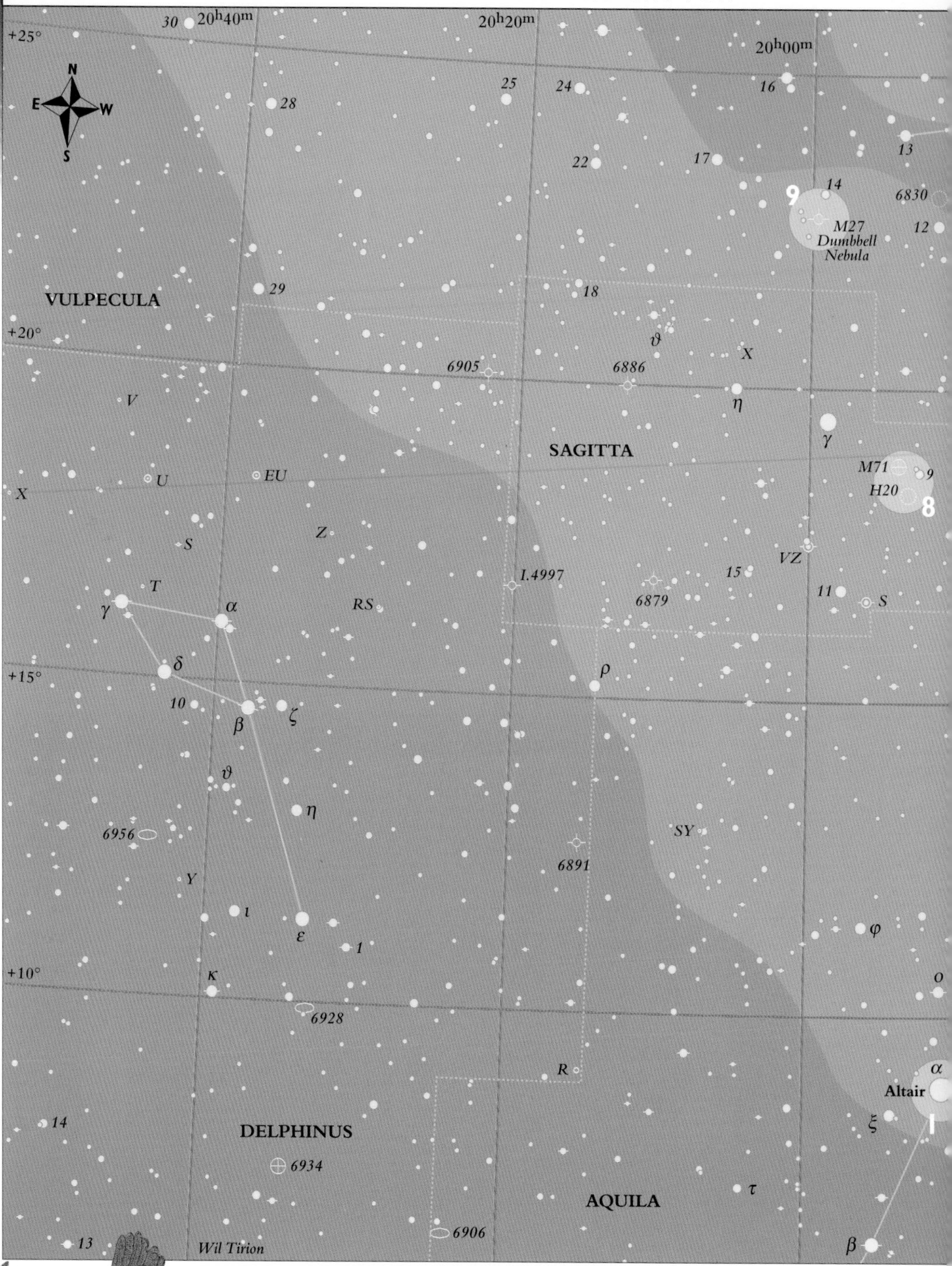
20h40m
20h20m
20h00m
+25°
+20°
+15°
+10°
N
E
W
S
VULPECULA
SAGITTA
DELPHINUS
AQUILA
M27
Dumbbell
Nebula
M71
H20
Altair
6830
6905
6886
6879
6956
6891
6928
6934
6906
I.4997
9
8
1
Wil Tirion

Hunting with the Eagle and the Fox

This part of the night sky is visible to skywatchers in both hemispheres from July through September. The starhop zooms in on parts of the constellations of Aquila, Vulpecula, and Delphinus, the Dolphin, and includes the third smallest constellation of all—Sagitta, the Arrow. Aquila's name is derived from the Arabic *Al Nasr al Tair*, meaning the Flying Eagle. Today, it is simply known as the Eagle, while Vulpecula is generally referred to as the Fox.

This region features rich fields containing thousands of sparkling bright stars, and coal-black dark nebulas. Wide swaths of dust clouds, known as the Great Rift, divide the Milky Way into two parallel bands of stars through the center of this starhop, as they do through nearby Cygnus, the Swan (see p. 264).

In Bayer's seventeenth-century star atlas, Uranometria, *he depicted Aquila as a bird belonging to the Greek God Zeus.*

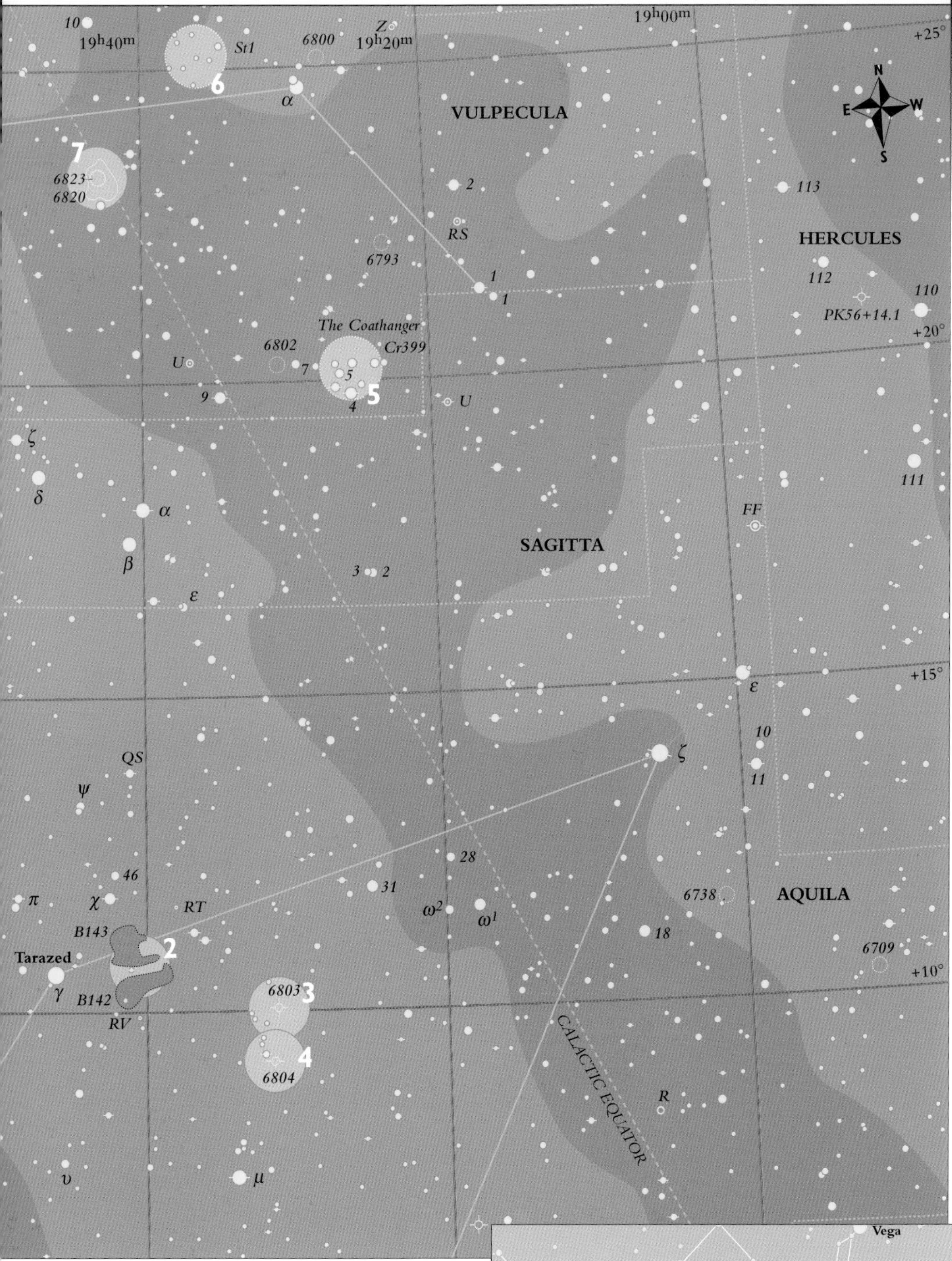
19h40m
19h20m
19h00m
+25°
+20°
+15°
+10°
N
E
W
S
VULPECULA
HERCULES
SAGITTA
AQUILA
St1
6800
6823-6820
6793
RS
113
112
110
PK56+14.1
The Coathanger
Cr399
6802
111
FF
QS
B143
RT
Tarazed
B142
RV
6803
6804
31
28
6738
18
6709
GALACTIC EQUATOR
R
1
2
3
4
5
6
7

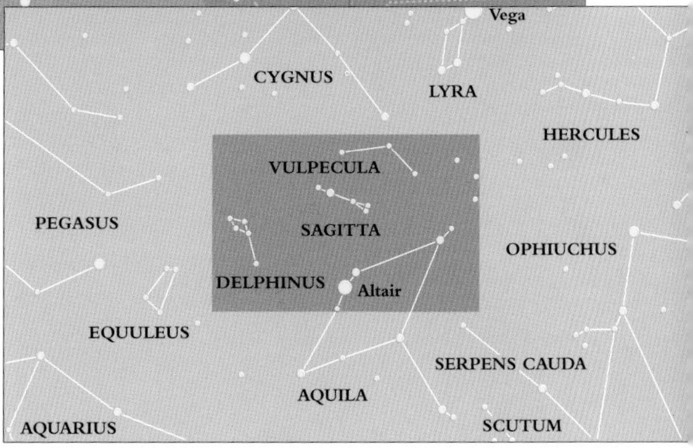
Vega
CYGNUS
LYRA
HERCULES
VULPECULA
SAGITTA
PEGASUS
OPHIUCHUS
DELPHINUS
Altair
EQUULEUS
SERPENS CAUDA
AQUILA
AQUARIUS
SCUTUM

❶ Altair (53 Alpha [α] Aquilae) Our jumping-off point for this starhop is the magnitude 1 white star Altair, the brightest star in Aquila. It is also the 12th brightest star in the night sky, marking the point where the Eagle's wing attaches to its body. At some 16 light-years away, Altair is also one of our closest neighbors among all the brighter stars in the night sky.

Spectroscopic studies show that Altair is rotating at an incredible speed of 160 miles per second (258 km/s), a truly remarkable feature. One Altair "day" is completed in approximately 6½ hours. By comparison, the Sun takes more than 25 Earth days to complete one rotation.

To find the next object on our starhop, first nudge your scope about 2 degrees northwest of Altair, to **Tarazed (Gamma [γ] Aquilae)**, then hop 1½ degrees west to the Double Dark Nebula.

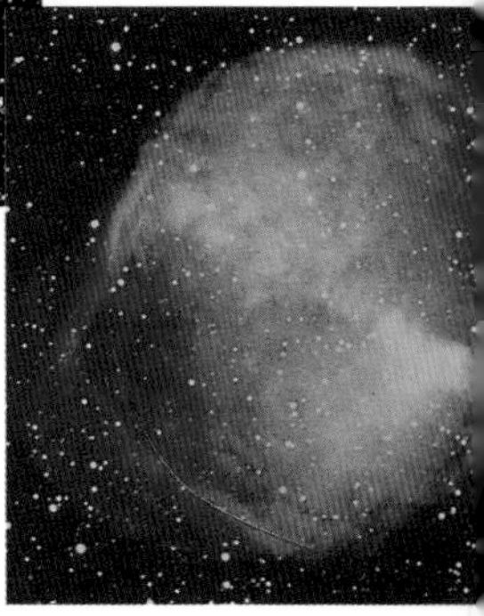

NGC 6820 and NGC 6823 (above), B142 and B143 (left), and M27 (below) all lie within the same patch of the northern Milky Way.

❷ **The Double Dark Nebula (B142 & B143)** The U-shaped nebula B143 and the elongated oval nebula B142 are made up of clouds of non-luminous gas and dust and are fine examples of dark nebulas. These clouds completely blot out the grainy star fields that are thought to be located some 5,000 light-years beyond. An 8 inch (200 mm) scope or larger and a dark sky are usually needed to study the edges of these clouds in detail.

❸ **NGC 6803** About 2 degrees west and a little south of the Double Dark Nebula is a fine planetary nebula, NGC 6803, glowing at magnitude 11. At magnitude 14, its central star is visible in scopes larger than 16 inches (400 mm). NGC 6803 is only 4 arcseconds in diameter, so it is essential to use high power when observing it.

❹ **NGC 6804** Hop 1 degree south to NGC 6804, another planetary nebula, making a fine pair with NGC 6803. At 60 arcseconds in diameter, NGC 6804 is an easier target to locate than its close neighbor. It is dimmer overall, at magnitude 13, but its central star is a little brighter.

Both NGC 6803 and NGC 6804 look like smaller and much fainter versions of the Ring Nebula (M57) in Lyra (see p. 266).

❺ **The Coathanger (Collinder 399)** Sweep some 10 degrees north of NGC 6804 to the spectacular open star cluster known as the Coathanger. It includes about 40 stars forming an arc of 6 bright stars running in an east-west direction, with a hook on the southern side. The Coathanger is best viewed with binoculars or a finderscope, because you can more easily orient yourself within a wider field. The magnitude 5.1 red giant at the bottom of the hook is the multiple star **4 Vulpeculae**.

❻ **Stock 1** Hop 5 degrees northeast to the large, very loose open star cluster Stock 1. It contains about 40 stars that are scattered amorphously, making it a challenge to resolve individual stars from the surrounding Milky Way star field. Stock 1 is about 1 degree in diameter, so observe it with binoculars or a finderscope, instead of with a telescope eyepiece.

❼ **NGC 6823** Swing your scope about 5 degrees southeast of Stock 1 to the open star cluster NGC 6823, surrounded by the elusive glow of the faint emission nebula **NGC 6820**.

M71 (above), in the constellation of Sagittae, is most often classified as a loosely-packed globular cluster.

The 30-member cluster itself, with a combined magnitude of 7.1, is an easy binocular target.

8 M71 (NGC 6838) Locate the 3rd magnitude stars **7 Delta [δ] Sagittae** and **12 Gamma [γ] Sagittae**, both red giant stars, about 4 degrees east of NGC 6823, to help find M71, lying midway between them. This compact cluster glows at magnitude 8.3, with many faint stars forming a triangular shape. Most astronomers consider M71 to be a loose globular rather than a compact open cluster because it contains red giant stars. However, no short-period variables, which are common to all globular clusters, have ever been detected in M71, leaving its true nature somewhat in doubt.

Harvard 20 Only 30 arcminutes southwest is the faint, sparse open cluster Harvard 20. An elongated box-like formation of 15 members, Harvard 20 is a challenge to pick out from the rich star field behind it.

9 The Dumbbell Nebula (NGC 6853, M27) Return to Gamma Sagittae before hopping about 4 degrees north to the magnitude 7.6 Dumbbell Nebula. Considered by many to be one of the finest planetary nebulas, it is also one of the closest to us at 1,000 light-years away. This softly glowing, greenish, hourglass-shaped planetary nebula is best viewed with a telescope, although, at 8 arcminutes across, it is also worth a look with binoculars or a finderscope.

There is so much more to see in this part of the night sky, so plan to return and slowly scan the wonderful Milky Way star fields and dark nebulas, again and again. RG

PLANETARY NEBULAS

Seen with early scopes, planetary nebulas were described as somewhat like the disks of planets, when actually, they are the last stage of a star's life.

When a star like the Sun gets old, it swells to become a red giant, soon running out of hydrogen and helium to burn in its core. Instead, it continues its nuclear reactions in shells around the core. As it does this, the star begins to pulsate, culminating in a shedding of its outer layers, which form a planetary nebula. This process reveals the hot central core which, in turn, pours ultraviolet radiation out into the expanding planetary shell, causing it to glow. This ghost-like shell survives for only about 50,000 years before it dissipates, leaving the core behind as a slowly cooling white dwarf star.

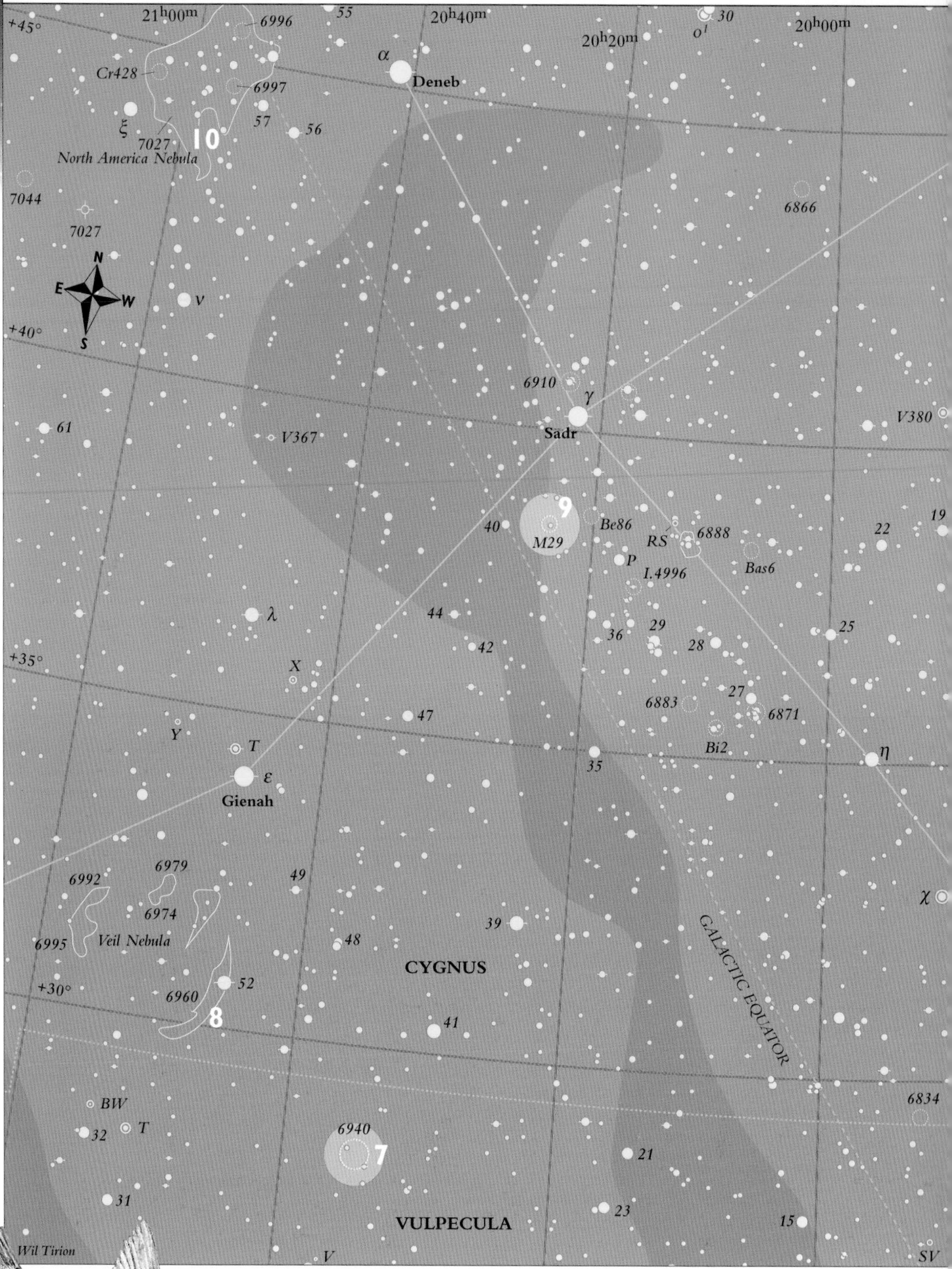
21h00m
20h40m
20h20m
20h00m
+45°
+40°
+35°
+30°
Deneb
Sadr
Gienah
North America Nebula
Veil Nebula
CYGNUS
VULPECULA
GALACTIC EQUATOR
M29
6996
6997
Cr428
7027
7044
6866
6910
Be86
6888
I.4996
Bas6
6883
6871
Bi2
6992
6979
6974
6995
6960
6940
6834
V367
V380
N
E
W
S
Wil Tirion

Cygnus in the Milky Way

For mid-northern observers, the two constellations traversing this region of sky hover near the zenith on July and August nights, making these the best months for viewing. This particular part of the Milky Way is rich with a variety of celestial objects in the constellations of Cygnus, the Swan, and Lyra, the Lyre.

This image of Cygnus, the Swan, was painted as part of a larger fresco on a ceiling of the Villa Farnese in Caprarola, Italy, in 1575.

Cygnus forms a prominent cross shape that straddles the Milky Way at the point where the galaxy is split by a dark dust lane. It is also known as the Northern Cross—the Northern Hemisphere's answer to Crux, the Southern Cross (see p. 246). The constellation of Lyra appears as a small box lying to the west of Cygnus.

Vega in Lyra, Deneb in Cygnus, and Altair in Aquila (south of this area), are the three 1st magnitude stars that form what is known in the Northern Hemisphere as the Summer Triangle.

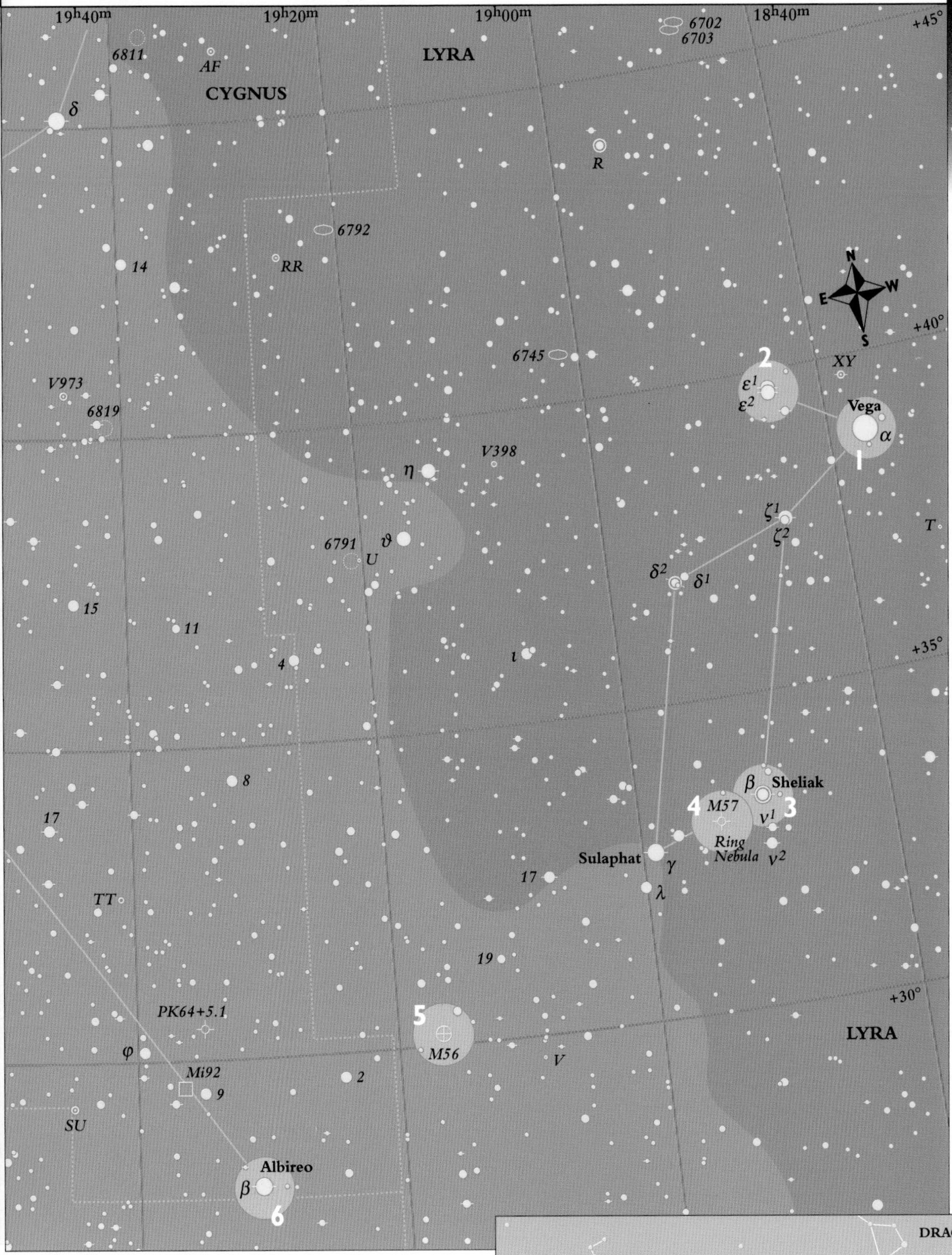
19h40m
19h20m
19h00m
18h40m
+45°
+40°
+35°
+30°
CYGNUS
LYRA
LYRA
6811
AF
δ
6702
6703
R
6792
RR
14
N
E
W
S
6745
XY
2
ε1
ε2
Vega
α
1
V973
6819
V398
η
ζ1
ζ2
T
6791
ϑ
U
δ2
δ1
15
11
4
ι
8
β
Sheliak
17
4
M57
ν1
3
Ring Nebula
ν2
Sulaphat
γ
λ
17
TT
19
PK64+5.1
5
M56
φ
V
Mi92
9
2
SU
Albireo
β
6

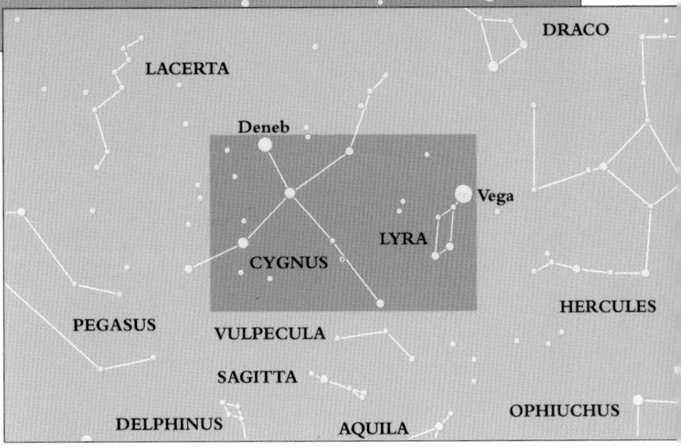
DRACO
LACERTA
Deneb
Vega
LYRA
CYGNUS
PEGASUS
VULPECULA
HERCULES
SAGITTA
DELPHINUS
AQUILA
OPHIUCHUS

1 Vega (3 Alpha [α] Lyrae) The starting point for this tour of celestial gems in the Milky Way is the brightest star in the Northern Hemisphere's summer sky—Vega. This bright, blue-white star is very close to zero magnitude, being listed at magnitude 0.03. Vega is the fifth brightest star in the entire sky, and is estimated to be intrinsically about 58 times brighter than the Sun. It is 25 light-years away from Earth.

Vega appears in many ancient legends. Its common name comes from the Arabic *Al Nasar al Waki*, meaning "the swooping eagle." In one Greek legend, it symbolizes the lyre of Hermes, so it is sometimes referred to as "the harp star."

Because Vega's magnitude does not vary significantly, the star is often used by astronomers when calibrating photometric equipment that is used to measure stellar brightness.

❷ The Double Double Star (4 Epsilon[1] & 5 Epsilon[2] [ε[1], ε[2]] Lyrae) The famous "double double" star, lying about 1½ degrees northeast of Vega, consists of four white stars. The two primary stars are 208 arcseconds apart, and each is itself a binary star. The magnitude 6 companion of Epsilon[1] is 2.8 arcseconds from its magnitude 5.1 primary. The Epsilon[2] system is composed of two 5th magnitude stars lying 2.6 arcseconds apart.

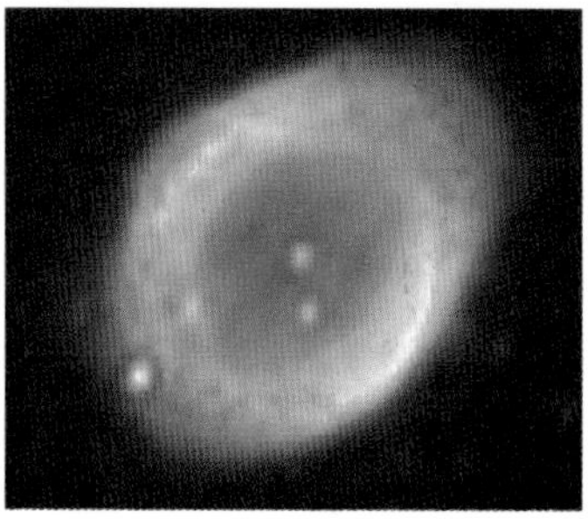

The Cross shape of Cygnus (above) with Vega at center right. The Ring Nebula (left).

❸ Sheliak (10 Beta [β] Lyrae) Hop about 6 degrees south to Sheliak, a bright eclipsing binary variable, whose magnitude fluctuates from 3.3 down to 4.4. The two unequally bright stars in this system orbit about a common center of gravity over a period of 12.9 days. The stars seem to be constantly transferring massive amounts of gas between their atmospheres. Being only about 22 million miles (35 million km) apart, the components of Sheliak have never been visually separated.

❹ The Ring Nebula (M57, NCG 6720) Located about midway between Sheliak and **Sulaphat (14 Gamma [γ] Lyrae)** is the famous bright planetary nebula M57. It is often referred to as the "smoke ring" or "doughnut" of the sky. M57 is the result of a star that blew off a massive amount of gas thousands of years ago. The shell of gas is easy to see through an 8 inch (200 mm) scope and has a slightly greenish cast to it. From a dark site, with a 15 inch (380 mm) scope, you can glimpse the magnitude 17 dwarf star from which the gas was expelled.

❺ M56 (NGC 6779) Some 4½ degrees southeast from Sulaphat is the globular cluster M56. This small, circular 8th magnitude mass of stars was discovered by Messier in 1779. Many faint stars around the edges of the cluster can be resolved with scopes smaller than 8 inches (200 mm).

❻ Albireo (6 Beta [β] Cygni) Continue southeast about 3½ degrees from M56 to Albireo. This is probably the finest color-contrast double star and a real showpiece at any star party. This magnitude 3.1 star appears bright yellow or topaz with a sapphire-blue companion shining at magnitude 5.1. The stars are 34.3 arcseconds apart, making them an easy pair to separate in any size telescope. While near Albireo, scan the rich star clouds with binoculars, looking for the myriad clusters and dark nebulas.

❼ NGC 6940 From Albireo shift about 14 degrees east to the open cluster NGC 6940, across the border in Vulpecula. About 60 to 80 stars make up this large, scattered cluster. Nine bright stars appear centered within the cluster, while other stars seem to form chain-like strings just north of the center of the cluster.

The North American Nebula (above) is a mix of bright gas and dark dust lanes. The Pelican Nebula appears to its right (west).

8 The Veil Nebula (NGC 6960) Hop about 6 degrees northeast of NGC 6940 to magnitude 2.5 **Epsilon [ε] Cygni**. Three degrees due south of Epsilon is the magnitude 4.2 star **52 Cygni**, straddling the long filaments of the western side of the Veil Nebula. The nebula is the expanding shell of gas from a supernova explosion. Your best views will be through an Oxygen III or other high-contrast filter (see p. 56). To fully explore this intriguing object, trace the intertwining filaments from end to end with a 6 inch (150 mm) scope or larger.

9 M29 (NGC 6913) Some 10 degrees northwest of the Veil Nebula is magnitude 2.1 **Sadr (37 Gamma [γ] Cygni)**. Use this star to find the open star cluster M29, about 2 degrees south. This attractive open cluster lies on the edge of one of the obscuring bands of the Great Rift dust clouds that seem to cut through the Milky Way in Cygnus. The seven brighter stars in M29 form a dipper-like asterism.

10 The North American Nebula (NGC 7000) Hop about 6 degrees from Sadr to the bright magnitude 1.3 star **Deneb (50 Alpha [α] Cygni)**, "the tail of the swan." Another 3 degrees east of Deneb is NGC 7000. From a very dark site, this giant, diffuse emission nebula can just be seen with the naked eye, glowing softly at magnitude 5.9. Its common name arises from its shape, which is very similar to that of the North American continent, including Mexico.

Take your time to fly with the Swan and seek out and explore the multitude of open clusters and double stars that are embedded in the star fields of Cygnus in the Milky Way. RG

THE VISUAL MILKY WAY

One of the joys of stargazing is to sit out under a very dark sky and watch the broad swath of the faintly glowing Milky Way slowly pass overhead. Take the time to study the Milky Way without binoculars or a telescope. Look for the delicate interweaving of the bands of obscuring dust clouds and the hazy star fields. After observing the Milky Way with the naked-eye, scan the band with binoculars, stopping to study the rich star fields and coal-black dark nebulas throughout it. The plane of the Milky Way Galaxy is made up of vast clouds of glowing gas and stars cut by dark lanes of dust (left). Billions of faint stars, many that are far too faint for our eyes to separate into individual points of light, merge like a mottled, grayish fog.

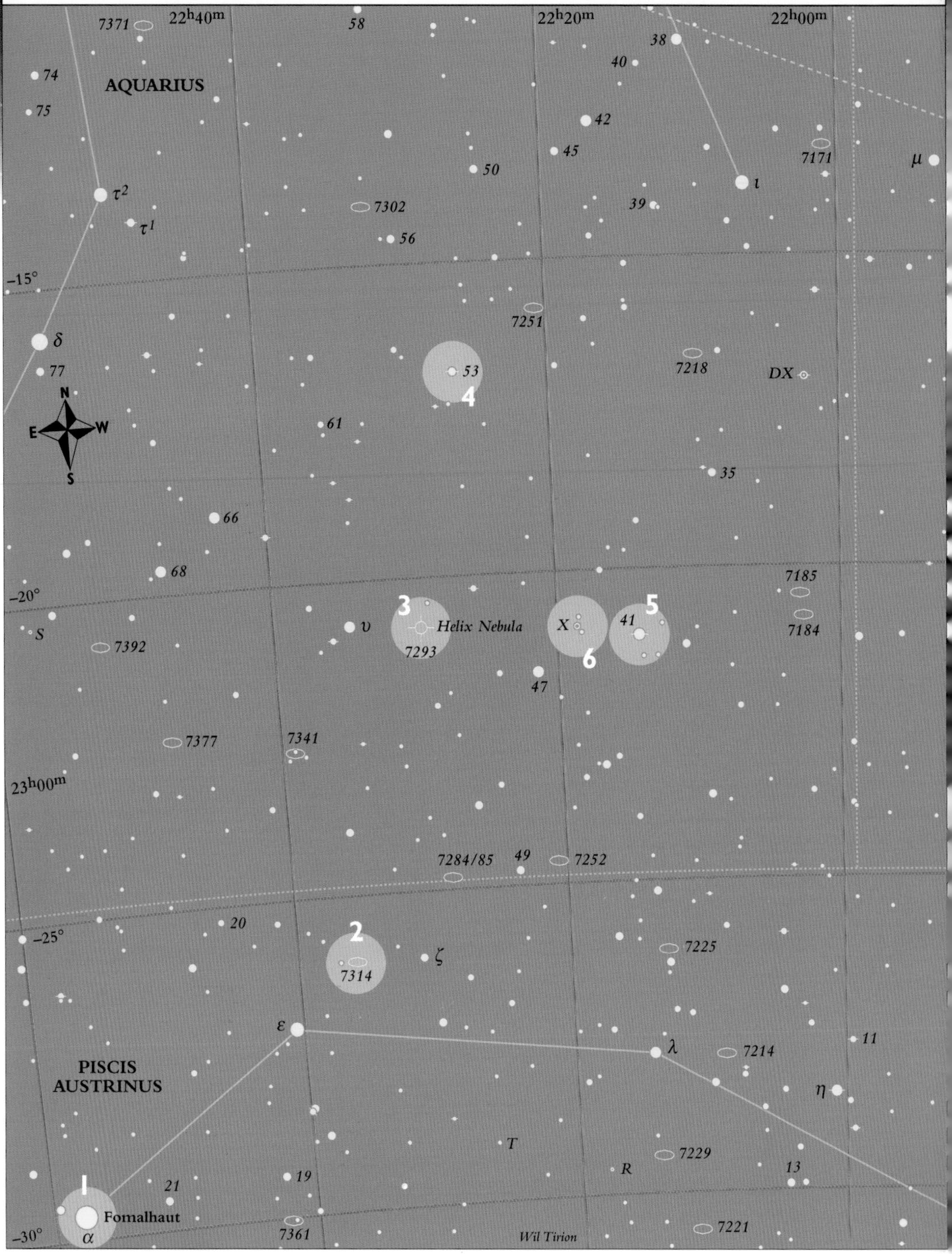
22h40m
22h20m
22h00m
23h00m
AQUARIUS
PISCIS AUSTRINUS
-15°
-20°
-25°
-30°
N
E
W
S
7371
58
74
75
38
40
42
45
50
7171
μ
ι
τ2
τ1
7302
56
39
δ
77
7251
53
4
7218
DX
61
35
66
68
7185
7184
3
Helix Nebula
7293
υ
X
41
5
6
47
S
7392
7377
7341
7284/85
49
7252
20
2
7314
ζ
7225
ε
λ
7214
11
η
T
7229
R
13
19
21
1
Fomalhaut
α
7361
7221
Wil Tirion

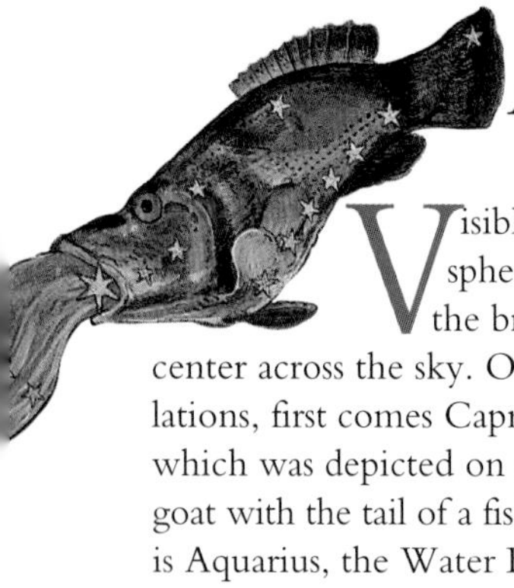

Among the Watery Constellations

Visible from both hemispheres, this region follows the brilliant area of the galactic center across the sky. Of several water constellations, first comes Capricornus, the Sea Goat, which was depicted on charts of long ago as a goat with the tail of a fish. Following Capricornus is Aquarius, the Water Bearer, and Pisces, the Fishes, partly shown on the locater map (right). All three are zodiacal constellations, which the planets regularly pass. Also in this region are the constellations of Cetus, the Whale, and Piscis Austrinus, the Southern Fish—the parent of Pisces, according to ancient legend.

This starhop begins at Fomalhaut, near the southeastern corner of the main sky chart, before going on to browse around the celestial delights of parts of three watery constellations—Aquarius, Capricornus, and Piscis Austrinus.

Piscis Austrinus, the Southern Fish, is often depicted as a fish drinking the flow of water poured from the urn of Aquarius.

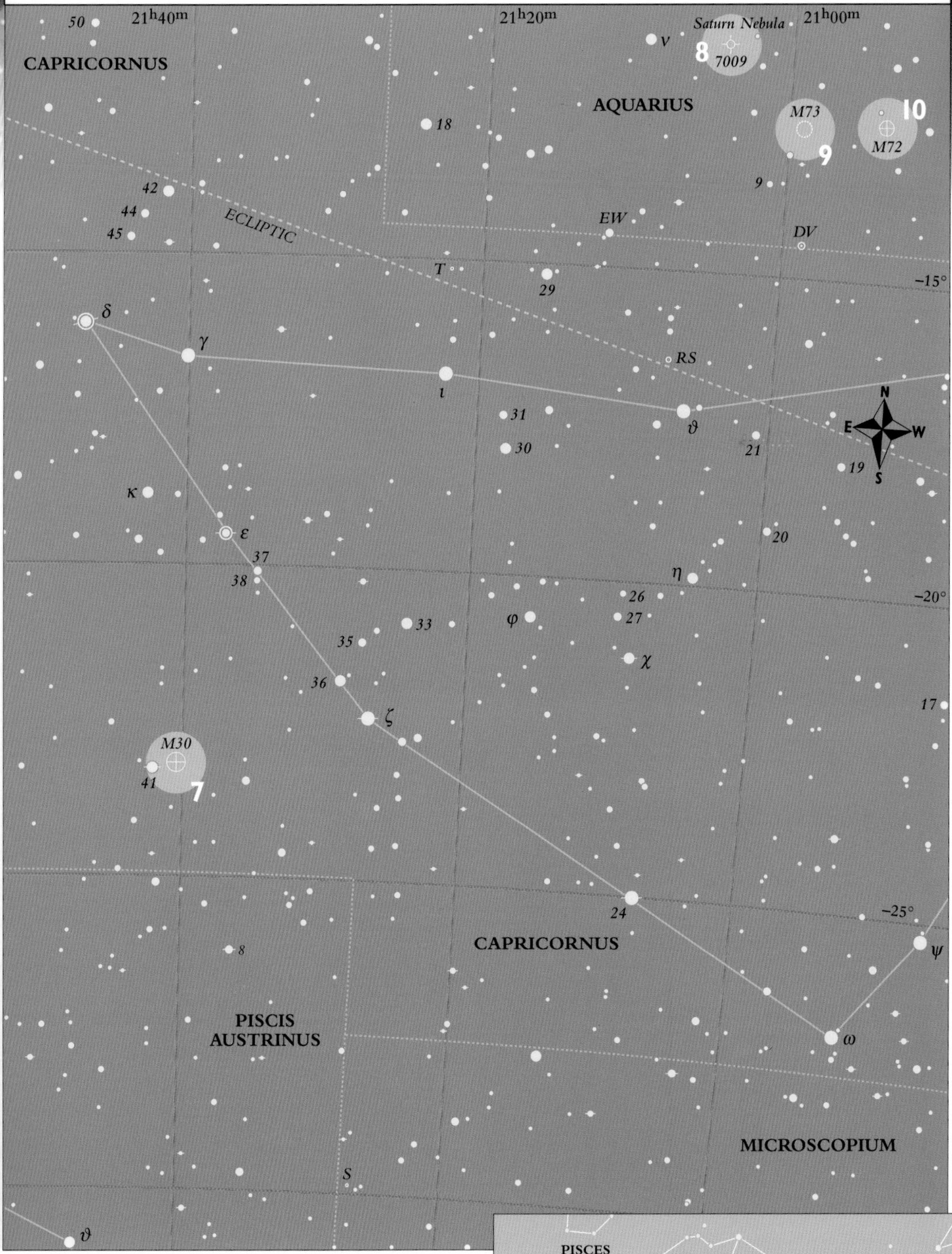
21h40m
21h20m
21h00m
CAPRICORNUS
AQUARIUS
Saturn Nebula
7009
M73
M72
ECLIPTIC
EW
DV
T
RS
−15°
−20°
−25°
N
E
W
S
M30
CAPRICORNUS
PISCIS AUSTRINUS
MICROSCOPIUM

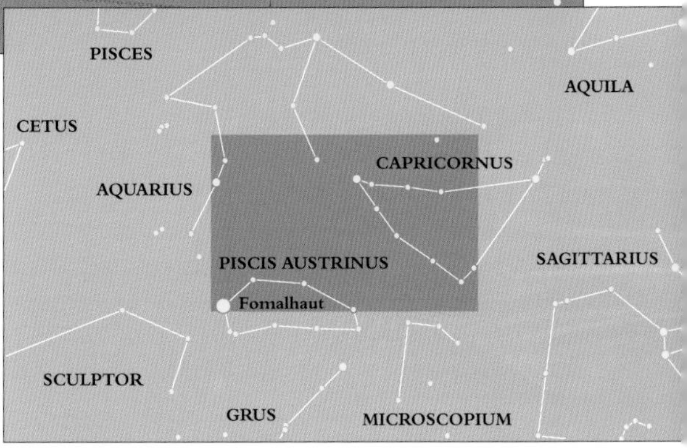
PISCES
AQUILA
CETUS
CAPRICORNUS
AQUARIUS
PISCIS AUSTRINUS
SAGITTARIUS
Fomalhaut
SCULPTOR
GRUS
MICROSCOPIUM

❶ Fomalhaut (24 Alpha [α] Piscis Austrinus) Our starting point for this star-hop is the brilliant magnitude 1.2 blue-white star Fomalhaut, the 18th brightest star in the night sky. This star lies in quite a barren region of the southern sky and because of this has become known as "The Solitary One." Its isolation makes it very easy to find. Fomalhaut lies some 23 light-years distant from Earth and is estimated to have about twice the diameter of the Sun and to be some 14 times more luminous.

Fomalhaut's name comes from the Arabic *Fum al Hut*, meaning "mouth of the fish," as it marks the position of the Fish's mouth in the constellation of Piscis Austrinus. The Greek poet Aratos described Fomalhaut as "One large and bright by the Pourer's feet," with the pourer, of course, being Aquarius, the Water Bearer.

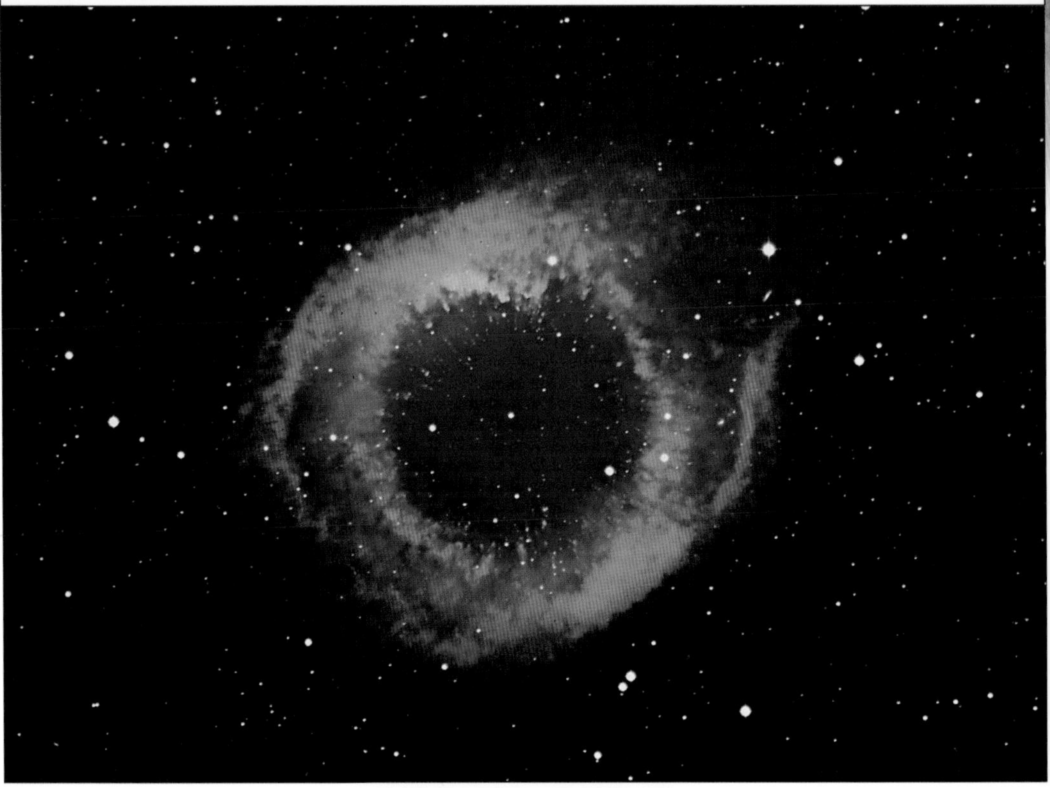

The Helix Nebula (above) is a highlight in Aquarius. Fomalhaut, a beacon of light (left).

❷ **NGC 7314** To find the next object on this hop, head just over 4 degrees northwest, to **18 Epsilon [ε] Piscis Austrius**. Another 1½ degrees farther northwest will bring you to NGC 7314. In a 4 inch (100 mm) scope, this 11th magnitude spiral galaxy can be seen as a fuzzy patch of light with a small, but fairly bright, nucleus. A larger scope reveals considerably more detail surrounding its center.

❸ **The Helix Nebula (NGC 7293)** Cross the constellation border into Aquarius and aim your scope some 5 degrees north to find the star **59 Upsilon [υ] Aquarii**. About 1 degree to the west of Upsilon lies the famed Helix Nebula (see Box). The Helix is a planetary nebula—one of the best known of its type. You should just be able to capture a view of it in a good pair of binoculars on a clear, dark night, appearing as a circular patch of light about a ¼ degree in diameter. When using a scope, you need a low-power, wide-field eyepiece to really appreciate it. If you have an Oxygen III filter, you can treat yourself to a fantastic view.

❹ **53 Aquarii** Continue swinging your scope northward, about 4 degrees, until you find 53 Aquarii. This pretty, yellowish pair of 6th magnitude stars was much easier to separate when measured early in the last century. They were then about 10 arcseconds apart, but now the distance has shrunk to a mere 3 arcseconds. Even so, you should easily separate them with a 3 inch (75 mm) scope on a steady night.

❺ **41 Aquarii** Direct your scope 5 degrees southwest of 53 Aquarii to the next double star, 41 Aquarii. The brighter of this pair, at magnitude 5.6, has a distinctive yellow color, while its 7th magnitude companion is yellow-white. At 5 arcseconds apart, they make a fine double for small scopes.

❻ **X Aquarii** Shift your telescope 1 degree east of 41 Aquarii, then nudge it just a few arcminutes north to view the field containing X Aquarii. A long-period Mira-type variable, X is also one of the rare S-type variables, which show zirconium oxide in their spectra. This orange-red star varies between magnitude 7.5 and 14.8 over a period of 312 days, so you need to catch it near maximum to see it clearly.

❼ **M30 (NGC 7099)** To find M30, make another border crossing, into Capricornus, 8 degrees west and 2 degrees south of X Aquarii, to find 5th magnitude **41 Capricorni**. About ½ degree west of it lies the small but attractive globular cluster M30, which is just visible with binoculars. A 3 inch (75 mm) telescope will begin to resolve some of its stars, but the view through a 6 inch (150 mm) telescope is without doubt the more attractive one.

The globular cluster M30 (above) is the only bright deep-sky object to be found in the constellation of Capricornus.

8 The Saturn Nebula (NGC 7009) Get ready to hop quite a distance to our next target. First, stop off at 4th magnitude **Zeta [ζ] Capricorni**, then turn northwest where, 7 degrees away, you will spot the less bright **Theta [θ] Capricorni**. Finally, cross back into Aquarius by moving 6 degrees farther north, to the Saturn Nebula. This delightful little 8th magnitude planetary nebula, a little more than 1 degree west of **Nu [ν] Aquarii**, glows a beautiful blue color. A 3 inch (75 mm) scope with moderate magnification shows the interesting elliptical shape of this object. It was given the name of the Saturn Nebula because of its almost ring-like projections that can be seen with 12 inch (300 mm) telescopes or larger.

9 M73 (NGC 6994) Aim your scope 2 degrees southwest of the Saturn Nebula to see M73. While not exactly a spectacular sight, it has historic significance, since Charles Messier included M73 in his catalog. It is merely a tight group of four faint stars between 10th and 12th magnitude.

10 M72 (NGC 6981) About 1 degree west of M73 is M72. Although it is moderately concentrated, 9th magnitude M72 is a relatively dim globular cluster, and one which is, unfortunately, quite difficult to resolve. A 10 inch (250 mm) scope is needed to show a hint of some of its multitude of stars, and only observers with large scopes will obtain a good view of it.

Although this patch of sky is relatively sparse, our starhop around some of the watery constellations has shown that even here there are many fascinating objects waiting to be found. MG

THE HELIX NEBULA

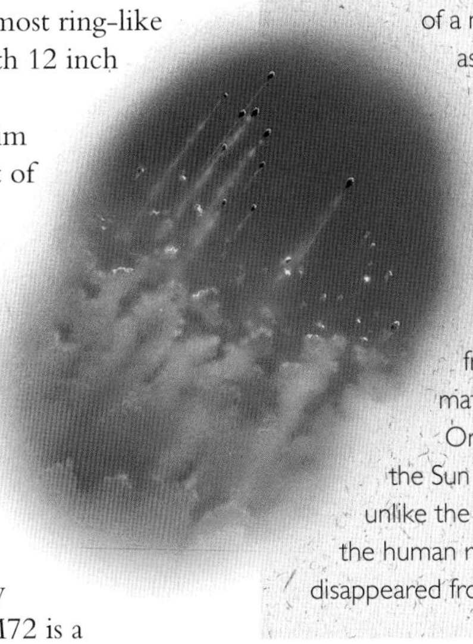

Lying at a distance of about 450 light-years from Earth, the Helix Nebula is the closest planetary nebula to us, and therefore appears as the largest in the sky. It has been the subject of a number of serious studies. In 1996, astronomers researching images of the Helix from the Hubble Space Telescope found thousands of comet-like gaseous fragments, especially near the inner edge of the nebula's ring. It is thought that these "knots" may have formed when new material from the star interacted with material that had been ejected earlier. One day, billions of years from now, the Sun will have a shell, probably not unlike the Helix Nebula, but by that time, the human race will probably have long since disappeared from Earth.

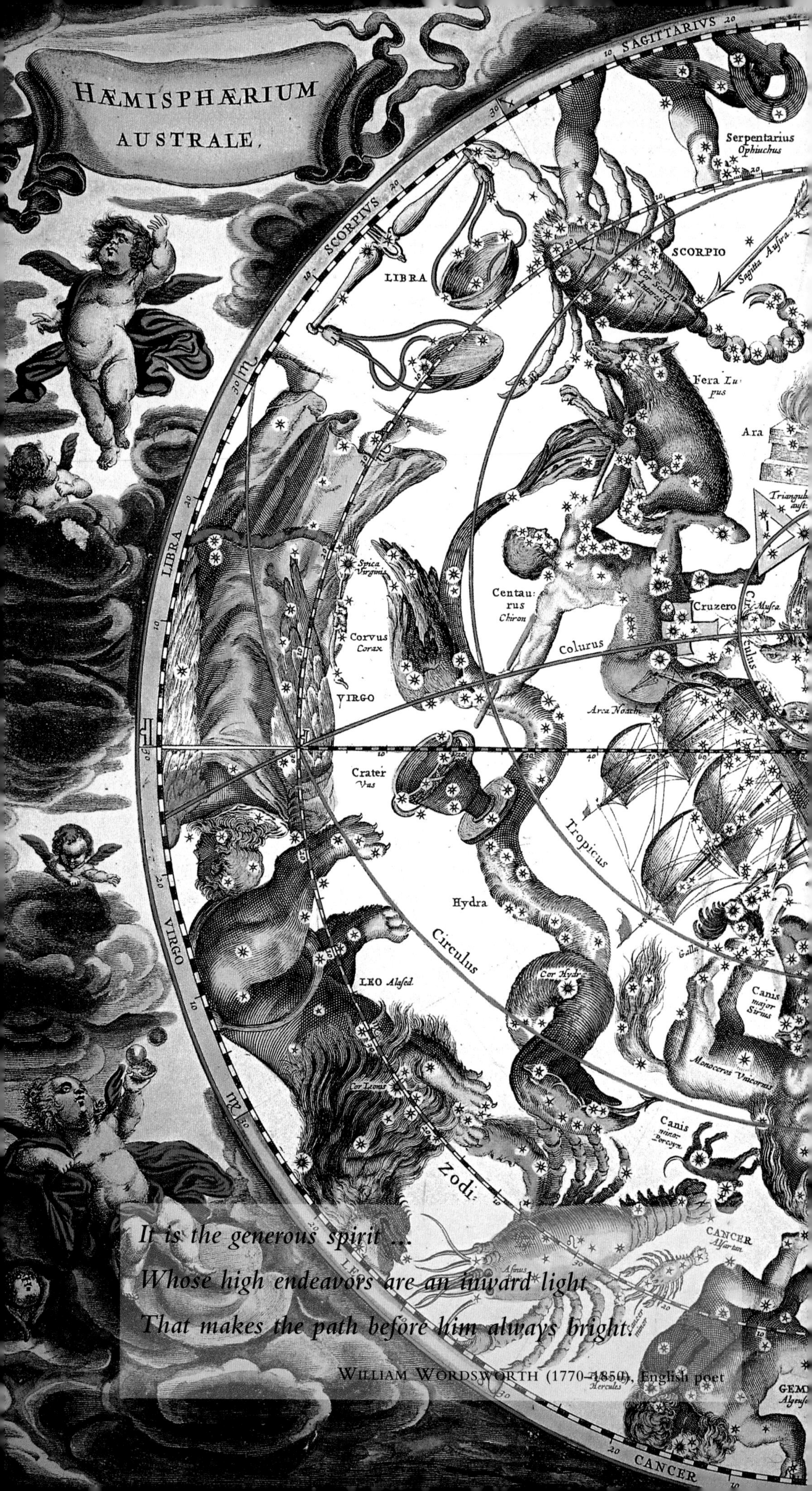
HÆMISPHÆRIUM AUSTRALE.
SAGITTARIUS
SCORPIUS
LIBRA
VIRGO
CANCER
Serpentarius Ophiuchus
SCORPIO
Sagitta Australis
LIBRA
Cor Scorpii Antares
Fera Lupus
Ara
Spica Virginis
Centaurus Chiron
Cruzero
Musca
Corvus Corax
Colurus
VIRGO
Arca Noach
Crater Vas
Tropicus
Hydra
Circulus
LEO Alased
Cor Hydræ
Canis major Sirius
Monoceros Unicornis
Cor Leonis
Zodi
Canis minor Procyon
CANCER Alsartan

It is the generous spirit …
Whose high endeavors are an inward light
That makes the path before him always bright.

WILLIAM WORDSWORTH (1770–1850), English poet

Resources Directory

Indus
Grus
Toucan
Phœnix
Dorado
Cetus
Balena
PISCES
Alsemcha
Capricorni
Æquinoctialis
Lepus
Orion
TAURUS
Aldebaran
PISCES
ARIES
TAURUS
GEMINI

FURTHER INFORMATION

Armchair Astronomy

The Cambridge Atlas of Astronomy, by Jean Audouze and Guy Israël (Cambridge University Press, 1994). A massive illustrated encyclopedia of all areas of astronomy, from cosmology to Earth science. (INTERMEDIATE)

A Concise Dictionary of Astronomy, by Jacqueline Mitton (Oxford University Press, 1991). A dictionary is one of those essentials no backyard sky-watcher can do without. (BEGINNER)

Exploration of the Universe, by George O. Abell, David Morrison, and Sidney C. Wolff (Saunders College Publishing, 1993). This is one of the best general reference textbooks, updated often. (INTERMEDIATE)

Hubble Vision, by Carolyn Collins Petersen and John C. Brandt (Cambridge University Press, 1995). As its subtitle says, this well-illustrated book is about "doing astronomy with the Hubble Space Telescope." (BEGINNER–INTERMEDIATE)

The New Astronomy, by Nigel Henbest and Michael Marten (Cambridge University Press, 1996). A colorful guided tour of astronomers' current understanding of the universe. (BEGINNER)

One Giant Leap (Discovery Channel, 1994). A 90-minute video looking at the race to the Moon. Discover the untold story of the Apollo project and the adventure of the space race. (BEGINNER)

History & Mythology

History of Astronomy, by Antonie Pannekoek (Dover, 1989). This is still the best general history of astronomy from antiquity up to about 1930. (INTERMEDIATE)

The Norton History of Astronomy and Cosmology, by John North (W. W. Norton, 1994). A comprehensive study of celestial knowledge from pre-history to the modern era. (INTERMEDIATE)

Star Names: Their Lore and Meaning, by Richard H. Allen (Dover, 1963). A comprehensive look at the history and legends behind star and constellation names. First published in 1899. (BEGINNER–INTERMEDIATE)

Stargazers (Discovery Channel, 1994). A 50-minute video that takes a look at the notions of Percival Lowell and explores his 100-year-old observatory; provides a wider view of natural phenomena. (BEGINNER)

Celestial Atlases

Atlas of the Moon, by Antonín Rükl (Kalmbach, 1990). No moonwatcher should be without these beautiful and detailed charts of the Moon's near side. (INTERMEDIATE)

The Cambridge Star Atlas, by Wil Tirion (Cambridge University Press, 1996). Shows the whole sky in 20 maps, with a list of deep-sky objects on each facing page. (BEGINNER)

Norton's 2000.0 Star Atlas and Reference Book, by Ian Ridpath (Longman Scientific and Technical, 1989). Contains plenty of information on telescopes and observing. Sixteen maps display the whole sky. (BEGINNER–INTERMEDIATE)

Sky Atlas 2000.0, by Wil Tirion (Cambridge University Press, 1981). A must for every amateur astronomer; covers the entire sky. (BEGINNER)

Uranometria 2000.0, by Wil Tirion, Barry Rappaport, and George Lovi (Willmann-Bell, 1987 & 1988). The next step after *Sky Atlas 2000.0*; goes almost 2 magnitudes fainter. A matching *Deep-Sky Field Guide to Uranometria 2000.0* (Willmann-Bell, 1992) lists basic data for non-stellar objects on the *Uranometria 2000.0* charts. (INTERMEDIATE–ADVANCED)

Sky Catalogs

NGC 2000.0, edited by Roger W. Sinnott (Cambridge University Press/Sky Publishing, 1988). This catalog of non-stellar objects updates and replaces Dreyer's *New General Catalogue* and *Index Catalogues*. (INTERMEDIATE)

Sky Catalogue 2000.0, edited by Alan Hirshfeld and Roger W. Sinnott (Cambridge University Press, 1991 & 1985). A two-volume work listing basic data, such as position, size, and brightness, for stars (8th magnitude or brighter) and non-stellar objects. (INTERMEDIATE)

Handbooks for Observers

Amateur Radio Astronomer's Handbook, by John Potter Shields (Crown Publishers, 1986). Just about the only guidebook on the subject of radio astronomy. Provides detailed how-to instructions for building and hooking up antennas and other equipment. (INTERMEDIATE–ADVANCED)

The Backyard Astronomer's Guide, by Terence Dickinson and Alan Dyer (Camden House, 1991). Thorough and authoritative handbook with reliable information on observing, and sound, clear-headed advice on equipment. (BEGINNER–INTERMEDIATE)

Binocular Astronomy, by Craig Crossen (Willmann-Bell, 1992). A guide to the seasonal sky sights visible with an ordinary pair of binoculars. (BEGINNER–INTERMEDIATE)

Celestial Objects for Common Telescopes, by Thomas W. Webb (Dover, 1962). An observer's classic from the early 1900s; delightful to read for its period flavor. (BEGINNER–INTERMEDIATE)

City Astronomy, by Robin Scagell (Sky Publishing, 1994). Good advice on how to make the best of light-polluted skies. (BEGINNER–INTERMEDIATE)

Skywatching, by David H. Levy (The Nature Company/ Time-Life Books, 1994). A comprehensive and beautifully illustrated guidebook covering all aspects of astronomy. (BEGINNER)

Astro-imaging

Astrophotography For the Amateur, by Michael A. Covington (Cambridge University Press, 1991). The first book to purchase if you want to try your hand at celestial picture-taking with camera and film. (BEGINNER–INTERMEDIATE)

Choosing and Using a CCD Camera, by Richard Berry (Willmann-Bell, 1992). Answers those "what-are-CCDs-all-about" questions. (INTERMEDIATE–ADVANCED)

Telescope Building

Build Your Own Telescope, by Richard Berry (Willmann-Bell, 1994). Includes detailed instructions for telescopes that are straightforward to build and will provide excellent performance. (INTERMEDIATE–ADVANCED)

The Dobsonian Telescope, by Richard Berry (Willmann-Bell, 1997). If you are even thinking about building a big telescope, this is the book to get. (INTERMEDIATE–ADVANCED)

The Solar System

Atlas of Neptune, by Garry E. Hunt and Patrick Moore (Cambridge University Press, 1994). Up-to-date guide to the planet, incorporating Voyager findings. (INTERMEDIATE)

The Giant Planet Jupiter, by John H. Rogers (Cambridge University Press, 1995). A detailed and beautifully illustrated guide to Jupiter. (INTERMEDIATE)

Introduction to Observing and Photographing the Solar System, by Thomas A. Dobbins, Donald C. Parker, and Charles F. Capen (Willmann-Bell, 1988). A practical guidebook to exploring the Sun's family with an amateur telescope. (INTERMEDIATE)

Mars, by Peter Cattermole (Chapman & Hall, 1992). An up-to-date overview of Mars as a planet—its volcanoes, craters, deserts, and ice caps. (INTERMEDIATE–ADVANCED)

Mercury: the Elusive Planet, by Robert G. Strom (Smithsonian Institution Press, 1987). A clear and detailed look at a planet that superficially resembles the Moon but whose interior is more like Earth's. (INTERMEDIATE)

The New Solar System, edited by J. Kelly Beatty and Andrew Chaikin (Cambridge University Press, 1990). All Solar System topics for the interested lay-reader. (INTERMEDIATE)

Observing Comets, Asteroids, Meteors, and the Zodiacal Light, by Stephen J. Edberg and David H. Levy (Cambridge University Press, 1994). Practical and thorough information. (INTERMEDIATE)

Observing the Sun, by Peter O. Taylor (Cambridge University Press, 1991). How you can safely view the Sun with any telescope. (INTERMEDIATE)

The Once and Future Moon, by Paul Spudis (Smithsonian Institution Press, 1996). A superb explanation of the Moon's history and features. (INTERMEDIATE)

Orbit, by Jay Apt, Michael Helfert, and Justin Wilkinson (National Geographic Society, 1996). An astronaut's-eye view of Earth combined with text explaining the amazing images. (BEGINNER)

The Planet Saturn, by A. F. O'D. Alexander (Dover, 1980). A historical survey of pre-Voyager information and a wonderful archive for a cloudy night. (INTERMEDIATE–ADVANCED)

Planets X and Pluto, by William G. Hoyt (University of Arizona Press, 1980). Retells the absorbing story of how off-the-track guesswork led astronomers to find the planet Pluto. (BEGINNER)

Rendezvous in Space, by John C. Brandt and Robert D. Chapman (W. H. Freeman, 1992). Explains what astronomers have learned about comets, including the much-studied return of Halley's Comet. (INTERMEDIATE)

Uranus, by Ellis Miner (Ellis Horwood, 1997). Brings the story up to date through the Voyager 2 encounter, with all its scientific findings presented clearly. (INTERMEDIATE)

Venus Revealed, by David Harry Grinspoon (Addison-Wesley, 1997). A highly readable account by a planetary scientist who has studied Venus. (INTERMEDIATE)

Lights in the Sky

The Aurora Watcher's Handbook, by Neil Davis (University of Alaska Press, 1992). Lots of how-to details on observing and photographing the northern and southern lights. (INTERMEDIATE)

Light and Color in the Outdoors, by M. G. J. Minnaert (Springer-Verlag, 1993). Looks at what makes rainbows, sun dogs, mirages, and other phenomena in the atmosphere. (BEGINNER–INTERMEDIATE)

Observing Earth Satellites, by Desmond King-Hele (Van Nostrand Reinhold, 1983). A detailed how-to book for satellite watchers. (INTERMEDIATE)

Thunderstones and Shooting Stars, by Robert T. Dodd (Harvard University Press, 1986). Short guide to meteorites and asteroids for the interested lay-reader. (INTERMEDIATE)

UFOs: The Public Deceived, by Phillip J. Klass (Prometheus Books, 1983). A clear-headed view of a topic rife with uncritical thinking. (BEGINNER)

Exploring Deep Space

Burnham's Celestial Handbook, by Robert Burnham, Jr. (Dover, 1977–78). Three handy volumes with detailed information on many celestial objects outside the Solar System. (INTERMEDIATE)

Hartung's Astronomical Objects for Southern Telescopes, by David Malin and David J. Frew (Cambridge University Press, 1995). Brings up to date a classic Southern Hemisphere guide. (INTERMEDIATE)

Messier Marathon Observer's Guide, by Don Machholz (Makewood Productions, 1994). Explains how to plan your observing session to see all 110 Messier objects in a single night. (ADVANCED)

Observing Variable Stars, by David H. Levy (Cambridge University Press, 1989). An easy introduction explaining how amateurs with backyard telescopes can follow variable stars. (INTERMEDIATE)

Star-Hopping; Your Visa to Viewing the Universe, by Robert A. Garfinkle (Cambridge University Press, 1997). Round-the-year tours for starhopping through all areas of the sky visible from middle northern latitudes. Also describes observing techniques and equipment. (INTERMEDIATE)

The Supernova Search Charts and Handbook, by Gregg Thompson and James Bryan, Jr. (Cambridge University Press, 1989). Charts for more than 300 galaxies to help observers hunt for supernovae within them. (ADVANCED)

Webb Society Deep-Sky Observer's Handbooks (Enslow Publishers, 1979–). An eight-volume series on specific deep-sky subjects. (INTERMEDIATE–ADVANCED)

Almanacs & Annuals

Astronomical Almanac (US Government Printing Office, 1997). Planet positions and data for each year, published annually. (INTERMEDIATE)

Astronomical Calendar, by Guy Ottewell (Astronomical Workshop, 1997). A wonderfully graphic way to explore the astronomical events of each year. (BEGINNER–INTERMEDIATE)

Observer's Handbook, edited by Roy L. Bishop (Royal Astronomical Society of Canada, 1997). If you want a single compendium of data and annual events, this is the one to get. (INTERMEDIATE)

Cosmology and Beyond

Are We Alone?, by Paul Davies (Basic Books, 1995). In the words of the subtitle, this is about the philosophical implications for the discovery of extraterrestrial life. (BEGINNER)

The Big Bang, by Joseph Silk (W. H. Freeman, 1989). Explains how the Big Bang model emerged from nearly a century's research and why it has proved so robust. (INTERMEDIATE–ADVANCED)

Web Sites

Astronomical Image Library (a searchable index of astronomical images). http://www.syz.com/images/

Astronomical Society of the Pacific (includes an image collection of objects listed in the *New General Catalogue*). http://www.aspsky.org/html/resources/ngc.html

AstroWeb (a searchable guide to resources on the Internet). http://www.cv.nrao.edu/fits/www/astronomy.html

Automated Telescopes (links to telescopes you can operate via the Internet). http://www.eia.brad.ac.uk/rti/automated.html

Central Bureau for Astronomical Telegrams (information about transient phenomena such as comets and supernovae). http://cfa-www.harvard.edu/cfa/ps/cbat.html

European Space Agency. http://www.esrin.esa.it/

Hubble Space Telescope. http://www.stsci.edu/

The Messier Catalog (images and information about Messier and the objects he listed). http://www.seds.org/messier/

NASA (links to many areas of exploration). http://www.nasa.gov/

Sky Online (links to lots of sites). http://www.skypub.com/

Sky & Telescope reviews of astronomical products. http//www.skypub.com/testrpt/testrpt.shtml

Visual Satellite Observers. http://www.satellite.eu.org/sat/vsohp/satintro.html

Software

Skywatching, by David H. Levy (The Nature Company/Time-Life Books, 1994). Excellent CD-ROM guide to the whole subject of astronomy; available in both IBM and Macintosh formats. (BEGINNER)

Starry Night, by Sienna Software (1996), 105 Pears Avenue, Toronto, Ontario M5R 1S9 Canada. Outstanding Macintosh program with an exceptionally easy-to-use interface and a realistic sky. (INTERMEDIATE) tel. (416) 926-2174 http://www.siennasoft.com/sienna/

TheSky, by Software Bisque (1996), 912 12th St, Suite A, Golden, CO 80401, USA. Highly capable IBM program, with many add-ons allowing telescope control from the keyboard. (INTERMEDIATE) tel. (303) 278-4478 http://www.bisque.com/thesky/

Check the computer columns in astronomy magazines for the latest versions and new offerings in computer software.

Periodicals

USA

Astronomy, 21027 Crossroads Circle, Waukesha, WI 53187. A colorful monthly magazine devoted to presenting the science of astronomy and backyard observing for all kinds of astronomers. http://www.kalmbach.com/astro/astronomy.html

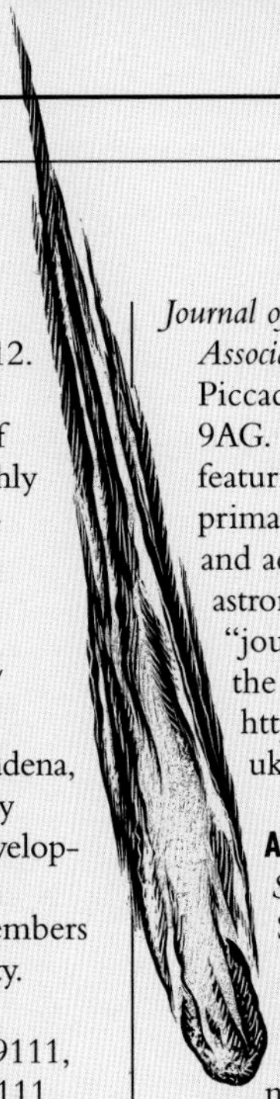

Mercury, 390 Ashton Ave, San Francisco, CA 94112. Sent to members of the Astronomical Society of the Pacific, this bimonthly publication contains interesting articles on astronomy written at a lay-reader level. http://www.aspsky.org/

The Planetary Report, 65 N. Catalina Ave, Pasadena, CA 91106. A bimonthly publication covering developments in Solar System exploration. Sent to members of The Planetary Society. http://planetary.org/

Sky & Telescope, PO Box 9111, Belmont, MA 02178-9111. A monthly magazine covering both the science and the hobby of astronomy, aiming features toward the more experienced amateur astronomer and the professional. http://www.skypub.com/

The Strolling Astronomer, PO Box 171302, Memphis, TN 38187-1302. A bimonthly published by the Association of Lunar and Planetary Observers, appealing primarily to dedicated amateur astronomers with strong interests in planetary viewing. http://www.lpl.arizona.edu/alpo/

CANADA

Journal of the Royal Astronomical Society of Canada, 136 Dupont St, Toronto, Ontario M5R 1V2. Publishes scientific papers bimonthly on historical, biographical, or educational topics of interest to the astronomical community. http://www.rasc.ca/journal/

SkyNews, National Museum of Science and Technology, PO Box 9724, Stn. T, Ottawa, Ontario K1G 5A3. A popular-level bimonthly for Canadian amateur astronomers, written in an engaging style. http://www.cmpa.ca/o9.html

BRITAIN

Astronomy Now, PO Box 175, Tonbridge, Kent TN10 4ZY. Monthly publication combining popular-level articles about the science with how-to articles for observers. http://www.astronomynow.com/

Journal of the British Astronomical Association, Burlington House, Piccadilly, London W1V 9AG. Bimonthly publication featuring articles that appeal primarily to the intermediate and advanced amateur astronomer. Serves as the "journal of record" for the BAA. http://www.ast.cam.ac.uk/~baa/index.html

AUSTRALIA

Sky & Space, Sky & Space Publishing, PO Box 1233, Bondi Junction, NSW 2022. A heavily illustrated monthly magazine presenting the science of astronomy and how-to features from a Southern Hemisphere perspective. http://www.skyandspace.com.au

Organizations

American Association of Variable Star Observers (AAVSO), 25 Birch St, Cambridge, MA 02138, USA. A group consisting mainly of amateur astronomers who keep thousands of variable stars under watch. http://www.aavso.org/

Association of Lunar and Planetary Observers (ALPO), PO Box 171302, Memphis, TN 38187-1302, USA. An amateur group that carries out planetary patrol observations. http://www.lpl.arizona.edu/alpo/

Astronomical Society of New South Wales (ASNSW), PO Box 1123, Sydney, NSW 1043, Australia. One of Australia's larger amateur societies. http://www.ozemail.com.au/~asnsw/

Astronomical Society of the Pacific (ASP), 390 Ashton Ave, San Francisco, CA 94112, USA. An international organization of amateur and professional astronomers. http://www.aspsky.org/

Belgian Working Group Satellites (BWG), Bart De Pontieu, MPE, Giessenbachstr., D-85740 Garching-bei-München, Germany. http://www2.plasma.mpe-garching.mpg.de/sat/bwgs.html

British Astronomical Association (BAA), Burlington House, Piccadilly, London W1V 9AG, England. The national society for amateur astronomers. http://www.ast.cam.ac.uk/~baa/index.html

International Dark-Sky Association, 3545 N. Stewart St, Tucson, AZ 85716, USA. Dedicated to helping astronomers of all kinds preserve dark skies for observing. http://www.darksky.org/ida/

International Meteor Association (IMO), Robert Lunsford, 161 Vance St, Chula Vista, CA 91910, USA, or Ina Rendtel, Mehlbeerenweg 5, D-14469 Potsdam, Germany. http://www.imo.net/

International Occultation Timing Association (IOTA), 2760 SW. Jewel Ave, Topeka, KS 66611-1614, USA. http://www.sky.net/~robinson/iotandx.htm

The Planetary Society, 65 N. Catalina Ave, Pasadena, CA 91106, USA. A space-advocacy group promoting spacecraft exploration of the Solar System. http://planetary.org/

Royal Astronomical Society of Canada (RASC), 136 Dupont St, Toronto, Ontario M5R 1V2, Canada. The national organization for professional and amateur astronomers. http://www.rasc.ca/

Royal Astronomical Society of New Zealand (RASNZ), PO Box 3181, Wellington, New Zealand.

Royal Greenwich Observatory, Satellite Laser Ranging Team, Herstmonceux Castle, Hailsham, East Sussex BN27 1RP, England. For information on making position measurements of satellites.

The Webb Society, Don Miles, 96 Marmion Rd, Southsea, Hants PO5 2BB, England. Members are interested in deep-sky observing. http://www.ukindex.co.uk/ukastro/webbmain.html

To find out if there is a planetarium or astronomy club in your area, ask at the public library or a local college or university, or check in the annual special issues published by *Astronomy* and *Sky & Telescope*.

INDEX *and* GLOSSARY

In this combined index and glossary, bold page numbers indicate the main reference, and italics indicate illustrations and photographs.

C

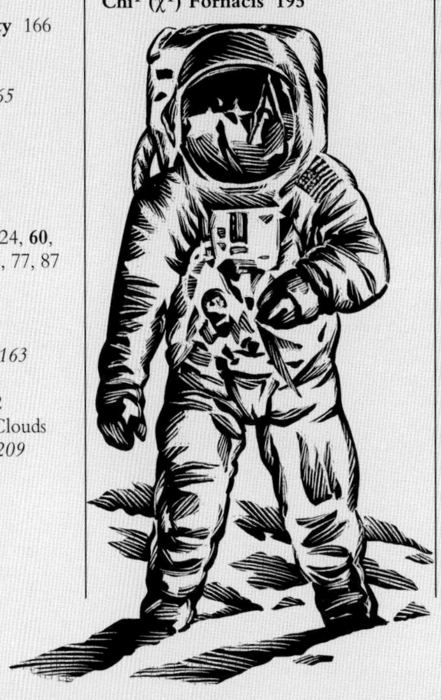

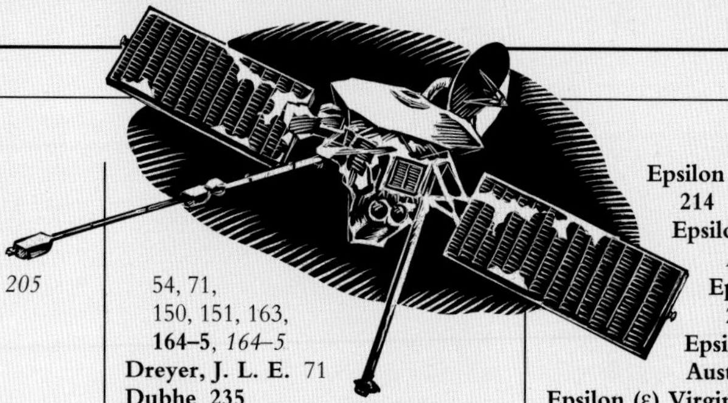

G

H

I

J

K

L

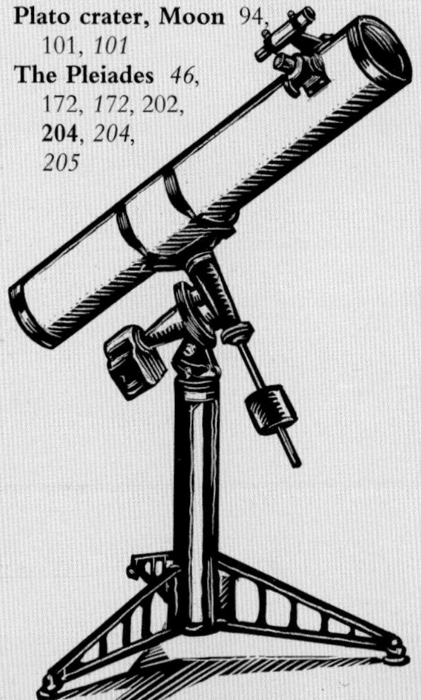

CONTRIBUTORS

Dr. John O'Byrne has been interested in astronomy all his life, having been given his first telescope when he was 12 years old. He is now Senior Lecturer in Physics at the University of Sydney and a secretary of the Astronomical Society of Australia. His main area of expertise is in optical systems used in astronomy, but he is also particularly interested in astronomy education and regularly conducts courses and telescope-viewing nights for adults, schools, and clubs.

Alan Dyer has been an amateur astronomer and astrophotographer for 25 years, and is one of Canada's leading astronomy writers. He is co-author of *The Backyard Astronomer's Guide* and is a regular contributor to the *Observer's Handbook*, published annually by the Royal Astronomical Society of Canada. He is a former associate editor of *Astronomy* magazine and is currently a contributing editor for *Sky & Telescope* and *SkyNews*.

Robert A. Garfinkle has been an avid amateur astronomer for more than 40 years, writing astronomy books, book reviews, and articles since the mid-1980s. He is the author of *Star-Hopping: Your Visa to Viewing the Universe* and is currently writing a lunar observer's handbook, as well as the monthly SkyChart and SkyTalk pages for the Astronomical Society of the Pacific's magazine, *Mercury*.

Robert Burnham is the author of *Comet Hale-Bopp: Find and Enjoy the Great Comet* and *The Star Book*. A lifetime amateur astronomer, he is a former editor-in-chief of *Astronomy* magazine, where he founded *The Observer's Guide*, an annual publication aimed at budding amateur astronomers. His main interests in astronomy revolve around observing the Moon and planets. He is a member of the American Astronomical Society, and minor planet 4153 Roburnham was named after him.

Jeff Kanipe has been a science journalist for more than 15 years. Various positions held include editor of *StarDate* magazine, columnist and feature writer for *Astronomy* magazine, and a contributing writer for the Time-Life book series, *Voyage Through the Universe*. A former managing editor at *Astronomy* magazine, he is currently writing for *New Scientist* magazine and contributes to the *Earth and Sky* radio program.

David H. Levy is one of the world's best known amateur astronomers. Even though he has never taken a course in astronomy, he has discovered 21 comets, 8 of them using a telescope in a backyard observatory, and 13 of which he shares with Gene and Carolyn Shoemaker. David has had 18 books published, lectures internationally, and lives in Vail, Arizona, with his wife and two dogs.

Martin George is Curator of the Launceston Planetarium in Tasmania, Australia. His interest in astronomy began when he was six years old. Since studying physics at the University of Tasmania, he has spent much time popularizing astronomy and introducing others to the night sky. Martin conducts a regular radio program on astronomy for the Australian Broadcasting Corporation and is a contributing editor of *Sky & Space* magazine.

ACKNOWLEDGMENTS

The publishers wish to thank the following people for their assistance in the production of this book: Coral Cooksley (AAO), Garry Cousins, Simon Gilchrist, Greg Hassall, Ionas Kaltenbach, Lena Lowe, Margaret McPhee, Meade Instruments Corporation, Catherine O'Byrne, Mary O'Byrne, Mike Smith and Don Whiteman (Binocular and Telescope Shop, Sydney), Tele Vue Optics Inc., Pasang Tenzing, Tashi Tenzing, Patrick Terrett, Jane Tonkin (State Library of South Australia).

PICTURE AND ILLUSTRATION CREDITS

t = top, c = center, b = bottom, l = left, r = right, i = inset, co = chapter opener
A = Auscape International; AAO = David Malin/Anglo-Australian Observatory; AF = Akira Fujii; APL = Australian Picture Library; ASP = Astronomical Society of the Pacific; AV = Astro Visuals; BAL = The Bridgeman Art Library, London; ESA = European Space Agency; Granger = The Granger Collection, New York; IS = Image Select; LC = Lee C. Coombs; JPL = Jet Propulsion Laboratory; Mendillo = Mendillo Collection of Antiquarian Astronomical Prints; MIC = Meade Instruments Corporation; NC = Newell Color; NOAO = National Optical Astronomy Observatories; NRAO = US National Radio Astronomy Observatory; NW = North Wind Picture Archives; OS = Oliver Strewe; PE = Planet Earth Pictures; ROE = Royal Observatory, Edinburgh; SPL = Science Photo Library; SF = Space Frontiers Collection, London; STScI = Space Telescope Science Institute; TPL = The Photo Library, Sydney; TS = Tom Stack and Associates; UCO = University of California Observatory; USGS = US Geological Survey

1 University Library Istanbul/ET Archive/APL; **2** NC/NASA; **3** NC/NASA; **4–5** NASA/SPL/TPL; **6–7** David Nunuk/SPL/TPL; **8–9** Pekka Parviainen/SPL/TPL; **10-11** Chlaus Lotscher/Peter Arnold Inc./A; **12–13**co & i TSM-Mark M Lawrence/Stock Photos; **14**t Ann Ronan/IS; c Hulton-Deutsch/TPL; b NASA/TPL; **15**t SF/PE; c Max Planck Institute/SPL/TPL; b NC/NASA; **16**c NW; b SPL/TPL; **17**t Baum & Henbest/SPL/TPL; c Mt Wilson & Las Campanas Observatories/ASP slide set "Astronomers of the Past"; **18**t NOAO/AV; c & b ROE/SPL/TPL; **19**t Mary Lea Shane Archives, Lick Observatory/ASP slide set "Astronomers of the Past"; b ESA/SPL/TPL; **20**cl Jodrell Bank/SPL/TPL; cr & b SPL/TPL; **21**t Ann Ronan/IS; **22**t Seth Shostak/SPL/TPL; b Dornier Space/SPL/TPL; **23**t SF/PE; c NASA; b NASA/SPL/TPL; **24**c NC/NASA; bl NASA; br C. R. O'Dell/Rice University/NASA; **25**t, c & b NC/NASA; **26**tl NASA/AV; tr SPL/TPL; c Novosti Press Agency/SPL/TPL; **27**c NASA; b NASA/Peter Arnold Inc./A; **28**t NASA/Airworks/TS; cl IS; cr SF/PE; **29**tl IS; tr, SF/PE; b SPL/NASA/TPL; **30**c SF/PE; b NASA/SPL/TPL; **31**tl, tr & b SF/PE; **32**tl Novosti/SPL/TPL; cl JPL/TSADO/TS; cr NASA; **33**tl & tr NASA; bl & br NASA/SPL/TPL; **34**c & b SF/PE; **35**tl & b SF/PE; tr David Parker/SPL/TPL; **36**tl Julian Baum/SPL/TPL; tr NASA/AV; c PE; b JPL/NASA; **37**t SF/PE; b Seth Shostak/SPL/TPL; **38**c F. Espenak/SPL/TPL; b Ray Nelson/Phototake/Stock Photos; **39**t ESA; bl Ray Pfortner/Peter Arnold Inc./A; br ROE/SPL/TPL; **40–41**co & i Roger-Viollet; **42**t & b OS; **43**tl AF; tr SPL/TPL; b Tony and Daphne Hallas/Astro Photo; **44**t Hulton-Deutsch/TPL; bl Bill and Sally Fletcher; br Alan Dyer; **45**t & c OS; b Tenmon Guide, Japan; **47**c OS; b Alan Dyer; **48**t Museo della Scienza, Firenze/Scala; b MIC; **49**t & b OS; **51**t Don Whiteman; c AF; b Yerkes Observatory; **52**tl Roger-Viollet; c OS; b Alan Dyer; **53**cr & b OS; **54**tl & c OS; **55**b OS; **56**tl & cr MIC; **57**t Tele Vue Optics Inc.; **58**t & b MIC; **59**b OS; **60**t OS; cl MIC; cr Luke Dodd; **61**t & b Yerkes Observatory; **62**t OS; cl & cr David Miller; **63**t Pekka Parviainen/SPL/TPL; b Alan Dyer; **64**t Alan Dyer; b Luke Dodd; **65**t Luke Dodd; **66**t MIC; c AF; cr C. Arsidi/MIC; **67**b MIC; br Tele Vue Optics Inc.; **68**t OS; b Gregory Terrance; **69**t MIC; c Jack Newton/MIC; b Jack Newton; **70**t Mendillo; c Corbis/Bettman/APL; b OS; **71**t OS; b NW; **72**t OS;

bl Maris Multimedia; br MIC; **73**t & c NASA; b Alan Dyer; **74–75**co & i Sun and Moon © Copyright Paul Klee/Reproduced By Permission Of BILD-KUNST (Germany) and VI$COPY Limited 1997/BAL; **76**t Egyptian Museum, Cairo/Werner Forman Archive; c F. Espenak; b Tele Vue Optics Inc.; **77**t F. Espenak; b SPL/TPL; **78**t AKG, London; **79**t NASA/TSADO/TS; c JPL/TSADO/TS; **80**t Mary Evans Picture Library; c NASA; b Photri Inc./APL; **81**t & c NOAO; b Dr. Dan Gezari/NASA/SPL/TPL; **82**t NRAO/AUI; **83**c *Sky & Telescope* Magazine; b Guildhall Library, London/BAL; **84**bl Francisco Arias/Sipa Press/TPL;br F. Espenak; **85**tr F. Espenak; b SPL/TPL; **86**t Pacific Stock/APL; c F. Espenak; b Roger-Viollet; **87**t Tony and Daphne Hallas/Astro Photo; b F. Espenak; **88**t Granger; bl Rev. Ronald Royer/SPL/TPL; br AF; **90**t Mary Evans Picture Library; cl Chad Ehlers/TPL; cr F. Espenak; **91**t Ann Ronan/IS; **92**t Wallace & Gromit/Aardman Animations Ltd. 1989; c NASA; b NASA/Airworks/TS; **93**t & c NASA; **94**t Zuber et al./SPL/TPL; c Granger; b NASA/JPL/TSADO/TS; **95**t UCO/Lick Observatory; b NC/NASA; **96–103** UCO/Lick Observatory; **104–105**co & i NASA/SPL/TPL; **106**t BAL; **108**t **108**t Erich Lessing/AKG, London; c NASA/SPL/TPL; b JPL/TSADO/TS; **109**t NASA/SPL/TPL; b AF; **110**t Museum of Mankind, London/BAL; c & b AF; **111**t & cr NASA; **112**t & c NASA; b Image Library/State Library of NSW; **113**t NASA; c & b NASA/SPL/TPL; **114**t Archaeological Museum, Aleppo/ET Archive/APL; **115**tl Arthur Pengelly/Sipa Press/TPL; tr Sylvain Grandadam/TPL; bl Adam Jones/TPL; br Peter Knowles/TPL; **116**tl NASA/SPL/TPL; tr Worldsat/Knighton/SPL/TPl; **117**t NASA; b Penny Gentieu/TPL; **118**t Prado, Madrid/BAL; b USGS/TSADO/TS; **119**t NASA; b Mary Lea Shane Archives, Lick Observatory/ASP slide set "Astronomers of the Past"; **120**tl USGS/TSADO/TS; tr USGS/SPL/TPL; b NASA; **121**br Don Parker; **122**t NASA; c SPL/TPL; b NASA; **123**tl NASA/SPL/TPL; tr NASA; b David Hardy/SPL/TPL; **124**t NW; b NASA/SPL/TPL; **125**tl & tr NASA/SPL/TPL; **126**t Don Parker; **127**t NRAO/AUI; r1, r2, & r3 LC; **128**t & br SF/PE; bl NW **129**tl & tr NASA/SPL/TPL; c PE; b SPL/TPL; **130**t Palazzo Vecchio, Florence/BAL; b SF/PE; **131**tl NASA/SPL/TPL; b NW; **132**t & c NASA/SPL/TPL; bl Peter H Smith/University of Arizona Lunar and Planetary Laboratory/NASA; br Hermitage, St Petersburg/BAL; **133**t Don Parker; **134**t Granger; r NASA; **135**ct NASA/SPL/TPL; tr NASA; b USGS/SPL/TPL; **136**tl Giraudon/BAL; tr NASA; bl NASA/SPL/TPL; br Mary Evans Picture Library; **137**t SF/PE; c NASA/SPL/TPL; **138**tl Palazzo Vecchio, Florence/BAL; c NC/NASA; cr Corbis/Bettman/APL; **139**tr NC/NASA; b NASA/SPL/TPL; **140**t Ann Ronan/IS; br Jayaraman & Dermott/SPL/TPL; **141**t R.S. Hudson and S. J. Ostro/NASA; c NASA/SPL/TPL; b Photofest/Retna/APL; **142**t Musee de la Tapisserie, Bayeux/BAL; cl AF; cr SPL/TPL; b ESA/SPL/TPL;**143**b Gordon Garradd/SPL/TPL; **144**c AF; b SPL/TPL; **145**t Index Stock/Phototake/Stock Photos; c Alan Dyer; b Observatoire de Paris/Bulloz; **146**br Hubble Space Telescope Comet Team/STScI/NASA; b Dr. H. A. Weaver and T. E. Smith/STScI/NASA; **147**t NASA/SPL/TPL; c Jonathan Blair/Woodfin Camp/APL; **148–149**co & i I M House/Tony Stone/TPL; **150**t Mary Evans Picture Library; c Dave Parkhurst/The Aurora Collection; b AAVSO; **151**t & cl AAO; cr ROE/AAO; **152**t & c Dave Parkhurst/The Aurora Collection; b NASA; **153**t Dave Parkhurst/The Aurora Collection; b John Sanford/SPL/TPL; **154**t Ann Ronan/IS; b Pekka Parviainen/SPL/TPL; **155**t Dave Parkhurst/The Aurora Collection; c Pekka Parviainen/SPL/TPL; b David Miller; **156**t Boyer-Viollet; c & b Mineralogy Dept./Nature Focus/Australian Museum; **157**t Pekka Parviainen/SPL/TPL; c SPL/TPL; b Bill Bachman; **158**t Dennis Hilon/SPL/TPL; b Ann Ronan/IS; **159**c SF/PE; b NW; **160**c Paul D. Maley; b AAO; **161**tl European Satellite Observers Group; tr Alan Dyer; b FPS/TPL; **162**tl St Madelein, France/BAL; c Alan Dyer; **163**t NOAO/AAO & AURA © 1993-1995; b Harvard College Observatory/Whitin Observatory/ASP slide set "Astronomers of the Past"; **164**t Michael Carroll; **165**t & c Mullard Radio Astronomy Lab./SPL/TPL; bl & br NW; **166**t AAVSO; **167**b NASA; **168**t Ann Ronan/IS; c SPL/TPL; b F. Paresce and R. Jedrzejewski/STScI/NASA/ESA; **169**t AAO; c Malin/Pasachoff/Caltech/AAO; b AAO; **171**tl Bill and Sally Fletcher; tr NASA/TSADO/TS; b The Webb Society; **172**bl AAO; br Bill and Sally Fletcher; **173**t NOAO/TS; c Bill and Sally Fletcher/TS; **174**t Ann Ronan/IS; cl John Chumack/TPL; cr AAO; b NOAO/TSADO/TS; **175**t AAO; b AAO; **176–177**co & i AAO; **Ch 6** banding AAO; **180**t Ben Davidson; c AF; b AAO; **181**c OS; **182**t Mendillo; **184**b Lauros-Giraudon/BAL; **186**t Bill and Sally Fletcher/TS; c AF; **187**t IAC/RGO/David Malin; **188**b Southampton City Art Gallery/BAL; **190**t AF; **191**t George Greaney; c LC; **192**b Granger; **194**t ROE/AAO; c Dr. Jean Lorre/SPL/TPL; **195**t AAO; **196**b Mountford-Sheard Collection (#1642)/Special Collections, State Library of South Australia; **198**t AF; b AAO; **199**t ROE/SPL/TPL; c ROE/AAO; **200**t AAO; **201**t SPL/TPL; c AAO; **202**b Granger; **204**t AF; b Mountford-Sheard Collection (#1480)/Special Collections, State Library of South Australia/Reproduced with the permission of the family of Wongu; **205**t & b SPL/TPL; **206**b Mendillo; **208**t AF; c John Sanford/SPL/TPL; **209**t Tony and Daphne Hallas/Astro Photo; **210**b Granger; **212**t John Chumack/TPL; c AF; **213**t AAO; **214**t AF; c John Chumack/TPL; b AAO; **215**t AAO; b AF; **216**b Granger; **218**t Luke Dodd/SPL/TPL; **219**t Luke Dodd/SPL/TPL; c AF; **220**b Bibliotheque Nationale, Paris/BAL; **222**t Jack Newton; c George Greaney; **223**t LC; **224**b Scala; **226**t ROE/AAO; c AAO; **227**t AF; **228**b Granger; **230**t AF; c AAO; **231**t AAO; **232**b Scala; **234**t AF; c Jack Newton; **235**t Jack Newton; **236**b Scala; **238**t ROE/AAO; c AAO; **239**t NOAO; c AF; b Dr. Rudolph Schild/SPL/TPL; **240**t AAO; **241**t Dr Rudolph Schild/SPL/TPL; b Jodrell Bank/SPL/TPL; **242**b Mendillo; **244**t AAO; **245**t AF; c AAO; **246**b Mountford-Sheard Collection (#1470)/Special Collections, State Library of South Australia/Reproduced with the permission of the family of Wongu; **248**t & c AF; **249**t AAO; **250**b Mendillo; **252**t AF; c ROE/AAO; **253**t Luke Dodd; **254**b Bibliotheque Nationale, Paris/BAL; **256**t & c AAO; **257**t AAO; b ROE/AAO; **258**t SPL/TPL; **259**t AF; c AAO; **260**b Mendillo; **262**t George Greaney; c Gordon Garrad; cr IAC/AAO; **263**t Jack Newton; **264**b Granger; **266**t AF; c Jack Newton; **267**t Bill and Sally Fletcher; **268**b Granger; **270**t AAO; c AF; **271**t LC; **272–3**co & i Granger.

ILLUSTRATION CREDITS

Illustrations by **Rod Burchett** 57; **Karen Clarke** 82-3 (graphs), 85, 89t; **Lynette Cook** 16, 107, 143; **Les Dalrymple** 170 (galaxy sketch); **Chris Forsey** 116; **Clare Forte** 164c & b, 166, 167 (graphs); **Connell Lee** 160; **Martin Macrae/Folio** 117; **Rob Mancini** 20, 24, 30, 46t, 55, 56, all sketch pads (82, 91, 121, 126, 133, 170), 170t, 172, 187, 190, 195, 200, 209, 213, 218, 223, 227, 231, 235, 240, 244, 249, 253, 258, 263, 267, 271, 274–286; **Mick McCullagh** 82 (solar sketch); **James McKinnon** 147; **Paulineke Nolan** 91 (Moon sketch), 126 (Jupiter sketch), 133 (Saturn sketch); **Oliver Rennert** 114 (Earth), 146; **Trevor Ruth** 34; **Kevin Stead** 140; **Wil Tirion** 59t, 65, 182–3 (Key Maps), all maps in the Starhopping Guide; **Chris Toohey** 121 (Mars sketch); **Simon Williams/Garden Studio** 78; **David Wood** 19, 21, 46b, 50, 54, 59c, 77, 84, 89b, 106, 109, 111, 114 (Jupiter), 118, 125, 131, 135, 137, 139, 162.

JACKET

Front: AAO; ASP/Mt Wilson and Las Campanas Observatories; Lynette Cook; Alan Dyer; AF; Tony Hallas/Astro Photo; Rob Mancini; MIC; NASA; Wil Tirion **Back:** AKG London; Granger; MIC; NASA; OS; PE **Front flap:** F. Espenak; Granger; NASA **Back flap:** NASA/Peter Arnold/A; David Wood

CAPTIONS

Page 1: A miniature sixteenth-century painting of astronomers observing the sky from Galata Tower, Constantinople.

Page 2: A young planetary nebula, known as the Hourglass Nebula, blown out from a central star.

Page 3: Comet-like gaseous fragments are left behind as the Helix Nebula's ring expands out behind it.

Pages 4–5: Astronaut Bruce McCandless floats above Earth in a maneuvring unit.

Pages 6–7: A long-exposure photograph showing a Utah rock formation silhouetted against star trails.

Pages 8–9: A time-lapse montage of a partial solar eclipse in five stages.

Pages 10–1: The observatory domes at Mauna Kea, Hawaii, USA.

Pages 12–3: The space shuttle *Atlantis* launch, Florida, USA.

Pages 40–1: A group of observers viewing an eclipse of the Sun, Paris, 17 April 1912.

Pages 74–5: A detail from *Sun and Moon* © Copyright Paul Klee (1879–1940). Reproduced by permission of BILD-KUNST (Germany) and VI$COPY Limited, 1997.

Pages 104–5: A detail of the surface of Jupiter, photographed by the Voyager 2 spacecraft.

Pages 148–9: A Hubble Space Telescope image of the interstellar gas pillars within the Eagle Nebula.

Pages 176–7: The Eagle Nebula (M16).

Pages 272–3: A detail from Andreas Cellarius's *Atlas Coelestis seu Harmonia Macrocosmica* (1660).